D0207744

AN INTRODUCTION TO
PARTICLE PHYSICS AND THE STANDARD MODEL

AN INTRODUCTION TO
PARTICLE PHYSICS AND THE STANDARD MODEL

ROBERT MANN
UNIVERSITY OF WATERLOO
ONTARIO, CANADA

CRC Press
Taylor & Francis Group
Boca Raton London New York

CRC Press is an imprint of the
Taylor & Francis Group, an **informa** business

A TAYLOR & FRANCIS BOOK

CRC Press
Taylor & Francis Group
6000 Broken Sound Parkway NW, Suite 300
Boca Raton, FL 33487-2742

Printed in the United States of America on acid-free paper
10 9 8 7 6 5 4 3 2

International Standard Book Number: 978-1-4200-8298-2 (Hardback)

Library of Congress Cataloging-in-Publication Data

Mann, Robert, 1955-
   An introduction to particle physics and the standard model / Robert Mann.
      p. cm.
   Includes bibliographical references and index.
   ISBN 978-1-4200-8298-2 (hardcover : alk. paper)
   1. Particles (Nuclear physics) 2. Quark models. 3. String models. I. Title.

QC793.2.M36 2010
539.7'2--dc22
                                      2009026775

Visit the Taylor & Francis Web site at
http://www.taylorandfrancis.com

and the CRC Press Web site at
http://www.crcpress.com

Contents

Preface

The purpose of this book is to introduce 4th-year or senior undergraduate students to what is known as the Standard Model of Particle Physics, the model that presently encompasses all of our empirical knowledge about the subject.

Particle physics was in a near-continual state of flux for several decades, finally settling down around the mid 1990s when the mass of the Z boson had been accurately measured, the number of light quarks and leptons had been established, and the top quark had been discovered. The Standard Model has since then faced pretty much every experimental challenge to its authority with flying colors, and today it stands as the established fundamental theory of the non-gravitational interactions, describing all known forms of subatomic matter that we have observed.

The goal of this book is to familiarize students with the Standard Model and in so doing, with particle physics in general. It grew out of a one-term course I have taught at the University of Waterloo nearly every year over the past two decades. It was an interesting course to teach because the subject matter would change as particle physics continued to develop, with new results coming out from LEP, Fermilab, Super-K, SNO and more on the experimental side, and from supersymmetry, string theory, and lattice gauge theory on the theoretical side. Students taking the course typically had taken at least one course in quantum mechanics (in which they would have seen the solution to the hydrogen atom from Schroedinger's equation), one in mathematical physics (covering vector calculus, Fourier transforms, and complex functions), and had a solid background in special relativity (having encountered the basic phenomena of length contraction and time dilation).

This book assumes that students have a good working knowledge of special relativity, quantum mechanics, and electromagnetism. From this basis students who work through the material will develop a solid command of the subject and a good working knowledge of the basics of particle physics, in terms of mathematical foundations, experimental methods, and basic processes. Each chapter has a number of questions, and there is a solutions manual available that has complete answers to all of the questions.

I have taken the approach of describing the Standard Model in terms of its Electromagnetic, Strong, and Weak components, so that students can understand the subject from the perspective of the reigning paradigm. Throughout I have endeavored to show why this paradigm does indeed reign – in other words, how and why the different parts of the Standard Model came to be what they are today, particularly pointing out and describing the experiments that were essential in arriving at these conclusions. I have also made efforts to show where the Model is in need of improvement and what possible physics might lie beyond what it describes. This is particularly addressed in the last chapter, but also appears throughout the book in a number of places. Our understanding of particle physics is by no means a finished project, and I hope that students will catch the excitement of the ongoing nature of research in this subject.

Particle physics is both mathematically and conceptually challenging, and many have thought that it can only be taught in a very superficial way at the undergraduate level, if it is taught at all. In my 20 years of teaching this subject I have found that students can indeed rise to the challenge, if both the formalism and background are carefully explained to them in a manner that allows them to connect with the physics they have already learned. I have taken that approach in this book, beginning (after a review of relativity) with some basic concepts in group theory and classical mechanics that lead into the subjects of symmetries, conservation laws, and particle classification. Three chapters following that are devoted to the experimental tools and methods, and analysis of particle physics. The next three chapters introduce students to Feynman diagrams, wave equations, and gauge invariance, building up to the theory of Quantum Electrodynamics. The remainder of the book then deals with the three pillars of the Standard Model: QED in Chapters 13 and 14, the strong interactions and QCD in Chapters 15 – 18, and Electroweak interactions in Chapters 19 – 24, with the final chapter devoted to what might lie beyond the Standard Model. I have also taken an historical approach to the development of the subject wherever possible, showing how it emerged from the physics that most students have learned about in other courses.

The book is designed to be used in a single course over one term, essentially twelve weeks of lectures in a three-hour lecture week. Though I would typically cover two chapters per week, there is a bit too much in the book for one term, and so a few topics inevitably get scant attention. I suggest that students read the first chapter on their own, and that instructors need cover only the formalism in chapter 2 that may be unfamiliar to students. Instructors may also wish to omit the material on the Higgs mechanism in Chapter 23, and perhaps the material on QCD in Chapter 18 if time does not permit.

Theoretical Particle Physics rests on the foundation of Quantum Field Theory (QFT), that subject combining both special relativity and quantum mechanics into a unified whole. I have found that students can learn and make use of the basic results of QFT – Feynman diagrams, scattering amplitudes, antiparticles, decay processes – without having to go through a full discussion of path integrals, Wick's theorem, Interaction pictures, and the like. I have avoided the use of the language of quantum fields, preferring to use the term wavefunction so that students can make better contact with what they are familiar with. Throughout the book I acknowledge the quantum field theoretic foundations on which the subject stands where appropriate. My goal is that students see both the forest and the trees, and not get too bogged down in formalism.

That said, the subject is one requiring serious mathematical and intellectual effort. I have attempted to cater to the more mathematically inclined students by putting into appendices mathematically challenging material that enriches but is not essential to the understanding of the material in a given chapter. Any appendix can be avoided in a first reading of the book, and most students will probably wish to do this. However, calculational derivations are made explicit wherever possible, and students willing to work through the appendices will be rewarded with an enriched understanding of the material and a set of formidable technical skills.

It is my hope that undergraduate students reading this book or taking a course that makes use of this book will be inspired by the subject of particle physics. I also hope that beginning graduate students may be able to make use of the book as preparation for more advanced courses they might take or as a resource for basic calculations and background material. I have tended to err on the side of completeness in my discussion to ensure that students are able to make use of the book in as broad a range of applications as possible.

This book was written while I was at the University of Waterloo in Ontario, Canada, and completed while I was on sabbatical at the Kavli Institute for Theoretical Physics at the University of California, Santa Barbara, California U.S.A., for whose hospitality I am most grateful. I am also grateful to Don Marolf and Martin Einhorn for their efforts in ensuring that I could be hosted there.

Acknowledgments

My first round of thanks is to the many students who have taken the particle physics course that I taught at the University of Waterloo over the years. They were always a source of challenging inspiration and provocative delight, keeping me on my toes with interesting questions about the subject. A number of them have gone on to make a career in particle physics at various laboratories and universities throughout the world, and I feel privileged to have had some part in that process.

I am also grateful for the guidance from mentors and teachers during my formative years at McMaster University and the University of Toronto. I am grateful to John Cameron and Bob Summersgill, who helped me learn my way around an accelerator lab, and to Rajat Bhadhuri and Yuki Nogami, who first stimulated my interest in the theoretical aspects of the subject. Thanks are due as well to Nathan Isgur, Bob Pugh, and Pat O'Donnell, who introduced me to the quark model, and especially to John Moffat, who taught me the value of considering things from a different angle.

My next round of thanks goes to the many colleagues, postdoctoral fellows, and students that I have collaborated with in various research projects in particle physics over the years: Catalina Alvarez, David Asner, Emilio Bagan, Harry Blundell, Biswajoy Brahmachari, Alex Buchel, Faroukh Chishtie, Debajyoti Choudhury, Mansour Chowdhury, Lou Culumovic, Rainer Dick, Victor Elias, Katherine Freese, Nicholas Hill, Marcia Knutt, Blaine Little, Martin Leblanc, Gerry McKeon, Jonas Mureika, Tim Rudy, Manoj Samal, Utpal Sarkar, Brad Shadwick, Tom Sherry, Tom Steele, Lev Tarasov, Ted Treml, and Kai Wunderlie. Each of them contributed in part – small or large – to my understanding of the subject. I am particularly grateful to Utpal Sarkar, who provided me with valuable help with some LaTex files in the final stages of preparation of this book. My ongoing collaborations with him, and with Victor Elias, Gerry McKeon, and Tom Steele, have always been enjoyable and fruitful. I am also grateful to all of my graduate students, summer students, and postdoctoral fellows – whether working in particle physics or not – for their contributions to the subject of physics and my understanding of it. The financial support of the Natural Sciences & Engineering Research Council of Canada has been invaluable throughout it all.

I would also like to thank Rudy Baergen, George Ellis, Mark Gedcke, Russ Howell, David Humphreys, Esther Martin, Don McNally, John Morris, Don Page, Judy Toronchuk, Alan Wiseman, and Bob Wright for discussions over the years on science and its broader meaning for our place in the scheme of things, and on the Intelligence behind the intelligibility of our universe.

Finally, I would like to thank my friends and family for the relationships that are part of what make life worth living, and my daughter, Heather, for the special part she will always play in that respect. Last – but certainly not least – I am most grateful to my wife, Nancy, who inspired me to write this book in the first place, and whose partnership over the decades has been an exciting adventure.

Further Reading

The most importance resource in particle physics is the *Review of Particle Physics* [1], published every other year since 1960 by the Particle Data Group, or PDG, at Berkeley in the United States. This document – over 1000 pages in length – contains all of the current empirical information that exists about the subject, along with the most up-to-date reviews on every aspect of particle physics relevant to the discipline. While it is not a place for beginners to learn about particle physics, it is truly the bible of the subject insofar as the information it contains is concerned. No student nor practitioner of the subject should be without it. A summary version appears in booklet form, but even that is now over 300 pages. I think a better way to access its information is via the PDG Web site:

http://pdg.lbl.gov/

From this Web site you can obtain all known information about any particle or process you want with only a couple of clicks of the mouse, along with any review article you like.

If you want to read further in particle physics, I recommend the following three books, which I have found particularly helpful in preparing this book:

A. Bettini, *Introduction to Elementary Particle Physics*, Cambridge University Press, 2008.

D. Griffiths, *Introduction to Elementary Particles*, 2nd edition, Wiley VCH, 2008.

D. Perkins, *Introduction to High Energy Physics*, 4th edition, Cambridge University Press, 2000.

There are other particle physics books that I can recommend that are accessible to the undergraduate student. These are of varying levels of difficulty, with some accessible to the non-specialist, and others of more difficulty. All of them will have substantive parts that are accessible to someone in their final year of an undergraduate physics program:

F. Close, M. Marten, and C. Sutton, *The Particle Explosion*, Oxford University Press, 1987.

P. C. W. Davies and J.R. Brown, *Superstrings: A Theory of Everything?*, Cambridge University Press, 1992.

R.P. Feynman, *The Character of Physical Law*, BBC Publications, 1965.

H. Frauenfelder and E.M. Henley, *Subatomic Physics*, Englewood Cliffs, N.J. Prentice-Hall, 1991.

F. Halzen and A.D. Martin, *Quarks and Leptons*, John Wiley & Sons, New York, 1984.

G.L. Kane, *Supersymmetry: Unveiling the Ultimate Laws of Nature*, Perseus, Cambridge, MA, 2000.

M. F. L'Annunziata, *Radioactivity: Introduction and History*, Elsevier Science, 2007.

H. J. Lipkin, *Lie Groups for Pedestrians*, Dover, New York, 2002.

A. Pais, *Inward Bound*, Clarendon Press, Oxford, 1986.

J.C. Polkinghorne, *Rochester Roundabout: The Story of High Energy Physics*, W.H. Freeman & Co. (Sd), 1989.

S.S. Schweber, *QED and the Men who Made it: Dyson, Feynman, Schwinger, and Tomonaga*, Princeton University Press, 1994.

A. Seiden, *Particle Physics: A Comprehensive Introduction*, Addison Wesley, San Francisco, 2005.

E.J.N. Wilson, *An Introduction to Particle Accelerators*, Oxford University Press, 2001.

S. Weinberg, *Dreams of a Final Theory The Scientist's Search for the Ultimate Laws of Nature*, Vintage, 1994.

S. Weinberg, *Subatomic Particles*, Scientific American Library, New York, 1983.

Finally, for those who may want to dig deeper into more advanced treatments at the quantum field theory level, or at a deeper level of model building, I can recommend:

C. Burgess and G. Moore, *The Standard Model: A Primer*, Cambridge University Press, 2006.

H. Enge, *Introduction to Nuclear Physics*, Addison Wesley, 1966.

M.E.Peskin and H.D.Schroeder, *An Introduction to Quantum Field Theory*, Addison-Wesley, 1995.

J.J. Sakurai, *Advanced Quantum Mechanics*, Addison Wesley, 1967.

U. Sarkar, *Particle and Astroparticle Physics*, Taylor & Francis, 2007.

S. Weinberg, *The Quantum Theory of Fields, Modern Applications* vol I, Cambridge University Press, Cambridge, 1996.

Be warned: these are challenging books at an advanced level!

1

Introduction and Overview

It is remarkable to realize that at the beginning of the 20th century – little more than 100 years ago – the structure of the atom was unknown. The electron had only been discovered a few years earlier, and its behavior and properties were still not well understood. Nobody knew anything about nuclei, protons, quarks, neutrinos, photons, gluons, and any of the many subatomic particles that we know about today. Quantum mechanics and special relativity were unknown conceptual frameworks for describing the physical world.

As the first decade of the 21st century draws to a close, the world will see the Large Hadron Collider (or LHC) at CERN turn on*. The thousands of scientists making use of this enormous machine – 27 kilometers in circumference and 24 stories underground – pivot their efforts around a key goal: to experimentally observe the Higgs particle and to measure its mass. If this experiment is successful, then we will have full empirical confirmation of the model – known as the Standard Model – that summarizes everything we know about the subatomic world at this point in history. Such confirmation would represent both a triumph of the human intellect and a gift of understanding that would ennoble humankind. Yet if the Higgs particle is not found, then the situation will be even more exciting. It will mean that something is wrong with our current understanding of particle physics, something that will be superseded by – it is hoped – more fundamental knowledge.

So what is particle physics? Particle physics is the study of nature at the most reductionist level possible: **it is the study of the ultimate constituents of matter and the laws governing their interactions.** The idea that matter ultimately consisted of small indivisible particles is an old idea, going back 2500 years to Democritus and Leucippus of Abdera, a town on the seacoast of Thrace in Greece [2]. These philosophers proposed that all of matter was made of $\alpha\tau o\mu o\sigma$, or atoms (a Greek word meaning "uncuttable") and empty space.

This idea survived through the centuries, and was used by scientists such as Newton, Dalton, Maxwell, and Mendeleyev to explain the behavior of gases and chemical compounds. It grew into the subjects we now call chemistry and

---

*The LHC has attracted a lot of attention worldwide, in part because of the fundamental questions it addresses and in part because of its large cost. There are many Web sites about it with information, novels have been written in which the LHC is a principal setting, a movie, *Angels and Demons*, in which the LHC plays a role, was released in May 2009, and there is even a rap (reproduced in Appendix J) about the LHC on YouTube!

physics, each of which has further subdivided into a variety of subdisciplines, that in turn have a healthy synergy with one another.

Particle physics can be regarded as the subdiscipline that pushes the atomic idea as far as possible. Simply put, it proceeds from two basic observations about our world, common to everyday experience:

1. Things exist   (i.e. there is matter)

2. Things happen   (i.e. interactions occur)

The goal of particle physics, then, is to reduce to as elementary a level as possible our understanding of these two observations.

1.1   Methods of Study

One of things that distinguishes particle physics from most other subdisciplines in physics is in its approach to the natural world. In most other subdisciplines – optics, condensed matter physics, acoustics, biophysics – the basic (or effective) physical laws and constituents are known, and one works out the consequences of these laws[†]. However, in particle physics the goal is to discover what the laws and constituents are – one cannot take them as given.

So how does one study particle physics? As with all of science, research proceeds on two fronts: experimental and theoretical. Each has a broad range of intellectual activity, with theoretical efforts often appearing to be nothing more than abstract mathematics, and experimental work seeming at times indistinguishable from engineering. Don't be fooled by superficial appearances though! Each of these activities plays a vital role in advancing the subject, and the two approaches have a healthy and vibrant interplay. Conceptually we can categorize each approach, as summarized in table 1.1. The two columns in table 1.1 form the primary conceptual categories in each area. There is a lot of overlap both vertically down the columns and horizontally across the rows. Let's look briefly at each category.

1.1.1   Large Accelerators

Most of our experimental knowledge of particle physics comes from colliding particles together at very high speeds, resulting in very energetic collisions. For this reason, particle physics is sometimes called *high energy physics*. Very

---

[†]This by no means makes such subdisciplines less intellectually challenging, less valuable, or less important. Indeed, they have led to an understanding of many novel phenomena and applications, including vortices, superfluids, photonic band gaps, and more.

TABLE 1.1

Approaches to Particle Physics

Experimental	Theoretical
Large Accelerators	Empirical Analysis
Detectors	Model Building
Precision Measurements	Numerical Computation
Cosmological Data	Mathematical Foundations

large machines called accelerators –kilometers in length – are needed to do this [3]. When we wish to examine very tiny systems (i.e. very short distances and/or very short times) we must cope with limitations imposed by the uncertainty principle :

$$\triangle p \triangle x \geq \hbar \Longrightarrow \triangle p \geq \hbar/\triangle x \Rightarrow \quad \triangle p \text{ is large for small } \triangle x$$
$$\triangle E \triangle t \geq \hbar \Longrightarrow \triangle E \geq \hbar/\triangle t \Rightarrow \quad \triangle E \text{ is large for small } \triangle t$$

Also, since from relativity $E = mc^2$, large mass particles need high energies to be created. Hence we need accelerators that can attain very high energies in order to study such short distance effects. In order to implement this, the accelerators need to be quite large in size – we'll see why in Chapter 7. The LHC at CERN is currently the largest such machine in the world, and is capable of accelerating particles to almost the speed of light [4].

1.1.2 Detectors

It does no good to smash particles together unless you can see what happens. A detector is a machine designed to do just that. There are many kinds of detectors, as we'll see in Chapter 8, with the main job of each one being that of measuring as much physical information about the particles emerging from a collision as possible: their momenta, their masses, their spins, their charges, their energies, and so on. These detectors are typically of enormous size – the ATLAS detector at CERN is as high as a 5-story building (see figure 1.1) – because of the large amount of sophisticated apparatus needed to ensure that all of these measurements can take place.

However, large detectors are not the only kinds of detectors employed in particle physics, nor are all detectors deployed in high-energy collision experiments. This brings us to our next approach.

1.1.3 Precision Measurements

Not all of what we know about particle physics comes from smashing particles together. Sometimes we need to measure very subtle properties about particles that cannot be observed in high-energy collisions. For example interactions of neutrinos with other kinds matter (electrons, nuclei) do not require

FIGURE 1.1
Installation of the beam pipe in the Atlas detector at the LHC in June 2008.
(Photograph: Maximilien Brice, copyright CERN; used with permission).

high energies. Furthermore, they are very infrequent and unlikely to occur.
Hence there is a need for sensitive detectors to pick out the signal from the
noisy background of the everyday world. The Sudbury Neutrino Observatory was an example of a large-scale precision measurement facility designed
to detect the properties of solar neutrinos [5]. Other experiments – searches
for dark matter, axions, and other exotic phenomena – employ detectors of
all shapes and sizes, custom-made to seek out (or place limits on) the phenomenon of interest.

1.1.4 Cosmological Data

The early universe was an environment of a hot plasma of all kinds of particles
[6]. The average temperature – and hence the average collision energy –
was very high, much higher than can be attained in controlled terrestrial
experiments. This means that observations from cosmology can provide us
with useful and important information about particle physics. An example of
this was a cosmological limit on the number of kinds of low-mass neutrinos,
which had to be less than 4 from the corroboration of Big Bang nucleosynthesis
with observation. The limit was later confirmed by experiments on the Z-
particle [7], which showed that there were only three kinds of neutrinos that
were lighter than half the mass of the Z.

It is common today for particle physicists and cosmologists to interact and

FIGURE 1.2
View of the Sudbury Neutrino Observatory detector after installation of the
bottom photomultiplier tube panels, but before cabling (Photo courtesy of
Ernest Orlando Lawrence Berkeley National Laboratory; used with permission).

collaborate with one another, with findings from each subdiscipline shedding light on findings from the other. In fact, astroparticle physics has pretty much become a separate subdiscipline of its own, with a community of theorists and experimentalists actively seeking to further our understanding of this interesting interdisciplinary subject [8].

1.1.5 Empirical Analysis

You might expect that a key job of a particle physicist is to analyze the data, and you would be right. The experimentalists do this first, converting raw data into usable information, such as measurements of the masses or lifetimes of particles. Theorists make use of this information to seek new patterns in the data, to critique existing analysis, and to suggest new experiments.

The analysis of the data itself makes use of a variety of mathematical techniques of some sophistication, and today typically require vast amounts of computer processing. The LHC will produce a data volume of 1 trillion bytes per second, equivalent to 10,000 sets of an *Encylopedia Brittanica* each second [4]. During its expected lifespan the LHC should produce an amount of data equivalent in volume to that contained in all of the words ever spoken by humankind in its existence on earth. Such an enormous volume of data per unit time must be supplemented by a large computational infrastructure, as well as a very sophisticated level of data processing and programming skill.

1.1.6 Model Building

A very common activity for a particle theorist is to propose a model for how nature works at the subatomic level. This involves making a clear set of assumptions about the particle content, the interactions between the particles, and the basic symmetries respected by each, all with an eye toward making a falsifiable prediction that an experimentalist could check. For example, a theorist might suggest that electrons and muons are themselves made of simpler particles that bind together according to some new force[‡].

The difference between a theory and a model is often confusing to newcomers to the subject. The distinction between the two is rather subtle, and perhaps can best be understood in the following way. A *theory* is a basic mathematical framework used for describing physics. Quantum mechanics, Yang-Mills theory, and Special Relativity are all examples of these. A *model* is a particularization of a theory to a specific context – it still very much has a mathematical character, but also has a specificity designed to describe a particular system or situation. For example the quark model is a particular

---

[‡]Such models were indeed proposed, with the constituents known by names such as preons and rishons, and became known as substructure models [9].

description of the underlying structure of particles such as the proton, neutron, and pion (more generally of all hadrons), that makes use of quantum mechanics, group theory, and Yang-Mills theory to elucidate its key features.

1.1.7 Numerical Computation

Computers have gone from playing a supplementary role in analyzing data to an essential role in working out the consequences of physical theory. Many of the problems in particle physics cannot be analyzed from a theoretical standpoint without the use of computers. The calculations are simply too big or too long for any person (or group of persons) to carry out in a reasonable amount of time.

Lattice gauge theory is a good example [10]. In this approach to understanding the behavior of quarks and gluons, many theorists work on attempting to solve the basic equations of Quantum Chromodynamics (QCD) on a computer, where spacetime is approximated as a lattice of discrete points. The goal here is to solve the equations with as few approximations as possible, something that has eluded formal theoretical analysis thus far.

1.1.8 Mathematical Foundations

This type of work involves a basic exploration of the mathematical structure of particle physics and its models. It is highly mathematical, and involves examining the basic foundations of current theory, as well as its proposed extensions. Here the theorist attempts to prove/refute certain properties of broad classes of models, with secondary regard as to their empirical content.

String theory is perhaps the best-known example of this kind of work [11]. Over the past 25 years it has given birth to new mathematical methods, new conceptual frameworks, and new calculational techniques in particle physics, a number of which could have interesting implications for the subject in the years to come.

1.1.9 Units

Particle physics is commonly concerned with understanding highly energetic processes at very short distances. This is a regime where special relativity and quantum mechanics are both important, as noted earlier, and so both Planck's constant $\hbar$ and the speed of light c, which have the values [1]

$$\hbar = 1.05457266 \times 10^{-34} \text{ Js}$$
$$c = 2.99792458 \times 10^8 \text{ ms}^{-1}$$

must be taken into account. Retention of these constants in every expression can often be a cumbersome nuisance, so most particle physicists prefer to work in what are called *natural units* where $\hbar = \frac{h}{2\pi} = 1$ and $c = 1$. This also allows one to set the permittivity of free space, $\varepsilon_0 = 1$, provided all charges are rescaled in units of $(\hbar c)^{-1/2}$. I will typically adopt these conventions, except in certain cases where it is useful to illustrate the explicit units. This will typically be when I display a result that can be directly compared to experiment (such as a decay rate or a cross-section), in which case the factors of $\hbar$ and c are useful.

With a bit of practice it is not hard to convert an expression in natural units to one with the proper powers of $\hbar$ and c. The general prescription for any given expression is to (a) express all velocities as a fraction of the speed of light and all times in terms of the light-travel distance (b) convert distances into units of inverse energy (or vice versa) as appropriate, using the conversion factor $\hbar c = 197$ MeV-fm, and (c) express charges, masses and momenta in units of energy.

TABLE 1.2
Natural Units

Physical Quantity	Notation	Units	Natural→ Physical
velocity	$\vec{\beta}$	unitless	$\vec{\beta} \rightarrow \frac{\vec{v}}{c}$
time	t	$\hbar$/MeV	$t \rightarrow t/\hbar$
length	d	$\hbar c$/MeV	$d \rightarrow d/\hbar c$
mass	m	MeV/c^2	$m \rightarrow mc^2$
momentum	$\vec{p}$	MeV/c	$\vec{p} \rightarrow \vec{p}c$
charge	q	unitless	$q \rightarrow \frac{q}{\sqrt{\hbar c}}$
energy	E	MeV	

So for a given expression that depends on mass, time, momentum, energy and charge, to convert it to standard units just apply the conversion factors in the right-hand column of table 1.2. The resultant expression will be in terms of physical quantities with respective units of kilograms, seconds, kg-meters/second, Joules, and electrostatic units. There will also be a number of factors of $\hbar\,c$ that will cancel out to leave an appropriate resultant expression.

For example in natural units the Compton wavelength $\lambda = 1/m$. To convert this to physical units we set $\lambda \rightarrow \lambda/\hbar c$ and $m \rightarrow mc^2$, giving $\lambda \rightarrow \lambda/\hbar c = 1/mc^2$ or $\lambda = \hbar/mc$, which is the standard formula.

1.2 Overview

The picture of particle physics circa 1940 was that everything in the universe was made of 4 particles:

e^-	the electron
p	the proton
n	the neutron
ν	the neutrino

All known chemical elements were made of the first three of these in some combination, and it was generally believed at that time that all known extraterrestrial matter – stars, planets, and interstellar dust – were made of the same three particles. The last of these, the neutrino, was a hypothetical particle needed to ensure that radioactive processes respected conservation of energy and angular momentum, but had no direct observational confirmation at that time.

The following 60 years saw a radical modification in our understanding of the subatomic world. As experimental energies increased, *hundreds* of new particles (almost all of them unstable) were discovered. For a period of time the subject was in a considerable amount of confusion, but by the end of the 1970s a general understanding of the situation had emerged, along with a model – now called the Standard Model – that could describe all current knowledge of the subject [12].

1.2.1 Bosons and Fermions

Today we know that all matter and its physical interactions can be described in terms of two basic kinds of particles: bosons and fermions. *Bosons* are particles of integer spin in units of Planck's constant $\hbar$ – they are the elementary particles that govern what we describe as a force in the everyday world. *Fermions* are particles of half-integer spin in the same units, and are the elementary constituents of what we call matter.

Bosons and fermions are distinguished by their collective properties under the interchange of two particles. Suppose we have a system consisting of two identical particles. If we enclose the system in a box, then the probability of finding one particle in a given position and the other in another position – let's call this $P(1,2)$ – must be equal to the probability of finding the particles interchanged in position (in other words we have $P(1,2) = P(2,1)$) because they are identical and so we can't tell them apart. Since quantum mechanics implies that probabilities are given by square of wavefunction amplitudes, we have

$$P(1,2) = P(2,1) \Rightarrow |\Psi(1,2)|^2 = |\Psi(2,1)|^2 \tag{1.1}$$

However, this doesn't mean that the wavefunction $\Psi(1,2) = \Psi(2,1)$. Instead we have the more general possibility that $\Psi(1,2) = e^{i\phi}\Psi(2,1)$ where ϕ is some phase. Applying the switch again would give $\Psi(1,2) = e^{2i\phi}\Psi(1,2)$, implying that $e^{i\phi} = \pm 1$, or

$$\Psi(1,2) = \begin{cases} +\Psi(2,1) & \text{boson} \\ -\Psi(2,1) & \text{fermion} \end{cases} \tag{1.2}$$

and we mathematically define bosons to be particles whose wavefunctions maintain sign under particle interchange, whereas fermions are particles whose wavefunctions flip sign under particle interchange. The *spin-statistics theorem* states that all fermions have half-integer spin and all bosons have integer spin [13]. In this text I will assume as valid the conditions that render the theorem true[§].

Note that indistinguishable means just that – two elementary particles of the same type are perfect duplicates of one another. It is simply not possible by any measurement we can make to tell one electron apart from another, or put a label on one π^+ to distinguish it from another π^+. This property of elementary particles is unlike anything in our everyday experience in the macroscopic world, where we are used to things that are similar – such as identical twins, or computers off of an assembly line – but not exactly the same. The elementary microscopic constituents of our universe are huge in number – about 10^{80} particles in all – but are of only 38 elementary types (including antiparticles), as we shall see shortly. It could have been otherwise, in which case it is hard to imagine how a coherent physical description of the universe would be possible. Just imagine trying to construct a theory with 10^{80} different kinds of particles, each of which had its own distinct properties!

Of course the existence of these 38 elementary types is our state of knowledge at the present time, and we now know that it is almost certainly incomplete. Over the past three decades we have discovered from observations in cosmology and astronomy that only 4% of the total energy of the universe is made of known matter (i.e. the matter that makes up the elements in the periodic table). Another 23% of the this energy budget is *dark matter*, whose presence is known to us only by the gravitational attraction it exerts on galaxies and clusters of galaxies [6]. Its composition in terms of elementary particles remains unknown to us at this point in history. The remaining part of the energy – 73% – is called *dark energy*, which is even more mysterious since it is causing the universe to *accelerate* in its expansion, whereas ordinary mass/energy (and the dark matter) would exert a decelerating influence. The ultimate composition of this form of energy is not at all clear, though the simplest explanation would appear to be that it is the vacuum energy

---

[§]A generalization of the result that all bosons have integer spin and all fermion half-integer spin (in units of $\hbar$) occurs in theories with only two space dimensions. In this case it is possible to have particles that have any possible spin, and such particles are called "anyons." These kinds of wavefunctions have applications in condensed matter physics [14].

of the universe, i.e. the ground-state energy of the aggregate of all bosons and fermions [15]. The problem with this interpretation is that although the cosmological vacuum energy is the largest fraction of the total energy of the universe (about 3/4), theoretical calculations indicate that it should be much larger than the value we observe – about 10^{120} times larger!

Nobody knows how to resolve the puzzles of dark energy and dark matter, and much effort is currently being expended by cosmologists, astronomers, and particle physicists to find out what these things are and how they behave. For the most part I will ignore these interesting issues thoughout most of this textbook, concentrating on elucidating the structure of the 4% of matter that we do know something about, and which is described by the Standard Model of particle physics. But there is one thing that we can be fairly confident about – whatever the dark stuff is, its elementary constituents will be bosons and/or fermions.

1.2.2   Forces

As far as experiment has been able to tell us, all known interactions in the world are governed by some combination of four basic forces: gravity, electromagnetism, nuclear (called the strong force), and radioactive (called the weak force). These forces have very different properties and manifestations, as illustrated in table 1.3.

Note that each force is associated with something called a mediator. What does this mean? Suppose we have a source of electric charge Q and we want to know what force a small test charge q experiences in its vicinity. In classical physics the answer is well known: we express the effect of Q on q in terms of something called an electric field $\vec{E}$:

$$\vec{F}_{\text{on } q} = q\vec{E} = \frac{Qq}{4\pi r^2}\hat{r} + \cdots = (\text{monopole}) + (\text{dipole}) + (\text{quadrupole, etc.})$$

which we say is due to the source Q. It is a vector, each component of which is a continuous function of the distance r. Of course the electric field itself is not directly observable; only the force $\vec{F}$ is.

TABLE 1.3
The Four Forces

	Gravity	Electromagnetism	Weak	Strong
Example	Planetary motion	Lightning	Radioactivity	Stars
Range	Long	(Long)	Short	Short
Strength	10^{-40}	10^{-2}	10^{-5}	1
Mediators	Graviton	Photon	W^+, W^-, Z^0	8 gluons
Helicity	2	1	1	1
Symmetry Group	Lorentz	Abelian Gauge	Non-Abelian Gauge	Non-Abelian Gauge
at 0 GeV	$\mathbf{SO}(3,1)$	$\mathbf{U}(1)$	Broken	$\mathbf{SU}(3)$
at 100 GeV	$\mathbf{SO}(3,1)$	$\mathbf{SU}(2) \times \mathbf{U}(1)_Y$		$\mathbf{SU}(3)$
at 10^{15} GeV	$\mathbf{SO}(3,1)$	Grand Unified Theory????		
at 10^{19} GeV	Theory of Everything (Superstring Theory?)			

This classical notion of force is modified at short distances due to quantum effects [16]. As Planck and Einstein noted in the last century, in order to describe certain phenomena (such as blackbody radiation or the photoelectric effect) the electromagnetic field should be "lumpy" – that is, it should come in distinct quanta called *photons*. In this picture, the source Q influences q by exchanging a photon with it, as shown in figure 1.3.

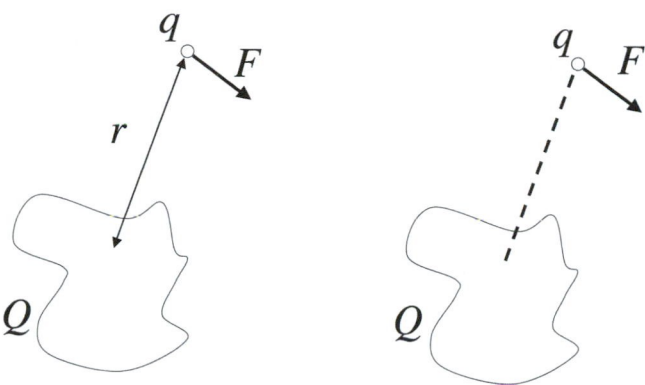

FIGURE 1.3
The left side is the classical picture of forces and fields, in which an element of charge Q a displacement $\vec{r}$ away exerts a force $\vec{F}$ on a test charge q. The right side is the quantum picture, in which the force is mediated by photons exchanged between Q and q, here represented by the dashed line.

We say that the exchanged photon is *virtual*, which means that it is not directly observed. But this means that we shouldn't see it transmit any net energy or momentum. In other words, the exchange of this virtual photon can only satisfy the requirements of energy and momentum conservation provided that the energy ΔE and momentum $\Delta \vec{p}$ exchanged in the process cannot be detected. This will be true if these quantities are bounded within the limits of what the uncertainty principle permits (so that any putative detection of the photon is washed out by quantum uncertainty). Specifically this means that

$$\Delta E \Delta t \leq \hbar \quad \text{and} \quad |\Delta \vec{p}| |\Delta \vec{r}| \leq \hbar \tag{1.3}$$

where

$$\Delta t = r/c = \text{time scale for photon exchange to take place} \tag{1.4}$$
$$|\Delta \vec{r}| = r = \text{distance scale for photon exchange to take place}$$

So the force experienced by q due to N virtual photons each transferring momentum $\Delta \vec{p}$ is

$$\vec{F} = N\frac{\Delta \vec{p}}{\Delta t} = N\frac{(\hbar/r)}{(r/c)}\hat{r} = N\hbar c\frac{\hat{r}}{r^2} = \frac{Qq}{4\pi r^2}\hat{r} \qquad (1.5)$$

provided we normalize $N = Qq/(4\pi\hbar c)$ – in other words, the number of virtual photons emitted/absorbed should be proportional to the product of the charges. In this sense the photon *mediates* the electromagnetic force!

This concept of a mediator is how all forces are understood in the context of modern particle physics. The basic idea is that any two subatomic particles exert forces on each other by exchanging other virtual quanta of subatomic particles – the mediators.

There are 12 mediators for all the non-gravitational forces: 1 photon, 8 gluons, and 3 weak vector bosons, more commonly referred to as the W^+, W^- and the Z. So far, experiment has indicated that these mediators have helicity 1 (i.e. integer spin $\hbar$). This is important, since our descriptions of bosons entail the following properties of bosons listed in table 1.4. I'm not

TABLE 1.4
Attractive/Repulsive Character of Forces

ODD-INTEGER SPIN PARTICLES:	mediate forces that are *both* attractive *and* repulsive
EVEN-INTEGER SPIN PARTICLES:	mediate forces that are *either* attractive *or* repulsive

going to prove the results in table 1.4 here – they can be demonstrated from some basic properties in quantum field theory. What this means is that since all non-gravitational forces are mediated by spin-1 bosons, they all have both an attractive and a repulsive character.

What if we include gravity? The fact that gravity is always attractive means that it must be mediated by an even-integer spin particle; we call this particle the *graviton* and it has spin-2. The Higgs boson (should it be found) has spin-0 and so will have a purely attractive character as well. This will add two more elementary particle types to the list, for a total of 14 out of the 38 elementary types¶.

¶It could be argued that I shouldn't include these two because, strictly speaking, we don't have direct observational confirmation of their existence. However, there is little doubt that the graviton is present, and the Higgs particle is foundational to our understanding of the Standard Model. I may be skating on thin ice with this one though: more Higgs particles than one may be discovered at the LHC or, more radically, no Higgs particle may be discovered.

1.2.3   Matter

So far we have been exploring bosons: the things that mediate forces. Now let's look at particles: the stuff that things are made of.

All known matter (excluding dark matter and dark energy) is made of spin-1/2 particles called *fermions* that have anticommuting statistics:

$$\Psi(1,2) = -\Psi(2,1)$$

where $\Psi(1,2)$ is the wavefunction for a system of two identical fermions. The proton and the electron are the best known of these, with the electron being the first fermion to be discovered (in 1897) [17]. The nucleus was discovered by Rutherford in 1910 [18], but it was not until 1932 that the nucleus was understood to be a composite object consisting of two kinds of fermions, known as protons and neutrons.

Uhlenbeck and Goudsmit postulated that the electron had spin $\hbar/2$ [19] in order to properly account for the Zeeman effect [20] and the Stern-Gerlach experiments [21]. This meant that Schroedinger's wave equation could not be used to describe the behavior of electrons, since it did not have take spin into account. In 1927 Dirac wrote down a relativistic wave equation that he required to be LINEAR in E and $\vec{p}$ (i.e. linear in the operators $\partial/\partial t$ and $\vec{\nabla}$). This equation predicted that a charged particle of $-e$ had spin $\hbar/2$. Remarkably, it also predicted that there was another particle of spin $\hbar/2$ with identical mass but charge $+e$. This particle is the *antiparticle* of the electron and is called the *positron*. Just as remarkably, it was discovered by Anderson in 1933.

All spin-$\hbar/2$ (or spin-1/2) particles obey Dirac's equation; you just adjust the mass and charge in the equation to describe the particle of interest. Hence each spin-1/2 particle has a corresponding antiparticle. This is also true for most bosons, but for some bosons (the photon being the best known example) the antiparticle is the particle itself $^{\|}$. In general all quantum numbers of an antiparticle are the negative of that of its corresponding particle except for its mass, which remains unchanged.

As far as experiment has indicated, all spin-1/2 particles in the world come in two basic types, leptons and quarks, listed in table 1.5. We can also construct a table listing the basic features of each of these kinds. In table 1.6 the quantity "color" labels the three distinct kinds of strong charge that a quark can have. These colors – called red, green and blue – have nothing to do with actual colors we can see – instead the term "color" is shorthand for "strong charge." Antiquarks have strong charges antired, antigreen and antiblue. Since the leptons do not experience the strong interactions, they have no color charge (they are color neutral) and so that entry in table 1.6 is blank.

$^{\|}$It is also possible for a fermion to be its own antiparticle; in this case it is called a Majorana fermion instead of a Dirac fermion

TABLE 1.5
The Kinds of Matter

LEPTONS	particles that *do not* experience strong interactions
QUARKS	particles that *do* experience strong interactions

TABLE 1.6
Basic Properties of Quarks and Leptons

	Flavor	EM Charge	Color	Helicity	Mass (MeV)
L E P T O N S	electron e e-neutrino ν_e muon μ μ-neutrino ν_μ tau τ τ-neutrino ν_τ	-1 0 -1 0 -1 0		L,R L,? L,R L,? L,R L,?	0.511 $< 2 \times 10^{-6}$ 106 < 0.19 1777 < 18
Q U A R K S	up u down d charm c strange s top t bottom b	+2/3 -1/3 +2/3 -1/3 +2/3 -1/3	R,G,B R,G,B R,G,B R,G,B R,G,B R,G,B	L,R L,R L,R L,R L,R L,R	2 5 1200 100 171,000 4,200

The quantity "flavor" is actually the charge experienced by the weak interactions. We will see that a strong interaction has the effect of changing a quark of one color (red, say) into a quark of another color (blue, say). Likewise, a weak interaction has the effect of changing a particle of one flavor (for example an electron) into another flavor (its corresponding neutrino). So in this sense flavor is to weak what electric-charge is to electromagnetism and what color is to strong (and what mass is to gravity). Note that flavor is the property by which we distinguish different types (or species) or particles. This is often the source of some confusion when one first tries to learn about the weak force, and so I have deferred the discussion of weak interactions to appear in Chapter 20 after the electromagnetic and strong forces in order that you can become more comfortable with a number of other concepts first. So there are 6 flavors (or types) of leptons, plus each of their antiparticles, and 6 flavors of quarks, plus each of their antiparticles, for a total of 24 fermion particle types in all**. Including the 14 types of bosons, this brings the list of elementary particle types to 38.

An additional complication in understanding particles has to do with quarks. Despite considerable effort over the years, it has not been possible to directly observe quarks. What we actually observe are bound states of quarks. It is generally believed (but not yet proven!) that quarks can bind together in only two possible ways, given in table 1.7. We call a bound state of quarks a

TABLE 1.7
Quark Bound States

BARYONS:	qqq	(a 3-quark bound state)
MESONS:	$q \neq q$	(a quark-antiquark bound state) } HADRONS

hadron. So far all observed hadrons have come in one of two types: *baryons* (which are bound states of three quarks) and *mesons* (which are bound states of a quark with some other antiquark). All hadrons are color-neutral, and the proton is the only stable hadron known (a free neutron will decay in less than 15 minutes). Understanding this structure will be one of our main tasks when we come to the strong interactions beginning in Chapter 15.

Note that the masses of the particles vary quite widely and (unlike the quantities in the rest of table 1.6) appear to have no pattern. It is generally believed that the masses of all elementary particles are acquired through some symmetry breaking-effect with a field called a Higgs field. The basic idea is that the Higgs field causes a kind of "drag" on all elementary particles of varying strength, and this drag is what we perceive as inertial mass. A consequence of this idea is that there must exist at least one spin-0 boson, which is called a Higgs particle. In October 2000 tantalizing evidence was presented from LEPII at CERN that the Higgs particle has been observed with a mass of approximately 114 GeV. However, subsequent analysis of the data has failed to confirm this, and the best that we can say at this point in time is that the Higgs mass is not less than 114 GeV. One of the main purposes of the LHC is to find the Higgs particle if it indeed exists.

Finally, note that both the leptons and the quarks come in three groupings according to mass: a light group, a medium group, and a heavy group. We refer to these groupings as *generations* (or sometimes as *families*). The lightest generation consists of the up and down quarks, and the electron and its neutrino (plus all their antiparticles). As noted above, all matter in the peri-

---

**It might be argued that I shouldn't count the antiparticles, since they are just like the particles except for a reversal of the signs of all the charges. I have made the distinction because a particle and its antiparticle counterpart are not identical – they can easily be distinguished in experiment.

odic table is made of combinations of these 4 particles (with the up and down quarks combining into protons and neutrons). Each generation is identical to the others in structure except for the differing masses. At present it is not known why this structure exists, and why the pattern has this particular form.

1.3   The Standard Model

The Standard Model of particle physics refers to the sum total of our knowledge of all the forces and particles described above. It is a particular kind of quantum theory, called a quantum field theory, that has a particular particle content and a particular symmetry group. There are infinitely many other models of the same general type that one could construct, but with different particle content and/or symmetries, as well as different values for the parameters. The Standard Model is the one that describes what we actually observe.

This particularity sounds reasonable – after all, isn't it the job of physics to describe what is observed? Furthermore, so far there are no experiments known that disagree with the predictions of this model. However, the model has a number of parameters that must be input from experiment before any further predictions can be made, summarized in table 1.8.

TABLE 1.8
Standard Model Parameters

masses of all fermions	12
coupling constants	4
mixing angles	8
vacuum angle	1
Higgs mass and coupling	2
Vacuum Energy	1

We see from table 1.8 that there are a total of 28 parameters in all. I've included the possibility that neutrinos have nonzero mass, though strictly speaking the Standard Model assumes they have zero mass. I've also included gravity (which accounts for one of the coupling constants of the four forces) and the strength of the dark energy (the vacuum energy) as two of the parameters. Together with General Relativity, the Standard Model provides us with 38 coupled differential equations for each of the particle types, dependent on the 28 parameters above. Once these 28 parameters are given we can predict the outcome of any process in particle physics from this model – and so far there are no experiments that disagree with any predictions which have been made. Not all of these parameters have been accurately measured, and their precise measurement is one of the current tasks of experimental particle physics.

The origin of these parameters and their empirical values is an unexplained mystery. It is hoped that a unified theory of physics (if one can ever be constructed) will explain both their values and their interrelationships, and that the set of 38 equations will actually be derived from one more unified theoretical structure, ideally one master equation. At present we don't know what that might be, so we must make do with the model that we have if we want to explain our observations of the subatomic world. You might think that 38 equations is a bit much to deal with. It is, but in practice we don't have to deal with them all at once – instead we can work with subsets of these equations, depending on the system of interest. Furthermore, the different equations have some common (and elegant) features amongst them due to the particular symmetries of Lorentz invariance and gauge invariance, subjects we will explore in chapters 2 and 12 respectively.

The purpose of this book is to teach you about the Standard Model: to fully acquaint you with its basic structure and features, to show you how to compute simple predictions from it, and to inform you of its empirical underpinnings. So let's get started!

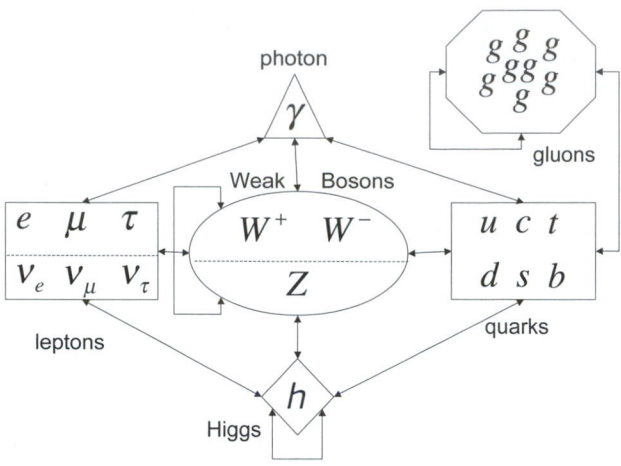

FIGURE 1.4

A pictorial representation of the different particles in the Standard Model and their interactions. A line between different shapes indicates that all particles within a given shape interact with all others in the connected shape, except for the photon, which does not interact with particles below the dashed lines (i.e. the neutrinos and the Z). Lines joining a shape to itself indicate self-interactions amongst the particles in the shape.

1.4 Questions

1. (a) It is possible to understand the strong force between a neutron and a proton as being effectively mediated by a boson called the pion, whose mass is 139 MeV. Use this to estimate the range of the strong force.

   (b) Similarly the weak force is mediated by the W-boson, whose mass is about 80 GeV. What range do you estimate this force has?

2. Before the neutron was discovered, beta-decay (the emission of an electron from a nucleus) seemed to support the idea that a nucleus consisted of protons with electrons trapped inside. The atomic number of a nucleus was given by the excess of protons over electrons.

(a) Using this picture, estimate the minimum momentum that a trapped electron must have.

(b) Use the relativistic relation $E^2 = (pc)^2 + (mc^2)^2$ to estimate its corresponding energy.

(c) How does this energy compare to that for an electron emitted in a typical beta-decay process? Does beta-decay tend to support or refute the trapped-electron model of the nucleus?

3. An experimentalist wants to probe distances of $d \leq 10^{-20}$ cm. How much collision energy must the machine be able to produce? How does this compare with the maximum energy of the LHC? If the size of the machine scales with the energy, how large would this machine have to be?

4. Suppose the electron had a mass 10 times its mass of 511 KeV. What particles would you expect to populate the universe?

5. Suppose that fermions were found to have different masses from their antiparticles. How many new parameters would there be in the Standard Model?

2

A Review of Special Relativity

Particle physics is concerned with understanding the behavior of the natural world in the most reductionistic manner possible. This means we need to probe interactions over very short timescales and very tiny distances. This in turn requires, among other things, an ability to collide particles at very high energies, energies that are far larger than the rest mass energies of the particles in the collision, where relativistic effects are significant.

So for particle physics an understanding of special relativity is crucial. Here I will review the basics of special relativity [22]. You probably have seen this material elsewhere – but what we do here might be in a repackaged notation that you may not be familiar with [23].

2.1 Basic Review of Relativity

The basic postulates of special relativity are twofold:

1) The laws of physics are equally valid in all inertial reference systems, where inertial means that Newton's first law holds.

2) The speed of light is constant and of the same value in all inertial systems.

The first of these postulates is familiar from Newtonian mechanics. It means that the laws of physics do not depend upon either the location, the orientation, or the constant velocity of the reference frame that we use to describe motion.

It is the second postulate that is counterintuitive, forcing a conceptual leap in our understanding to less than familiar territory. So let's explore this postulate in more detail.

Suppose we have 2 inertial systems (frames) S and S', where S' moves at velocity $\vec{v}$ with respect to S. If we orient the axes of both systems so that they are identical at $t = t' = 0$ and so that the relative motion is along the x/x' direction, then we know that

$$
\begin{aligned}
x' &= \gamma(x - vt) \\
y' &= y \\
z' &= z \\
t' &= \gamma(t - vx/c^2)
\end{aligned}
\qquad \text{where } \gamma = \frac{1}{\sqrt{1 - v^2/c^2}}
\qquad (2.1)
$$

a transformation known as a *boost transformation.* To invert this transformation (i.e. to go from S' to S), just change $\vec{v} \rightarrow -\vec{v}$:

$$
\begin{aligned}
x &= \gamma(x' + vt') \\
y &= y' \\
z &= z' \\
t &= \gamma(t' + vx'/c^2)
\end{aligned}
\qquad (2.2)
$$

You can easily check that solving eq. (2.1) for x' and t' in terms of x and t results in eqs. (2.2).

These transformations have several important physical consequences. Let's look briefly at what they are.

2.1.1 Relativity of Simultaneity

An *event* is defined as something that occurs at a specific place and time, along with all necessary preconditions and unavoidable consequences. It is something that happens in the real world, such as a supernova, an earthquake, or the fall of a raindrop. Usually we think of events taking place in a particular chronological order, regardless of who is observing them. The first surprise of special relativity is that this is not the case: *events occurring at the same time in S but at different locations* do NOT *occur at the same time in S'.*

Let's look at this more carefully. Suppose an observer in frame S records two events A and B located at the respective positions x_A and x_B as taking place at the same time: $t_A = t_B$. The above transformations then give

$$
\Rightarrow t'_A - t'_B = \gamma(t_A - vx_A/c^2) - \gamma(t_B - vx_B/c^2) = \gamma\frac{v}{c^2}(x_B - x_A) \neq 0 \quad (2.3)
$$

and so the observer in S' would record them as happening at different times. As an example consider an event (fig. 2.1) in which lightning strikes both ends of a moving train. There is an observer on a flatcar in the middle of the train that is moving (Mona), and another observer (Stan) standing on the ground beside the train as it moves by. Let **F** be the event "Lightning strikes the front end," and **B** be the event "Lightning strikes the back end."

Stan receives ("sees") the light emitted from events F and B simultaneously, and concludes that: "since I am midway between the marks on the tracks, and since the speed of light is a constant, then events F and B are simultaneous." However, Mona receives ("sees") the light emitted from event F first, then from B, and concludes that: "since I am midway between the marks on the train, and since the speed of light is a constant, then events F and B are NOT simultaneous: F occurred before B." See fig. 2.2 for a diagram in space and time of this event.

The reason this happens is that light has a finite speed, and so the information conveyed in the light from the bolt in the back has to "catch up" to Mona, who is riding in the middle of the train. Similarly Mona is "catching

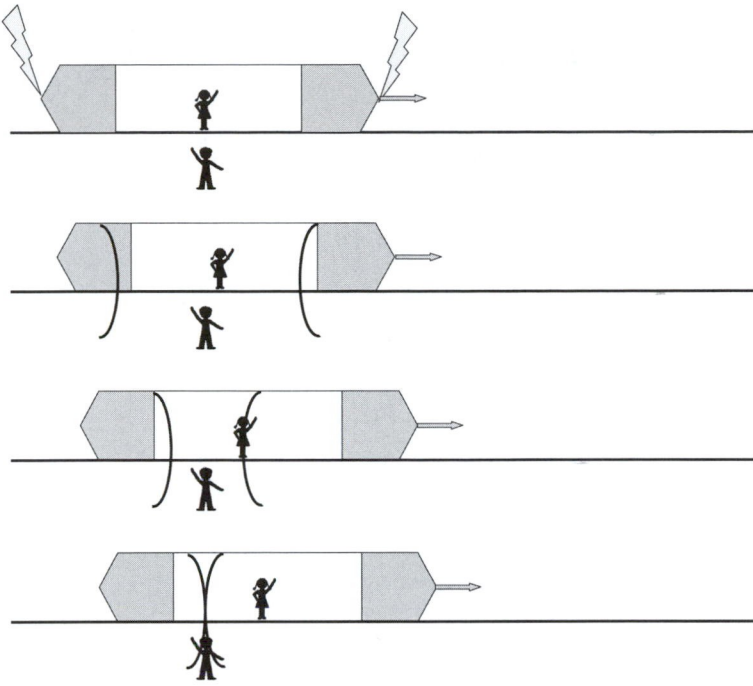

FIGURE 2.1

Lightning strikes the front and back ends of a moving train. The lightning strikes leave marks on the train and on the tracks, and the pulse of light is large enough to be detected by both Stan (on the ground) and Mona (on the train). Mona sees the front signal before the back signal; Stan sees both at the same time.

up" to the light coming toward her from the bolt at the front of the train. Stan, on the other hand, is not moving and so sees the bolts at the same time.

As we will see, since the second postulate implies that nothing travels faster than the speed of light, it will in turn imply that the transmission of all forms of information cannot travel faster than light. Unlike a Newtonian universe (in which an infinite speed of information transmission is possible in principle), finite transmission speeds in a relativistic universe force us to abandon a notion of universal simultaneity.

2.1.2 Length Contraction

This phenomenon refers to the fact that an object at rest in S' is shortened with respect to S. For example a rod of length L at rest in S' can be positioned so that one end is at $x' = 0$, the other at $x' = L$. Since it is moving in S, we

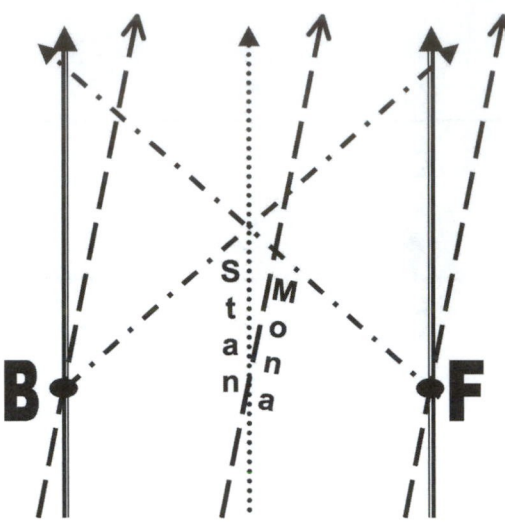

FIGURE 2.2

Paths of Stan (dotted), Mona and the ends of the train (dash), train tracks (solid), and the light pulses (dot-dash) from the lightning bolts as they move forward in time.

must record its length by measuring its ends at the same time ($t = 0$, say). So its length in S is, using (2.1) with $t = 0$:

$$x_{\text{one end}} - x_{\text{other end}} = \frac{1}{\gamma}\left(x'_{\text{one end}} - x'_{\text{other end}}\right) = L/\gamma < L \qquad (2.4)$$

==So *moving objects are shortened along the direction of motion.*==

Consider observers on a spaceship that have a poster of Einstein on the wall of their ship. From their perspective the picture is 4 ft. × 3 ft. However, an observer standing on an asteroid looking through the window of the ship as it goes by at 77% of the speed of light will see that the poster is only 4 ft. × 1.5 ft., as shown in fig. 2.3.

The importance of this phenomenon in particle physics occurs in the construction of accelerators – as the particles move at high velocities down an accelerator tube the effective length of the pipe is shortened in their direction of motion, and must be taken into account in the design parameters of the machine.

What observers on the ship see   What observers on the asteroid see

FIGURE 2.3
Length contraction demonstrated pictorially. Apic/Hulton archive/Getty Images.

2.1.3 Time Dilation

Suppose there is a clock ticking off an interval in S' – a consequence of relativity is that it ticks off a longer interval in S. If the clock runs from $t' = 0$ to $t' = T$, then using eq. (2.2) above we have

$$\Delta t_{\text{in } S} = \gamma(\Delta t' + v\Delta x'/c^2) = \gamma\Delta t' = \gamma T > T \tag{2.5}$$

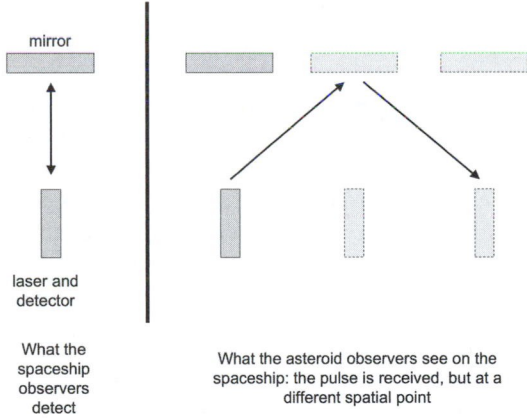

mirror

laser and
detector

What the
spaceship
observers
detect

What the asteroid observers see on the
spaceship: the pulse is received, but at a
different spatial point

FIGURE 2.4
A laser clock for measuring time dilation.

where $\Delta x' = 0$ because the clock is always at the same location in S'. So *moving clocks run slow.*

Consider a "clock" that consists of a laser capable of sending out very short pulses and a mirror to reflect those pulses back to a detector (fig. 2.4). The proper time will be measured by an observer at rest with respect to this "clock." Suppose this clock is on that same spaceship. Then an observer on the asteroid can watch as this "clock" goes by him/her at some speed (from his/her point of view). This observer will see the light pulse traveling a greater distance before it returns to the detector (see fig. 2.4). Since according to the postulates of special relativity the speed of light is the same for both inertial observers, the observer on the asteroid will claim that the spaceship clock is running slowly with respect to asteroid clocks. If the spaceship is moving at 77% of the speed of light, then for every three hours of time on the asteroid, the spaceship clocks register that only a little more than two hours have passed.

In particle physics we see this phenomenon manifest in decay rates of unstable particles: they take longer to decay the faster they are moving.

2.1.4   Velocity Addition

If a particle has velocity $\vec{u}'$ in the x' direction in S', then its velocity in S is

$$u = \frac{\Delta x}{\Delta t} = \frac{\gamma(\Delta x' + v\Delta t')}{\gamma(\Delta t' + v\Delta x'/c^2)} = \frac{\left(\frac{\Delta x'}{\Delta t'} + v\right)}{\left(1 + \frac{v}{c^2}\frac{\Delta x'}{\Delta t'}\right)} = \frac{u' + v}{1 + \frac{u'v}{c^2}} \tag{2.6}$$

showing that *velocities do not simply add.* Notice, though that if the speed of light is large compared to either of the velocities, then

$$u = \frac{u' + v}{1 + \frac{u'v}{c^2}} = (u' + v)\left(1 - \frac{u'v}{c^2} + \cdots\right) \simeq u' + v \tag{2.7}$$

and we see that velocities add as they do in the non-relativistic everyday world of experience.

This is one of the more counterintuitive things to understand about relativity, but it follows from (and is consistent with) the second postulate. To see how, suppose the moving observer in S' is shining a flashlight pointed in the direction of motion. The light will leave with speed c as seen by the moving observer. This means that we must set $|\vec{u}'| = u' = c$. The observer in S then measures this speed to be

$$u = \frac{c + v}{1 + \frac{cv}{c^2}} = c\frac{c + v}{c + v} = c \tag{2.8}$$

or in other words, the speed of the light is seen to be exactly the same, in accord with postulate 2!

2.2   4-Vector Notation

In the examples in section 2.1 the two frames were neatly aligned along the direction of motion. Of course this need not be the case; in general they could be moving in any direction relative to each other. So it's useful to introduce simplifying notation that unifies time and space components together.

This anticipated union suggests that vectors will have to be generalized from their 3-component form to having 4-components: one for each spatial direction and one for time. We call such objects 4-vectors, and can write one of them as, say, $a^\mu = (a^0, \vec{a})$. The 0th component is the "time" component, and the others are the spatial (3-vector) components.

Let's define

$$x^0 = ct \qquad x^1 = x \qquad x^2 = y \qquad x^3 = z \tag{2.9}$$

to be the components of a 4-vector x^μ that denotes the location in space and the time elapsed at some event (relative to a fixed choice of origin). Multiplication of the time by c ensures that each coordinate has units of length. In other words, we measure the time elapsed in terms of the light-travel-time: how far light would travel in a given time. Since all observers agree on what the speed of light is, the difference in ct as measured by various observers is equivalent to the difference in time t.

Now we can write the particular Lorentz transformation given in eq. (2.1) as a matrix equation

$$x'^\mu = \sum_{\nu=0}^{3} \Lambda^\mu{}_\nu x^\nu \text{ where } \Lambda^\mu{}_\nu = \begin{pmatrix} \gamma & -\beta\gamma & 0 & 0 \\ -\beta\gamma & \gamma & 0 & 0 \\ 0 & 0 & 1 & 0 \\ 0 & 0 & 0 & 1 \end{pmatrix} \tag{2.10}$$

(with "row" labeling the upper index and "column" the lower index)

with $\gamma = \frac{1}{\sqrt{1-\beta^2}}$ and $\beta = v/c$. So for the transformation (2.1) all components of the matrix $\Lambda^\mu{}_\nu$ are zero, except for

$$\Lambda^0{}_0 = \Lambda^1{}_1 = \gamma \qquad \Lambda^0{}_1 = \Lambda^1{}_0 = -\beta\gamma \qquad \Lambda^2{}_2 = \Lambda^3{}_3 = 1 \tag{2.11}$$

Before continuing, let's get a few conventions straight: I will use Greek letters to indicate spacetime indices that run over the values 0,1,2,3, and I will use Latin letters from the middle of the alphabet $(i, j, k, ...)$ to denote spatial indices of the type familiar from vector calculus; these will take on the values 1,2,3. I'll also use the *summation convention*: repeated indices are summed over (unless otherwise indicated). This means I can replace $\sum_{\nu=0}^{3} \Lambda^\mu{}_\nu x^\nu$ with $\Lambda^\mu{}_\nu x^\nu$ in eq. (2.10) since the ν index is repeated. Repeated indices almost always occur in pairs, so it is always clear what indices to sum over – I'll make it clear whenever this is not the case. Note that the repeated index is arbitrary, since we are summing over all of its values; in the sum in eq. (2.10), instead of ν we could have used ρ, σ, ς, or any Greek letter we wanted! We can use the summation convention for any repeated index (Greek or Latin). So Greek indices are summed from 0 to 4, and Latin indices from 1 to 3.

Now by postulate (2), a pulse of light emitted at $t = 0$ and $\vec{x} = 0$ in S has the same velocity in S'. Hence the distance ds to the wavefront from any point $\Delta\vec{x}$ in S at time Δt is the same as the distance ds' measured from any point $\Delta\vec{x}'$ in S' at time $\Delta t'$. Using Pythagoras' theorem we have

$$(c\Delta t)^2 - \left(\Delta x^1\right)^2 - \left(\Delta x^2\right)^2 - \left(\Delta x^3\right)^2 = ds^2 = ds'^2$$
$$= \left(c\Delta t'\right)^2 - \left(\Delta x'^1\right)^2 - \left(\Delta x'^2\right)^2 - \left(\Delta x'^3\right)^2 \qquad (2.12)$$

and you can easily check that (2.1) and (2.2) satisfy this relation.

The quantity is called an *invariant*; it has the same value in any inertial system. Invariants are nothing new – this concept is present in non-relativistic Newtonian mechanics. For example the length of a vector is invariant under rotations in Newtonian mechanics. What is new in relativity is that invariants typically involve a mixture of spatial quantities with temporal quantities.

The invariant quantity above in eq. (2.12) is a bit cumbersome, and it would be nice to write it in the form of a sum. The problem is we have 3 minus signs. We deal with this by writing

$$ds^2 = g_{\mu\nu}\Delta x^\mu \Delta x^\nu = g_{\mu\nu}\Delta x'^\mu \Delta x'^\nu \qquad (2.13)$$

where $g_{\mu\nu}$ is a matrix called the *metric*

$$g_{\mu\nu} = \begin{pmatrix} +1 & & & \\ & -1 & & \\ & & -1 & \\ & & & -1 \end{pmatrix} \qquad \text{i.e.} \begin{array}{l} g_{00} = +1 \\ g_{ij} = -\delta_{ij} \\ g_{0i} = 0 \end{array} \qquad (2.14)$$

which generalizes the 3-dimensional Kronecker-delta function used in vector calculus as you can see from eq. (2.14).

Notice that, since $x'^\mu = \Lambda^\mu{}_\nu x^\nu$, we have

$$g_{\mu\nu}x'^\mu x'^\nu = g_{\mu\nu}\left(\Lambda^\mu{}_\alpha x^\alpha\right)\left(\Lambda^\mu{}_\beta x^\beta\right) = \left(g_{\mu\nu}\Lambda^\mu{}_\alpha \Lambda^\mu{}_\beta\right)x^\alpha x^\beta \qquad (2.15)$$

(handwritten: $\mu = \nu$ because of the construction of $g_{\mu\nu}$)

Since this relation has to hold for all possible x^α and x'^μ, we must have

$$\boxed{g_{\mu\nu}\Lambda^\mu{}_\alpha \Lambda^\mu{}_\beta = g_{\alpha\beta}} \qquad (2.16)$$

This "boxed" relation is very important: it *must* be obeyed by *all* Lorentz transformations $\Lambda^\mu{}_\nu$. Alternatively, any matrix obeying this "boxed" equation is a valid Lorentz transformation. We take the boxed relation to mathematically define a Lorentz transformation.

While there are infinitely many possible Lorentz transformations, it is possible (and useful) to categorize them into six kinds. Since there are three distinct directions (along x, y, and z) for a moving observer to travel, three kinds of Lorentz transformations are transformations from the stationary to the moving observer along each of these directions. These are called *boosts.*

What are the other three kinds of transformations? Suppose we don't make any boosts. This means that Λ must have the form

$$\Lambda^\mu{}_\nu = \begin{pmatrix} 1 & 0 \\ 0 & R_{ij} \end{pmatrix} \tag{2.17}$$

where R_{ij} is some 3×3 matrix. It's easy to show from the boxed relation that this matrix must obey $R_{ik}R_{jk} = \delta_{ij}$, which in matrix form is $R^T R = I$. This is just the defining relation for a rotation (or rather an orthogonal transformation, to put it in mathematical terms), and so we see that rotations are a subset of the Lorentz transformations! It's clear then that there are 3 kinds of *rotations* (about x, y and z), and putting these together with the 3 boosts makes a total of 6 kinds of continuous Lorentz transformations.

Lorentz transformations are actually the relativistic analog of rotations. They transform 4-vectors analogous to the way that rotations transform 3-vectors. In this sense they define for us 4-vectors (and 4-tensors – mathematical objects with more than one Lorentz index) just in the same way that rotations are used to define 3-vectors (and 3-tensors – mathematical objects with more than one spatial index). We can make the following comparative table:

TABLE 2.1
Rotations compared to Lorentz transformations

Rotations $R_{ik}R_{jk} = \delta_{ij}$		Lorentz Transformations $g_{\mu\nu}\Lambda^\mu{}_\alpha\Lambda^\mu{}_\beta = g_{\alpha\beta}$
$\phi'(\vec{x}) = \phi(R^{-1}\vec{x})$	SCALAR	$\phi'(x^\mu) = \phi((\Lambda^{-1})^\mu{}_\nu x^\nu)$
$V_i'(\vec{x}) = R_i^j V_j(R^{-1}\vec{x})$	VECTOR	$V'^\alpha(\vec{x}) = \Lambda^\alpha{}_\mu V^\mu((\Lambda^{-1})^\mu{}_\nu x^\nu)$
$T_{ij}'(\vec{x}) = R_i^k R_i^l T_{kl}(R^{-1}\vec{x})$	TENSOR	$T'^{\alpha\beta}(\vec{x}) = \Lambda^\alpha{}_\mu \Lambda^\beta{}_\nu T^{\mu\nu}((\Lambda^{-1})^\mu{}_\nu x^\nu)$

2.3 Spacetime Structure

We can use the metric to define a new vector x_ν from our event-vector x^ν and vice versa:

$$g_{\mu\nu} = (g^{\mu\nu})^{-1} = g^{\mu\nu}$$

$$x_\mu = g_{\mu\nu}x^\nu \qquad \text{and} \qquad x^\mu = g^{\mu\nu}x_\nu \tag{2.18}$$

$$x_0 = ct \qquad\qquad x^0 = ct$$
$$x_1 = -x \qquad\qquad x^1 = x$$
$$x_2 = -y \qquad\qquad x^2 = y$$
$$x_3 = -z \qquad\qquad x^3 = z$$

where $g^{\mu\nu}$ is the inverse of $g_{\mu\nu}$. So using this we can write (2.13) as

$$ds^2 = \Delta x_\mu \Delta x^\mu = \Delta x'_\mu \Delta x'^\mu \tag{2.19}$$

"Raised" indices are called *contravariant*; "lowered" indices are called *covariant*. This distinction – unimportant in the three-dimensional flat spatial world of vectorial quantities in Newtonian mechanics – is of crucial importance in our description of the four-dimensional relativistic world. A given covariant vector A_μ and its contravariant counterpart A^μ contain the same information – it is how this information is expressed and manipulated compared to other physical quantities that makes the covariant/contravariant distinction important.

Here's an example. Notice that for *any* 2 four-vectors A_μ and B_μ we have

$$A'_\mu B'^\mu = g_{\mu\nu} A'^\mu B'^\nu = g_{\mu\nu} \left(\Lambda^\mu{}_\alpha A^\alpha \right) \left(\Lambda^\mu{}_\beta B^\beta \right)$$
$$= \left(g_{\mu\nu} \Lambda^\mu{}_\alpha \Lambda^\mu{}_\beta \right) A^\alpha B^\beta = g_{\alpha\beta} A^\alpha B^\beta = A_\beta B^\beta \tag{2.20}$$

i.e. $A_\beta B^\beta$ is *invariant* (it is the same in any inertial frame). It is the four-dimensional analog of a dot-product. Sometimes we will write $A \cdot B$ for the dot product:

$$A \cdot B = A_\beta B^\beta = A_0 B^0 - A_j B^j = A_0 B^0 - \vec{A} \cdot \vec{B} \tag{2.21}$$

where 3-vectors will always have "arrows" on top (or have Latin indices attached to them) so we know that's what they are*.

Notice that the "square" of a 4-vector need not be positive:

$$A^2 = A \cdot A = A_\beta A^\beta = A_0 A^0 - \vec{A} \cdot \vec{A} \tag{2.22}$$

This means that our concept of magnitude needs to be extended in special relativity. To do this we define:

$$A^2 > 0 \qquad \text{timelike}$$
$$A^2 = 0 \qquad \text{null or lightlike}$$
$$A^2 < 0 \qquad \text{spacelike}$$

Why this terminology? It's descriptive of how spacetime (i.e. space and time) appear to a given observer. Consider an observer at a position x^μ relative to the origin. We thus have either $x^2 > 0$, $x^2 = 0$, or $x^2 < 0$. This splits spacetime up into three regions about any given point, as illustrated in fig. 2.5. A classification is given in table 2.2.

---

*I will write $A_j B^j = A^j B^j = A_j B_j = A^j B_j = \vec{A} \cdot \vec{B}$ – these all mean the same thing for 3-vectors. Note that the quantity $A^\beta B^\beta = A^0 B^0 + A^j B^j$ is not invariant – it will look different in different Lorentz frames.

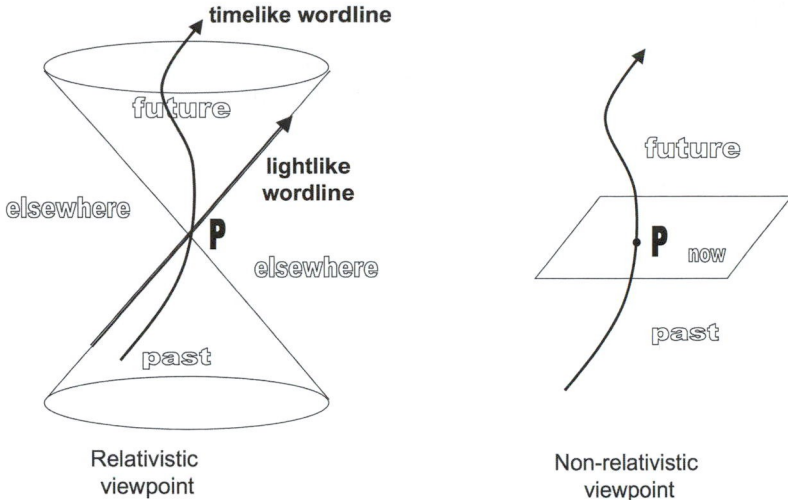

FIGURE 2.5

The diagram on the left shows the causal structure of space time in special relativity. The light cone about a point "**P**" plays a fundamental role in determining the events that can and cannot affect (or be affected by) **P**. The non-relativistic version of this diagram is at the right: all events are either to the future or past of **P** unless they are simultaneous with **P**, i.e. are in the "now" of **P**.

2.4 Momentum and Energy

Because x^μ is a position 4-vector, it cannot remain fixed – even if the spatial components remain constant, the time x^0 must increase, since the observer cannot remain frozen in time. In the most general situation, all the components of x^μ will be changing with time. We can parametrize this change by making each component a function of some parameter τ. As we shall see, we are typically more interested in how these functions change as τ changes, so let's define

$$u^\mu = \lim_{\Delta\tau \to 0} \frac{\Delta x^\mu}{\Delta\tau} = \frac{dx^\mu}{d\tau} \tag{2.23}$$

which is the rate of change of the position 4-vector with respect to this parameter.

What is τ? If we choose it to be the time coordinate, then we would have $u^\mu = \left(\frac{d(ct)}{dt}, \frac{d\vec{x}}{dt}\right) = \left(c, \frac{d\vec{x}}{dt}\right)$. This looks like a velocity in the spatial part, but leaves us with a problem: which clock is measuring the time? In Newtonian

TABLE 2.2
Past, Present, and Elsewhere

$x^2 > 0 \Rightarrow \lvert ct\rvert > \lvert\vec{x}\rvert \begin{cases} t > 0 \\ t < 0 \end{cases}$	Future	everything the observer potentially will influence
	Past	everything that potentially influenced the observer
$x^2 < 0 \Rightarrow \lvert ct\rvert < \lvert\vec{x}\rvert$	Elsewhere	no influence between observer and environment
$x^2 = 0 \Rightarrow \lvert ct\rvert < \lvert\vec{x}\rvert$	Lightcone	Boundary between elsewhere past and future

mechanics time is absolute, but in Einsteinian mechanics, moving clocks run slow, as we saw in section 2.1.3.

What we want is for τ to be a time that all observers can agree on. There is one such definition of time – namely the time measured by a clock that is at rest with respect to the observer whose position 4-vector is $x^\mu(\tau)$. For example, suppose the observer is on a plane. The clocks on the plane (and everything else: the flight attendants, the heart rate of the passengers, the movie) tick slow relative to you standing on the ground. If your clock ticks off an interval Δt, the plane's time is $\Delta\tau$ where

$$\Delta\tau = \Delta t/\gamma \tag{2.24}$$

Even though the clocks on the plane are moving slower, everybody will agree on what those clocks say. So both you and observers on the plane (and anyone else moving in some other inertial frame) will agree that the clocks on the plane tick off an interval $\Delta\tau$. We call $\Delta\tau$ the *proper time* of the observer. It is the shortest time any observer can measure, and it is invariant, since all observers (moving or not) agree with what it is.

Note that if we have a collection of N observers located at x^μ_A, each moving at different speeds, then each will have their own (invariant) proper time τ_A $(A = 1, ..., N)$. At ordinary speeds $\gamma = 1$ to about 1 part in 10^{20}, so we never notice this effect, and each observer perceives time as absolute. This is why the non-relativistic picture at the right of fig. 2.5 is so useful, and so much in accord with our everyday intuition. However, in particle accelerators $\gamma \gg 1$ and these time-dilation effects are crucial. For example the lifetimes of unstable subatomic particles can dramatically increase at very high velocities.

Since τ is invariant, u^μ is covariant:

$$u'^\mu = \frac{dx'^\mu}{d\tau'} = \frac{dx'^\mu}{d\tau} = \frac{d}{d\tau}\left(\Lambda^\mu{}_\nu x^\nu\right) = \Lambda^\mu{}_\nu \frac{dx^\nu}{d\tau} = \Lambda^\mu{}_\nu u^\nu \tag{2.25}$$

u^μ is covariant.

$u \cdot u$ is invariant.

and the components of u^μ are

$$u^\mu = \left(c\frac{dt}{d\tau}, \frac{d\vec{x}}{d\tau} \right) = \frac{dt}{d\tau} \left(c, \frac{d\vec{x}}{dt} \right) = \Lambda^0{}_0 \left(c, \frac{d\vec{x}}{dt} \right) = \gamma \left(c, \frac{d\vec{x}}{dt} \right) = \gamma(c, \vec{v})$$

$$\gamma = \frac{1}{\sqrt{1 - \frac{v^2}{c^2}}}$$

$$(2.26)$$

We see that the spatial component of the 4-velocity is related to the usual 3-velocity by $u^i = \gamma v^i \simeq v^i$, where the approximation is good whenever $|\vec{v}| \ll c$. The magnitude of the 4-velocity is

$$u \cdot u = \gamma^2 \left(c^2 - \vec{v} \cdot \vec{v} \right) = c^2 \gamma^2 \left(1 - \vec{v} \cdot \vec{v}/c^2 \right) = c^2 \tag{2.27}$$

which is invariant!

Note that $u^2 > 0$: the 4-velocity of a particle is always timelike. In terms of the causal structure in figure 2.5, this means that the trajectory of any physical object must remain within its future light cone, and must have emerged from its past light cone at any event P. This trajectory is called the worldline of the particle, and if the particle has mass the worldline must be timelike as shown in figure 2.5.

Now let's consider momentum. Since this is (mass)×(velocity) in Newtonian mechanics, let's define

$$p^\mu = mu^\mu \tag{2.28}$$

where we identify the constant m with the mass of the body whose 4-velocity is u^μ since

$$p^i = mu^i = \gamma mv^i \quad (\simeq mv^i \text{ for small } v^i) \tag{2.29}$$

This seems like a reasonable definition of the spatial components of the momentum. What, then, is p^0? We have

$$p^0 = mu^0 = \gamma mc = \frac{mc}{1 - v^2/c^2}$$
$$= \frac{1}{c} \left[mc^2 + \frac{1}{2}mv^2 + \frac{3}{8}m\frac{v^4}{c^2} + \cdots \right] \tag{2.30}$$

upon expanding in powers of $\frac{v}{c}$. The 2nd term in the series is easily recognized as the non-relativistic kinetic energy, and so we define

$$p^0 = E/c \quad \text{where} \quad E = \gamma mc^2 = \frac{mc^2}{\sqrt{1 - v^2/c^2}} \tag{2.31}$$

Remember that in non-relativistic mechanics it is only *changes* in energy that are physically meaningful. However, in relativity, every body has a minimum constant energy mc^2 called its *rest energy* (or *rest mass*):

$$R = mc^2 \tag{2.32}$$

Note that since m is a constant it is an invariant: all observers agree on what it is. The kinetic energy is the difference between the full energy and the rest energy

$$T = E - R = (\gamma - 1)mc^2 = \frac{1}{2}mv^2 + \frac{3}{8}m\frac{v^4}{c^2} + \cdots \tag{2.33}$$

which is the usual non-relativistic term plus an infinite series of speed-dependent corrections.

What is the magnitude of p^μ? It's easy to compute it:

$$p^2 = p \cdot p = m^2 u \cdot u = m^2 c^2 \tag{2.34}$$

and so p^2 is an invariant. However, we also have $p \cdot p = (E/c)^2 - \vec{p} \cdot \vec{p}$, and so

$$E^2 = (mc^2)^2 + |c\vec{p}|^2 \Rightarrow E = \sqrt{(mc^2)^2 + c^2 |\vec{p}|^2} \tag{2.35}$$

which gives an expression for the energy of a particle in terms of its mass and its momentum, analogous to the non-relativistic relation $E = \frac{p^2}{2m}$.

The interesting thing about the relation (2.35) is that it holds even for a massless object! Setting $m = 0$ in eq. (2.35) we find

$$E = |\vec{p}| c \quad \text{and} \quad p^2 = 0 \tag{2.36}$$

which means that a massless particle has a 4-momentum of zero magnitude! We can't make sense out of this from our earlier expressions $E = \frac{mc^2}{\sqrt{1-v^2/c^2}}$ and $\vec{p} = \frac{m\vec{v}}{\sqrt{1-v^2/c^2}}$ because both would give zero – unless the massless body travels at the speed of light (i.e. $|\vec{v}| = c$). In this case we have a zero-over-zero limit, and the preceding expressions become ambiguous.

Rather than work with such ambiguity, we regard equation (2.34) as being the equation that fundamentally defines the 4-momentum of a particle of rest mass m. This equation is valid for all $m \geq 0$, and yields equation (2.35) for $m \neq 0$, and equation (2.36) for $m = 0$.

2.5 Collisions

Energy and momentum are conserved in any process (as we'll see in Chapter 4), which is why they are useful quantities to deal with. In fact our picture of collisions in relativity is quite similar to our non-relativistic picture! Each has both elastic and inelastic collisions, and energy and momentum are always conserved. The key difference between the non-relativistic and relativistic cases has to do with the conservation of mass, something not true in relativistic physics. Table 2.3 illustrates the parallels between the two situations.

In solving collision problems, it is generally a matter of ensuring that the relation $\sum p^\mu_{\text{in}} = \sum p^\mu_{\text{out}}$ is satisfied. In principle this can be done component-by-component. However, in practice such an approach is rather cumbersome. A better strategy is to search for invariants, exploiting relationships between them to simplify the situation and solve the problem. This is typically done

TABLE 2.3
Collisions

Collisions	
Non-relativistic	Relativistic

Momentum is conserved
$$\sum \vec{p}_{in} = \sum \vec{p}_{out}$$
Energy is conserved
$$\sum E_{in} = \sum E_{out}$$

$4 -$ momentum is conserved
$$\sum p^{\mu}_{in} = \sum p^{\mu}_{out} \Rightarrow \begin{cases} \sum E_{in} = \sum E_{out} \\ \sum \vec{p}_{in} = \sum \vec{p}_{out} \end{cases}$$

Mass is conserved
$$\sum m_{in} = \sum m_{out}$$

Mass is NOT necessarily conserved
$$\sum m_{in} \neq \sum m_{out}$$

Kinetic energy may
or may not be conserved

Kinetic energy may
or may not be conserved

(by isolating the quantities of interest in the problem, and then taking dot-products and squares to obtain the desired answer.)

Let's look at some examples of how to use this formalism.

2.5.1 Broadside Collision

Two particles each of mass m and speed v, collide at right angles, forming a new body of mass M. What is the mass of M?

Answer
The best way to solve this problem is to recognize that the squares of the 4-momenta of both the incoming and outgoing particles are invariants.

$$\text{First conserve momentum}: \quad p^{\mu}_1 + p^{\mu}_2 = p^{\mu}$$
$$\text{Square both sides}: \quad p^2_1 + p^2_2 + 2p_1 \cdot p_2 = p^2$$

Each of the squared momenta is the rest mass of its corresponding particle, and so

$$(mc)^2 + (mc)^2 + 2\left(\frac{E_1 E_2}{c^2} - \vec{p}_1 \cdot \vec{p}_2\right) = (Mc)^2 \qquad (2.37)$$

Now we use the fact that the spatial momenta are orthogonal ($\vec{p}_1 \perp \vec{p}_2$) and that the energies are related to the rest masses via $E_A = \gamma_A mc^2$, where A=1,2 labels which particle we are interested in. Hence

$$2m^2 (1 + \gamma_1 \gamma_2) = M^2$$
$$\Rightarrow M^2 = 2m^2 \left(1 + \frac{1}{1 - v^2/c^2}\right) = 2m^2 \left[\frac{2 - v^2/c^2}{1 - v^2/c^2}\right] \qquad (2.38)$$

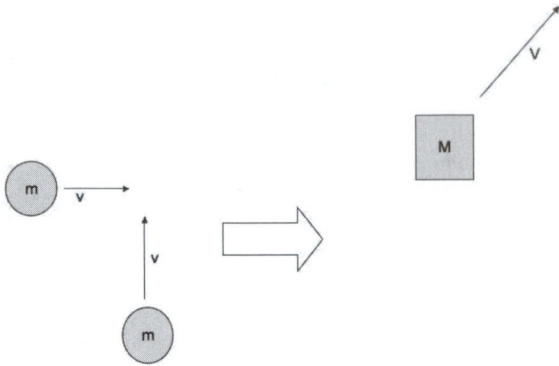

FIGURE 2.6
A Broadside collision between two equal mass particles moving at right-angles to each other.

Note that $M > 2m$. This is a sticky collision and so the mass increases because kinetic energy is converted into rest energy.

2.5.2 Compton Scattering

A massless particle elastically collides with a massive one (mass M) at rest. What is the final energy of the massless particle if it is scattered at an angle θ?

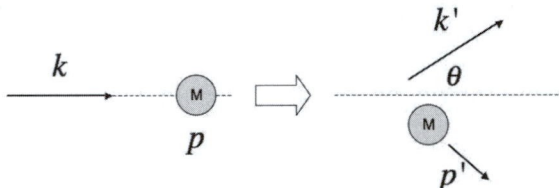

FIGURE 2.7
Compton Scattering of a massless particle from a massive one

<u>Answer</u>

Setting $c = 1$ for simplicity, as before we conserve momentum and find invariants. The most complicated momentum is the momentum of the massive scattered particle, since we don't know the angle it scatters to, nor do we know its final speed. So we isolate p'^μ and then square both sides:

$$\text{First conserve momentum}: \quad p'^\mu = p^\mu + k^\mu - k'^\mu$$

$$\begin{aligned} \text{Isolate the most complicated} \\ \text{momenta and square}: \quad & p'^2 = (p + k - k')^2 \\ \text{Simplify}: \quad & M^2 = M^2 + k^2 + k'^2 \\ & \quad + 2\left(p \cdot (k - k') - k \cdot k'\right) \end{aligned}$$

$$\text{Rearrange (remember } k^2 = 0 = k'^2): \quad p \cdot k = k' \cdot (k + p)$$

Now we need to insert the given information in the problem, namely that $k^\mu = E\left(1, \hat{k}\right)$, where $\hat{k}$ is a unit vector that points along the direction of motion of the initial massless particle. We know that this is the form k^μ takes because it is null: we must have $k^\mu k_\mu = 0$. Similarly $k'^\mu = E'\left(1, \hat{k}'\right)$, where we know from the setup of the problem that $\hat{k} \cdot \hat{k}' = \cos\theta$. Hence we have $k' \cdot k = EE' - EE'\hat{k} \cdot \hat{k}' = EE'(1 - \cos\theta)$. We also know the initial 4-momentum of the massive particle because it is at rest: $p^\mu = \left(M, \vec{0}\right)$. Putting this into the above gives

$$\begin{aligned} \text{Write in terms of components}: \quad ME &= E'(E + M) - E'E\left(\hat{k} \cdot \hat{k}'\right) \\ &= EE'(1 - \cos\theta) + ME' \\ \text{Solve for the final energy}: \quad E' &= \frac{ME}{E(1 - \cos\theta) + M} \\ &= \frac{E}{1 + \frac{E}{M}(1 - \cos\theta)} \quad (2.39) \end{aligned}$$

Note that the only variable in the problem is the scattering angle θ: the energy E' of the scattered massless particle is fully determined by this angle and all other given variables.

2.5.3 3-Body Decay

A particle of mass M explodes into 3 identical particles each of mass m, which move away from each other at equal angles. What speed(s) do the particles move at?

<u>Answer</u>

Again, we conserve momenta. The relation is

$$p^\mu = p_1^\mu + p_2^\mu + p_3^\mu \quad (2.40)$$

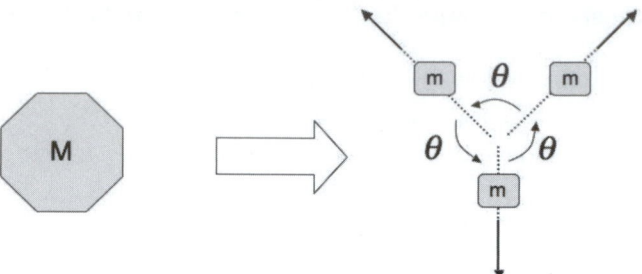

FIGURE 2.8

Diagram for the three-body explosion

Here we have to make clever use of the fact that all 3 particles come out at the same angle. This means that

$$\vec{p}_i \cdot \vec{p}_j = |\vec{p}_i| \cdot |\vec{p}_j| \cos\theta \quad \text{for every } i, j = 1, 2, 3 \tag{2.41}$$

In this example it is more useful to look at the components of the conservation law. First let's check the spatial components

$$\text{Spatial momentum conservation} \Rightarrow \vec{p}_1 = -(\vec{p}_2 + \vec{p}_3)$$
$$\Rightarrow |\vec{p}_1|^2 = -\vec{p}_1 \cdot (\vec{p}_2 + \vec{p}_3) = \cos\theta \, |\vec{p}_1| \, (|\vec{p}_2| + |\vec{p}_3|)$$

Doing this for all three particles, we find

$$\cos\theta = \frac{|\vec{p}_1|}{|\vec{p}_2| + |\vec{p}_3|} = \frac{|\vec{p}_2|}{|\vec{p}_3| + |\vec{p}_1|} = \frac{|\vec{p}_3|}{|\vec{p}_2| + |\vec{p}_1|}$$

and so we see that $|\vec{p}_1| = |\vec{p}_2| = |\vec{p}_3|$.

Now let's use energy conservation. Since $p^\mu = \left(Mc, \vec{0}\right)$, we have

$$\text{Energy conservation}: \quad Mc = \frac{E_1}{c} + \frac{E_2}{c} + \frac{E_3}{c} = 3\frac{E}{c} = 3\frac{mc}{\sqrt{1 - v^2/c^2}}$$

$$P_i^2 = \frac{E_i^2}{c^2} - |\vec{P}_i|^2 = P_2^2 = P_3^2 = m^2 c^2$$

where each energy must be equal since the momenta and rest masses of the final particles are all equal. Hence

$$v = c\sqrt{1 - 3\,(m/M)^2} \tag{2.42}$$

2.6 Questions

1. Show that the product $A_\beta B^\beta$ is Lorentz-invariant but the products $A^\beta B^\beta$ and $A_\beta B_\beta$ are not.

2. Suppose we write the velocity v of a particle as $v = c\tanh\eta$, where η is a parameter called the rapidity.

   (a) Find the form of a Lorentz transformation for a particle moving in the x direction in terms of this rapidity parameter.

   (b) Consider a succession of two boosts, both in the x direction, with velocities v_1 and v_2. What is the value of the rapidity parameter for the combination of these transformations in terms of the rapidity parameters η_1 and η_2 for each?

3. A particle moving at speed v collides with an identical particle at rest. What is the center-of-mass frame speed of this particle?

4. The lifetime of the muon is 2.2×10^{-6} seconds. Muons are produced high in the atmosphere (10,000 meters above the surface of the earth) from pions in cosmic rays moving at 99.9% of the speed of light. Once produced the muons move at the same very high speed.

   (a) How far will the muon travel according to non-relativistic physics? Will it make it to the surface of the earth?

   (b) How far will the muon travel according to relativistic physics? Will it make it to the surface of the earth?

   (c) The pion lifetime is 2.6×10^{-8} seconds. Is it possible for the pions to reach the surface of the earth?

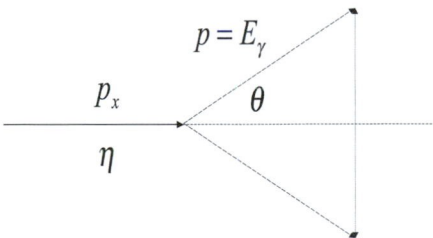

FIGURE 2.9
Diagram for question #5

5. (a) An η moving at $3/4$ the speed of light decays into two photons of equal energy E, which are each emitted at an angle θ relative to the direction of the η's motion. Find θ.

   (b) Consider the same process, but now with the photons emitted along the direction of motion. What is the frequency difference between the two photons?

6. It is possible to define the force on a particle in special relativity as the proper-time derivative of its spatial momentum

$$F^\mu = \frac{dp^\mu}{d\tau}$$

   where $p^\mu = m u^\mu$.

   (a) Show that the acceleration $a^\mu = \frac{du^\mu}{d\tau}$ of a particle is always orthogonal to its 4-velocity u^μ.

   (b) Express the relationship between the spatial acceleration $\vec{a} = \frac{d\vec{v}}{dt}$ and the spatial force $\vec{F}$. Do they always have the same direction?

   (c) What is the meaning of the 0th component F^0?

7. Neutral pions can be produced from the collision of a photon with a proton $p + \gamma \rightarrow p + \pi^0$.

   (a) What is the minimum energy a photon must have for this process to take place when the proton is at rest?

   (b) The largest energy a photon in the cosmic microwave background can have is about 1 meV. What is the minimum energy that a cosmic ray proton must have in order to produce pions by scattering off of the microwave background?

8. A pion traveling at speed v decays into an lepton of mass m_ℓ and its correspond antineutrino $\bar{\nu}_\ell$. Suppose the antineutrino is emitted at right angles to the direction of motion of the pion.

   (a) Find an expression for the angle that the lepton is emitted relative to the orginal direction of motion.

   (b) Suppose a pion of speed v emits a muon at angle θ, and another pion at speed v' emits an electron at the same angle, each having emitted the antineutrinos at right angles to the direction of motion. How much larger or smaller is v compared to v'?

9. Particle A decays into two particles: $A \longrightarrow B + C$.

   (a) Find an expression for the energy of each outgoing particle in terms of the various rest masses.

   (b) Find the magnitudes of the outgoing momenta of B and C.

   (c) Under what circumstances can your answer in part (b) equal zero?

10. Find the energies of the outgoing particles for the following decays

$$\pi^+ \longrightarrow e^- + \bar{\nu}_e$$
$$K^- \longrightarrow \pi^- + \pi^0$$
$$\pi^0 \longrightarrow 2\gamma$$
$$\Lambda \longrightarrow p + \pi^-$$
$$\rho \longrightarrow \pi^+ + \pi^-$$

11. Particle A collides into particle B, which is at rest, and three or more particles are produced as a result: $A + B \longrightarrow C_1 + C_2 + \cdots + C_N$.

    Find the threshold energy for this reaction to take place in terms of the rest masses of the particles.

12. Find the threshold energies for the following reactions

$$p + p \rightarrow p + p + \pi^0$$
$$\pi^- + p \rightarrow K^0 + \Sigma^0$$
$$p + p \rightarrow p + K^0 + \Sigma^+$$
$$p + p \rightarrow p + p + \pi^+ + \pi^-$$
$$p + p \rightarrow p + p + p + \bar{p}$$

13. Particle A decays into three or more particles: $A \longrightarrow B + C + D + \cdots$

    (a) Find the maximum and minimum energies that particle B can have in terms of the rest masses of the other particles in the problem.

    (b) For the decay $D^0 \longrightarrow K^- + \pi^- + e^+ + \nu_e$, find the maximum and minimum energies the e^+ can have.

14. Consider a two-body scattering event $A + B \longrightarrow C + D$. Define the following quantities

$$s = (p_A + p_B)^2 \qquad t = (p_A - p_C)^2 \qquad u = (p_A - p_D)^2$$

    which are a set of variables called *Mandelstam variables*.

    (a) Show that the sum of the Mandelstam variables is a Lorentz invariant quantity and compute its value.

    (b) Find the energy of A in the lab frame, where particle B is at rest in terms of the Mandelstam variables and the rest masses of the particles.

    (c) Find the center-of-mass energy of A in terms of the Mandelstam variables and the rest masses of the particles. What is the total energy in this frame?

3

Symmetries

One of the most fundamental notions of physics is that of symmetry: the idea that certain systems, or properties of a system, or laws governing a system, remain unchanged when you do something to them. For example, the gravitational field of a homogeneous spherical object is rotationally symmetric: it is the same no matter how you orient yourself about the sphere. Another example is a crystal: an arrangement of atoms that looks the same if you shift it in space in a certain direction. You can learn a lot about a physical system (and the mathematics that describes it) just by knowing what symmetries it has. In physical theories symmetries manifest themselves in terms of two basic notions: *invariance* and *covariance*.

Invariance is the term used to describe properties of a system that do not change when a symmetry transformation is performed. If a quantity is invariant, all observers will agree on its value. We've already seen that the proper time of an observer is an invariant under the Lorentz transformations: all observers agree on what the proper time is for any given observer. Rest mass is another such invariant.

Covariance is the term used to describe properties of a system that change in accord with the changes induced by the symmetry transformation. For example, if the equations of motion describing a given system have well-defined transformation properties when a given symmetry operation is performed, we say that these equations are covariant (i.e. they vary along, or co-vary, with the transformation). In the above example, the equations describing the motion of a body around a homogeneous spherical object are covariant with respect to rotations – they will transform in a manner consistent with rotational symmetry.

We'll see later that symmetries necessarily result in conservation laws. These laws may be used to obtain new information about a given system, which in turn may yield further laws. In general a given interaction respects many conservation laws: conservation of momentum, angular momentum, charge etc. This stringently constrains the possible mathematical description of the interaction.

Symmetries can be either continuous or discrete. For example, time-reversal is a discrete symmetry (all clocks either advance or retreat), whereas time-translation is a continuous one (the zero of time is a completely arbitrary choice). Discrete symmetries result in multiplicative conservation laws, whereas continuous symmetries yield additive laws.

3.1 Groups

Mathematically, symmetries are properly described by making use of two concepts: that of a *group* and that of an *algebra*. Let's look at both of these, with some examples [24].

3.1.1 Axioms of a Group

When we think of a symmetry, we think of performing certain operations, or transformations, on a system that leave some of its basic properties unchanged. We can collect all of the relevant transformations into a set, $\mathcal{G} = \{g_1, g_2, ...\}$ (which may be infinite in size), and then ask what basic properties the set should have in order for it to describe a symmetry. Mathematically the answer is given in terms of what we call a group.

A *group* is a set of $\mathcal{G} = \{g_1, g_2, ...\}$ of objects with a binary operation "$\diamond$" – some operation that allows us to combine two elements in the set – that has four properties:

TABLE 3.1
Properties of a Group

CLOSURE	If $g_1, g_2 \in G \Rightarrow g_1 \diamond g_2 \in G$	(combinations remain in the set)
IDENTITY	There exists $I \in G$ $\Rightarrow I \diamond g_i = g_i$ for every $g_i \in G$	(one element does nothing)
INVERSE	Every $g_i \in G$ has a $g_i^{-1} \in G$ such that $g_1 \diamond g_i^{-1} = I$	(combinations can be undone)
ASSOCIATIVITY	If $g_1, g_2, g_3 \in G$ $\Rightarrow (g_1 \diamond g_2) \diamond g_3 = g_1 \diamond (g_2 \diamond g_3)$	(combinational groupings can be interchanged)

These properties are (in most cases) the minimum ones needed for transformations to be meaningful in physics, and for the most part are motivated by common-sense considerations. If the elements g_i are transformations, then we'd like a combination of two transformations to be the same kind of transformation (closure). We also want to undo transformations (inverse), and not do transformations (identity). Finally, we'd like our answer to be independent of how we combine any three transformations together, provided we don't change their sequential order (associativity). Note the order of applying the

"$\diamond$" is relevant: in general $g_1 \diamond g_2 \neq g_2 \diamond g_1$ for any two elements*.

How do we know if something is a group or not? The only sure way is to check all four properties. Let's look at the integers as an example, shown in table 3.2. A simple check shows that the integers form a group under addition,

TABLE 3.2

Integers as a Group

	Integers: $G = Z$			
	Addition $\quad(\diamond = +)$		Multiplication $\quad(\diamond = \times)$	
Closure	$Z_1 + Z_2 \in Z$	✓	$Z_1 \times Z_2 \in Z$	✓
Identity	$0 + Z_i = Z_i$	✓	$1 \times Z_i = Z_i$	✓
Inverse	$Z_i + (-Z_i) = 0$	✓	$Z_i \times 1/(Z_i) = 1$ but $1/(Z_i) \notin Z$	✗
Associativity	$(Z_1 + Z_2) + Z_3$ $= Z_1 + (Z_2 + Z_3)$	✓	$(Z_1 \times Z_2) \times Z_3$ $= Z_1 \times (Z_2 \times Z_3)$	✓

but not under multiplication. However, in this latter case, if we modify the set of integers $\mathbb{Z}$ to include all fractions and exclude 0, we get a group $\mathbb{Q}$: the non-zero rational numbers under multiplication.

This latter example is a common feature: if we have a set of elements that don't form a group under some combining operation, we can sometimes get a group by either modifying the set, or modifying the combining operation (or both). It is quite common in physics to have a set of transformations (typically implied by experiment) that almost satisfies the properties of a group, but not quite. Often it is obvious how to generalize the set so that the group properties are satisfied, although there is no general prescription as to how to do this. Part of the job of a theoretical particle physicist is to make intelligent guesses as to how to find such generalizations to ensure that a given system has a desired symmetry. This symmetry can then be exploited to help make further predictions about the system that can then be tested against experiment.

Table 3.3 provides an example of another kind of group. This group is called **U**(1), the **U**nitary group of complex 1×1 matrices. Unlike the group

---

*If $g_1 \diamond g_2 = g_2 \diamond g_1$ then we say that the group is abelian. Transformations that shift location in space (translations) yield the same result no matter what order they are applied in, and so they are abelian. Rotations in space about a fixed point are not abelian – a rotation about the x-axis followed by one about the y axis will yield a different result than performing these operations in the opposite order.

TABLE 3.3

Complex Phases as a Group

Complex Phases ($G = \mathbf{U}(1) = \left\{e^{i\theta}\right\}$, $\diamond = \times =$ multiplication; $\theta \in R$)	
Closure	$e^{i\theta_1} \times e^{i\theta_2} \in \mathbf{U}(1)$ ✓
Identity	$e^0 \times e^{i\theta_i} = e^{i\theta_i}$ ✓
Inverse	$e^{i\theta_i} \times e^{-i\theta_i} = e^0 = 1$ ✓
Associativity	$\left(e^{i\theta_1} \times e^{i\theta_2}\right) \times e^{i\theta_3} = e^{i\theta_1} \times \left(e^{i\theta_1} \times e^{i\theta_3}\right)$ ✓

of the integers under addition, which has countably infinitely many elements, this group has an uncountably infinite number of elements since the elements depend continuously on a real parameter θ. As we'll see in Chapter 12, this is the symmetry group of the electromagnetic interactions.

3.1.2   Representations

So far we've discussed properties and examples of sets of transformations. But how are groups of transformations used in particle physics? What is it that the transformations act on?

In particle physics the most common thing that is transformed under a symmetry operation is the wavefunction of a particle or set of particles. In general the wavefunction can be written down as a multicomponent single column matrix with complex entries (a complex column vector) in some (abstract) multi-dimensional space[†]. The symmetries of the system are transformations that act on the wavefunction via multiplication by complex matrices.

In (almost) all cases in particle physics, the groups we are interested in are sets of complex matrices, and the combining operation is matrix multiplication. So what we need is a way of understanding group elements as matrices. This leads us to the concept of *representations.*

If we have a symmetry group, how do we write down the elements of the group in terms of matrices? Or alternatively, if we have a set of matrices, how do we know that these matrices correspond to the group elements of the symmetry that we are interested in?

It is a result of the mathematics of group theory [25] that every group $\mathcal{G}$ can be *represented* by a set of matrices: in other words

For every $g \in \mathcal{G}$ there is a corresponding matrix M_g such that

$$g_1 \diamond g_2 = g_3 \Rightarrow M_{g_1} M_{g_2} = M_{g_3}$$

which means that the matrix multiplication table of our set of matrices must correspond to the multiplication table of the group. A set of matrices $\{M_{g_1}\}$

[†]In introductory quantum mechanics, the wavefunction has only a single component, i.e., it is one complex function. In particle physics it will commonly be necessary to generalize it to multiple components.

that has this property is called a *representation*[‡] of the group $\mathcal{G}$.

The main advantage of using a matrix representation of a symmetry is that you can carry out explicit calculations. Since matrix multiplication is associative, the associativity property is automatically satisfied for any representation. Furthermore, the inverse property demands that all matrices in a representation must be invertible, which significantly reduces the possible forms the matrix representation of a group can have. Finally, the identity group element is easily represented by the identity matrix.

There are several somewhat counterintuitive points about representations that are worth noting:

- Representations are not unique. Since the only constraint on a set of matrices to represent a group is that it reproduce the multiplication table as described above, it is clear that a group can have many different matrix representations, since there are many ways of doing this. In fact, a given group $\mathcal{G}$ has infinitely many representations!

- Representations in general are not faithful for the same reason: typically, more than one element of a given group is represented by one matrix. The most extreme case of this is to represent all elements by the identity (the $n \times n$ unit matrix)[§]. If the representation is 1-1 (i.e. if every group element is represented by one and only one matrix – and vice versa) then we say the representation is *faithful*[¶]. A faithful representation provides the maximum amount of information that we can have about the group.

- A group of matrices is already a faithful representation of itself: this is called the *fundamental* representation. In general it is possible to use other matrices to form different representations of the same matrix group. For example the group of 2×2 unitary matrices (called $\mathbf{SU}(2)$) can be represented by matrices of every possible dimension: $1 \times 1, 2 \times 2, 3 \times 3, 4 \times 4, ...$

[‡]Strictly speaking, a d-dimensional representation is a map ρ (called a homomorphism) from the group to a set of $d{\times}d$ matrices $\mathcal{M} = \{M_g\}$:

$$\rho : \mathcal{G} \to \mathcal{M}$$

[§]Note that this representation causes us to lose all information about the multiplication properties of the group.

[¶]So why not only use faithful representations? Why would we be interested in any representation that was less than faithful? One reason is that we often don't need all of the information about the group. A good example of this is to represent every group element by either +(Identity) or −(Identity) – this will tell us about the basic even/odd properties of the group, and hence something about the basic physics of the system. For example, if the group is the group of rotations, then this kind of representation tells us whether or not we have a reflection, i.e. a transformation of odd parity.

3.1.3   Irreducible Representations

How do we know what all the representations are for a particular group? This is a major problem in group theory, one complicated by the fact that you can combine old representations together to form new ones:

$$\text{Suppose } M_{g_1} M_{g_2} = M_{g_3} \text{ and } \quad N_{g_1} N_{g_2} = N_{g_3}$$

$$\text{where } \left\{ \begin{array}{l} M_{g_i} \text{ is an } m \times m \text{ representation} \\ N_{g_i} \text{ is an } n \times n \text{ representation} \end{array} \right\} \text{ of } \mathcal{G}$$

$$\text{Then } \ L_{g_i} = \begin{pmatrix} M_{g_i} & 0 \\ 0 & N_{g_i} \end{pmatrix} \text{ is also a representation: } L_{g_1} L_{g_2} = L_{g_3}$$

$$\text{and is an } (m + n) \times (m + n) \text{ representation}$$

This shows why a group has infinitely many representations: we can always keep enlarging the matrices from smaller representations in this block-diagonal way.

But this is redundant! Clearly we shouldn't count the set $\{L_{g_i}\}$ as a separate representation, because it is a trivial combination of smaller, block-diagonal representations. What we really want to know are the non-redundant sets of matrices that can represent a group. We call such sets *irreducible* representations. Specifically, any representation whose matrices simultaneously cannot all be decomposed into block-diagonal form is called an irreducible representation[||]. Once the irreducible representations (irreps) are known for a given group, then all possible representations are known, because they can be constructed out of the irreps.

So in particle physics (and other kinds of physics), if we have a system **S** that is believed to have some symmetry $\mathcal{S}$, we conceptualize our treatment of the system as follows:

1. Find the group $\mathcal{G}$ associated with the symmetry $\mathcal{S}$.

2. Find all the irreps of the group $\mathcal{G}$.

3. The wavefunctions that transform under these different irreps are the only wavefunctions that are mathematically permitted – and hence physically realizable** – to describe the physics of the system **S**.

For example, the group **SU**(3) is the symmetry group under which the color charges of quarks transform (we'll see why later). This group has representations of dimension $3, 6, 8, 10, \ldots$. Hence quark wavefunctions must be complex column vectors with either $3, 6, 8, 10$, etc. entries. These are the only kinds of quark wavefunctions that can exist if **SU**(3) is the symmetry –

[||]Mathematically we say that irreducible representations have no invariant subspaces.
**These wavefunctions are commonly called *multiplets*, where the "multi" part refers to how many components the wavefunction has. A doublet has 2 components, a triplet 3, etc.

the math won't allow anything else! Likewise, bound states of quarks must have wavefunctions that are one or more of these same dimensionalities.

How do we find (step 2) the different irreps of a given group $\mathcal{G}$? This is a major task for group theory [26], one beyond the scope of what I'm able to present here. However, the problem has been solved (and a well-defined procedure exists) for a certain class of groups that are of interest in particle physics: Lie Groups!

3.1.4 Multiplication Tables

You are no doubt familiar with the multiplication table

$\times$	1	2	3
1	1	2	3
2	2	4	6
3	3	6	9

where I've only written down the first few entries for all the natural (counting) numbers. The table tells us to combine two numbers under multiplication to get another number: by choosing a number x in the leftmost column and another number y topmost row, we can find out what the answer is when we multiply these two numbers together simply by looking at where the row that x is in intersects the column that y is in. In a similar manner, a multiplication table provides a straightforward way of writing down all of the information about a given group.

For groups of very large or infinite size this is generally not very practical. But for groups of a reasonable finite size it can provide a handy way of letting us know what kind of information the group contains, and what kinds of relationships exist between its various elements.

One of the interesting things to consider in a finite group is what happens when we take a given element and keep combining it with itself. Consider the sequence of elements obtained by taking successive "powers" of some element $x \in \mathcal{G}$ of order n: $x,$ $x^2 = x \diamond x,$ $x^3 = x \diamond x \diamond x, \ldots$ We can show that eventually this sequence repeats itself in a cyclic manner. This is not too hard to see. Suppose that the first repeated element in the group is x^p and that it is repeated after $q + 1$ steps. Therefore $x^{q+1} = x^p$ for $p \leq q + 1$. Consequently

$$ x^{-1} \diamond x^{q+1} = x^{-1} \diamond x^p \Rightarrow x^q = x^{p-1} \Rightarrow p = 1 $$

because we can't have x^{p-1} appear in the sequence before x^p unless $p = 1$.

3.2 Lie Groups

A *Lie Group* is a group in which the group elements are smooth (continuous and differentiable) functions of some finite set of parameters $\theta_\mathfrak{a} \in \mathbb{R}$, and in which the "$\diamond$" operation depends smoothly on those parameters. A typical element $g \in \mathcal{G}$ is written as

$$g = g(\theta_1, \ldots, \theta_N) = \exp\left[i\theta_\mathfrak{a}\mathbf{T}^\mathfrak{a}\right] = \exp\left[i\vec{\theta} \cdot \vec{\mathbf{T}}\right] \qquad (3.1)$$

$$\text{where } \mathfrak{a} = 1, \ldots, N \text{ and } \{\theta_\mathfrak{a}\} \text{ are continuous parameters}$$

The $\mathbf{T}^\mathfrak{a}$'s are mathematical objects called the *generators* of the group. There are N of them – they are called generators because if you know what they are, you can construct any element of the group that you want by exponentiating as above – in other words, they generate the group. So the $\mathbf{T}^\mathfrak{a}$'s contain all the relevant information[††] of the group in a very economical form!

Lie groups are important in physics because most of the symmetries we consider are continuous symmetries. For example, rotations depend continuously on parameters called angles. A given angle specifies how much rotation has been carried out around a given axis.

Almost all of the important symmetry groups in particle physics are Lie groups [27]. The most important of these are the UNITARY and ORTHOGONAL groups. In particular, $\mathbf{SO}(3)$ is the group that describes rotational symmetry, a symmetry we believe to be true of all the laws of nature[‡‡]. It is almost identical in structure to another group, $\mathbf{SU}(2)$, which is the most important *internal* symmetry group. In general, in particle physics, we refer to symmetries as being either *spacetime* symmetries (they transform space and time co-ordinates in some way) or *internal* symmetries (they transform wavefunction components and/or charges amongst themselves).

The action of any group on a wavefunction multiplet will result in a rearrangement of multiplet components. If the group is a symmetry, then we mean that the physics of the system is insensitive to this rearrangement.

Lie groups have been completely classified in terms of their transformation properties. Table 3.4 lists all the kinds and their uses in particle physics.

[††]Actually this is not quite true. The $\mathbf{T}^\mathfrak{a}$'s contain all *local* information about the group – in other words, if you know all N of the $\mathbf{T}^\mathfrak{a}$'s then you can construct any group element that is continuously connected to the identity element. However, other group elements – those not continuously connected to the identity – cannot be constructed in this way. This property is relevant in constructing the vacuum state in quantum electrodynamics as discussed in Chapter 25.

[‡‡]Of course this assertion must be tested by experiment, which so far has provided no indication that rotational symmetry is not a symmetry of the laws of physics at their most fundamental level.

TABLE 3.4
Lie Groups in Particle Physics

Group Name	Defining Property	Applications
$\mathbf{U}(n)$	$n \times n$ unitary ($U^\dagger U = 1$)	$\mathbf{U}(1)$ electromagnetism
$\mathbf{SU}(n)$	$n \times n$ unitary ($U^\dagger U = 1$) with $\det U = 1$	$\mathbf{SU}(3)$ strong interactions $\mathbf{SU}(2)$ weak interactions
$\mathbf{O}(n)$	$n \times n$ orthogonal ($O^{\mathrm{T}}O = 1$)	$\mathbf{O}(3)$ rotations and reflections
$\mathbf{SO}(n)$	$n \times n$ orthogonal ($O^{\mathrm{T}}O = 1$) with $\det O = 1$	$\mathbf{SO}(3)$ rotations $\mathbf{SO}(3,1)$ Lorentz transformations
$\mathbf{Sp}(n)$	$n \times n$ symplectic ($S^{\mathrm{T}}\alpha S = \alpha$) with $\alpha = \begin{pmatrix} 0 & I \\ -I & 0 \end{pmatrix}$	$\mathbf{Sp}(6)$ Hamiltonian structure of phase space
G_2	Exceptional	G_2: 5-Dimensional Chern-Simons Supergravity
F_4	Exceptional	F_4: 6-Dimensional Chern-Simons Supergravity
E_6	Exceptional	E_6 : Grand unified theory
E_7	Exceptional	E_7 : Grand unified theory?
E_8	Exceptional	E_8 : Superstring theory

3.3 Algebras

Closely related to the notion of a group is that of an algebra. An *algebra* is a vector space $V = \{v_I\}$ with a binary (combining) operation "∘" such that

$$v^I \circ v^J = \sum_K C^{IJ}{}_K v^K = C^{IJ}{}_K v^K \qquad (3.2)$$

where the summation convention was used in the last step. This rule ensures that when vectors in V are combined, they always give a vector in V, which can be expressed as a linear combination of the other vectors in V ; the $C^{IJ}{}_K$'s are the coefficients of this combination, known as the *structure constants.* Note that all we care about here is closure – there doesn't have to be either an inverse or an identity! The relation (3.2) implies that knowledge of all possible $C^{IJ}{}_K$'s is equivalent to knowledge of the combining operation "∘" – in this sense the $C^{IJ}{}_K$'s characterize the algebra under consideration. The I, J in the $C^{IJ}{}_K$ are not raised using the metric, but rather are placed there for ease of notation.

A familiar example of an algebra is the vector cross-product. The set of elements in the algebra is the set of vectors in 3-dimensional space, and the combining operation ∘ = ×, the cross-product. Any two vectors $\vec{a}$ and $\vec{b}$ combine under the cross-product to give a third vector $\vec{a} \times \vec{b}$, which is also a vector in the 3-dimensional space. In terms of the unit vectors $\{\hat{x}, \hat{y}, \hat{z}\}$ of a Cartesian coordinate system we have

$$\hat{x} \times \hat{y} = \hat{z} \qquad \hat{y} \times \hat{z} = \hat{x} \qquad \hat{z} \times \hat{x} = \hat{y}$$

with all other cross-products zero. In this example $C^{IJ}{}_K$ turns out to be the Levi-Civita symbol ϵ^{IJK}, which shall be defined below.

3.3.1 Lie Algebras

Why is an algebra important? The reason is that the Taylor-series expansion of a Lie group $\mathcal{G}$ gives us an algebra called a *Lie Algebra* (denoted $\mathfrak{G}$). The demonstration of this is given in the appendix. The elements of $\mathfrak{G}$ are the generators $\mathbf{T}^a$ of the Lie group, and the combining operation ∘ = [,] is the commutator

$$\left[\mathbf{T}^a, \mathbf{T}^b\right] = i f^{ab}{}_c \mathbf{T}^c \qquad (3.3)$$

where the C^{IJ}_K's are denoted by the $f^{ab}{}_c$'s. The group associativity law implies the relation

$$\left[\left[\mathbf{T}^a, \mathbf{T}^b\right], \mathbf{T}^c\right] + \left[\left[\mathbf{T}^b, \mathbf{T}^c\right], \mathbf{T}^a\right] + \left[\left[\mathbf{T}^c, \mathbf{T}^a\right], \mathbf{T}^b\right] = 0 \qquad (3.4)$$

which is called the *Jacobi identity*. Any algebra for which (3.3) and (3.4) hold is a Lie Algebra.

As noted previously, we can write the group elements of any Lie Group in the form $g(\theta_1, ..., \theta_N) = \exp(i\theta_a \mathbf{T}^a)$. Consequently, if we know what all of the $\mathbf{T}^a$'s are, we can produce any element of the group that we want. As stated above, we say that a Lie group is *generated* by the $\mathbf{T}^a$'s, which are called the *generators* of the group. Hence we can study (most of) the symmetries of physics simply by studying their infinitesimal structure using Lie Algebras [28], instead of the more complicated structure of Lie Groups.

3.4 The Rotation Group SO(3)

A rotation about, say, the $\hat{z}$-axis is given by:

$$\begin{aligned} x' &= \cos\theta\, x + \sin\theta\, y \\ y' &= -\sin\theta\, x + \cos\theta\, y \end{aligned} \xrightarrow{\text{small } \theta} \begin{aligned} x' &= x + \theta\, y = x - (\theta\hat{z} \times \vec{x}) \cdot \hat{x} \\ y' &= -\theta\, x + y = y - (\theta\hat{z} \times \vec{x}) \cdot \hat{y} \end{aligned} \tag{3.5}$$

$$(\text{ More generally } \quad \vec{x}' = \vec{x} - \left(\vec{\theta} \times \vec{x}\right) \text{ where } \quad \vec{\theta} = \theta\hat{z})$$

Do the set of all possible rotations form a group? It's clear that any two rotations yields a rotation (closure), that rotation by an angle $\theta = 0$ is the identity and that every rotation can be undone simply by performing the same rotation by the negative of the angle. So if we use matrices to represent the rotations, we'll have a group (since associativity will automatically hold). Moreover, we know that any rotation can be constructed by taking products of matrices from the set

$$\mathcal{R} = \left\{ \begin{pmatrix} 1 & 0 & 0 \\ 0 & \cos\theta_x & \sin\theta_x \\ 0 & -\sin\theta_x & \cos\theta_x \end{pmatrix}, \begin{pmatrix} \cos\theta_y & 0 & -\sin\theta_y \\ 0 & 1 & 0 \\ \sin\theta_y & 0 & \cos\theta_y \end{pmatrix}, \begin{pmatrix} \cos\theta_z & \sin\theta_z & 0 \\ -\sin\theta_z & \cos\theta_z & 0 \\ 0 & 0 & 1 \end{pmatrix} \right\} \tag{3.6}$$

and so our group $\mathcal{G} = \mathcal{R}$ is the set of all matrices that are products of matrices in the above set, where $\{\theta_x, \theta_y, \theta_z\}$ are three arbitrary parameters. A general group element $\mathfrak{r} \in \mathcal{R}$ will be a function of these parameters, i.e. $\mathfrak{r} = \mathfrak{r}(\theta_x, \theta_y, \theta_z)$. This group is the orthogonal group of 3×3 matrices of determinant one, so it is **SO**(3).

Let's construct the Lie Algebra of **SO**(3). We can do this by examining the elements of $\mathcal{R}$, rewriting them in the form $\exp\left[i\theta_{\mathfrak{l}}\mathbf{T}^{\mathfrak{l}}\right] \simeq I + i\theta_{\mathfrak{l}}\mathbf{T}^{\mathfrak{l}}$ for $\theta_{\mathfrak{l}} \ll 1$. For example

$$\begin{pmatrix} \cos\theta_z & \sin\theta_z & 0 \\ -\sin\theta_z & \cos\theta_z & 0 \\ 0 & 0 & 1 \end{pmatrix} \simeq \begin{pmatrix} 1 & \theta_z & 0 \\ -\theta_z & 1 & 0 \\ 0 & 0 & 1 \end{pmatrix} + \mathcal{O}\left(\theta_z^2\right)$$

$$= \begin{pmatrix} 1\,0\,0 \\ 0\,1\,0 \\ 0\,0\,1 \end{pmatrix} + \theta_z \begin{pmatrix} 0\,\,1\,0 \\ -1\,0\,0 \\ 0\,\,0\,0 \end{pmatrix}$$

$$= I + i\theta_z \mathbf{T}^z \tag{3.7}$$

which means that $\mathbf{T}^z = -i \begin{pmatrix} 0\,\,1\,0 \\ -1\,0\,0 \\ 0\,\,0\,0 \end{pmatrix}$. This is the generator of rotations about the $\hat{z}$-axis. Performing a similar exercise on the other two matrices, we obtain the set of generators

$$\{\mathbf{T}^x, \mathbf{T}^y, \mathbf{T}^z\} = \left\{ -i \begin{pmatrix} 0\,\,0\,\,0 \\ 0\,\,0\,\,1 \\ 0\,-1\,0 \end{pmatrix}, -i \begin{pmatrix} 0\,0\,-1 \\ 0\,0\,\,\,0 \\ 1\,0\,\,\,0 \end{pmatrix}, -i \begin{pmatrix} 0\,\,1\,0 \\ -1\,0\,0 \\ 0\,\,0\,0 \end{pmatrix} \right\} \tag{3.8}$$

which form the basis of the Lie algebra of $\mathbf{SO}(3)$, which we call $\mathbf{so}(3)$. We can check the commutation relations

$$[\mathbf{T}^x, \mathbf{T}^y] = (-i)^2 \begin{pmatrix} 0\,\,0\,\,0 \\ 0\,\,0\,\,1 \\ 0\,-1\,0 \end{pmatrix} \begin{pmatrix} 0\,0\,-1 \\ 0\,0\,\,\,0 \\ 1\,0\,\,\,0 \end{pmatrix} - (-i)^2 \begin{pmatrix} 0\,0\,-1 \\ 0\,0\,\,\,0 \\ 1\,0\,\,\,0 \end{pmatrix} \begin{pmatrix} 0\,\,0\,\,0 \\ 0\,\,0\,\,1 \\ 0\,-1\,0 \end{pmatrix}$$

$$= \begin{pmatrix} 0\,\,1\,0 \\ -1\,0\,0 \\ 0\,\,0\,0 \end{pmatrix} = i\mathbf{T}^z \tag{3.9}$$

with the rest being similar: $[\mathbf{T}^y, \mathbf{T}^z] = i\mathbf{T}^x$ and $[\mathbf{T}^z, \mathbf{T}^x] = i\mathbf{T}^y$. So altogether we have

$$[\mathbf{T}^a, \mathbf{T}^b] = i\epsilon^{ab}{}_c \mathbf{T}^c \quad (\text{i.e. } f^{ab}{}_c = \epsilon^{ab}{}_c) \tag{3.10}$$

which are the commutation relations of the rotation group familiar from quantum mechanics (e.g. in solving for the wavefunction of the Hydrogen atom).

The object ϵ^{ab}_c is called the Levi-Civita symbol or (more commonly) the epsilon tensor. We can write it in several equivalent ways: $\epsilon^{ab}_c = \epsilon^{cab} = \epsilon_{cab}$. It has the following properties

$$\epsilon^{abc} = \begin{cases} 0 & \text{if any of } c, a, b \text{ are equal} \\ -\epsilon^{acb} = +\epsilon^{cab} = -\epsilon^{cba} & \text{switches sign under interchange} \\ \quad = +\epsilon^{bca} = -\epsilon^{bac} & \text{of any pair of indices} \\ 1 & \text{if } a = 1, b = 2, c = 3 \end{cases} \tag{3.11}$$

The epsilon tensor plays a role in the cross-product analogous to the role that the Kronecker-delta symbol plays in dot product. We have

$$\vec{A} \cdot \vec{B} = \delta^{ij} A^j B^k \qquad \qquad \left(\vec{A} \times \vec{B}\right)^i = \epsilon^{ijk} A^j B^k \tag{3.12}$$

or more explicitly

$$\vec{A} \cdot \vec{B} = A^1 B^1 + A^2 B^2 + A^3 B^3$$

$$\left(\vec{A} \times \vec{B}\right)^1 = A^2 B^3 - A^3 B^2 \qquad \left(\vec{A} \times \vec{B}\right)^2 = A^3 B^1 - A^1 B^3$$

$$\left(\vec{A} \times \vec{B}\right)^3 = A^1 B^2 - A^2 B^1$$

Just in the same way that δ^{ij} can be defined in n-dimensions (by letting the indices i, j take on all values from 1 to n), so also can the epsilon tensor be defined in n-dimensions, with the key distinction being that it has as many indices as there are dimensions. So in 2 dimensions, the epsilon tensor has two indices (ϵ^{ij}), in 3 dimensions three indices (ϵ^{ijk}), in 4 dimensions four indices ($\epsilon^{\mu\nu\alpha\beta}$) and so on. We will be making use of all of these dimensional versions of the epsilon tensor in this text***.

The algebra $\mathbf{so}(3)$ (and its Lie group $\mathbf{SO}(3)$) is so commonly used in physics, we give its generators a special notation: $\mathbf{T^a} = \mathbf{J^a}$, so that the set of generators $\{\mathbf{J^x}, \mathbf{J^y}, \mathbf{J^z}\}$ obeys

$$\left[\mathbf{J^a}, \mathbf{J^b}\right] = i\epsilon^{abc}\mathbf{J^c} \tag{3.13}$$

and using (3.11) it's not hard to show that

$$\epsilon^{abd}\epsilon^{dce} + \epsilon^{bcd}\epsilon^{dae} + \epsilon^{cad}\epsilon^{dbe} = 0 \tag{3.14}$$

and so from (3.13) we have

$$\left[\left[\mathbf{J^a}, \mathbf{J^b}\right], \mathbf{J^c}\right] + \left[\left[\mathbf{J^b}, \mathbf{J^c}\right], \mathbf{J^a}\right] + \left[\left[\mathbf{J^c}, \mathbf{J^a}\right], \mathbf{J^b}\right] = 0 \tag{3.15}$$

which means the Jacobi identity (3.4) holds. So $\mathbf{so}(3)$ is indeed a Lie algebra.

As noted above, each group element of $\mathbf{SO}(3)$ can be written in the form $\exp\left[i\vec{\theta} \cdot \vec{\mathbf{J}}\right]$, which is the matrix that corresponds to a rotation of angle $\left|\vec{\theta}\right|$ about the $\hat{\theta}$ direction. For example. if $\vec{\theta} = \theta\hat{y}$:

$$
\begin{aligned}
\exp\left[i\vec{\theta} \cdot \vec{\mathbf{J}}\right] = \exp\left[i\theta\mathbf{J^y}\right] &= \sum_{n=0}^{\infty} \frac{(i\theta\mathbf{J^y})^n}{n!} \\
&= \begin{pmatrix} 1 & 0 & 0 \\ 0 & 1 & 0 \\ 0 & 0 & 1 \end{pmatrix} + \sum_{n=1}^{\infty} \frac{(-1)^n (\theta)^{2n}}{(2n)!} \begin{pmatrix} 1 & 0 & 0 \\ 0 & 0 & 0 \\ 0 & 0 & 1 \end{pmatrix} \\
&\quad + \sum_{n=0}^{\infty} \frac{(-1)^n (\theta)^{2n+1}}{(2n+1)!} \begin{pmatrix} 0 & 0 & -1 \\ 0 & 0 & 0 \\ 1 & 0 & 0 \end{pmatrix}
\end{aligned} \tag{3.16}
$$

*** There is an important distinction in 4 dimensions compared to 3 and 2, having to do with the raising and lowering of the indices. In 3 spatial dimensions we have $\epsilon_c^{ab} = \epsilon^{cab} = \epsilon_{cab}$. However, it is NOT true that $\epsilon^{\mu\nu\alpha\beta} = \epsilon_\mu^{\ \nu\alpha\beta} = \epsilon^\mu_{\ \nu}^{\ \alpha\beta}$ etc. Instead we must raise/lower indices of $\epsilon^{\mu\nu\alpha\beta}$ using the metric $g_{\mu\nu}$.

$$
= \begin{pmatrix} 1 + \sum_{n=1}^{\infty} \frac{(-1)^n (\theta)^{2n}}{(2n)!} & 0 & -\sum_{n=0}^{\infty} \frac{(-1)^n (\theta)^{2n+1}}{(2n+1)!} \\ 0 & 1 & 0 \\ \sum_{n=0}^{\infty} \frac{(-1)^n (\theta)^{2n+1}}{(2n+1)!} & 0 & 1 + \sum_{n=1}^{\infty} \frac{(-1)^n (\theta)^{2n}}{(2n)!} \end{pmatrix}
$$

$$
= \begin{pmatrix} \cos\theta & 0 & -\sin\theta \\ 0 & 1 & 0 \\ \sin\theta & 0 & \cos\theta \end{pmatrix} \tag{3.17}
$$

which is a rotation of angle θ about the $\hat{y}$−axis. From this we see that $\mathbf{J}^y$ generates rotations about the y-axis. Using similar methods you can show that $\mathbf{J}^x$ generates rotations about the x-axis and $\mathbf{J}^z$ generates rotations about the z-axis.

The above formalism means that we can rewrite a general element in the set of rotational transformations (3.6) in the form

$$
x_i' = \left(\exp\left[i\vec{\alpha} \cdot \vec{J} \right] \right)_i^j x_j \tag{3.18}
$$

which for small angles $|\alpha_k| << 1$ gives

$$
x_i' \simeq \left(\delta_i^{\ j} - \epsilon_i^{\ jk} \alpha_k \right) x_j = x_i - (\vec{\alpha} \times \vec{x})_i \tag{3.19}
$$

of which eqs (3.5) are a special case.

3.5   Appendix: Lie Algebras from Lie Groups

As I noted earlier in this chapter, a Lie group, when "Taylor-series expanded," becomes a Lie Algebra. Here I will outline in more detail how this works.

Closure of the Lie group implies

$$
\exp\left(i\Theta_a \mathbf{T}^a \right) \exp\left(i\Upsilon_b \mathbf{T}^b \right) = \exp\left(i\Xi_c(\Theta, \Upsilon) \mathbf{T}^c \right) \tag{3.20}
$$

where $\Xi_c(\Theta, \Upsilon)$ must be an analytic function of (Θ, Υ). Clearly $\Xi_c(\Theta, 0) = \Theta_c$ and $\Xi_c(0, \Upsilon) = \Upsilon_c$. In fact, Ξ_c can only depend on odd powers of Θ and Υ; otherwise $\Xi_c(\Theta, -\Theta)$ won't equal zero, which it must equal when $\Upsilon_c = -\Theta_c$. Hence

$$
\Xi_c(\Theta, \Upsilon) = \Theta_c + \Upsilon_c - \frac{1}{2} f^{ab}_{\ c} \Theta_a \Upsilon_b + \mathcal{O}(\Theta \Upsilon^3 \text{ or } \Theta^3 \Upsilon) + \cdots \tag{3.21}
$$

where the $f^{ab}_{\ c}$ are coefficients in the series expansion.

Expanding the exponentials on both sides of equation (3.20)

$$
1 + i\left(\Theta_a + \Upsilon_a \right) \mathbf{T}^a + \frac{(i)^2}{2!} \left(\Theta_a \Theta_b + \Upsilon_a \Upsilon_b \right) \mathbf{T}^a \mathbf{T}^b + (i)^2 \Theta_a \Upsilon_b \mathbf{T}^a \mathbf{T}^b + \cdots
$$

$$
= 1 + i\left(\Theta_c + \Upsilon_c \right) \mathbf{T}^c - \frac{i}{2} f^{ab}_{\ c} \Theta_a \Upsilon_b \mathbf{T}^c + \frac{(i)^2}{2!} \left(\Theta_a + \Upsilon_a \right) \left(\Theta_b + \Upsilon_b \right) \mathbf{T}^a \mathbf{T}^b + \cdots
$$

gives upon cancellation

$$-\Theta_a\Upsilon_b\mathbf{T}^a\mathbf{T}^b = -\frac{1}{2}\Theta_a\Upsilon_b\left(\mathbf{T}^a\mathbf{T}^b + \mathbf{T}^b\mathbf{T}^a\right) - \frac{i}{2}f^{ab}_{\ \ c}\Theta_a\Upsilon_b\mathbf{T}^c \quad (3.22)$$

$$\Rightarrow -\frac{1}{2}\Theta_a\Upsilon_b\left(\mathbf{T}^a\mathbf{T}^b - \mathbf{T}^b\mathbf{T}^a\right) = -\frac{i}{2}f^{ab}_{\ \ c}\Theta_a\Upsilon_b\mathbf{T}^c \quad (3.23)$$

This must be true for any Θ and Υ, so

$$\left[\mathbf{T}^a, \mathbf{T}^b\right] = if^{ab}_{\ \ c}\mathbf{T}^c \quad (3.24)$$

which is (3.3).

3.6 Questions

1. Consider an equilateral triangle, with the following set of symmetry operations

I	identity
R_+	Positive rotation of $\frac{2\pi}{3}$ around center
R_-	Negative rotation of $\frac{2\pi}{3}$ around center
A	Reflection through a line joining vertex A to the midpoint of line BC
B	Reflection through a line joining vertex B to the midpoint of line CA
C	Reflection through a line joining vertex C to the midpoint of line AB

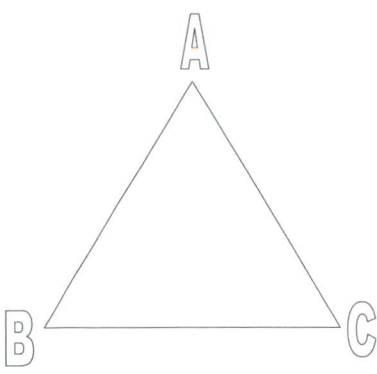

FIGURE 3.1
Diagram for question #1.

(a) Show that these six elements form a group.

(b) Construct the multiplication table of this group.

2. Find a non-trivial one-dimensional representation of the triangle group in question #1.

3. (a) Find multiplication tables for the groups $\mathcal{G}_2 = \{I, a\}$ and $\mathcal{G}_3 = \{I, a, b\}$ respectively consisting of two and three elements each.

(b) Are the multiplication tables unique? Why or why not?

(c) Find a multiplication table for the group $\mathcal{G}_3 = \{I, a, b, c\}$ consisting of four distinct elements. Is it unique?

(d) Find the multiplication table for the group Δ consisting of symmetry transformations on the equilateral triangle.

4. An orthogonal matrix is defined by the relation $R^T R = 1$, where R is an $N \times N$ matrix, where T refers to the transpose, $\left(R^T\right)_{ij} = R_{ji}$

(a) Show that the set of all orthogonal $N \times N$ matrices forms a group.

(b) Show that the set of all orthogonal $N \times N$ matrices of determinant 1 forms a group.

5. A unitary matrix is defined by the relation $U^\dagger U = 1$, where U is an $N \times N$ matrix, and the $\dagger$ operation means take the Hermitian conjugate (the complex-conjugate of the matrix transpose), i.e. $\left(U^\dagger\right)_{ij} = U^*_{ji}$.

(a) Show that the set of all unitary $N \times N$ matrices forms a group.

(b) Show that the set of all unitary $N \times N$ matrices of determinant 1 forms a group.

6. A symplectic matrix is defined by the relation $S^T \kappa S = \kappa$, where S is an $2N \times 2N$ matrix and where κ is a matrix of the form

$$\kappa = \begin{pmatrix} 0 & I_N \\ -I_N & 0 \end{pmatrix}$$

and where I_N is an $N \times N$ identity matrix. Show that the set of all symplectic $N \times N$ matrices forms a group.

7. Consider a set of three objects $\{i, j, k\}$ with the following properties

$$ij = k \quad jk = i \quad ki = j$$

and where $i^2 = j^2 = k^2 = -1$. Show that the set $\{1, i, j, k\}$ forms a group under multiplication using these rules.

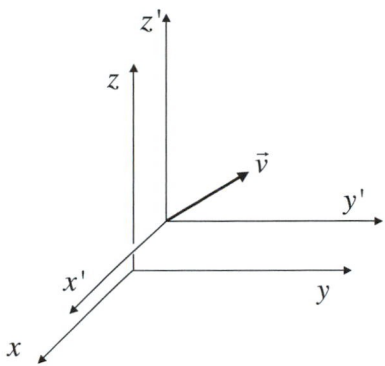

FIGURE 3.2
Diagram for question #8

8. (a) Find the general form of a Lorentz transformation for an arbitrary velocity $\vec{v}$ as shown in 3.2.

   (b) From your answer in part (a), work out the general form for velocity addition for two velocities $\vec{v}$ and $\vec{u}$.

   (c) Under the velocity addition formula in part (b), does the set of velocities form a group? Why or why not?

9. For matrices $\mathbf{A}$ and $\mathbf{B}$, show

$$\exp\left[-\mathbf{A}\right]\mathbf{B}\exp\mathbf{A} = \exp(-ad_{\mathbf{A}})\mathbf{B} \quad \text{where} \quad ad_{\mathbf{A}}\mathbf{B} \equiv [\mathbf{A}, \mathbf{B}]$$

10. (a) Show that the Jacobi identity implies

$$f^{ab}_{\ \ \partial}f^{\partial c}_{\ \ e} + f^{bc}_{\ \ \partial}f^{\partial a}_{\ \ e} + f^{ca}_{\ \ \partial}f^{\partial b}_{\ \ e} = 0$$

   (b) Show that the structure constants of a Lie group generate a representation of the group, i.e. that $(\mathbf{T}^a)^b_{\ c} = if^{ab}_{\ \ c}$ generates a representation. This is called the adjoint representation of the group.

4

Conservation Laws

In the previous chapter I mentioned that symmetries necessarily result in conservation laws. In fact, it turns out that a conservation law exists for every symmetry. This is very useful, because conservation laws are something that can be checked experimentally. Even better, from an experimental perspective we can turn the reasoning around, and inductively conclude that if a conservation law is present, then the underlying theory should have a related symmetry.

So what is the connection between symmetries and conservation laws? Recall that a symmetry is an operation on a system that leaves some (or all) of its properties unchanged. Rotations of a sphere are an excellent example, in this case, of a symmetry of shape. A crystal forms another example, in this case, a symmetry of structure, in which the atoms forming the crystal can be displaced in a certain way that replicates the crystal structure.

In particle physics, fundamental symmetries are revealed not so much in terms of shape or physical structure, or even in terms of the motions of particular objects or systems. Rather they are revealed in terms of the *set of all possible motions* a system can have. In other words, symmetries are in general manifest in the EQUATIONS OF MOTION of a system, rather than in particular solutions to these equations*. To see how this works let's recall the basic formulation of the equations of motion from Lagrangian dynamics.

4.1 The Action Principle

To describe the motion of a body in classical physics, we assign it a set of coordinates $\mathbf{q(t)}$, that in general are functions of time, since the body will be

---

*Of course it is possible for a particular solution of a given theory to have a high degree of symmetry. But when we talk about a symmetry of nature, we are referring to general properties of the equations and not to particular solutions.

moving around[†]. To deduce its equations of motion we require that the action

$$S[\mathbf{q}] \equiv \int dt\, L[\mathbf{q}, \dot{\mathbf{q}}] \qquad \text{where} \qquad \begin{aligned} \mathbf{q} &= \{q_1, q_2, q_3, \cdots q_n\} \\ \dot{\mathbf{q}}_i &= \tfrac{dq_i}{dt} \end{aligned} \qquad (4.1)$$

be stationary with respect to variations of the trajectory. Conceptually, this means that we modify the actual (and at this point unknown) path of motion that the particle takes and change it in some small but arbitrary manner (see fig. 4.1). This is done point-by-point along the trajectory, so we replace in the action $\mathbf{q(t)}$ with $\mathbf{q(t)} + \delta\mathbf{q(t)}$, where $\delta\mathbf{q(t)}$ is the arbitrary small change. By demanding that the action be stationary with respect to small changes of the trajectory, we in fact require that the action is minimized (or more generally, extremized) with respect to these small changes. Hence we require

$$0 = \delta_{q_i} S = S[\mathbf{q} + \delta\mathbf{q}] - S[\mathbf{q}]$$

$$= \int_{t_I}^{t_F} dt \left[\frac{\partial L}{\partial q_i} \delta q_i + \frac{\partial L}{\partial \dot{q}_i} \delta \dot{q}_i \right] + \cdots \qquad \frac{\partial L}{\partial \dot{q}_i} \delta \dot{q}_i = \frac{d}{dt}\left(\frac{\partial L}{\partial \dot{q}_i} \delta q_i \right)$$

$$= \int_{t_I}^{t_F} dt \left[\frac{\partial L}{\partial q_i} - \frac{d}{dt}\left(\frac{\partial L}{\partial \dot{q}_i} \right) \right] \delta q_i + \frac{\partial L}{\partial \dot{q}_i} \delta q_i \Big|_{t_I}^{t_F} \qquad - \frac{d}{dt}\left(\frac{\partial L}{\partial \dot{q}_i} \right) \delta q_i \qquad (4.2)$$

The "$+\cdots$" in eq. (4.2) means that I neglect any terms proportional to $(\delta q_i)^2$. The third line follows from the second via an integration by parts, where $\delta\dot{q}_i = \delta\frac{dq_i}{dt} = \frac{d}{dt}\delta q_i$.

To proceed further we need to incorporate additional information: namely we assume $\delta q_i = 0$ at the endpoints of all the trajectories. The reason for this is that we want to find out what equations of motion take our body from a given set of initial conditions to another given set of final conditions. The only way this comparison will be meaningful is if we require that all the trajectories in our variation have the same initial (and final) conditions, i.e., the same endpoints. This means that $\delta q_i = 0$ at these endpoints. This eliminates the last term in the preceding expression. Since δq_i is otherwise arbitrary, eq. (4.2) can only be satisfied if   *except at the endpoints.*

$$\frac{\partial L}{\partial q_i} - \frac{d}{dt}\left(\frac{\partial L}{\partial \dot{q}_i} \right) = 0 \qquad \text{Euler-Lagrange equations} \qquad (4.3)$$

where I have written the general name these equations get due to their special significance. We can alternatively write them as

$$\frac{dp_i}{dt} = \frac{\partial L}{\partial q_i} \qquad (4.4)$$

[†]What we are doing here is considering the body to be a point mass. However, if we were interested in the detailed structure of the body we could assign each of its constituents a coordinate, idealizing each constituent as a point mass.

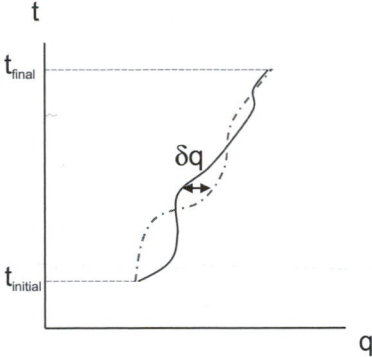

FIGURE 4.1

A sample trajectory (solid), and a variation of this trajectory (dot-dash).

where

$$p_i \equiv \frac{\partial L}{\partial \dot{q}_i}$$

a quantity known as the *canonical momentum p_i.*

So what I have shown is that solutions to the Euler-Lagrange equations are trajectories $\mathbf{q}(t)$ that extremize the action. For a given set of initial conditions, any solution $\mathbf{q}(t)$ to the equations (4.3) is regarded as a possible motion of the system. Conversely, functions $\mathbf{q}(t)$ that do not obey equations (4.3) will not yield an extremal action.

The variational principle is a very powerful principle – it implies that equations of motion governing a system are determined by extremization of a scalar functional (the action) [29]. It generalizes to quantum mechanics, general relativity and quantum field theory. Pretty much all of the equations of fundamental physics are founded on an action principle. The scalar character of the action makes it ideal for introducing symmetries.

When we consider a symmetry of the system, we are contemplating a group of transformations $\mathcal{G} = \{g_i\}$ such that

$$q_i' = g_i[\mathbf{q}] \quad \text{where invariance} \Rightarrow S[\mathbf{q}'] = S[g[\mathbf{q}]] = S[\mathbf{q}] \quad (4.5)$$

Suppose we choose to describe the motion in terms of the transformed variables $\mathbf{q}'(t')$. What equations of motion will they obey? We can find them by computing the variation

$$0 = \delta_{q_i} S[\mathbf{q}] = \frac{\delta q_j'}{\delta q_i} \delta_{q_j'} \left(S[\mathbf{q}] \right) = \frac{\delta q_j'}{\delta q_i} \delta_{q_j'} \left(S[g[\mathbf{q}]] \right) = \frac{\delta q_j'}{\delta q_i} \delta_{q_j'} S[\mathbf{q}']$$

$$\Rightarrow \delta_{q_i'} S[\mathbf{q}'] = 0 \quad (4.6)$$

where the first equality holds provided $\mathbf{q}(t)$ solves (4.3) and the last holds because $\frac{\delta q_j'}{\delta q_i} \neq 0$.

Hence if $\mathbf{q}(t)$ solves (4.3), then the transformed quantities $\mathbf{q}'(t')$ will also be a solution to a set of equations of motion. These will not in general be identical to the set that the $\mathbf{q}(t)$ obey, but they will be obtainable from this set using the same transformations: an *invariant* action yields *covariant* equations of motion!

In other words both $q_i(t)$ and $q_i'(t') = g[\mathbf{q}(t)]$ extremize the action if $S[g[\mathbf{q}]] = S[\mathbf{q}]$, provided $\mathcal{G}$ is a group. This last requirement ensures that $\det \frac{\delta q_j}{\delta q_i'} \neq 0$ (all group elements must have an inverse, ensuring $\frac{\delta q_j'}{\delta q_i} \neq 0$); were it not to hold then we could not ensure that $\delta_{q'} S[\mathbf{q}'] = 0$, but only that some linear combination of variations of $S[\mathbf{q}']$ would vanish.

4.2 Noether's Theorem

An even more general principle holds whenever we have a symmetry, namely,

For every symmetry of the action there is a conservation law

which is called Noether's theorem, obtained by Emmy Noether in 1917.

To see why this is true, notice that for a *fully arbitrary* variation (one for which neither the equations of motion (4.3) hold nor for which the endpoints are fixed) we have

$$\delta_{q_i} S = \int_{t_I}^{t_F} dt\, [\delta_{q_i} L] = \int_{t_I}^{t_F} dt\, \left[\frac{\partial L}{\partial q_i} - \frac{d}{dt}\left(\frac{\partial L}{\partial \dot{q}_i}\right)\right] \delta q_i + \left.\frac{\partial L}{\partial \dot{q}_i} \delta q_i\right|_{t_I}^{t_F} \quad (4.7)$$

so that

$$\delta_{q_i} L = \left[\frac{\partial L}{\partial q_i} - \frac{d}{dt}\left(\frac{\partial L}{\partial \dot{q}_i}\right)\right] \delta q_i + \frac{d}{dt}\, [p_i \delta q_i] \quad (4.8)$$

Now let's consider a (small) symmetry transformation, parametrized by a set of parameters $\alpha = \{\alpha_1, \alpha_2, \alpha_3, \ldots, \alpha_m\}$ so that

$$q_i \to q_i' = q_i'[\alpha] = g_i[\alpha, \mathbf{q}]$$

$$\Rightarrow \delta_{\alpha_j} q_i = q_i' - q_i = \left.\frac{\partial q_i'}{\partial \alpha_j}\right|_{\alpha=0} \delta\alpha_j + \cdots$$

and where we define $\alpha = 0$ to be the identity transformation ($q_i'[\alpha = \mathbf{0}] = g_i[\mathbf{0}, \mathbf{q}] = q_i$).

Let's see what happens if set $\delta_j q_i = \delta_{\alpha_j} q_i$, neglecting terms of order $(\delta\alpha)^2$ and higher. First we see that equation (4.8) becomes

$$\frac{d}{dt}\left[p_i \frac{\partial q_i'}{\partial \alpha_j}\right] = \frac{\partial L}{\partial \alpha_j} - \left[\frac{\partial L}{\partial q_i} - \frac{d}{dt}\left(\frac{\partial L}{\partial \dot{q}_i}\right)\right] \frac{\partial q_i'}{\partial \alpha_j} \quad (4.9)$$

to leading order. Next consider the action: invariance of the action means $S[q'] \equiv S[q + \delta_{\alpha_j} q] = S[q]$, or alternatively $\delta_{\alpha_j} S = 0$. This can only be true if

$$0 = \delta_{\alpha_j} S = \int_{t_I}^{t_F} dt \, [\delta_{\alpha_j} L] = \int_{t_I}^{t_F} dt \, \frac{\partial L}{\partial \alpha_j}\bigg|_{\alpha} = 0 + \cdots \qquad (4.10)$$

The most general condition under which this holds is

$$\frac{\partial L}{\partial \alpha_j}\bigg|_{\alpha=0} = \frac{d}{dt}\left[G\left(\mathbf{q}, \frac{\partial \mathbf{q}'}{\partial \alpha_j}\right)\right]_{\alpha=0} \qquad (4.11)$$

provided the function G is some function of $\mathbf{q}$ and $\delta_{\alpha_j} \mathbf{q}$ that vanishes at the endpoints. This is the most general form $\delta_{\alpha_j} L$ can take that vanishes upon integration. The function G is some function that must be calculated case-by-case for each given symmetry.

The important thing to note here is that this equality follows because we have a symmetry transformation. For a symmetry transformation, $\delta_{\alpha_j} L$ *must be a total time derivative*, whereas for an arbitrary variation it is not in general a total time derivative (unless the equations of motion are satisfied), as is clear from eq. (4.8) above. We can form an object

$$J_\ell = \left(p_i \frac{\partial q_i'}{\partial \alpha_\ell} - G\left(\mathbf{q}, \frac{\partial \mathbf{q}'}{\partial \alpha_\ell}\right)\right)\bigg|_{\alpha_a=0} \qquad (4.12)$$

and easily show using (4.9) that

$$\begin{aligned}
\frac{d}{dt} J_\ell &= \left(\frac{d}{dt}\left[p_i \frac{\partial q_i'}{\partial \alpha_\ell}\right] - \frac{d}{dt} G\right)\bigg|_{\alpha_a=0} \\
&= \left(\frac{\partial L}{\partial \alpha_\ell} - \left[\frac{\partial L}{\partial q_i} - \frac{d}{dt}\left(\frac{\partial L}{\partial \dot{q}_i}\right)\right] \frac{\partial q_i'}{\partial \alpha_\ell} - \frac{d}{dt} G\right)\bigg|_{\alpha_a=0} \quad \text{“} 0 \text{ if the equations of motion} \\
&= \left(\frac{\partial L}{\partial \alpha_\ell} - \frac{d}{dt} G\right)_{\alpha_a=0} \qquad\qquad\qquad\qquad\qquad\qquad \text{are satisfied.”} \\
&= 0 \qquad\qquad\qquad\qquad\qquad\qquad\qquad\qquad\qquad\qquad\qquad\quad (4.13)
\end{aligned}$$

which means that J_ℓ is conserved: it stays the same at all times, provided the equations of motion are satisfied. Note that we impose the Euler-Lagrange equations to get the third line; the last line follows from (4.11).

The current J_ℓ is called the *Noether current* associated with the symmetry α. Equation (4.13) is Noether's theorem: the symmetry parametrized by the α's ensures that J_ℓ does not change with time. In other words J_ℓ is conserved. Noether's theorem is very far-reaching – I've only proved it for classical mechanics, but it can be shown to hold in both general relativity [30] and quantum field theory [31]. It is a very general result that holds for any theory of physics based on action principle.

The meaning of J_ℓ depends on what the symmetry is, or in other words, what the α's are. Let's look at some examples.

4.3 Spacetime Symmetries and Their Noether Currents

The foundational symmetries of particle physics (indeed all of physics) are spacetime symmetries: symmetries that transform the coordinates of space and time. Let's find the associated Noether currents for these.

4.3.1 Spatial Translations

We expect the fundamental laws of nature to be insensitive to a special location in space. This means that there is no special choice for the origin of coordinates. Under a spatial translation, the choice of origin is shifted, so this symmetry requirement means that the action should be translationally invariant. A translationally invariant action will have

$$L[\mathbf{q} + \alpha, \dot{\mathbf{q}}] = L[\mathbf{q}, \dot{\mathbf{q}}] \qquad \text{where} \quad \mathbf{q}' = \mathbf{q} + \alpha \quad \text{and} \quad \dot{\mathbf{q}}' = \dot{\mathbf{q}} \qquad (4.14)$$

where there are three symmetry parameters $\{\alpha_1, \alpha_2, \alpha_3\}$ that parametrize the translation distances. Equation (4.14) implies $\frac{\partial L}{\partial \alpha^i}\big|_{\alpha=0} = 0$ and so we can set $G = 0$ (or a constant vector; it doesn't matter). We also have

$$\frac{\partial q_i'}{\partial \alpha_k} = \delta_{ik} \qquad (4.15)$$

and so the Noether current $\mathsf{J}_k^{\text{sp-trans}}$ is

$$\mathsf{J}_k^{\text{sp-trans}} = p_i \frac{\partial q_i'}{\partial \alpha_k} - G\left(\mathbf{q}, \frac{\partial \mathbf{q}'}{\partial \alpha}\right) = p_i\left(\delta_{ik}\right) - 0 = p_k \qquad (4.16)$$

This means that[‡]

$$\frac{d}{dt}\mathsf{J}_k^{\text{sp-trans}} = 0 \Rightarrow \frac{d\vec{p}}{dt} = 0 \qquad (4.17)$$

which is the conservation of linear momentum! Hence

space translation invariance $\Rightarrow$ conservation of linear momentum

4.3.2 Rotations

What happens if we insist that our system be rotationally invariant? Rotational symmetries are described by the orthogonal group $\mathbf{SO}(3)$. In Chapter 3 we saw that these take the form

$$q_i' = \left(\exp\left[i\vec{\alpha} \cdot \vec{J}\right]\right)_i^{\ j} q_j \simeq \left(\delta_i^{\ j} - \epsilon_i^{\ jk}\alpha_k\right) q_j = q_i - (\vec{\alpha} \times \vec{q})_i \qquad (4.18)$$

[‡]Note that if G were set to be a constant vector, then this would have the effect of shifting the definition of conserved momentum by some constant (and irrelevant) value; momentum would still be conserved. This is why we choose $G = 0$.

where the (small) symmetry parameters $\{\alpha_1, \alpha_2, \alpha_3\}$ are the angles of rotation around each orthogonal coordinate axis, the J's are the rotation generators we looked at before, and $\{q_1, q_2, q_3\}$ are the 3 spatial coordinates. Hence

$$\Rightarrow \frac{\partial q_i'}{\partial \alpha_k} = -\epsilon_i{}^{jk} q_j = +\epsilon_i{}^{kj} q_j = \epsilon_{ikj} q_j \tag{4.19}$$

Since the magnitude of a vector is invariant under rotations, a rotationally invariant Lagrangian will only depend on the magnitude of $\vec{q}$, (i.e. $L = L(|\vec{q}|)$), so we will have $L(|\vec{q}'|) \equiv L(|\vec{q}|)$ (i.e. $\left.\frac{\partial L}{\partial \alpha^i}\right|_{\alpha=\mathbf{0}} = 0$) and we can once again choose $G = 0$.

The Noether current J_k^{rot} is thus

$$J_k^{\text{rot}} = p_i \frac{\partial q_i'}{\partial \alpha_k} = p_i (\epsilon_{ikj} q_j) = -\epsilon_{kij} p_i q_j = -(\vec{p} \times \vec{q})_k = (\vec{q} \times \vec{p})_k \tag{4.20}$$

and we obtain

$$\frac{d}{dt} J_k^{\text{rot}} = 0 \Rightarrow \frac{d}{dt}(\vec{q} \times \vec{p}) = 0 \tag{4.21}$$

which is the conservation of angular momentum! Hence

$$\boxed{\text{rotational invariance} \Rightarrow \text{conservation of angular momentum}}$$

4.3.3 Time Translations

We expect that the same laws of physics were operative yesterday as they are today, so it is reasonable to require invariance of the action under time translations. Under a small time translation $t' = t + \delta\alpha$,

$$\delta_\alpha \mathbf{q} = \mathbf{q}(t + \delta\alpha) - \mathbf{q}(t) \Rightarrow \left.\frac{\partial q_i'}{\partial \alpha}\right|_{\alpha=0} = \dot{q}_i \tag{4.22}$$

There is now just one symmetry parameter $\{\alpha\}$, which parametrizes how much the origin of time has shifted. We will also have a non-zero G because

$$\delta_\alpha L = L(t') - L(t) = \frac{dL}{dt}\delta\alpha + \cdots \Rightarrow \left.\frac{\partial L}{\partial \alpha^i}\right|_{\alpha=0} = \frac{dL}{dt} \tag{4.23}$$

which tells us that $G = L - L_0$, where L_0 is an irrelevant constant that we can set to zero. Hence the Noether current $J^{\text{t-trans}}$ is

$$J^{\text{t-trans}} = \left.\left(p_i \frac{\partial q_i'}{\partial \alpha} - L\right)\right|_{\alpha=0} = p_i \dot{q}_i - L \equiv H \tag{4.24}$$

$$\frac{dJ^{\text{t-trans}}}{dt} = 0 \Rightarrow \frac{dH}{dt} = 0 \tag{4.25}$$

which is the conservation of energy, since $H = p_i \dot{q}_i - L$ is the Hamiltonian. Hence

$$\boxed{\text{time translation invariance} \Rightarrow \text{conservation of energy}}$$

Time-translation will be a symmetry of any system whose Lagrangian does not explicitly depend on the time coordinate. After all, if the Lagrangian did depend on the time coordinate, then the action would not be invariant under a time translation. This is a feature of all closed systems: that there is no possibility of the system being externally modified at some later time(s). Consequently conservation of energy will hold for any closed system.

4.4   Symmetries and Quantum Mechanics

The preceding discussion was valid for systems of non-relativistic classical particles. The transition to quantum mechanics changes the situation in two ways[§]. First, it necessitates a reformulation of the symmetry-principle/conservation-law relationship. Second, it allows for a broader class of symmetries called internal symmetries: symmetries that rearrange the charges and internal structure of wavefunctions without transforming space and time.

A physical system (or particle) in non-relativistic quantum mechanics is represented by a wavefunction Ψ that obeys Schroedinger's equation:

$$i\hbar\frac{\partial\Psi}{\partial t} = \mathbf{H}\Psi(t) \tag{4.26}$$

where $\mathbf{H}$ is a quantum operator that is constructed from the classical Hamiltonian $H = p_i\dot{q}_i - L$. Already we see a difference from the classical discussion in that $\mathbf{H}$ (and not the Lagrangian) is now playing a fundamental role. For a stationary state $\Psi_s(t) = \exp\left(-iEt/\hbar\right)\varphi\left(\vec{q}\right)$, the eigenvalues of $\mathbf{H}$ correspond to the allowed energies (the allowed values of E) that the system can have.

In passing from classical to quantum mechanics, we replace $\mathbf{q}(t) \longrightarrow \Psi(t, \mathbf{q})$, so we expect that a symmetry operation is described as

$$\mathbf{q}_i' = g_i[\alpha, \mathbf{q}] \longrightarrow \Psi'(t; \mathbf{q}') = \mathbf{U}(\alpha)\Psi(t; \mathbf{q}) \tag{4.27}$$

where $\mathbf{U}(\alpha)$ is a matrix that acts on the (possibly multicomponent) wavefunction Ψ. Since $\mathbf{U}(\alpha)$ represents a symmetry of the system, the normalization of the wavefunction should not change and so

$$\int d\mathbf{q}'\Psi'^{\dagger}(t; \mathbf{q}')\Psi'(t; \mathbf{q}') = 1$$

$$\Rightarrow \int d\mathbf{q}\Psi^{\dagger}(t; \mathbf{q})\mathbf{U}^{\dagger}(\alpha)\mathbf{U}(\alpha)\Psi(t; \mathbf{q}) = \int d\mathbf{q}\Psi^{\dagger}(t; \mathbf{q})\Psi(t; \mathbf{q}) = 1 \tag{4.28}$$

[§]The transition to relativity, while preserving the basic conservation laws in section 4.3, introduces a change of another kind that we will examine in Chapter 11.

which means for any symmetry operation represented $\mathbf{U}(\alpha)$ we must have

$$\mathbf{U}^\dagger\mathbf{U} = \mathbf{I} \tag{4.29}$$

so that the normalization is preserved. Hence a symmetry in quantum mechanics must be represented by a unitary transformation on the wavefunction: $\mathbf{U}(\alpha)$ must be a unitary matrix.

How are conserved quantities associated with symmetries in quantum mechanics? Based on the classical discussion we expect that a symmetry operator $\mathbf{U}(\alpha)$ corresponds to a conserved quantity ω that is observable. Specifically, consider

$$\omega = \int d\mathbf{q}\,\Psi^\dagger(t;\mathbf{q})\mathbf{U}(\alpha)\Psi(t;\mathbf{q}) \tag{4.30}$$

and compute its time derivative

$$\frac{d\omega}{dt} = \int d\mathbf{q}\left(\frac{\partial\Psi^\dagger}{\partial t}\mathbf{U}\Psi + \Psi^\dagger\mathbf{U}\frac{\partial\Psi}{\partial t}\right) = \frac{i}{\hbar}\int d\mathbf{q}\,(\Psi^\dagger\mathbf{H}\mathbf{U}\Psi - \Psi^\dagger\mathbf{U}\mathbf{H}\Psi)$$

$$= \frac{i}{\hbar}\int d\mathbf{q}\Psi^\dagger\,[\mathbf{H},\mathbf{U}]\,\Psi \tag{4.31}$$

where equation (4.26) was used. We see that the quantum number ω will be conserved (i.e. $\frac{d\omega}{dt} = 0$) provided $[\mathbf{H},\mathbf{U}] = 0$. So the quantum version of Noether's theorem is

Conserved quantum number $\Leftrightarrow [\mathbf{U},\mathbf{H}] = 0 \Leftrightarrow \mathbf{U}$ is a symmetry of the system

The set of matrices $\{\mathbf{U}(\alpha_i)\}$ that form a group is called the symmetry group of the Hamiltonian.

Note that the matrix $\mathbf{U}$ need not be connected with any spacetime transformation (e.g. rotations, translations) – it might simply represent a symmetry that mixes up the wavefunction components. An example of this would be a symmetry that rearranges the color charges of quarks. We will see later that such internal symmetries underlie all the non-gravitational forces of nature.

Let's look again at our spacetime symmetries in this context.

4.4.1 Spatial Translations

We expect that since the classical Noether current is $\vec{p}$, the symmetry operator associated with space translations is $(-i\vec{\nabla})$ once we make the identification $\vec{p} \to (-i\hbar\vec{\nabla})$ from standard quantum mechanics. This is indeed correct. Invariance under a translation means that *omit " ' " (primes)*

$$\mathbf{q}' = \mathbf{q} + \alpha \ \to \Psi'(\mathbf{q}') = \Psi(\mathbf{q}) \Rightarrow \Psi'(\mathbf{q}) = \Psi(\mathbf{q} - \alpha) \tag{4.32}$$

and expanding this for small α yields

$$\Psi'(q_i) = \Psi(q_i - \alpha_i) \simeq \Psi(q_i) - \left(\vec{\alpha}\cdot\vec{\nabla}\right)\Psi(q_i) = \left(1 - i\vec{\alpha}\cdot(-i\vec{\nabla})\right)\Psi(q_i) \tag{4.33}$$

and so the symmetry operator is $\vec{\mathbf{P}} = -i\hbar\vec{\nabla}$ (where Planck's constant $\hbar$ ensures that the units are correct on both sides of this relation). We thus must have

$$\left[\vec{\alpha} \cdot \vec{\mathbf{P}}, \mathbf{H}\right] = 0 \tag{4.34}$$

for a translationally invariant system. This is conservation of linear momentum in the direction $\vec{\alpha}$.

Since $\vec{\alpha}$ is arbitrary, we can choose it to point in any one of the three orthogonal directions, in which case we have

$$[\mathbf{P}_i, \mathbf{H}] = 0 \tag{4.35}$$

for any given direction. Since the various components of $\vec{\mathbf{P}}$ commute with each other (because $\nabla_i\nabla_j = \nabla_j\nabla_i$), in a translationally invariant quantum-mechanical system components of the linear momentum can be defined along all of the measurement axes.

4.4.2   Rotations

Under a rotation of angle α around an axis $\hat{n}$ $(\vec{\alpha} = \alpha\hat{n})$, a one-component wavefunction Ψ transforms as

$$q'_i = \left(\exp\left[i\vec{\alpha} \cdot \vec{J}\right]\right)_i^{\ j} q_j = \mathbf{R}_i^{\ j} q_j \longrightarrow \Psi'(\mathbf{q}') = \Psi(\mathbf{q}) \Rightarrow \Psi'(\mathbf{q}) = \Psi(\mathbf{R}^{-1}\mathbf{q}) \tag{4.36}$$

and expanding this for small $\vec{\alpha}$ gives

$$\Psi'(q_i) \simeq \Psi(q_i + (\vec{\alpha} \times \vec{q})_i) \simeq \Psi(q_i) + \left((\vec{\alpha} \times \vec{q}) \cdot \vec{\nabla}\right)\Psi(q_i)$$
$$= \left(1 + i\vec{\alpha} \cdot \left(\vec{q} \times (-i\vec{\nabla})\right)\right)\Psi(q_i) \tag{4.37}$$

and so the symmetry operator associated with rotations is $\vec{\mathbf{L}} = \vec{q} \times (-i\hbar\vec{\nabla})$. Note that this is equivalent to the Noether current $\vec{q} \times \vec{p}$ once we make the identification that $\vec{p} \to (-i\hbar\vec{\nabla}) = \vec{\mathbf{P}}$ from standard quantum mechanics as before. Hence a rotationally invariant quantum system must obey the relation

$$\left[\vec{\alpha} \cdot \vec{\mathbf{L}}, \mathbf{H}\right] = 0 \tag{4.38}$$

which we recognize from 3rd year quantum mechanics as conservation of angular momentum $\vec{\mathbf{L}}$ around the axis $\vec{\alpha}$.

Unlike the situation for translations, the different components of $\vec{\mathbf{L}}$ do *not* commute with each other. This leads to a subtle but important distinction from the classical case. In the classical case, we saw that $\vec{q} \times \vec{p}$ was conserved, and so angular momentum around any axis is classically conserved in a rotationally invariant system. Since in quantum mechanics the different components of $\vec{\mathbf{L}}$ do not commute with each other (even though any specific component commutes with the Hamiltonian **H**), in a rotationally invariant system the angular momentum can only be defined about a chosen axis. The choice of this axis is arbitrary (reflecting the classical conservation about any axis) but – once chosen – the remaining components are rendered inaccessible to measurement.

Let's look at the implications of this non-commutativity in a bit more detail. For any function $f(x, y, z)$

$$[\mathbf{L}_x, \mathbf{L}_y] f(x, y, z) = \left[\left(\vec{q} \times (-i\hbar \vec{\nabla}) \right)_x , \left(\vec{q} \times (-i\hbar \vec{\nabla}) \right)_y \right] f(x, y, z)$$

$$= i^2 \hbar^2 \left[\left(y \frac{\partial}{\partial z} - z \frac{\partial}{\partial y} \right), \left(z \frac{\partial}{\partial x} - x \frac{\partial}{\partial z} \right) \right] f$$

$$= -\hbar^2 \left(y \frac{\partial}{\partial z} - z \frac{\partial}{\partial y} \right) \left(z \frac{\partial f}{\partial x} - x \frac{\partial f}{\partial z} \right)$$

$$+ \hbar^2 \left(z \frac{\partial}{\partial x} - x \frac{\partial}{\partial z} \right) \left(y \frac{\partial f}{\partial z} - z \frac{\partial f}{\partial y} \right)$$

$$= \hbar^2 \left[zy \left(\frac{\partial^2 f}{\partial x \partial z} - \frac{\partial^2 f}{\partial z \partial x} \right) + xy \left(\frac{\partial^2 f}{\partial z^2} - \frac{\partial^2 f}{\partial z^2} \right) \right.$$

$$+ z^2 \left(\frac{\partial^2 f}{\partial y \partial x} - \frac{\partial^2 f}{\partial x \partial y} \right)$$

$$\left. + zx \left(\frac{\partial^2 f}{\partial y \partial z} - \frac{\partial^2 f}{\partial z \partial y} \right) - y \frac{\partial f}{\partial x} + x \frac{\partial f}{\partial y} \right]$$

$$= \hbar^2 \left(x \frac{\partial}{\partial y} - y \frac{\partial}{\partial x} \right) f$$

$$= i\hbar \mathbf{L}_z f \tag{4.39}$$

We can repeat this, cycling through (x, y, z) and arrive at the relation

$$\left[\mathbf{L}^a, \mathbf{L}^b \right] = i\hbar \epsilon^{abc} \mathbf{L}^c \tag{4.40}$$

which is the same kind of relationship between the **L**'s that we had for the **J**'s in Chapter 3 (see eq. (3.13)). We will make use of this in subsequent chapters.

4.4.3 Time Translations

This is a special case in quantum mechanics, since time plays the role of an ordering parameter. Under a time translation $t' = t + \alpha$,

$$\Psi'(t') = \Psi(t) \Rightarrow \Psi'(t) = \Psi(t - \alpha) \simeq \Psi(t) - \alpha \frac{\partial \Psi}{\partial t} = \left(1 + \frac{i}{\hbar} \alpha \mathbf{H} \right) \Psi \tag{4.41}$$

which means that the Hamiltonian **H** is the symmetry operator. This trivially commutes with itself, and so energy is always conserved in a quantum mechanical theory.

4.5 Summary

Under a symmetry $\mathbf{g}[\alpha]$:

INVARIANCE

of the ACTION $\quad \Rightarrow$ conservation of $\mathbf{J}_{\boldsymbol{\ell}} = \left(p_i \frac{\partial q_i'}{\partial \alpha_\ell} - G \right) \Big|_{\alpha_\ell = 0}$

COMMUTATION

$[\mathbf{U}(\alpha), \mathbf{H}] = 0 \quad \Rightarrow$ conservation of $\ \omega = \int d\mathbf{q} \, \Psi^\dagger(t; \mathbf{q}) \mathbf{U} \Psi(t; \mathbf{q})$

Specifically, for spacetime symmetries:

Rotational Invariance	$\Rightarrow$ conservation of angular momentum
Space-translation Invariance	$\Rightarrow$ conservation of linear momentum
Time-translation Invariance	$\Rightarrow$ conservation of energy

---

4.6 Questions

1. Suppose an operator F has real expectation values, i.e., $\langle \psi | F | \psi \rangle$ is real for any wavefunction ψ. Show that F is Hermitian, i.e., that $F^\dagger = F$.

2. Consider the following action in classical mechanics

$$S = \int dt \left[\frac{1}{2} m \left| \frac{d\vec{x}}{dt} \right|^2 - V(\vec{x}) \right]$$

(a) Suppose we want to make a transformation that rescales the coordinates by a constant factor of σ, i.e. $\vec{x}' = \sigma \vec{x}$. How must the time rescale in order that the action remains invariant if the potential $V(\vec{x}) = 0$?

(b) Under what circumstances is the action invariant under this transformation if $V(\vec{x}) \neq 0$? Find the general form of the potential.

(c) Find the Noether current associated with this transformation and show that it is conserved when the equations of motion are satisfied.

3. Consider a system of N particles, whose wavefunction is $\Psi\left(\mathbf{q}_1, \mathbf{q}_2, \ldots, \mathbf{q}_N\right)$. If this system is invariant under translations, find the associated Noether current.

4. Consider an algebra consisting of the set of operators $\{P_x, P_y, P_z, L_x, L_y, L_z\}$ with the combining operator being the commutator. Does this algebra close?

5. Show that the operators $\vec{\mathbf{P}} \cdot \vec{\mathbf{P}}$ and $\vec{\mathbf{L}} \cdot \vec{\mathbf{L}}$ commute with all elements of the algebra in question #4.

6. Consider the operator $U = \exp\left(-\frac{i}{\hbar}\vec{a} \cdot \vec{\mathbf{P}}\right)$ where $\vec{\mathbf{P}} = -i\hbar\vec{\nabla}$ is the momentum operator and $\vec{a}$ is a vector displacement from the origin. How does U act on a wavefunction $\Psi\left(\vec{x}, t\right)$?

7. Consider the operator $U = \exp\left(-\frac{i}{\hbar}\hat{n} \cdot \vec{\mathbf{L}}\right)$ where $\vec{\mathbf{L}} = \vec{x} \times \left(-i\hbar\vec{\nabla}\right)$ is the angular momentum operator and $\hat{n}$ is a unit vector. How does U act on a wavefunction $\Psi\left(\vec{x}, t\right)$?

5

Particle Classification

One of the uses of symmetries in particle physics is to classify the possible types of particles. Intuitively we conceive of a particle as a tiny, possibly indivisible bit of matter. However, we know from quantum mechanics that it is described by a wavefunction, and is in a certain sense delocalized over a region of space at any given instant.

But what kind of wavefunction should we use for a given particle? Clearly we can't use the same kind of wavefunction for each particle because they have very distinct properties. Electrons have small mass and negative charge, quarks have color and large mass, pions can be charged or neutral and have no spin, etc. A single type of wavefunction could not properly describe these distinct properties.

This is where symmetries come in. Symmetries provide a framework that constrains the types of wavefunctions we can use to describe particles [32]. Intuitively, the wavefunctions that we use should covariantly transform with respect to the fundamental symmetries of nature that we believe (on empirical grounds) to be valid. From this perspective we then ask given a symmetry, what particle wavefunctions can logically exist? And what are their characteristics? In this chapter we will consider these issues.

5.1 General Considerations

To answer these questions, the first thing we need to specify is the system. If we want to classify a particle, the system under consideration should be just the particle and nothing else. This means that the action for the system is just the kinetic energy of the particle – no other interactions or potentials are present!

Let's work non-relativistically to start with. In this case we have

$$\text{Classically}: \ S = \int d^3x \left[\frac{1}{2}m\left|\frac{d\vec{x}}{dt}\right|^2\right] = \frac{m}{2}\int d^3x \ |\dot{x}|^2 \quad (5.1)$$

$$\text{Quantum Mechanically}: \ i\hbar\frac{\partial\Psi}{\partial t} = -\frac{\nabla^2}{2m}\Psi(t) \quad (5.2)$$

for a free particle of mass m.

This system is clearly invariant under all spacetime symmetries: space translations, time translations, and rotations. Hence we can classify a free particle by its properties under these symmetries. Note that the particle might be composite (e.g. the pion, the proton, the Kaon, etc.) or elementary* (e.g. electron, up quark, photon, etc.). In either case, as long as the particle is represented non-relativistically by (5.1) or (5.2), our classification will be valid.

Consider first translations. Invariance under spatial translations means that the momentum of a free particle is always conserved, and invariance under time translations means that its energy is conserved (both classically and quantum mechanically). In the rest frame of the particle the momentum vanishes. Hence every free particle is characterized by its energy in the rest frame – which is its *mass*[†]. We say that the mass of the particle (by which we mean the inertial mass) is a good quantum number of the system.

Rotations are somewhat less trivial. We've already seen that a rotation is a transformation $\mathbf{R}$ for which $\mathbf{R}^{\mathsf{T}}\mathbf{R} = \mathbf{I}$, i.e., the group is the group of orthogonal transformations of 3×3 matrices, which we call $\mathbf{SO}(3)$. Such transformations preserve angles between vectors and their lengths. We saw earlier that they can be written in the form

$$\mathbf{R} = \exp\left[i\vec{\theta} \cdot \vec{\mathbf{J}}\right] \Rightarrow R_i^{\ j} = \left(\exp\left[i\vec{\theta} \cdot \vec{\mathbf{J}}\right]\right)_i^{\ j} \tag{5.3}$$

where the $\mathbf{J}$ operators obey

$$\left[\mathbf{J}^a, \mathbf{J}^b\right] = i\epsilon^{abc}\mathbf{J}^c \tag{5.4}$$

and also (as a consequence of (5.4))

$$\left[\left[\mathbf{J}^a, \mathbf{J}^b\right], \mathbf{J}^c\right] + \left[\left[\mathbf{J}^b, \mathbf{J}^c\right], \mathbf{J}^a\right] + \left[\left[\mathbf{J}^c, \mathbf{J}^a\right], \mathbf{J}^b\right] = 0 \tag{5.5}$$

with $a = 1, 2, 3$. We also saw in the previous chapter that $\vec{\mathbf{L}} = \vec{x} \times (-i\hbar\vec{\nabla})$ satisfies the relations (5.4) – that is, setting $\mathbf{J}^a = \mathbf{L}^a = \hbar\left(\vec{x} \times (-i\vec{\nabla})\right)^a$ for

---

*By definition elementary particles are indivisible – they are not made of smaller components. So how do we know if a particle is elementary or not? The answer is that we don't! What we can do theoretically is to assume a given particle is elementary and work out the consquences. To the extent that experiment is in agreement with theory, this assumption is valid. For example all experimental evidence to date indicates that the electron behaves like an elementary particle for distances no smaller than 10^{-20} cm [33]. Future experiments that can probe even shorter distances might uncover evidence that the electron is not elementary – if so, then our theory of the electron would have to be modified [34].

[†]Since we are working non-relativistically, we actually can't conclude that the energy in the rest frame is the mass. However, this result would follow were we to work relativistically (which we'll do later on). Note that even non-relativistically we must specify the parameter m in order to write down the action or Hamiltonian – since m commutes with $\mathbf{H}$, it can be used to characterize (i.e. classify) the particle.

$\mathfrak{a} = 1, 2, 3$ will satisfy (5.4), with an appropriate factor of $\hbar$ inserted. It is easy to check that (5.5) is also satisfied – indeed, it must be if (5.4) holds.

The nice thing is that any set of $N \times N$ matrices $\mathbf{J}^\mathfrak{a}$ that satisfy the relation (5.4) can be considered as symmetry operators of $\mathbf{SO}(3)$ that act on some wavefunction Ψ. Note that this turns the problem around. We originally began with the matrices

$$\{\mathbf{T}^x, \mathbf{T}^y, \mathbf{T}^z\} = \left\{ -i \begin{pmatrix} 0 & 0 & 0 \\ 0 & 0 & 1 \\ 0 & -1 & 0 \end{pmatrix}, -i \begin{pmatrix} 0 & 0 & -1 \\ 0 & 0 & 0 \\ 1 & 0 & 0 \end{pmatrix}, -i \begin{pmatrix} 0 & 1 & 0 \\ -1 & 0 & 0 \\ 0 & 0 & 0 \end{pmatrix} \right\} \quad (5.6)$$

and showed that they satisfied (5.4) by setting $\mathbf{J}^\mathfrak{a} = \mathbf{T}^\mathfrak{a}$ Now we want to consider (5.4) as the defining relation for rotational symmetry and find all possible matrices – of any dimensionality – that satisfy this defining relation.

Recalling the definition from group theory, any such set will be a representation of the rotation group. We only want the irreducible representations (the irreps). Each irrep will correspond to a possible way that a free quantum particle can manifest rotational symmetry. Hence the irreps of the rotation group classify free particles. Once we know these irreps, we know all the possible physically distinct particle wavefunctions!

It's a general problem in group theory to find these irreps, and I won't do that here [35]. Instead, setting $\hbar = 1$, I'll just write down the most general solution to (5.4):

$$(\mathbf{J}^\mathfrak{a})_I{}^K = \left(\vec{x} \times (-i\vec{\nabla}) \right)^\mathfrak{a} \delta_I{}^K + (\mathbf{S}^\mathfrak{a})_I{}^K$$
$$= \mathbf{L}^\mathfrak{a} \delta_I{}^K + (\mathbf{S}^\mathfrak{a})_I{}^K \quad (5.7)$$

or more succinctly, $\vec{\mathbf{J}} = \vec{\mathbf{L}} + \vec{\mathbf{S}}$. Non-relativistically this operator will act on an N-component wavefunction $\Psi(\vec{x}, t) = (\Psi_1(\vec{x}, t), \Psi_2(\vec{x}, t), \cdots, \Psi_N(\vec{x}, t))$, where the K-th component is

$$\Psi_K(\vec{x}, t) = \varphi(\vec{x}, t) \chi_K \quad (5.8)$$

i.e. the wavefunction is the product[‡] of a spatial part $\varphi(\vec{x}, t)$ and an N-component spin part χ. The orbital angular momentum operator $\vec{\mathbf{L}}$ acts only on φ and the spin angular momentum operator $\vec{\mathbf{S}}$ acts only on χ, i.e.

$$\vec{\mathbf{J}} \Psi = \left(\vec{\mathbf{L}} \varphi \right) \chi + \varphi \left(\vec{\mathbf{S}} \chi \right) \quad (5.9)$$

Note that rotational invariance implies via Noether's theorem only that

$$\left[\vec{\mathbf{J}}, \mathbf{H} \right] = 0 \quad (5.10)$$

[‡]Relativistically this does not hold – although it will hold in the non-relativistic limit and in the rest frame of the particle.

and that in general

$$\left[\vec{\mathbf{L}}, \mathbf{H}\right] \neq 0 \qquad \left[\vec{\mathbf{S}}, \mathbf{H}\right] \neq 0 \tag{5.11}$$

In other words $\vec{\mathbf{L}}$ and $\vec{\mathbf{S}}$ will in general not commute with the Hamiltonian, though they commute with each other.

In the rest frame of a particle, there is no orbital angular momentum and all eigenvalues of $\vec{\mathbf{L}}$ vanish. However, the eigenvalues of the spin (or intrinsic) angular momentum do *not* vanish. This gives us our second good quantum number for free particles: *spin.*

5.2 Basic Classification

So far we've seen that invariances of a free particle under time and space translations and rotations imply that its momentum, energy, and angular momentum are all well-defined (are "good quantum numbers"), and so can be used to classify a particle. In the rest frame of the particle these quantities reduce to the particle mass m, and its spin s.

The mass is specifiable simply by giving a numerical value for m in whatever the relevant units are. We know of no principle that determines the particular values that m might have for a given elementary particle. The best we can do is to determine the value of m from experiment on a case-by-case (or perhaps I should say particle-by-particle) basis and input this value into our theories.

The spin s, on the other hand, is not so freely specifiable; instead it is determined from one of the irreducible solutions to (5.4):

$$\left[\mathbf{S}^a, \mathbf{S}^b\right] = i\epsilon^{abc}\mathbf{S}^c \tag{5.12}$$

where $\vec{\mathbf{J}} = \vec{\mathbf{S}}$ since we are in the rest frame of the particle. Any irreducible set of 3 matrices $\{\mathbf{S}^x, \mathbf{S}^y, \mathbf{S}^z\}$ that solves (5.12) determines a possible s value of the particle. Hence knowledge of all irreps satisfying (5.12) (i.e. all possible $\vec{\mathbf{S}}$'s) is equivalent to knowledge of all allowed particle spins.

As noted above, finding these irreps is a problem in group theory. Table 5.1 lists the three simplest (and most commonly used) irreps, along with the general form.

The quantities σ^i in table 5.1 are

$$\{\sigma^x, \sigma^y, \sigma^z\} = \{\sigma^1, \sigma^2, \sigma^3\} = \left\{ \begin{pmatrix} 0 & 1 \\ 1 & 0 \end{pmatrix}, \begin{pmatrix} 0 & -i \\ i & 0 \end{pmatrix}, \begin{pmatrix} 1 & 0 \\ 0 & -1 \end{pmatrix} \right\} \tag{5.13}$$

and the quantities $\mathbf{T}^i$ in table 5.1 are

$$\{\mathbf{T}^x, \mathbf{T}^y, \mathbf{T}^z\} = \{\mathbf{T}^1, \mathbf{T}^2, \mathbf{T}^3\}$$

TABLE 5.1

Irreducible Representations of Spin

Spin	Irrep	Terminology	Transformation (Rest Frame)
$s = 0$	$\vec{S} = 0$	Scalar $\Psi = \phi$	$\phi' = \phi$
$s = \frac{1}{2}$	$\vec{S} = \frac{1}{2}\vec{\sigma}$	Spinor $\Psi = \psi$	$\psi'_a = \exp\left[\frac{i}{2}\vec{\theta}\cdot\vec{\sigma}\right]_a^b \psi_b$ $a, b = 1, 2$
$s = 1$	$\vec{S} = \vec{T}$	Vector $\Psi = \vec{A}$	$A'_i = \exp\left[i\vec{\theta}\cdot\vec{T}\right]_i^j A_j$ $i, j = 1, 2, 3$
$\vdots$	$\vdots$	$\vdots$	$\vdots$
s	$\vec{S}$	Tensor $\Psi = \chi$	$\chi'_M = \exp\left[i\vec{\theta}\cdot\vec{S}\right]_M^N \chi_N$ $M, N = 1, 2, \ldots, (2s+1)$

$$
= \left\{ -i\begin{pmatrix} 0 & 0 & 0 \\ 0 & 0 & 1 \\ 0 & -1 & 0 \end{pmatrix}, -i\begin{pmatrix} 0 & 0 & -1 \\ 0 & 0 & 0 \\ 1 & 0 & 0 \end{pmatrix}, -i\begin{pmatrix} 0 & 1 & 0 \\ -1 & 0 & 0 \\ 0 & 0 & 0 \end{pmatrix} \right\} \tag{5.14}
$$

which are the generators of spatial rotations of a vector that we encountered in Chapter 3. The general form of $\left(\vec{S}\right)_M^N$ is a square matrix of dimension $(2s+1) \times (2s+1)$ which obeys (5.12).

Note the implications of the results listed in table 5.1: the spin of an elementary particle must be either an integer or a half-integer (in units of $\hbar$). There can never be an elementary particle with $s = 9/7$ or $s = \sqrt{3}$ or some other real number that is not one of these two types – rotational symmetry forbids this possibility[§].

All representations have the following two features in common:

$$
\mathbf{S}^2\Psi = \left(\vec{\mathbf{S}}\cdot\vec{\mathbf{S}}\right)\Psi = s(s+1)\Psi \tag{5.15}
$$

$$
S^z\Psi = s_z\Psi \tag{5.16}
$$

where s_z is a number that can have any value in the set $\{-s, -s+1, -s+2, \ldots, s-1, s\}$, for a total of $2s+1$ values in all.

The $s = \frac{1}{2}$ representation (or the spinor representation) is of significant import in particle physics. It was found by Pauli [36], and the $\vec{\sigma}$'s are given the name Pauli matrices. You can find out more about their properties in

---

[§]Actually this statement depends on the number of spatial dimensions. In two spatial dimensions spin need not come in units of $\hbar/2$. An particle whose spin differs from some integer times $\hbar/2$ is called an *anyon*, as noted in Chapter 1.

appendix D. They obey (5.12), i.e.

$$\left[\frac{1}{2}\sigma^1, \frac{1}{2}\sigma^2\right] = \frac{i}{2}\sigma^3 + \text{cyclic} \tag{5.17}$$

and we see that the factor of $\frac{1}{2}$ is crucial to this end[¶]. This factor implies that under a 2π rotation (about, say, the z-axis) the spinor representation transforms as

$$\psi_a' = \exp\left[\frac{i}{2}(2\pi)\sigma^z\right]_a^b \psi_b = \exp\left[(i\pi)\begin{pmatrix} 1 & 0 \\ 0 & -1 \end{pmatrix}\right]_a^b \psi_b$$

$$= \begin{pmatrix} e^{i\pi} & 0 \\ 0 & e^{-i\pi} \end{pmatrix}_a^b \psi_b = \begin{pmatrix} -1 & 0 \\ 0 & -1 \end{pmatrix}_a^b \psi_b$$

$$= -\psi_b \tag{5.18}$$

So the wavefunction $\psi \to -\psi$ under a 2π rotation! Notice that for the same rotation the scalar and vector wavefunctions do not change sign ($\phi \to \phi$ and $\vec{A} \to \vec{A}$). It is this peculiar feature of half-integer spin particles that distinguishes them from integer-spin particles.

What does relativity do to all of this? We could repeat our analysis by replacing R_i^j with Λ_μ^ν. The preceding results do not change – the only kinds of wavefunctions allowed are those permitted by the rotation group. Including boosts adds nothing new to this. Instead, relativity has a different physical implication for the allowed wavefunctions, which we'll look at in Chapter 11.

Tables 5.2 and 5.3 summarize our current knowledge of which elementary and composite subatomic particles have which spins. Question marks appear beside particles hypothesized to exist, but which have not actually been observed [‖].

[¶] Note that $[\sigma^1, \sigma^2] = 2i\sigma^3$ & cyclic. The factor of 2 is what destroys agreement with eq. (5.12).

[‖] You might have noticed that there are no elementary particles with spins larger than 2, and that there appears to be no elementary particle of spin 3/2. There is a theoretical obstruction to writing down theories describing pointlike elementary particles spins larger than 2 – nobody knows how to get them to interact according to standard approaches in quantum field theory. Superstring theories do not have such obstructions – see chapter 25 for more details.

TABLE 5.2
Classification of Bosonic Particles

	Spin-0	Spin-1	Spin-2
Elementary	Higgs?	Photon, Gluon, W, Z	Graviton?
Composite	Pseudoscalar Mesons	Vector Mesons	??

TABLE 5.3
Classification of Fermionic Particles

	Spin-1/2	Spin-3/2
Elementary	Quarks, Leptons	Gravitino?
Composite	Baryon Octet	Baryon Decuplet

5.3 Spectroscopic Notation

It's common to write the general (non-relativistic) spin-s wavefunction as χ, where

$$\chi = |s\,s_z\rangle \quad \text{where } s_z = -s, ..., s \tag{5.19}$$

so that

$$\mathbf{S}^2 |s\,s_z\rangle = s(s+1) |s\,s_z\rangle \tag{5.20}$$

$$S^z |s\,s_z\rangle = s_z |s\,s_z\rangle \tag{5.21}$$

$$S^\pm |s\,s_z\rangle = \sqrt{s(s+1) - s_z(s_z \pm 1)} \, |s\,(s_z \pm 1)\rangle \tag{5.22}$$

where the matrices $\vec{\mathbf{S}}$ have been written as

$$S^\pm = S^x \pm iS^y = S^1 \pm iS^2 \quad \text{and} \quad S^z = S^3 \tag{5.23}$$

$$\mathbf{S}^2 = (S^x)^2 + (S^y)^2 + (S^z)^2 \tag{5.24}$$

In addition to its mass and spin a particle (e.g. if it is composite) may have other good quantum numbers ("good" because they are constants of the motion and so their associated operators commute with $\mathbf{H}$). For example, it is very common that

$$\left[\vec{\mathbf{L}} \cdot \vec{\mathbf{L}}, \mathbf{H}\right] = 0 \qquad \left[\vec{\mathbf{S}} \cdot \vec{\mathbf{S}}, \mathbf{H}\right] = 0 \tag{5.25}$$

which physically corresponds to assuming that there are no forces or interactions that can change the *magnitude* of either the spin or orbital angular momentum (although there may be forces or interactions that change their directions). So more generally we write

$$\mathbf{J}^2 |j\,m\rangle = j(j+1) |j\,m\rangle \tag{5.26}$$

$$J^z \,|j\,m\rangle = m\,|j\,m\rangle \qquad m = -j, -j+1, \ldots, j-1, j \qquad (5.27)$$

$$J^{\pm} \,|j\,m\rangle = \sqrt{j(j+1) - m(m \pm 1)}\,|j\,(m \pm 1)\rangle \qquad (5.28)$$

for a particle (composite or not) of total spin j and z-component m, with the J's defined analogously to the S's. When (5.25) holds, a particle is characterized by the quantum numbers given in table 5.4 in addition to the mass.

TABLE 5.4
Particle Quantum Numbers

Quantity	Operator	Eigenvalues
total angular momentum	$\vec{J} \cdot \vec{J}$	$j(j+1)$
total spin	$\vec{S} \cdot \vec{S}$	$s(s+1)$
total orbital ang. mom.	$\vec{L} \cdot \vec{L}$	$\ell(\ell+1)$
Axial component of ang. mom.	J^z	$j_z = m$

We summarize this information using spectroscopic notation[37]

$$^{2S+1}L_J$$

for a state of total angular momentum $j = J$ and spin $s = S$. The numerical values ℓ of L are often denoted by S,P,D,F (for historical reasons**) instead of 0,1,2,3. Note that L can have only integer (and not half-integer) values. Hence a particle in the state $^3S_1$ has $s = j = 1$ and $\ell = 0$; a particle in the state $^5D_3$ has $s = 2$, $j = 3$ and $\ell = 2$.

5.4 Adding Angular Momenta

When particles collide they produce resonances (short-lived bound states), which in turn decay into other particles. These resonances will have spins that will be determined by the intrinsic spins of the colliding particles and their relative angular momenta. For example, the vector mesons are bound states of a quark with an antiquark in a spin-1 combination.

Consequently a key general question in particle physics (as well as in atomic physics) is how, quantum mechanically, do we add two (or more) angular

---

**The letters, "S", "P", "D", and "F", for the first four values of ℓ respectively stand for "Sharp", "Principal", "Diffuse", and "Fundamental", based on the properties of the spectral series observed in alkali metals.

momenta? Of course we know the answer classically: we just add the components. Quantum mechanically this doesn't make sense, because we can't measure *all* components of $\vec{J}$ simultaneously. A measurement of J^x necessarily alters J^y. The best we can do is to simultaneously measure $\mathbf{J}^2 = \vec{J} \cdot \vec{J}$ (the magnitude $\vec{J}$) and one component of $\vec{J}$, since each component commutes with $\mathbf{J}^2$. By convention we choose this component to be J^z. This is why we often label angular momenta of wavefunctions in the form $|j \, j_z\rangle$ or, more commonly, as $|j \, m\rangle$.

So what can we do to describe how particles of different spin combine? The answer (as you might have guessed) is also found in group theory and is called the *Clebsch-Gordon Decomposition* (or CG decomposition[††]). Given two particles of angular momenta $|j_1 \, m_1\rangle$ and $|j_2 \, m_2\rangle$, they can combine in the following linear combination of angular momentum states

$$|j_1 \, m_1\rangle \otimes |j_2 \, m_2\rangle = \sum_{j=|j_1-j_2|}^{j_1+j_2} \mathcal{C}_m^j \, (j_1, \, j_2; m_1 \, m_2) \, |j \, m\rangle \qquad M = M_1 + M_2$$

$$= \mathcal{C}_{m_1+m_2}^{|j_1-j_2|} \, ||j_1 - j_2| \, , m_1 + m_2\rangle$$

$$+ \mathcal{C}_{m_1+m_2}^{|j_1-j_2|+1} \, ||j_1 - j_2| + 1, m_1 + m_2\rangle + \cdots$$

$$+ \mathcal{C}_{m_1+m_2}^{(j_1+j_2)} \, |(j_1 + j_2), m_1 + m_2\rangle \qquad (5.29)$$

where the value of m is always the sum of the incoming m's, i.e. $m = m_1 + m_2$. The left-hand side of eq. (5.29) is the product of two spin wavefunctions corresponding to angular momenta j_1 (with z-component m_1) and j_2 (with z-component m_2). The right-hand side of eq. (5.29) is a linear combination of all possible spin-wavefunctions that are permitted by the rules of quantum-mechanical angular momentum conservation.

This rather formidable looking notation is more easily understood by noting that the CG decomposition has three essential features.

1. Since we can always measure (by convention) J^z unambiguously, the z-components of angular momenta just add, as noted above, i.e. $m = m_1 + m_2$. So every wavefunction on the right-hand side of eq. (5.29) has z-component $m_1 + m_2$, as is clear from each term in the sum.

2. The magnitudes of the angular momenta do <u>not</u> add: the total magnitude depends on the relative orientation of the incoming angular momenta $\vec{J}_1$ and $\vec{J}_2$. Since this is empirically unknowable (because we can at best measure a component of each $\vec{J}$ along only one axis), we get all possible spin-wavefunctions in the linear combination on the right-hand side of eq. (5.29) that are group-theoretically allowed. The biggest

---

[††]The coefficients are named after two German mathematicians Alfred Clebsch (1833-1872) and Paul Gordan (1837-1912), who encountered an equivalent problem in invariant theory.

value for $\vec{\mathbf{J}}$ is when $\vec{\mathbf{J}}_1$ and $\vec{\mathbf{J}}_2$ are parallel (giving $j = j_1 + j_2$) – both angular momenta are aligned along the z-axis. The smallest is when $\vec{\mathbf{J}}_1$ and $\vec{\mathbf{J}}_2$ are antiparallel – both are anti-aligned along the z-axis, giving $j = |j_1 - j_2|$. All other possible values for $\vec{\mathbf{J}}$ are given by going between these values in integer steps

$$j = |j_1 - j_2|, |j_1 - j_2| + 1, |j_1 - j_2| + 2, \ldots, j_1 + j_2 - 1, j_1 + j_2 \quad (5.30)$$

3. The C_m^j's are numbers that depend on the input parameters j_1, j_2, m_1 and m_2. A book on quantum mechanics will tell you how to calculate them [38], and the numbers are listed in tables known as Clebsch-Gordon tables. They are included in the appendix to this chapter. Reading these tables takes a bit of practice. The total spins being combined are given in the upper left of one of the sub-tables. The m-values (or z-components) of these spins are given in the lower-left boxes in a subtable, and the possible output $|j\,m\rangle$ wavefunctions are in the upper right boxes in the same sub-table. The C_m^j's are the square roots of the numbers in the relevant middle boxes, where the minus sign (if there is one) goes outside of the square root.

5.4.1 Examples

5.4.1.1 Glueballs

A bound state of two gluons is called a glueball. What possible spins can the lowest-energy glueball states have?

Gluons have spin $s = 1$, so the possible total spin values range between $1 - 1 = 0$ and $1 + 1 = 2$. Since the states are of lowest energy, the gluons must have no orbital angular momenta. Hence the possible spins are $0, 1$ and 2.

5.4.1.2 Positronium

An electron and a positron can form a bound state called positronium. What are the possible spins of a positronium "atom" if the e^+ and e^- have relative orbital angular momentum 1?

The spin j of positronium will be given by combining the spins of e^+ and e^- and their relative orbital angular momentum. Each spin $s = \frac{1}{2}$, so the combined spin (without taking orbital angular momentum into account) is either 0 or 1. Since $\ell = 1$ we get $j = 0, 1, 2$ if the combined spin is 1. If the combined spin is 0 we get $j = \ell + 0 = 1$.

5.4.1.3 Capture

An electron is temporarily captured by an Ω^+ particle, forming a resonance. If the z-components of their spins are positive and equal, what is the probability of observing this resonance to have its maximal angular momentum allowed in a state of lowest energy? How does this answer change if the z-components of their spins are equal and opposite?

> For this problem we need the CG tables. The Ω^+ has spin $\frac{3}{2}$ and the electron has spin $\frac{1}{2}$ so we need the $\frac{3}{2} \otimes \frac{1}{2}$ sub-table. Since we are in a state of lowest energy, $\ell = 0$ and so orbital angular momentum makes no contribution. Since the z-components of the spins are positive and equal, we must have $m = \frac{1}{2}$ for each particle, because this is the only allowed positive z-value for the electron. Reading from the 3rd line of the lower-left box in the $\frac{3}{2} \otimes \frac{1}{2}$ table, we have
>
> $$\left|\frac{3}{2}\,\frac{1}{2}\right\rangle \otimes \left|\frac{1}{2}\,\frac{1}{2}\right\rangle = \frac{\sqrt{3}}{2}\,|2\,1\rangle - \frac{1}{2}\,|1\,1\rangle$$
>
> where the coefficients are found by taking square roots of the numbers in the central box to the right, and the wavefunction components are found from the box above this one, column by column. The maximal angular momentum state is therefore $|2\,1\rangle$, i.e. $j = 2$. The probability of observing it is $\left(\frac{\sqrt{3}}{2}\right)^2 = 75\%$. If the spins are equal and opposite we have, from the 5th line of the lower-left boxes in the $\frac{3}{2} \otimes \frac{1}{2}$ table,
>
> $$\left|\frac{3}{2}\,\frac{1}{2}\right\rangle \otimes \left|\frac{1}{2}\,-\frac{1}{2}\right\rangle = \frac{1}{\sqrt{2}}\,|2\,0\rangle - \frac{1}{\sqrt{2}}\,|1\,0\rangle$$
>
> and so the probability of observation of the $|2\,0\rangle$ state is now $\left(\frac{1}{\sqrt{2}}\right)^2 = 50\%$.

5.5 Appendix: Tools for Angular Momenta

5.5.1 Pauli-Matrices

The 3-Pauli matrices are

$$\{\sigma^x, \sigma^y, \sigma^z\} = \{\sigma^1, \sigma^2, \sigma^3\} = \left\{ \begin{pmatrix} 0 & 1 \\ 1 & 0 \end{pmatrix}, \begin{pmatrix} 0 & -i \\ i & 0 \end{pmatrix}, \begin{pmatrix} 1 & 0 \\ 0 & -1 \end{pmatrix} \right\} \tag{5.31}$$

and we don't distinguish between upper and lower indices, so that $\sigma^1 = \sigma_1, \sigma^2 = \sigma_2, \sigma^3 = \sigma_3$. We have the product rule

$$\sigma_i \sigma_j = \delta_{ij} I + i\epsilon_{ijk}\sigma_k = \delta_{ij} + i\epsilon_{ijk}\sigma_k \tag{5.32}$$

where the 2×2 unit matrix I is often suppressed, as in the 2nd part of the expression above. This rule implies

$$
\begin{aligned}
(\sigma_1)^2 = (\sigma_2)^2 = (\sigma_3)^2 &= 1 \\
\sigma_1 \sigma_2 = i\sigma_3 \ \text{ and cyclic} & \\
[\sigma_i, \sigma_j] = 2i\epsilon_{ijk}\sigma_k \qquad \{\sigma_i, \sigma_j\} &= 2\delta_{ij}
\end{aligned} \tag{5.33}
$$

and for any two vectors $\vec{a}$ and $\vec{b}$:

$$
(\vec{a} \cdot \vec{\sigma})\left(\vec{b} \cdot \vec{\sigma}\right) = \vec{a} \cdot \vec{b} + i\left(\vec{a} \times \vec{b}\right) \cdot \vec{\sigma} \tag{5.34}
$$

We also have the exponential relation

$$
\exp\left[\vec{\theta} \cdot \vec{\sigma}\right] = \sum_{n=0}^{\infty} \frac{\left(\vec{\theta} \cdot \vec{\sigma}\right)^n}{n!} = \cos\theta + i\widehat{\theta} \cdot \vec{\sigma} \sin\theta \tag{5.35}
$$

where $\vec{\theta} = \theta\widehat{\theta}$.

5.5.2   Clebsch-Gordon Tables

Clebsch-Gordon tables contain explicit formulae for all the $\mathcal{C}_m^j$ (j_1, j_2; $m_1 \, m_2$) coefficients given in eq. (5.29). I have reproduced them on the next two pages and in appendix F.

The total spins (j_1, j_2) being combined are given in the upper left of one of the sub-tables. The respective (m_1, m_2)-values (or z-components) of these spins are given in the lower-left boxes in a subtable, and the possible output $|j \, m\rangle$ wavefunctions are in the upper right boxes in the same sub-table. The $\mathcal{C}_m^j$'s are the square roots of the numbers in the relevant middle boxes, where the minus sign (if there is one) goes outside of the square root.

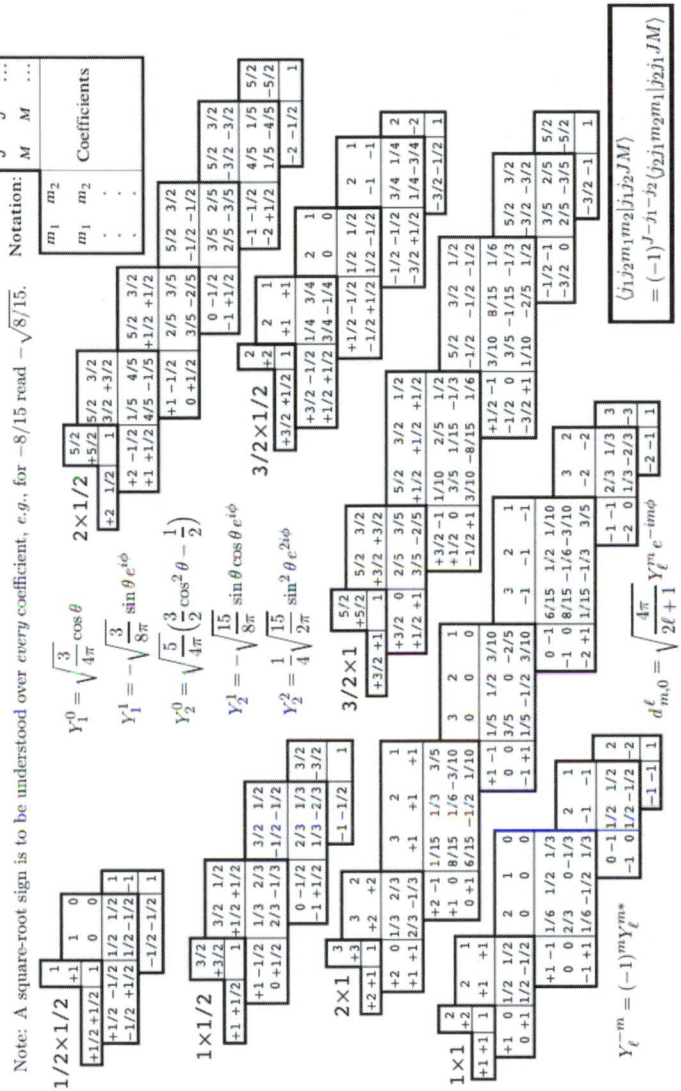

FIGURE 5.1

Table of Clebsch-Gordon Coefficients

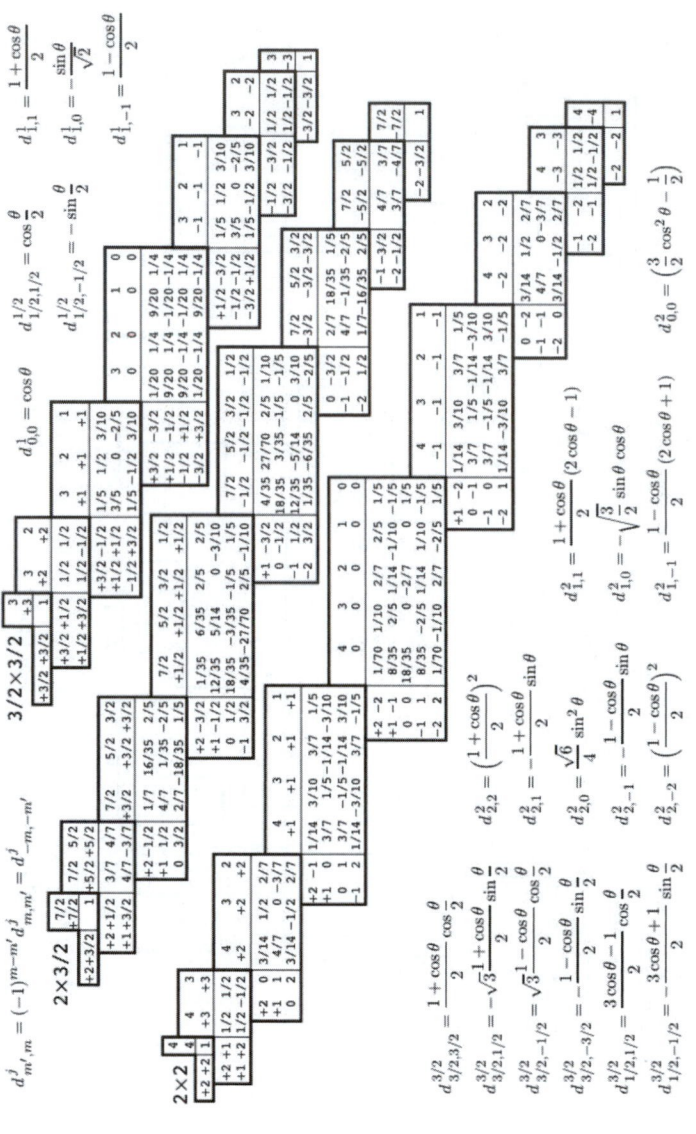

FIGURE 5.2
Table of Clebsch-Gordon Coefficients (cont'd)

5.6 Questions

1. (a) A Δ^+ with $j_z = -1/2$ collides with an Ω^- with $j_z = +1/2$ temporarily forming a resonance. What is the probability that the resonance is spinless? What is the probability that the resonance has spin-2?

   (b) A spin-2 glueball decays into a quark-antiquark pair. What are the possible values of the orbital angular momenta of the resultant $q\bar{q}$ system? What is the parity of the glueball?

2. A clever experimentalist figures out how to collide two Ω particles of opposite charge into a resonance (i.e., a short-lived bound state) called an omegaball.

   $L = 0$

   (a) What are the possible spins of an omegaball in its lowest-energy state?

   (b) The experimentalist then figures out how to make a supply of 10,000 omegaballs of lowest energy by repeatedly colliding Ω^+ and Ω^- together with opposite spins along the z-axis. What must these spin components be if more highest-spin omegaballs are made than any other kind? Find approximately how many of each spin are made in the sample of 10,000.

   (c) How does your answer to (b) change if the highest-spin omegaballs are fewest in number in the sample?

3. Λ-particles are produced by a pion beam in the reaction

   $$\pi^- + p \rightarrow K^0 + \Lambda$$

   and are observed via their decay $\Lambda \rightarrow \pi^- + p$. You are an experimentalist trying to determine the total spin $s(\Lambda)$ of the Λ. The angle of the decay products relative to the beam axis is θ.

   (a) If the Λ is produced exactly along the beam axis what are the possible values of $s_z(\Lambda)$?

   (b) A trustworthy theorist tells you that $s(\Lambda)$ can't be larger than 3. Given this constraint, what are the possible angular decay distributions for the forward-produced Λ's as a function of their spin?

4. Consider the Hamiltonian

   $$H = H_0 - g\frac{\mu_0}{\hbar}\vec{\mathbf{J}} \cdot \vec{B}$$

   where $\vec{B}$ is an external magnetic field, g, μ_0 are constants, and H_0 commutes with the angular momentum operator $\vec{\mathbf{J}}$. If the magnetic field is pointing along the z-axis, find the commutator $\left[H, \vec{\mathbf{J}}\right]$.

$|S|^2 |s, S_z\rangle = s(s+1) |s, S_z\rangle \qquad S_z | \quad \rangle = S_z | \quad \rangle . \Rightarrow S_x, S_y \quad 4 \times 4$

5. (a) Find the simplest irreducible representation of the gravitino, a spin-3/2 particle hypothesized to exist in superstring theory.

(b) Suppose two gravitini collide to form a bound state (a gravitiball). What possible spins could this bound state have?

(c) Suppose you had 10,000 gravitiballs of lowest energy, formed by repeatedly colliding two gravitini together with opposite spins along the z-axis. What must these spin components be if more highest-spin gravitiballs are made than any other kind? Find approximately how many of each spin are made in the sample of 10,000.

6. Verify the following:

(a) $\exp\left[\vec{\theta} \cdot \vec{\sigma}\right] = \sum_{n=0}^{\infty} \frac{(\hat{\theta} \cdot \vec{\sigma})^n}{n!} = \cos\theta + i\hat{\theta} \cdot \vec{\sigma} \sin\theta$

(b)

$$(\sigma_1)^2 = (\sigma_2)^2 = (\sigma_3)^2 = 1$$

$$\sigma_1 \sigma_2 = i\sigma_3 \text{ and cyclic}$$

$$[\sigma_i, \sigma_j] = 2i\epsilon_{ijk}\sigma_k \qquad \{\sigma_i, \sigma_j\} = 2\delta_{ij}$$

7. In 1932 the decay of the neutron into a proton and an electron was observed. What conservation laws did this decay violate, if any?

8. Suppose an electron is in the state $\psi = \alpha \begin{pmatrix} 1 \\ 0 \end{pmatrix} + \beta \begin{pmatrix} 0 \\ 1 \end{pmatrix}$.

(a) What relationship must α and β obey in order for ψ to be normalized?

(b) What values might be obtained upon measurement of S_x, and what is the probability of each?

(c) What values might be obtained upon measurement of S_y, and what is the probability of each?

(d) What values might be obtained upon measurement of S_z, and what is the probability of each?

6

Discrete Symmetries

Some symmetries in nature are not continuous, but discrete: you either perform them or you don't. In other words they can't be generated from infinitesimal transformations. Because of this the conservation laws associated with such symmetries are multiplicative instead of additive. In this chapter we will examine the notion of discrete symmetries and their implications for particle physics.

In general, a multicomponent wavefunction Ψ will transform under a discrete symmetry as

$$\Psi'_A(x') = (\mathbf{U_D})_A{}^B \Psi_B(x) \quad \text{where} \quad x'^\mu = \left(\hat{\mathbf{U}}_\mathbb{D}x\right)^\mu \tag{6.1}$$

$$\Rightarrow \quad \Psi'_A(x) = (\mathbf{U_D})_A{}^B \Psi_B(\hat{\mathbf{U}}_\mathbb{D}^{-1}x) \tag{6.2}$$

where $\mathbf{U_D}$ is some matrix (acting on the components of the wave function) that is to be determined for each discrete symmetry, and $\hat{\mathbf{U}}_\mathbb{D}$ is its representation when acting on the spacetime coordinates. For a continuous symmetry $\mathbf{U} = \mathbf{I} + \varepsilon(\mathbf{something})$ where ε is small. This kind of Taylor-series expansion is not possible for a discrete symmetry. The matrices $\mathbf{U_D}$ must be fully known in order for their action on wavefunctions to be explicitly computed.

Fortunately this is not difficult to do. I won't deal with all possible discrete symmetries here, but instead will concentrate on the three most important for particle physics: *parity*, *time-reversal*, and *charge conjugation*. Let's look at each.

6.1 Parity

Parity is the act of reflecting a system in a mirror. If the mirror-system has all the same physical properties as the original, then we say the system is invariant under parity.

Mathematically this kind of reflection involves specifying a plane for the mirror and then switching the signs of all the coordinates in the directions orthogonal to this plane. An example is shown in fig. 6.1. This is generally quite inconvenient, so usually a parity transformation is implemented by performing an *inversion* on the coordinates $\vec{x} \to -\vec{x}$: every point is carried to its diametrically opposite location through the origin. This is a combination of a reflection with a 180° rotation. Both transformations turn a right hand into a left hand and vice versa. Inversions are easier to work with, since we don't have to choose a plane for the mirror. We shall generally refer to inversions as parity transformations. The difference between the two is illustrated in figures 6.1 and 6.2.

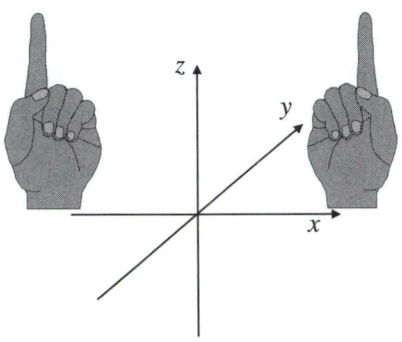

FIGURE 6.1
Reflection in the x-z plane: $(x, y, z) \longrightarrow (x, -y, z)$

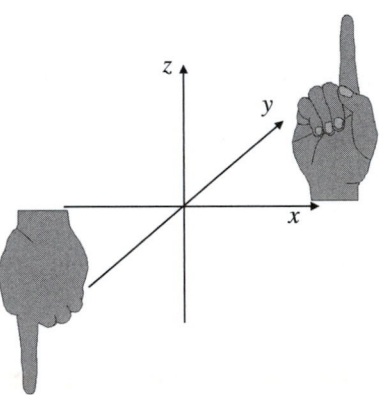

FIGURE 6.2
Inversion: $(x, y, z) \longrightarrow (-x, -y, -z)$

So if we call the parity operator $\mathbb{P}$, we have

$$\mathbb{P} : (t, x, y, z) \rightarrow (t, -x, -y, -z) \Rightarrow \mathbb{P}\left(\vec{V}\right) = -\vec{V} \qquad (6.3)$$

where $\vec{V}$ is any vector. Note that

$$\mathbb{P}\left(\vec{V} \cdot \vec{W}\right) = \mathbb{P}\left(\vec{V}\right) \cdot \mathbb{P}\left(\vec{W}\right) = \left(-\vec{V}\right) \cdot \left(-\vec{W}\right) = \vec{V} \cdot \vec{W} \qquad (6.4)$$

but

$$\mathbb{P}\left(\vec{V} \times \vec{W}\right) = \mathbb{P}\left(\vec{V}\right) \times \mathbb{P}\left(\vec{W}\right) = \left(-\vec{V}\right) \times \left(-\vec{W}\right) = +\vec{V} \times \vec{W} \qquad (6.5)$$

which we expect, since $\mathbb{P}$ changes right-handed coordinate systems to left-handed ones.

So we see that there are two kinds of vectors: those that reverse sign under $\mathbb{P}$ and those that do not. We call this 2nd kind of vector a *pseudovector*, since it transforms under parity opposite to the way a vector transforms. Note that if $\vec{V}$ is a vector and $\vec{A}$ is a pseudovector then

$$\mathbb{P}\left(\vec{V} \times \vec{A}\right) = \left(-\vec{V}\right) \times \left(+\vec{A}\right) = -\vec{V} \times \vec{A} \Rightarrow \text{ a vector} \qquad (6.6)$$

$$\mathbb{P}\left(\vec{V} \cdot \vec{A}\right) = \left(-\vec{V}\right) \cdot \left(+\vec{A}\right) = -\vec{V} \cdot \vec{A} \Rightarrow \text{ a pseudoscalar} \qquad (6.7)$$

and so tensor quantities (scalars, vectors, etc.) may or may not be pseudo, depending on how they transform under $\mathbb{P}$. The various possibilities are listed in table 6.1.

TABLE 6.1
Behavior of Scalars and
Vectors under Parity

Scalar	$\mathbb{P}(s) = +s$
Pseudoscalar	$\mathbb{P}(p) = -p$
Vector	$\mathbb{P}\left(\vec{V}\right) = -\vec{V}$
Pseudovector	$\mathbb{P}\left(\vec{A}\right) = +\vec{A}$

If $\mathbb{P}$ is applied twice we must get what we had originally, and so

$$\mathbb{P}^2 = I \qquad (6.8)$$

or in other words the eigenvalues of $\mathbb{P}$ are ± 1. Scalars and pseudovectors have eigenvalue $+1$, whereas pseudoscalars and vectors have eigenvalue -1.

Note also that we can write a representation of $\mathbb{P}$ as a matrix when it acts on spacetime coordinates:

$$\mathbb{P} : (t, x, y, z) \to (t, -x, -y, -z) \Rightarrow (\Lambda_{\mathbb{P}})^{\mu}{}_{\nu} = \begin{pmatrix} 1 & 0 & 0 & 0 \\ 0 & -1 & 0 & 0 \\ 0 & 0 & -1 & 0 \\ 0 & 0 & 0 & -1 \end{pmatrix} \quad (6.9)$$

This shows that $\Lambda_{\mathbb{P}}$ is also a Lorentz transformation, since $g_{\alpha\beta} (\Lambda_{\mathbb{P}})^{\alpha}{}_{\nu} (\Lambda_{\mathbb{P}})^{\beta}{}_{\mu} = g_{\mu\nu}$, something easily shown by explicit calculation for each component.

So in addition to their mass and spin, wavefunctions of elementary particles are also classified according to their parity eigenvalues. Let's look at a simple example.

6.1.1 Parity of the Photon

What is the parity of the photon? We can figure it out by recalling that in the hydrogen atom the wavefunctions (or rather, the spatial part of the wavefunctions) have the form

$$\Psi (\vec{x}) = \Psi (r, \theta, \phi) = \varphi(r)\mathcal{Y}_{\ell}^{m} (\theta, \phi) \quad (6.10)$$

where the $\mathcal{Y}_{\ell}^{m} (\theta, \phi)$ are the spherical harmonics, given in table ??. Now note that

$$\begin{aligned} \mathbb{P} \; &: \; (t, x, y, z) \to (t, -x, -y, -z) \\ &\Rightarrow \mathbb{P} : (t, r, \theta, \phi) \to (t, r, \pi - \theta, \pi + \phi) \quad (6.11) \\ &\Rightarrow \mathbb{P} [\Psi (\vec{x})] = \Psi (-\vec{x}) = \varphi(r)\mathcal{Y}_{\ell}^{m} (\pi - \theta, \pi + \phi) \\ &= (-1)^{\ell} \, \varphi(r)\mathcal{Y}_{\ell}^{m} (\theta, \phi) = (-1)^{\ell} \; \Psi (\vec{x}) \quad (6.12) \end{aligned}$$

and so we see that a state with orbital angular momentum ℓ has a parity eigenvalue of $(-1)^{\ell}$. This means that the S,D,G,... states of the Hydrogen atom (the ℓ =even ones) have even parity, whereas the P,F,H,... states have odd parity. In a transition where $\Delta \ell = \pm 1$, one photon is absorbed (or emitted). Hence (since electromagnetism is parity-conserving) the photon has negative parity.

6.1.2 Parity Conservation

We also assign positive parity to quarks and leptons (and negative parity to their antiparticles). Of course, this is a convention, and we could have chosen a reverse assignment; it won't matter as long as parity is conserved*.

---

*Thinking ahead to when we include relativity, we will find that we need to include antiparticles. It is possible to show that a fermion has a parity opposite to that of its antiparticle whereas a boson has a parity that is the same as that of its antiparticle. This result follows from quantum field theory, a subject beyond the scope of this text.

To the best of our experimental knowledge, parity *is* conserved in strong and electromagnetic interactions. However, it is *not* conserved in weak-interactions. Weak interactions affect all known particles including a neutral fermion of very small mass called a neutrino. All neutrinos are left-handed (and all antineutrinos are right-handed): their spins are always antiparallel to their momenta[†]. Figures 6.3 and 6.4 illustrate this concept. Nature is not mirror-symmetric – parity is violated!

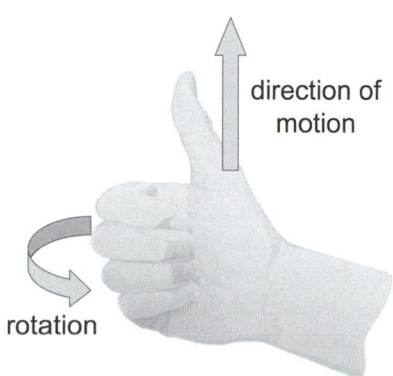

FIGURE 6.3

A right-handed particle is one that rotates in the direction of the fingers of the right hand while traveling in the direction of the thumb. Similarly, a left-handed particle rotates in the opposite direction.

6.2 Time-Reversal

Macroscopic physics is generally NOT invariant under time-reversal; for example an explosion does not look the same if a movie of it is run backwards! This is what we refer to as the arrow of time. Unlike the spatial coordinates, which can be traversed in any direction we like, the time coordinate seems

---

[†]This discovery was first made in the 1950s by Wu [39] and we'll look at it in more detail when we consider weak interactions in Chapter 19. Recent observations at Super-Kamionkande [41] and SNO [42] have revised the original understanding of neutrinos as zero-mass particles. There is now excellent evidence that they are very light-mass particles, in which case there are also right-handed neutrinos! Our best experimental information so far is that the right-handed neutrinos do not experience the weak interactions.

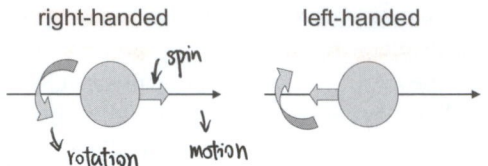

FIGURE 6.4

A diagram representing the relationship between momentum, spin, and handedness. The thick arrow points in the direction of the spin vector. The thin straight arrow is the momentum vector.

to be one for which only a one-way journey (from past to future) is not only allowed, but required. Understanding why this should be the case is one of the outstanding puzzles of modern physics [43].

At a macroscopic level the asymmetry associated with the arrow of time is a consequence of initial conditions, which according to the laws of thermodynamics always become less ordered overall as time increases (though local regions can become more ordered as in, say, the construction of a building). We say that a physical system in combination with its environment experiences an increase in entropy.

At the microscopic level of collisions between the fundamental particles in our example of the explosion, the situation is quite different. At this level we might expect time reversal invariance to hold because Newton's laws are time-reversal invariant. From this perspective the puzzle of the arrow of time reduces to the problem of the origin of initial conditions.

Common quantities transform under time reversal $\mathbb{T}$ and parity $\mathbb{P}$ are shown in table 6.2.

As with parity, we can represent $\mathbb{T}$ as a matrix when it acts on coordinates:

$$\mathbb{T} : (t, x, y, z) \to (-t, x, y, z) \Rightarrow (\Lambda_\mathbb{T})^\mu_{\ \nu} = \begin{pmatrix} -1 & 0 & 0 & 0 \\ 0 & 1 & 0 & 0 \\ 0 & 0 & 1 & 0 \\ 0 & 0 & 0 & 1 \end{pmatrix} \tag{6.13}$$

and it, too, is a Lorentz transformation: $g_{\alpha\beta} (\Lambda_\mathbb{T})^\alpha_{\ \nu} (\Lambda_\mathbb{T})^\beta_{\ \mu} = g_{\mu\nu}$. Both fermions and their antiparticles transform the same way under time-reversal[‡]; the same holds for bosons and their antiparticles.

6.2.1 Detailed Balance

Time-reversal is difficult to test – as noted above, since all physical systems experience time to move forward, we have no way of directly forcing a system

---

[‡]This is another result from quantum field theory.

TABLE 6.2

Transformations of Common Physical Quantites under Time-reversal and Parity

Quantity	$\mathbb{T}$	$\mathbb{P}$	Comments
Position $\vec{r}$	$\vec{r}$	$-\vec{r}$	
Momentum $\vec{p}$	$-\vec{p}$	$-\vec{p}$	Polar vector
Spin $\vec{s}$	$-\vec{s}$	$\vec{s}$	Axial vector (like $\vec{r} \times \vec{p}$)
Electric field $\vec{E}$	$\vec{E}$	$-\vec{E}$	$\vec{E} = -\vec{\nabla}V$
Magnetic field $\vec{B}$	$-\vec{B}$	$\vec{B}$	like a ring current
Magnetic dipole moment $\vec{\mu}$	$-\vec{\mu}$	$\vec{\mu}$	
Electric dipole moment $\vec{\mathbf{d}}$	$\vec{\mathbf{d}}$	$-\vec{\mathbf{d}}$	
Longitudinal polarization $\vec{s} \cdot \vec{p}$	$\vec{s} \cdot \vec{p}$	$-\vec{s} \cdot \vec{p}$	Chirality
Transverse polarization $\vec{s} \cdot (\vec{p_1} \times \vec{p_2})$	$-\vec{s} \cdot (\vec{p_1} \times \vec{p_2})$	$\vec{s} \cdot (\vec{p_1} \times \vec{p_2})$	

to reverse its trajectory in the time direction (i.e., we just can't "run the movie backward").

However, we can take a particular physical reaction and run it in reverse. Consider for example the two-body scattering of a neutron off of a proton to form a deuteron and a photon. Under time-reversal we would have

$$n + p \longrightarrow D + \gamma \quad \Leftarrow \mathbb{T} \Rightarrow \quad D + \gamma \longrightarrow n + p \tag{6.14}$$

$\mathbb{T}$-invariance, if it held, would force the rate for both processes to be the same for corresponding conditions of energy, momentum and angular momentum; this is called the *Principle of Detailed Balance* [44]. It is the most direct test of time reversal that we have.

So far all experiments have indicated that time-reversal invariance is a symmetry of the strong and electromagnetic interactions, with the principle of detailed balance holding for every known case for these interactions. Unfortunately the places where we expect to see $\mathbb{T}$-violation are in the weak interactions: after all, this is where parity is violated (and, as we will see, charge-conjugation invariance as well). Here the principle of detailed balance is extremely difficult to test. For example the weak decay of the Λ meson is $\Lambda \to p + \pi$, so we would expect to test detailed balance via

$$\Lambda \to p + \pi \quad \Leftarrow \mathbb{T} \Rightarrow p + \pi \longrightarrow \Lambda \tag{6.15}$$

This would in principle form a check, but in practice the pion and the proton form many other states besides the Λ when they collide. This is because the strong interaction between the pion and the proton totally dominates the feeble weak interaction that would form the Λ. It is therefore pretty much impossible in practice to pick out the time-reversed reaction on the right-hand side of eq. (6.15).

Can we do anything about this? Yes – we can precisely measure static quantities whose value should be *exactly zero* if $\mathbb{T}$ is a symmetry. Any empirical evidence that such quantities were nonzero would therefore be firm evidence that $\mathbb{T}$ is not a symmetry of nature.

One such quantity is the electric dipole moment $\vec{\mathbf{d}}$ of an elementary particle. If it were nonzero it would have to be either aligned or antialigned with the spin of the particle since there is no other direction available. But from table 6.2 we see that an electric dipole moment does not change sign under time-reversal whereas the spin does, so any nonzero-value of $\vec{\mathbf{d}}$ would be a signature of time-reversal violation. At present the best limits that we have are for the electron [45] and the neutron [46]:

$$|\vec{\mathbf{d}}_n| < 6 \times 10^{-26} e \text{ cm} \qquad |\vec{\mathbf{d}}_e| < 1.6 \times 10^{-27} e \text{ cm} \qquad (6.16)$$

setting stringent upper bounds on $\mathbb{T}$-violation. We'll look more at the situation for the neutron in Chapter 25.

6.3 Charge Conjugation

Charge conjugation, denoted by $\mathbb{C}$, transforms any state into a state with the same energy, momentum, spin and mass but with all other quantum numbers (the "charges") reversed. In other words, $\mathbb{C}$ transforms each particle into its antiparticle, e.g.

$$\mathbb{C}\,|p\rangle = |\bar{p}\rangle \qquad \mathbb{C}\,|\pi^+\rangle = |\pi^-\rangle \qquad (6.17)$$

Obviously $\mathbb{C}^2 = +1$ (just like $\mathbb{T}^2 = +1$ and $\mathbb{P}^2 = +1$), and so it also has eigenvalues ± 1. However,, unlike $\mathbb{P}$, most particles are *not* eigenstates of $\mathbb{C}$, because a particle is not the same state as its antiparticle (for example, a positron is not an electron). Any particle that is an eigenstate of $\mathbb{C}$ must be its own antiparticle, since the eigenvalue equation for $\mathbb{C}$ requires

$$\mathbb{C}\,|\text{particle}\rangle = |\text{antiparticle}\rangle = \pm\,|\text{particle}\rangle \qquad (6.18)$$

The photon is one such particle, and the π^0, η, η', ρ^0, ϕ, ω, and J/ψ mesons are also their own antiparticles. Consequently they can be assigned definite charge-conjugation quantum numbers. By convention, $\mathbb{C}$ does not change the mass, energy, momentum or spin of a particle[§].

6.3.1 Charge Conjugation of the Pion

If electromagnetic charge is reversed for all particles in a given system, then the sign of the electromagnetic field in the system must also be reversed, and so $\mathbb{C} = -1$ for the photon. We can use this to deduce the value of $\mathbb{C}$ for the pion. For the π^0, experiments have shown that

$$\pi^0 \to 2\gamma \tag{6.19}$$

which is an electromagnetic decay, because only photons are in the final state. Hence we must have $\mathbb{C} = +1$ for the π^0 since $\mathbb{C}$ is conserved in electromagnetism.

The conservation of $\mathbb{C}$ is therefore something we can check by searching for the decay of the π^0 into odd numbers of photons. Any positive evidence for such a decay would indicate that something was wrong with our understanding of electromagnetism and, by extension, the Standard Model. Perhaps our theory of electromagnetism would need revision, or perhaps the π^0 would have a structure that is different from our current understanding. Experimentally we have

$$\frac{\Gamma\left(\pi^0 \to 3\gamma\right)}{\Gamma\left(\pi^0 \to 2\gamma\right)} < 3.1 \times 10^{-8} \tag{6.20}$$

and so we see that $\mathbb{C}$ *is* conserved in electromagnetism, to better than one part in 10 million [47].

6.3.2 Charge Conjugation of Fermions

A fermion-antifermion bound state of orbital angular momentum ℓ and spin s must have $\mathbb{C} = (-1)^{\ell+s}$. This can be deduced by noting that the lowest-energy state, which has $s = \ell = 0$, can decay into two photons by energy conservation. Since $\mathbb{C} = -1$ for the photon, we must have $\mathbb{C} = (\mathbb{C}(\text{photon}))^2 = +1$ for this bound state. If the bound state has $s = 1$, then the fermion-antifermion pair have the same spins: a spin-flip transition to an $s = 0$ state results in the emission of one photon, so $\mathbb{C} = (-1)^s$ if $\ell = 0$. Finally, if $\ell \neq 0$, the excited bound state can decay electromagnetically with $\ell \to \ell - 1$ by emitting one photon. The only consistent choice that describes these observations is to set $\mathbb{C} = (-1)^{\ell+s}$ for a fermion-antifermion bound state.

[§]Strictly speaking, the antiparticle of an electron with spin $\vec{s}$, and momentum $\vec{p}$ is a positron of spin $-\vec{s}$, and momentum $-\vec{p}$. However, we are interested in how $\mathbb{C}$ changes the internal quantum numbers of a particle – the electric charge, the quark color, etc – and so we define $\mathbb{C}$ as an operator that reverses only internal quantum numbers of any given particle.

$\mathbb{C}$ is *not* a symmetry of the weak interactions: $\mathbb{C}$ applied to a left-handed neutrino gives a left-handed anti-neutrino (i.e. a neutrino with opposite "weak charge"). But left-handed anti-neutrinos do not couple to any known particles[¶], so $\mathbb{C}$ is not conserved for weak interactions.

6.4 Positronium

An ideal place to strengthen our understanding of discrete symmetries is with positronium. Positronium is an electromagnetically bound state of e^+e^- that can decay to photons:

$$e^+e^- \longrightarrow 2\gamma, 3\gamma, \ldots \tag{6.21}$$

It is bound together just like the Hydrogen atom, except that a positron is responsible for the binding instead of a proton. The wavefunction is

$$\Phi = \Psi\left(r, \theta, \phi; \mu\right)\Xi(s) = \Psi\left(r, \theta, \phi; \mu\right)\left(\psi_{e-} \otimes \psi_{e+}\right) \tag{6.22}$$

where Ψ is the wavefunction of the Hydrogen atom, but with reduced mass $\mu = \left(\frac{1}{m_e} + \frac{1}{m_e}\right)^{-1} = \frac{1}{2}m_e$. Its energy levels will be given by the Bohr formula $E_n = -\frac{\alpha^2 mc^2}{4n^2}$ for $n = 1, 2, 3, \ldots$ where $\alpha = \frac{e^2}{4\pi\hbar c} \simeq \frac{1}{137}$ is a dimensionless quantity called the *fine-structure constant*.

The wavefunction Ψ provides information about the relative spatial relationship between any two particles regardless of their structure. The object $\Xi(s) = (\psi_{e+} \otimes \psi_{e-})$ is the spin part of the wavefunction, composed of the spins of the electron and positron. Since each of these are spin-1/2, they can only combine to give a total spin of 0 or 1. Using the Clebsch-Gordon tables **??**, we have

$$\left.\begin{array}{c} \Xi(1,1) = \psi_{e-}^{\uparrow}\psi_{e+}^{\uparrow} \\[4pt] \Xi(1,0) = \frac{1}{\sqrt{2}}\left(\psi_{e-}^{\uparrow}\psi_{e+}^{\downarrow} + \psi_{e-}^{\downarrow}\psi_{e+}^{\uparrow}\right) \\[4pt] \Xi(1,-1) = \psi_{e-}^{\downarrow}\psi_{e+}^{\downarrow} \end{array}\right\} \quad \begin{array}{c} \text{triplet } S = 1 \\ \text{ORTHOPOSITRONIUM} \end{array} \tag{6.23}$$

$$\Xi(0,0) = \frac{1}{\sqrt{2}}\left(\psi_{e-}^{\uparrow}\psi_{e+}^{\downarrow} - \psi_{e-}^{\downarrow}\psi_{e+}^{\uparrow}\right) \quad \begin{array}{c} \text{singlet } S = 0 \\ \text{PARAPOSITRONIUM} \end{array} \tag{6.24}$$

Now recall that charge-conjugation changes a spin-up (which I'll write as spin-$\uparrow$) e^- into a spin-$\uparrow$ e^+. Hence under $\mathbb{C}$:

$$\mathbb{C}\left[\Xi(1,1)\right] = \mathbb{C}\left[\psi_{e-}^{\uparrow}\psi_{e+}^{\uparrow}\right] = \psi_{e+}^{\uparrow}\psi_{e-}^{\uparrow} = -\psi_{e-}^{\uparrow}\psi_{e+}^{\uparrow} = -\Xi(1,1) \tag{6.25}$$

[¶]It was once thought that they didn't exist, but experiments at the Sudbury Neutrino Observatory [42] strongly imply that neutrinos have mass, which means that both right-handed and left-handed neutrinos and antineutrinos exist. We will consider this subject in Chapter 25.

where the 2nd from last step follows because fermion wavefunctions anticommute due to the Pauli principle. Hence

$$\mathbb{C}\Phi^{ORTHO} = -\Phi^{ORTHO} \quad \text{and} \quad \mathbb{C}\Phi^{PARA} = \Phi^{PARA} \tag{6.26}$$

which you can easily check. Recalling that the photon is negative under $\mathbb{C}$, we consequently find that

γ – does not conserve momentum.

$$\Phi^{PARA} \longrightarrow 2\gamma + 4\gamma + 6\gamma + \cdots \tag{6.27}$$
$$\Phi^{ORTHO} \longrightarrow 3\gamma + 5\gamma + 7\gamma + \cdots \tag{6.28}$$

$\mathbb{C}$ conserves electromagnetism.

or in other words, PARA can only decay into even numbers of photons, and ORTHO only into odd numbers of photons$^{\parallel}$. A 2-Body decay process is "easier" to go to than a 3-Body one – the phase space is larger for decay into smaller numbers of objects – so we expect PARA to decay faster than ORTHO. Also, the amplitude for emission of one photon is proportional to the charge e of the electron. This means that the probability for PARA to decay will be proportional to $|e^2|^2 \sim \alpha^2$. Since ORTHO decays by emitting 3 photons instead of 2, we expect its decay rate to be smaller than PARA's by a factor of α.

Note that positronium decay depends on the electron and positron annihilating each other, a situation that can only occur if they are in the same place at the same time. Hence the decay rate must be proportional to $|\Psi(0)|^2$, i.e. the square of the wavefunction at the origin, which is where the electron and positron "collide." From atomic physics we know that

$$|\Psi(0)|^2 = \frac{1}{\pi a^3} = \frac{\alpha^6}{8\pi r_e^3} \tag{6.29}$$

where $a = \frac{2r_e}{\alpha^2}$ is the Bohr radius of the positronium atom and $r_e = \frac{e^2}{4\pi mc^2}$ is the classical electron radius. The actual theoretical calculations give [48, 49]

$$\Gamma(PARA \to 2\gamma) = 4\pi\hbar c r_e^2 |\Psi(0)|^2 = \frac{m_e c^2}{2}\alpha^5 = \left(1.252 \times 10^{-10}\text{s}\right)^{-1}$$
$$= 8.00\,(\text{nsec})^{-1} \tag{6.30}$$
$$\Gamma(ORTHO \to 3\gamma) = \frac{2}{9\pi}\left(\pi^2 - 9\right)m_e c^2 \alpha^6 = \left(1.374 \times 10^{-7}\text{s}\right)^{-1}$$
$$= 7.21\,(\mu\text{sec})^{-1} \tag{6.31}$$

for the decay of the ground states.

$^{\parallel}$The 2-photon and 3-photon decays are the dominant processes; decays into more photons are higher-order corrections.

6.4.1 A Puzzle with ORTHO

Until very recently more accurate calculations yielded a puzzle. While the agreement between theory and experiment for PARA was always in good shape, there existed a discrepancy between theory and experiment for ORTHO that went unexplained for decades. Experiments as recently as 1990 differed by at least 6 standard deviations from the theoretical calculation. This led theorists to propose that all kinds of exotic hypotheses that were sometimes rather bizarre extensions of the Standard model. These ideas included axions, $\mathbb{C}$-odd bosons, millicharged particles, forbidden numbers of gamma rays, and even a mirror universe!

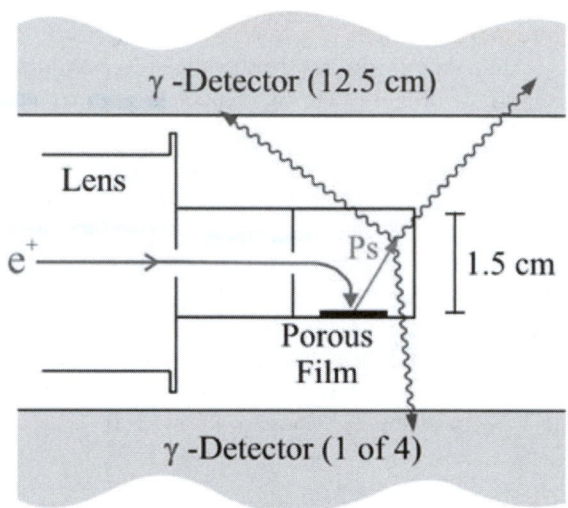

FIGURE 6.5

A schematic of the cavity used in the new precision orthopositronium decay rate measurement. Positrons are focused through two apertures of an aluminum cavity onto an porous silica film. The emitted thermal positronium decays in vacuum. Reprinted figure with permission from R. S. Vallery, P.W. Zitzewitz, and D.W. Gidley, *Phys. Rev. Lett.* **90** 203402 (2003) [50]. Copyright (2003) by the American Physical Society.

However, in May 2003, R.S. Vallery and colleagues published a paper [50] describing the results of a more careful experiment on orthopositronium. They created orthopositronium by firing a low-energy positron beam into a special micron-thick nanoporous silica film; orthopositronium formed from the slowed-down positrons as they captured electrons. Vallery and collaborators were able to measure how long this took by detecting the gamma rays after annihilation in a scintillator. This set-up, illustrated in fig. 6.5, overcame problems encountered in previous experiments to measure decay rates, which sometimes measured energetic positronium annihilating on the cavity walls of the detector. In the Vallery experiment, only positrons that annihilated their bound electrons were detected. They measured a lifetime for ORTHO in agreement with the current QED calculation that differed by only about 0.014% from the theoretical value!

The moral? Sometime a simple explanation – in this case, that something was wrong with the experiments – is the right one.

	Theory	Experiment
Γ (PARA)	$7.9852 \pm .010$ (nsec)$^{-1}$ [48]	$7.994 \pm .011$ (nsec)$^{-1}$ [51]
Γ (ORTHO)	$7.039979 \pm .000011$ (μsec)$^{-1}$ [49]	$\begin{cases} 7.0516 \pm .0013\,(\mu\text{sec})^{-1}\ [52] \\ 7.0514 \pm .0014\,(\mu\text{sec})^{-1}\ [53] \\ 7.0482 \pm .0016\,(\mu\text{sec})^{-1}\ [54] \\ 7.0404 \pm .0018\,(\mu\text{sec})^{-1}\ [50] \end{cases}$

$$\text{PARA} = \Xi(0,0) = \frac{1}{\sqrt{2}}\left(\psi_{e^-}^{\uparrow}\,\psi_{e^+}^{\downarrow} - \psi_{e^-}^{\downarrow}\,\psi_{e^+}^{\uparrow}\right)$$

6.4.2 Testing Fermion-Antifermion Parity

Consider next the parity of positronium. For PARA we have

$$\mathbb{P}\Phi^{\text{PARA}} = -(-1)^{\ell}\,\Phi^{\text{PARA}} \tag{6.32}$$

where the first minus sign is due to the opposite parity of the electron and positron, and the $(-1)^{\ell}$ comes from the parity of the spatial wavefunction. Since $\ell = 0$ for the ground state, the final state for the 2 emitted photons from PARA must have negative parity.

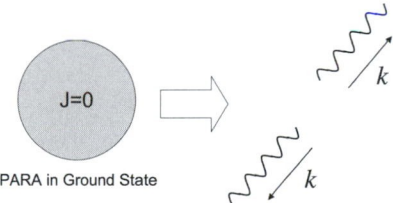

FIGURE 6.6
Schematic diagram of Positronium decay.

In the rest frame of PARA the 3-momenta of the photons must be $\vec{k}$ and $-\vec{k}$ as shown in figure 6.6. The initial state has no angular momentum ($J = 0$).

The final state wavefunction $|\gamma_1\gamma_2\rangle$ can only depend on the photon momenta and polarizations, and must also have $J = 0$ by angular momentum conservation. Therefore it must be a scalar function of the momenta and polarizations. Furthermore, since photons are bosons, we must have $|\gamma_1\gamma_2\rangle = +|\gamma_2\gamma_1\rangle$. Hence

$$|\gamma_1\gamma_2\rangle = \mathcal{A}\,\hat{e}_1 \cdot \hat{e}_2 + \mathcal{B}\,(\hat{e}_1 \times \hat{e}_2) \cdot \hat{k} \qquad (6.33)$$

where $\mathcal{A}$ and $\mathcal{B}$ are scalar functions of the momentum and polarizations. Now by conservation of parity we must have $\mathbb{P}\,|\gamma_1\gamma_2\rangle = -|\gamma_1\gamma_2\rangle$. Hence

$$|\gamma_1\gamma_2\rangle = \mathcal{B}\,(\hat{e}_1 \times \hat{e}_2) \cdot \hat{k} \qquad (6.34)$$

since $\mathbb{P}\,(\hat{e}_1 \times \hat{e}_2) \cdot \hat{k} = -(\hat{e}_1 \times \hat{e}_2) \cdot \hat{k}$ but $\mathbb{P}\hat{e}_1 \cdot \hat{e}_2 = \hat{e}_1 \cdot \hat{e}_2$. The amplitude $\langle \mathrm{PARA}\,|\gamma_1\gamma_2\rangle$ is largest for $\hat{e}_1 \perp \hat{e}_2$.

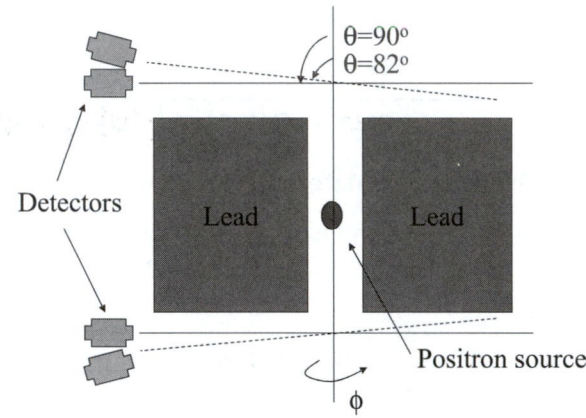

FIGURE 6.7
Schematic diagram of test of parity conservation in positronium decay. The polarization angle ϕ corresponds to a rotation perpendicular to the page.

If one emitted photon exhibits an X-polarization, the other always shows a Y-polarization, i.e., the planes of their polarizations must be perpendicular to each other. This may be confirmed experimentally by utilizing the feature that Compton-scattering cross sections for polarized photons are significantly greater for scattering into the plane at right angles to the E-vector of the incident photon, i.e., $90°$ to the direction of polarization. The setup is schematically shown in figure 6.7. The optical analog of the scattering material is the polarizing filter. The Klein-Nishina formula [56] shows the scattering cross section σ is proportional to:

$$\sigma = \frac{k}{k_0} + \frac{k_0}{k} - 2\sin^2\theta\cos^2\phi \qquad (6.35)$$

with k_0 and k representing the momenta of the incident and the scattered photons respectively, θ the angle of scattering, and ϕ the angle between the plane of scattering and the $\hat{e}$-vector of the incident photons.

This experiment was first done by Wu and Shaknov [57] with ^{64}Cu as the source. At $\theta = 90°$, annihilation radiation scattering turns out (for the energies relevant to the source ^{64}Cu) to be two times stronger for $\phi = 90°$ (when $\hat{e}_1 \perp \hat{e}_2$) than for $\phi = 0°$ (when $\hat{e}_1 \parallel \hat{e}_2$), thus providing an effective γ-ray polarization analyzer. They observed

$$\frac{\text{Rate}(\phi = 90°)}{\text{Rate}(\phi = 0°)} = 2.04 \pm 0.08 \qquad (6.36)$$

in agreement with the expected value of 2.00. We can regard this experiment as demonstrating that electrons and positrons have opposite parity.

6.5 The CPT Theorem

Strong and electromagnetic interactions are observed to separately conserve $\mathbb{C}$, $\mathbb{P}$ and $\mathbb{T}$ and our present theories of these interactions (QCD and QED) are constructed so that these symmetries are preserved**.

Weak interactions, however, violate $\mathbb{C}$ and $\mathbb{P}$ separately. Parity-violation in β-decay was first observed by Wu et al. [39] and has been seen directly in nuclear reactions [40] such as

$$\underset{J^P = 2^-}{^{16}\text{O}} \longrightarrow \underset{J^P = 2^+}{^{12}\text{C}} + {}^4\text{He} \qquad \Gamma = (1.0 \pm 0.3) \times 10^{-10} s \qquad (6.37)$$

Weak interactions also violate $\mathbb{C}$-invariance: no left-handed antineutrinos have ever been observed. And finally, Kaon decays violate both $\mathbb{C}$ and $\mathbb{P}$. We'll look at all of these processes in subsequent chapters.

There is a theorem by Schwinger, Lüders and Pauli [58] called the CPT-theorem which states that:

> ANY Lorentz-invariant Hermitian Lagrangian is invariant under CPT provided
>
> 1. The ground state is invariant.
>
> 2. The theory is local (i.e. has no action at a distance).

**There is actually a subtle exception to this for the strong interactions that we will look at in Chapter 25.

A given theory might violate any of $\mathbb{C}$, $\mathbb{P}$ or $\mathbb{T}$ separately, but it must respect the combined operation $\mathbb{CPT}$ if these two conditions hold. Experimentally this means

1. Particles and antiparticles must have the same mass

2. Particles and antiparticles must have the same lifetimes

3. Particles and antiparticles must have the same magnetic moments

Any discrepancy in such experiments would signal a breakdown of quantum mechanics! Hawking and Penrose have each separately suggested that quantum gravity forces such a breakdown [59]. For example, the presence of black holes might imply that pure states evolve into mixed states (something that can't happen in quantum mechanics) because part of the wavefunction is "absorbed" by the black hole is therefore irretrievably lost if the black hole evaporates in a purely thermal fashion[††].

Table 6.3 summarizes the results of a few key experiments that test the $\mathbb{CPT}$ theorem. So far $\mathbb{CPT}$ violation has not been observed – yet!

TABLE 6.3
Sample Tests of $\mathbb{CPT}$

Lifetime	$\tau_{\mu^+}/\tau_{\mu^-} = 1.00002 \pm .00008$
Mass	$|m_{K^0} - m_{\bar{K}^0}|/m_{\text{average}} < 10^{-18}$
Magnetic Moment	$\mu_{e^+}/\mu_{e^-} < 10^{-11}$

6.6 Questions

1. Are any of these processes allowed? Why or why not?

   (a) $\rho^0 \longrightarrow \pi^+ + \pi^-$ (b) $\eta^0 \longrightarrow 5\gamma$

   (c) $\rho^0 \longrightarrow 3\pi^0$ (d) $\eta^0 \longrightarrow \pi^+ + \pi^- + \pi^0$

2. According to the Standard Model, which of the following reactions are allowed and which are forbidden? State the reasons why if not. If allowed, state what interaction is responsible for the process.

[††]This is because no information is contained in the outgoing thermal radiation.

(a) $\bar{n} \longrightarrow \bar{p} + e^+ + \nu$

(b) $\gamma + Z^0 \longrightarrow \nu_e + \pi^0$

(c) $\mu^+ + \tau^- \longrightarrow \gamma + \gamma + \gamma$

(d) $Z^0 \longrightarrow \bar{\nu}_\mu + \nu_\mu$

(e) $D^0 \longrightarrow K^- + \rho^+$

(f) $e^- + \bar{\nu} \longrightarrow \bar{t} + b$

3. Consider the non-relativistic Schroedinger equation with the Hamiltonian

$$H = \frac{|\vec{p}|^2}{2m} + V(\vec{x})$$

satisfied by the wavefunction $\Psi(\vec{x})$. Under what circumstances will $\Psi(-\vec{x})$ satisfy the Schroedinger equation?

4. The deuteron is a $^3S_1$ bound state of a neutron and *hydrogen* a proton, and must have a wavefunction that is antisymmetric under proton-neutron interchange. What is the parity of the deuteron? $\Psi(n,p) = -\Psi(p,n)$

5. Show that Maxwell's equations are invariant under time-reversal.

6. Suppose the expectation value of $\mathbf{J} \cdot \mathbf{P}$ were found to be nonzero in some process.

   (a) What would this imply about parity conservation?

   (b) What would this imply about time-reversal invariance?

7. Time reversal interchanges initial and final states, so that if $\mathbb{T}|\chi\rangle = |\chi\rangle^t$ then

$$\mathbb{T}(\langle\vartheta|\chi\rangle) =^t \langle\vartheta|\chi\rangle^t = \langle\chi|\vartheta\rangle = (\langle\vartheta|\chi\rangle)^*$$

where the first equality is due to interchange of initial and final states and the second equality is the property of quantum-mechanical amplitudes.

   (a) Show that the above relation implies $\mathbb{T}c = c^*\mathbb{T}$ where c is any complex number.

   (b) Given that $\mathbb{T}\vec{J} = -\vec{J}\mathbb{T}$ where $\vec{J}$ is the angular momentum operator, show that

   time reversal → spin direction flips. S^2, S_z, S_+, S_-

$$|\chi_1\rangle \equiv -\mathbb{T}\left|\frac{1}{2}, -\frac{1}{2}\right\rangle \quad \text{and} \quad |\chi_2\rangle \equiv \mathbb{T}\left|\frac{1}{2}, \frac{1}{2}\right\rangle$$

   doublet
   form a spin-1/2 in the time-reversed system. $\mathbb{T}(\phi_1 \otimes \phi_2) = \mathbb{T}(\phi_1) \otimes \mathbb{T}(\phi_2)$.

   (c) What is $\mathbb{T}^2$ for a state $|\phi\rangle$ with an odd number of spin-1/2 particles? What is $\mathbb{T}^2$ for a state $|\varphi\rangle$ with an even number of spin-1/2 particles?

   (d) Suppose the state $|\phi\rangle$ in (c) is an eigenstate of the Hamiltonian. What is the degree of the degeneracy of this state? (i.e. how many distinct states are there at a given energy level?) This degeneracy is called *Kramer's degeneracy*.

(e) Is Kramer's degeneracy preserved in an electric field? Is it preserved in a magnetic field?

7

Accelerators

Experimental information in high-energy physics has historically most "commonly" been obtained from *accelerators*: machines that accelerate charged particles to very high speeds and then let them collide with other particles. They are very expensive to build and maintain, and they have a limited use in terms of the fundamental physics that they can reveal. Yet they provide us with a kind of information about the subatomic world that cannot be obtained in any other way.

Accelerators basically do two things. First, they provide us with information about the detailed structure of subatomic systems. They produce new interactions and/or bound states of known particles, whose characteristics provide us with further information about the laws of nature and the structure of matter. Second, they produce new particles. This is why they cost so much money – in order to produce a new particle you need at least as much energy as the rest mass of the particle. This might not sound like much at the subatomic level, but there is a lost of energy "wasted" in a collision process because we can't directly control the products of the collision that emerge nor the loss of energy due to other effects.

Accelerators have played an essential role in particle physics. Without them it would simply not be possible to check in any detailed way whether or not our theories were correct. They have led to the discovery of particles and interactions that nobody anticipated, and have provided us with a picture of the subatomic world that no one imagined as recently as a century ago. They probe the shortest distances humankind has ever measured by manipulating beams of particles traversing millions of kilometers in a few seconds to micron precision. They have been compared to the great cathedrals of Europe by Robert R. Wilson because of their immense size, intricate complexity, and symbolic representation of the human intellect [62].

The first accelerators that were constructed in the 1930s had particle beams whose energies were a few hundred keV. Seventy years later, in the first decade of the 21st century, the Large Hadron Collider will generate particle beams with energies of nearly 10^{13}eV – a factor 100 million times greater! The effective energy available to study new physics is even larger, about 10^{18}eV, since the LHC arranges for two beams of similar energy to collide. This will allow us to probe distances shorter than 10^{-18}cm, yielding the world's most powerful microscope.

Accelerators have grown out of their core purpose of providing experimental information for particle physics, and today are used in condensed matter physics, biomedical technologies, geophysics, electronics, food processing, and many other areas. Accelerator science has become a separate intellectual subdiscipline with both pure and applied aspects, providing yet another illustration of how basic research can foster positive economic and social development.

This chapter will provide a brief overview of accelerator physics. Following an historical path, we shall begin with the earliest machines, sketching the emergence of the more complex technologies as they developed over the past 70 years.

7.1   DC Voltage Machines

The simplest way to accelerate charged particles is with a high voltage DC source [61]. Such machines today can at best achieve beam energies of 20 MeV. For nuclear physics experiments this is useful, but for particle physics this is too low an energy. A picture is given in figure 7.1.

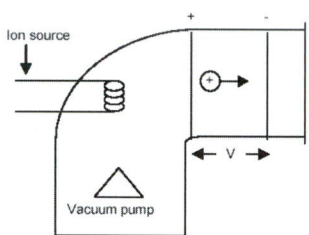

FIGURE 7.1
Prototype of the simplest accelerator.

A particle of charge q is generated by a source of ions. It is then accelerated in an electric field $\vec{E}$, which means it experiences a force $\vec{F}$, and consequently gains energy $\mathcal{E}$ as it travels a distance d

$$\vec{F} = q\vec{E} \Rightarrow \mathcal{E} = \left|\vec{F}\right| d = q\left|\vec{E}\right| d = qV \tag{7.1}$$

where $V = \left|\vec{E}\right| d$ is the voltage of the machine. The system must also be in as high a vacuum as is manageable – otherwise the accelerating particle will lose most of its energy in collisions with air molecules in the accelerator.

These elements – particle source, accelerating structure, and vacuum pump – appear in every accelerator.

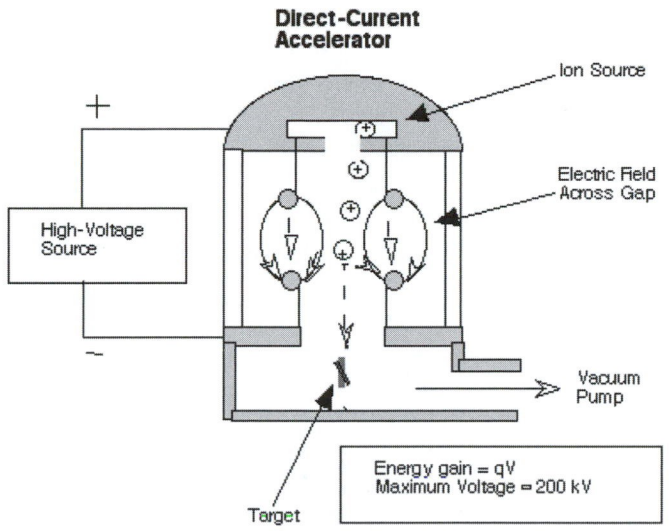

FIGURE 7.2

A Cockcroft-Walton Accelerator. Diagram courtesy of the Contemporary Physics Education Project (used with permission).

The earliest such machine of this type is called a Cockroft-Walton accelerator [63], shown in fig. 7.2. The ion source was hydrogen gas, and the original machine developed by John Cockroft and Ernest Walton was able to accelerate protons to 400 keV. Such machines today can reach a maximum voltage of 1 MeV due to voltage breakdown and discharge. These machines are often used as the first step of a multistage process that accelerates particles to much higher energies in more powerful accelerators.

A more sophisticated DC machine is a Van de Graff accelerator [64], in which a conveyor belt carries positive charge (in the form of ions that are sprayed onto the belt) to a collector which in turn transfers the positive charge to the dome. The principle at play here is that charge on any conductor resides on its outermost surface. If a conductor that is carrying charge touches another conductor whose surface surrounds it, then the charge on the first conductor will be transferred to the second one. Hence, as Robert van de Graff realized, one can "pile up" charge on a conductor to increase voltage by continually transferring charge to it via a conveyor belt, as shown in fig. 7.3.

This technique can yield voltages of up to 12 MeV. Positive charge (obtained

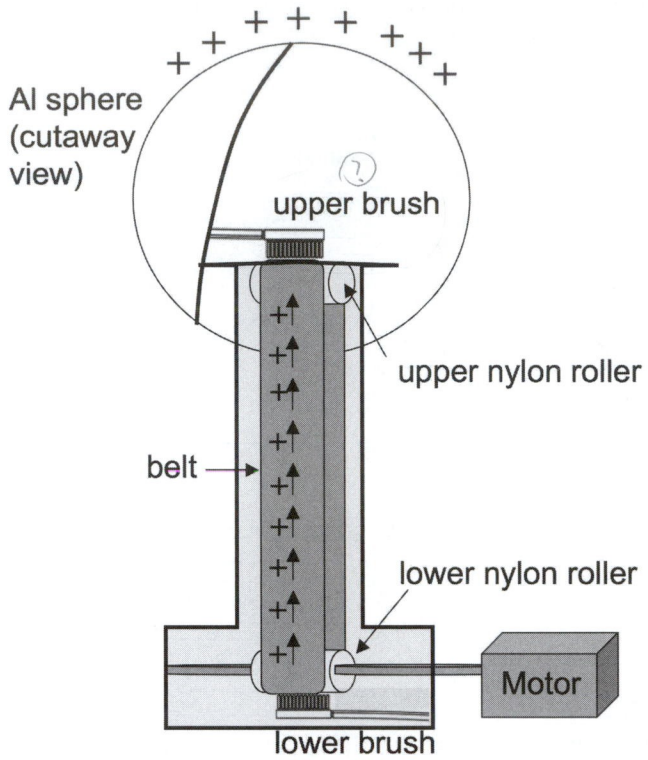

FIGURE 7.3

A diagram of a Van de Graff accelerator.

from gas ionized at high voltage) is sprayed or brushed onto a conducting conveyer belt, which continually rotates about two drums (somewhat like the fan belt in a car). One end of the belt is inside of a conducting dome. A wire brush is attached to the sphere and brushes against the belt. The charge on the belt will then travel through the brush and spread out on the sphere. The motors turning the drum provide the work needed to carry out this process. The points at which the charges are sprayed onto the belt are called *corona points*.

To complete the accelerator an ion source must be placed within the conducting dome near an evacuated tube. This tube leaves the dome and provides a conduit for the accelerated particles to eventually hit a target. If positive ions are emitted from the source, they will be accelerated away from the positively charged dome down the tube and toward the target. The tube is constructed with equipotential metallic rings, embedded within an insulating tube, and the entire structure is contained within a pressurized chamber of gas (typically SF_6) of about 15 atm. The energy limit that the machine can reach is constrained by the voltage at which the gas discharges.

Higher energies can be attained through the following trick. If negative ions are emitted from a source at one end of an acclerating tube that is outside of the dome, they will be attracted to it. A stripper inside the dome can remove some of their electrons and make them positive. These ions again accelerate down another tube, away from the terminal and toward a target. In this manner the maximal energy can be doubled to about 25 – 40 MeV. Such a machine is called a Tandem Van de Graff [65].

DC machines cannot achieve the requisite energies of modern particle physics. However, they have a high beam intensity (up to 100 microamperes) and a stable beam energy, and so are useful in nuclear physics research and (more recently) in solid state physics, where these machines are used to implant ions into materials to achieve a desired doping.

7.2   Linacs

Linear accelerators (LINACS) attempt to overcome the aforementioned limitation by giving the charged particle a series of "kicks" using an AC source. The basic idea is to repeatedly accelerate the particle many times over. A LINAC does this via a series of cylindrical tubes (called *drift tubes*), each of which is connected to a high-frequency oscillator. The succesive tubes are arranged to have opposite polarity. Inside the tubes the electric field is zero, but in the gaps in between, the electric field alternates with the generator frequency.

Suppose a particle of charge e enters this setup. The electric field at the first gap is set so that it attracts the particle, accelerating it to the first tube. The length of this tube is arranged so that when the particle arrives at the next gap the relative voltages of the tubes have flipped so as to provide another accelerating field in the gap. This further accelerates the particle and the process is repeated up to the tolerance voltages of the device, as illustrated in fig. 7.4. Each tube must increase in length because the speed of the electron rapidly increases as it moves down the tube. The length L of the tube must equal $\frac{1}{2}vT$, where T is the period of the oscillation and v is the speed of the particle.

Typically such machines gain a few MeV per meter in beam energy. Proton linacs typically reach about 50 MeV; the best in the world is the meson factory at Los Alamos that can reach 800 MeV. Electron linacs can reach much higher energies ($\sim$ 25 GeV maximum) since the electron is much lighter. The largest

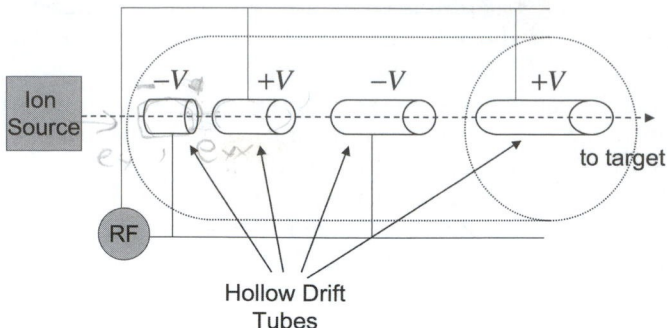

FIGURE 7.4
A schematic drawing of a linac.

electron linac in the world is at Stanford. It is 3 km long and has a 25 GeV electron beam which pulses 60 times per second.

7.2.1 Focusing the Beam

One of the problems with such big machines is in keeping the beam well-collimated. Light from a flashlight, say, will spread out, and the same kind of thing will happen with a particle beam. And, just as the spreading light can be refocused with lenses, the same sort of thing can be done for particle beams by using magnets*. The *Lorentz force equation* describes the motion of a particle of charge q in an electromagnetic field:

$$\vec{F} = q\left(\vec{E} + \frac{\vec{v}}{c} \times \vec{B}\right) = \frac{q}{c}\vec{v} \times \vec{B} \qquad (7.2)$$

where $\vec{v}$ is the velocity of the particle and we have set $\vec{E} = 0$. For $\vec{v} \perp \vec{B}$, the particle will experience a centripetal force $\left|\vec{F}\right| = \frac{mv^2}{r}$ normal to both of these directions, and so

$$\frac{mv^2}{r} = \frac{q}{c}vB \Rightarrow r = \frac{mvc}{qB} = \frac{pc}{qB} \qquad (7.3)$$

is the radius of curvature r through which the particle bends as it goes through the magnetic field.

To get an idea of how big the magnetic fields are that are required for focusing, consider an electron whose kinetic energy $\mathcal{E}_{kin}$ is 1 MeV. Its momentum

---

*Electric fields could also be used, but the field strength required for focusing high-energy particle beams is impractically large.

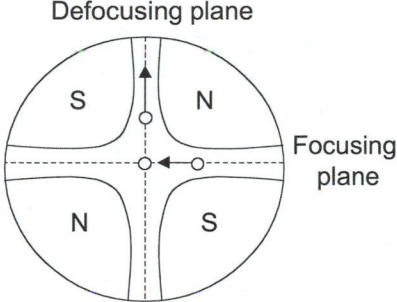

FIGURE 7.5

A schematic diagram of a quadrupole magnet.

is therefore given by solving

$$\mathcal{E}_{kin} = \sqrt{(pc)^2 + (mc^2)^2} - mc^2 \Rightarrow pc = 1.42 \text{ MeV} = 1.42 \times 10^6 \text{eV} \quad (7.4)$$

since the mass of the electron is 0.511 MeV. Because the charge of the electron is $1.60217733 \times 10^{-19}$ Coulombs $= 3 \times 10^9$ esu, we find from eq.(7.3)

$$Br = \frac{pc}{q} = \frac{1.42 \times 10^6}{3 \times 10^9} = 4.73 \times 10^3 \text{ Gauss-cm} \quad (7.5)$$

Unfortunately a magnet can bend particles only in one plane, and so can focus only in this plane, unlike an optical lens that can focus in more than one plane. How can a magnetic field be used to focus in two planes? The solution to this problem was found in 1950 by Christofilos [66] (and again independently in 1952 by Courant, Livingston and Snyder [67]): the *quadrupole magnet*! This magnet focuses in one plane, and defocuses in the orthogonal plane – see figure 7.5 for a conceptual representation of how this works. In the diagram, particles in the horizontal plane are deflected inward, while those in the vertical plane are deflected outward. Rotate this magnet by 90° and the opposite effect occurs: particles in the horizontal plane are deflected outward, and those in the vertical plane are deflected inward. Consequently two such magnets rotated 90° to one another around the beam axis behave as an optical lens, and a net focusing occurs. A picture of a quadrupole magnet that was used for PEP (Positron-Electron Project) at SLAC is shown in 7.6.

Fig. 7.7 is a diagram of the SLAC linac, which is the world's highest energy LINAC.

LINACs can achieve arbitrarily high energies in principle. However, the greater the desired energy, the larger the cost – a 500 GeV accelerator would have to be 75km long! This is prohibitively expensive and environmentally costly. A new solution for achieving higher energies is required.

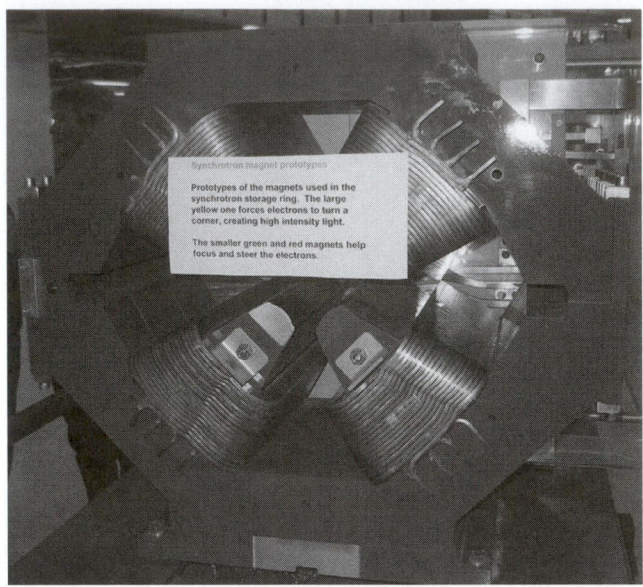

FIGURE 7.6

A quadrupole magnet once used in the storage ring at the Australian Synchrotron, Clayton, Victoria, Australia. Photo by John O'Neill; used with permission.

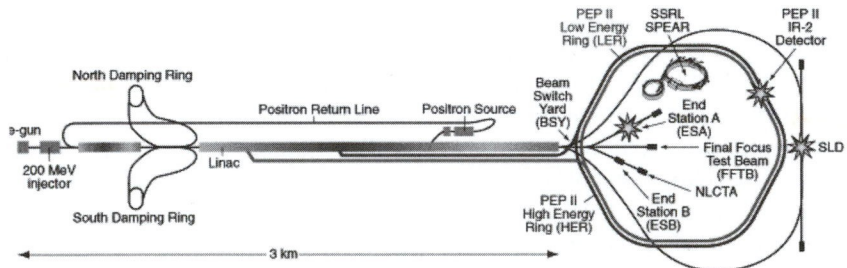

FIGURE 7.7

A diagram of the LINAC at SLAC. Image courtesy of SLAC National Accelerator Laboratory.

7.3   Synchrotrons

The idea of the synchrotron is to use one voltage source to repeatedly accelerate the particle instead of many sources as in the linac. Instead of the particles moving in a straight line, they move around in a circle, their motion synchronized with a changing voltage source and magnetic field that increases their energy. This idea of circularly accelerating particles was proposed in 1930 by Lawrence [68], and such machines were called cyclotrons.

A cyclotron consists of two hollowed out metal vaccum chambers, each in the shape of the letter D. These are placed side-by-side along their straight edges with a gap, and each is connected to an alternating high voltage source. This entire setup is then placed inside a magnetic field that is perpendicular to the D-shaped chambers. The high-voltage source produces an electric field only in the gap between the D's because the metal chambers shield the insides, where only the magnetic field pervades. If an ion source is placed in the gap, the electric field will accelerate ions toward one of the D-chambers. The magnetic field meanwhile causes the ion to move in circular motion. By appropriately setting the alternating frequency of the voltage source, the ion, as it leaves the first D-chamber, will be accelerated toward the second one. This process can be repeated many times, with the ion being accelerated across the gap each time it leaves a D-chamber. Its speed and its radial orbit will continue to get larger until one wishes to extract it (say by turning off the magnetic field) to have it impact upon a target.

A fixed-frequency cyclotron cannot accelerate particles to high energies, where relativistic effects must be taken into account. The maximum energy a proton cyclotron can obtain is 20 MeV. A more sophisticated machine is needed to attain higher energies. These machines are called synchrotrons.

A synchrotron consists of straight segments in accelerating cavities combined with circular segments that cause the particle to repeatedly traverse the same trajectory. The charged particles are first linearly accelerated into the ring, and then traverse a vacuum tube in a torus. A magnetic field keeps the particles moving in a circle. The straight segments have a RF field that turns on as the particles enter the cavity, accelerating them to higher speeds.

For a particle injected into a ring of radius R at speed v, the time for one full turn is

$$T = \frac{2\pi R}{v} = \frac{2\pi R}{pc^2}\mathcal{E} \tag{7.6}$$

since $p = m\gamma v$ and $\mathcal{E} = m\gamma c^2$. Hence the circular frequency is

$$\Omega = \frac{2\pi}{T} = \frac{pc^2}{\mathcal{E}R} \tag{7.7}$$

and the acceleration of the particle is

$$\frac{d\vec{p}}{dt} = \frac{q}{c}\vec{v} \times \vec{B} \tag{7.8}$$

where the magnetic field is orthogonal to the direction of motion. Recalling that $\vec{v} = v\hat{\theta}$ for circular motion, we have $\vec{v} \times \vec{B} = -vB\hat{r}$. Similarly, we have

$$\frac{d\vec{p}}{dt} = \frac{dp}{dt}\hat{\theta} + p\frac{d\hat{\theta}}{dt} = \frac{dp}{dt}\hat{\theta} - p\frac{v}{R}\hat{r} \tag{7.9}$$

which for constant $p = |\vec{p}|$ yields

$$-p\frac{v}{R}\hat{r} = -\frac{q}{c}vB\hat{r}. \tag{7.10}$$

or alternatively

$$B = \frac{pc}{qR} \tag{7.11}$$

Of course p is not constant: once per revolution, the particles are accelerated to increase v. Obviously B must increase in a synchronized manner (otherwise the particles will crash into the walls of the tube) – hence the name *synchrotron*. As the particles are accelerated by an RF generator of frequency ω, they will gain energy and momentum. The frequency ω must be an integer multiple of Ω to keep the beam within the tube. As the energy increases, we have $pc \to E$ and so

$$\omega = k\Omega = k\frac{pc^2}{\mathcal{E}R} \to \frac{kc}{R} \qquad \text{and} \qquad B = \frac{pc}{qR} \to \frac{\mathcal{E}}{qR} \tag{7.12}$$

since $pc \to \mathcal{E}$ at high energies. The magnetic field and the RF are increased from their initial values to final values chosen in such a manner as to always maintain the above relations. Clearly the limitation on the beam energy is B. The best superconducting magnets currently furnish magnetic fields slightly larger than 5 Tesla (50 kilogauss). A notable feature of large acclerators is that the particles cannot be accelerated from zero (or small) velocity into the machines – the range over which the RF and magnetic fields would have to operate is too big. Consequently such machines are built in stages, with smaller machines pre-accelerating the particles to speeds that the larger machines can handle. The Large Hadron Collider (LHC) at CERN is an excellent example of a machine that makes use of smaller linacs and synchrotrons to achieve high energies, as shown in fig. 7.8.

7.3.1   Focusing Beams at Synchrotrons

Particle beams in general do not consist of streams of charged particles. Instead, they occur in clusters, or bunches. The reason for this is that there is always some finite spread in the time of arrival of the particles as they enter an acceleration region. Consider a cyclotron. Particles arriving "on time" will experience just the right electric field to keep them moving in the correct orbit in the D-chambers. A particle arriving earlier will experience a stronger electric field, causing it to traverse an orbit of larger radius. This in turn causes it to return to the gap at a later time, closer to the return time of the original "on time" particles. A particle arriving later will experience a weaker electric field and thus traverse an orbit of smaller radius, causing it to return to the gap earlier, which again is closer to the return time of the original "on

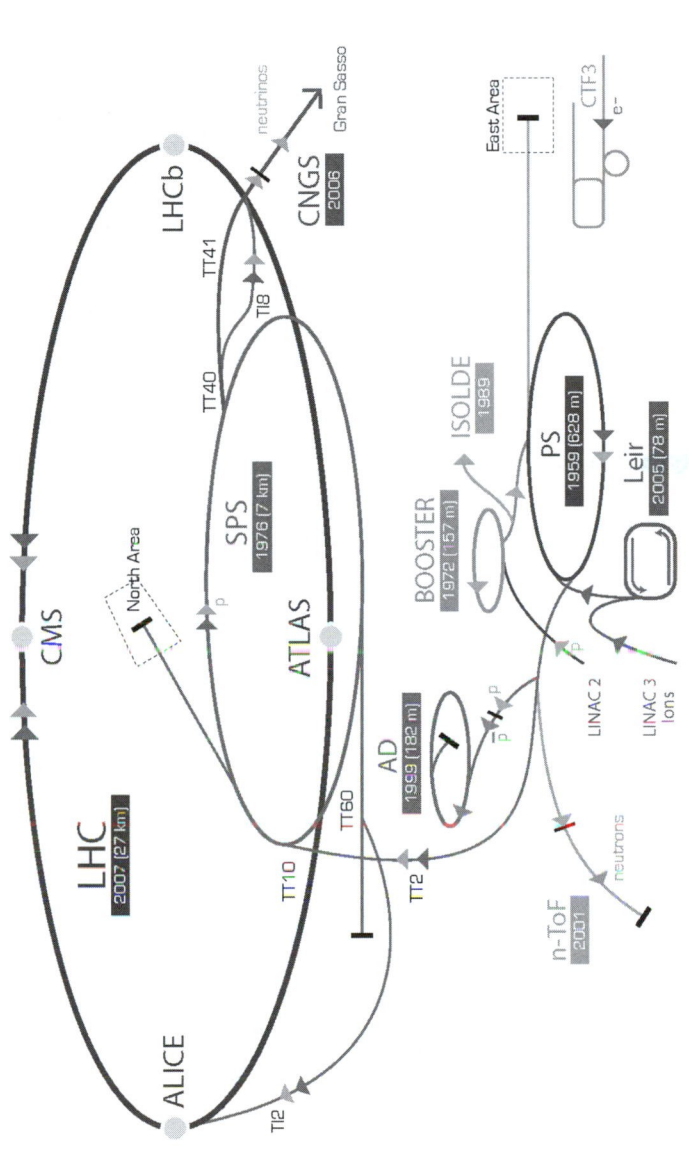

FIGURE 7.8

An overview of the CERN complex, indicating the Large Hadron Collider relative to the other accelerator rings (copyright CERN; used with permission)

time" particles. The second time around, the advanced particles will arrive less early and so be accelerated less, whereas the delayed particles will arrive less late and so be accelerated more. This continues for each orbit, resulting in the bunching of particles about the synchronous orbit. This bunching is the "RF structure" present in all such acclerators.

Ideally, the particle bunches would move uniformly, all accelerate in step, and hit the target precisely. In practice, small misalignments of the beam, magnet inhomogeneities, etc. cause the beam (or rather the bunches) to wander. Deviations from the ideal circular path are called *betatron oscillations.* With appropriate focusing using quadrupole magnet pairs (quadrupole doublets) as discussed above, these can be made quite small compared to the beam radius R. Longitudinal oscillations of a bunch are called *synchrotron oscillations.* Appropriate RF "kicking" stabilizes these bunches, which oscillate in size around the ideal equilibrium position. A typical beam thickness is ~ 1 mm. The concept of stabilizing the bunches via corrective field methods was developed independently by Vladimir Veksler [69] in 1944 and by Edwin McMillan [70] in 1945.

Proton synchrotrons operate by first accelerating the protons in a Cockroft-Walton machine to about 1 MeV, after which they are further accelerated into a linac before injection into the synchrotron, typically at an energy of several hundred MeV. The magnets are positioned in a ring along the circular path of the beam line. The world's largest proton synchrotron beam is the LHC at CERN.

Since electrons are lighter and so much easier to accelerate, why not build electron synchrotrons? This can be done, but there is a significant cost because all charged particles radiate energy as they accelerate. Circularly moving particles at higher and higher speeds accelerate more and more, losing energy ΔE with each revolution:

$$\Delta E = \frac{4\pi}{3} \frac{q^2 \beta^3 \gamma^4}{R} \longrightarrow \frac{4\pi}{3} \frac{q^2}{R} \left(\frac{E}{mc^2} \right)^4 \quad \text{as} \quad v \to c \qquad (7.13)$$

where the formula for energy loss is given in Jackson (eq. (14.31)) [71]. Hence the energy loss varies inversely with the fourth power of the mass, and so

In the circular case,
it costs more energy to
accelerate light mass particle .

$$\frac{(\Delta E)_{\text{electron}}}{(\Delta E)_{\text{proton}}} = \left(\frac{m_{\text{proton}}}{m_{\text{electron}}} \right)^4 \simeq 10^{13} \qquad (7.14)$$

which is a huge energy loss: at 20 GeV beam energy it is 16 MeV per turn. This makes electron synchrotrons a very expensive but excellent source of intense short-wavelength light, called synchrotron radiation. This emitted radiation permits unique research in a variety of scientific fields in physics chemistry and biology. The Canadian Light Source in Saskatoon, Canada, is a good example of a state-of-the-art machine of this type. However, for the purposes of particle physics, proton accelerators provide much more energy per unit cost.

7.4 Colliders

All of the aforementioned machines are *fixed-target* machines: the particle beam, after reaching its optimum energy, hits some target. Nearby detectors then measure what happens. This used to be the way that all high energy experiments were carried out. In the last 30 years, a new type of machine called a *collider* has become important. Here 2 beams are smashed into each other, with nearby detectors monitoring the collision.

Why is this better than a fixed-target machine? Consider 2 particles with 4 momenta p_b^μ and p_t^μ for the beam and target respectively. The total 4-momentum p^μ is, by momentum conservation

$$p^\mu = p_b^\mu + p_t^\mu \Rightarrow p^2 = (p_b + p_t)^2$$
$$\Rightarrow p^2 = p_b^2 + p_t^2 + 2p_b \cdot p_t$$
$$\Rightarrow p^2 c^2 = m_b^2 c^4 + m_t^2 c^4 + 2(E_b E_t - \vec{p}_b \cdot \vec{p}_t c^2) \quad (7.15)$$

where m_b and m_t are the masses of the beam and target particles respectively.

The center of momentum system (CMS) is defined to be that system in which $p^\mu = (E_{total}, \vec{0})$. If the target is at rest, then $\vec{p}_t = 0$ and so

$$E_{total}^2 = m_b^2 c^4 + m_t^2 c^4 + 2E_b m_t c^2 \rightarrow E_{total} \simeq \sqrt{2E_b m_t c^2} \quad (7.16)$$

for beam energies $E_b \gg m_b c^2, m_t c^2$. However, if both the beam and the target are moving toward each other so that $\vec{p}_t + \vec{p}_b = 0$, then

$$E_{total}^2 = m_b^2 c^4 + m_t^2 c^4 + 2(E_b E_t + |\vec{p}_b|^2 c^2) \rightarrow E_{total} \simeq 2E_b \quad (7.17)$$

for beam energies $E_b \simeq E_t \gg m_b c^2, m_t c^2$. Hence in fixed target machines, the total energy increases as the square root of the beam energy, whereas for colliders it increases linearly with the beam energy. Clearly much more total energy is available for particle creation in colliders!

Colliders have several disadvantages. The particles in the beam must be stable, unlike the previous machines we have considered, which can be used to produce secondary beams of unstable particles. Hence only protons, antiprotons, electrons and positrons can be used in colliders. The other (more serious) disadvantage is that the collision rate is low. The relationship between the rate $\mathcal{R}$ and the other parameters of the beam is:

$$\mathcal{R} = \sigma \mathcal{L} \quad \text{with } \mathcal{L} = fn\frac{N_1 N_2}{A}$$

where the quantities associated with the luminosity $\mathcal{L}$ are

$$f = \text{frequency of revolution of bunches}$$
$$n = \text{number of bunches per beam}$$
$$N_1, N_2 = \text{number of particles per bunch in beams 1,2}$$
$$A = \text{area of beam}$$
$$\sigma = \text{cross-section for interaction}$$
$$\text{(computed from underlying theory)}$$

For colliders, $\mathcal{L}$ is typically $\sim 10^{31}/\text{cm}^2/\text{s}$, whereas in fixed target machines $\mathcal{L} \sim 10^{37}/\text{cm}^2/\text{s}$. This is a necessary trade-off between the two kinds of machine – attaining higher energies comes at the cost of reduced luminosity.

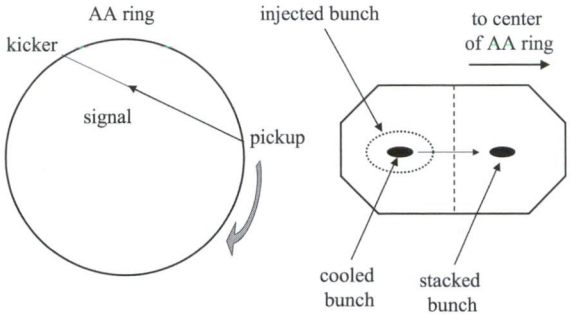

FIGURE 7.9
Schematic diagram of stochastic cooling. A pickup coil delivers a signal depending on deviation from the antiprotons from the ideal orbit, and this activates a kicker which deflects them toward the ideal orbit as they come around the ring. After the bunch is cooled, it is kicked into the inner half of the chamber and stacked together with previous bunches.

The cross-section, denoted by σ, is a quantity that is characteristic of the fundamental physics governing the interaction, and can be thought of as the effective area that one particle presents to another. We shall defer a discussion of its properties until chapter 9.

Non particle-antiparticle machines (e.g. *ep* or *pp*) need 2 separate beam pipes and 2 sets of magnets. Particle-antiparticle machines (e^+e^- or $p\bar{p}$) need only one pipe and set of magnets. $p\bar{p}$ machines pose unique problems in that obtaining a beam of antiprotons is much harder than obtaining a positron beam. Antiprotons are produced from fixed target proton-nucleus collisions: these have a low yield and give a hot gas of antiprotons.

To cool this gas a technique called *stochastic cooling* was developed by Simon van der Meer, illustrated in fig. 7.9. The antiprotons are placed in a

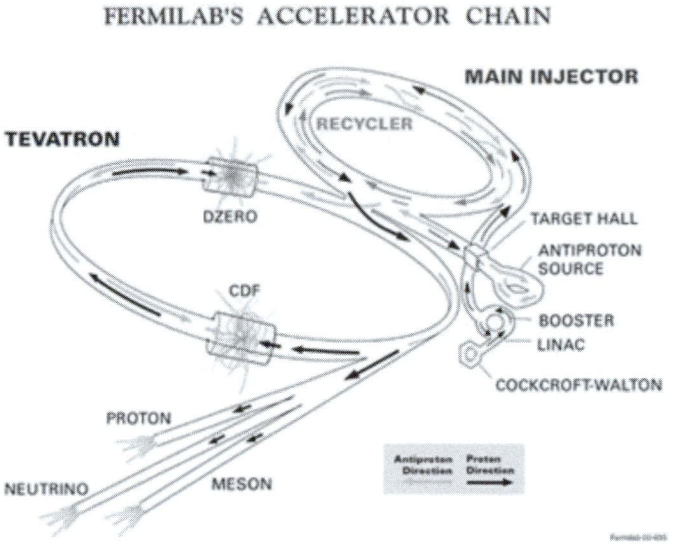

FERMILAB'S ACCELERATOR CHAIN

FIGURE 7.10

A diagram of the facilities at the Fermilab Tevatron. Image courtesy of Fermilab National Accelerator.

ring. A sensitive pickup coil in one part of the ring senses the deviation from an ideal path and sends a signal to a kicker in another part of the ring to deflect them to an ideal orbit. It takes about 2 seconds of circulation to cool the injected beam.

After cooling, the injected bunch is put into a stacking ring in the same tube. After a day or so about a trillion antiprotons exist in the beam and can be used for experimentation. This method was used in the CERN SPS collider to obtain proton-antiproton collisions which led to the discovery of the W and Z bosons [73].

The highest-energy machine currently operative is the Tevatron at Fermilab. A schematic illustration of the setup at Fermilab is shown in fig. 7.10. The TEVATRON at Fermilab is 1 km in radius and can achieve a beam energy of 1000 GeV = 1 TeV (tera electron volt). It is a $p\bar{p}$ machine, which uses a Cockroft-Walton, a Linac and a booster ring in order to get the protons to sufficiently high energy before they enter the main ring.

One machine that played a very important role in experimental particle physics in the 1990s was called LEP (for Large Electron-Positron machine). This machine was designed to run at collision energies (about 90-100 GeV) that would allow the Z-boson to be produced in large numbers. This worked very well, and an enormous amount of empirical information was obtained about the Standard Model from the experiments done at this facility. In April 2000 the energy of LEP's particle beams was boosted to be as high as possible – up to 209 GeV, well beyond the original design energy. Significant experimental data was accumulated at a center-of-mass energy in excess of 206 GeV, and a number of events compatible with a Higgs boson production with mass around 114-115 GeV were reported in the combined results of the four LEP experiments, ALEPH, DELPHI, L3 and OPAL. Unfortunately the topology of these events is also compatible with those originating from other known Standard Model processes, and so it is at present impossible either to rule out or confirm the existence of a 114-115 GeV Higgs boson [74]. LEP was shut down several years ago so that its facilities could be renovated to convert it to the LHC.

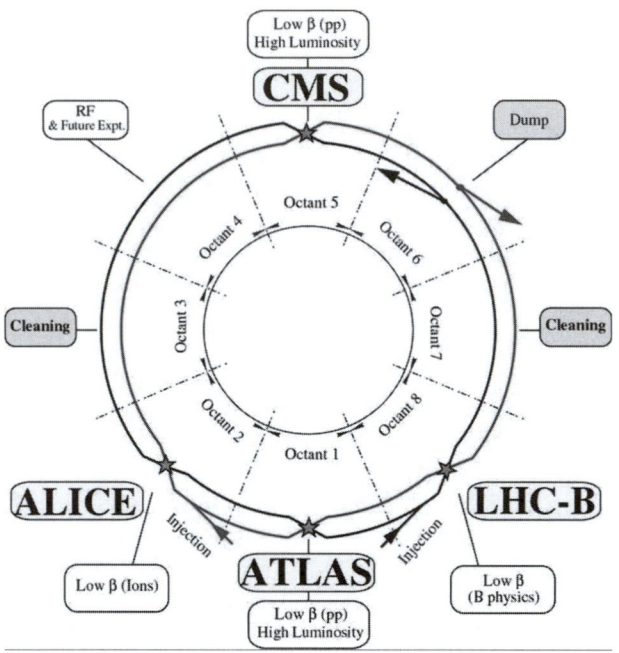

FIGURE 7.11
The four main experiments and the two ring structure of the LHC (copyright CERN; used with permission).

Higgs bosons of such a light mass are expected to be copiously produced at the LHC, the centerpiece of the future scientific program at CERN. The

LHC is housed in a circular tunnel 27 km in circumference, buried at a depth of 50 to 175 m underground on the Swiss/French border on the outskirts of Geneva, with a planned start-up in 2008. This machine will achieve energies of 7 TeV+7 TeV for a proton-proton beam (the highest ever attained), and luminosities of $10^{37}/\text{cm}^2/\text{s}$. For most of the ring, the beams travel in two separate vacuum pipes in opposite directions, but at four points they collide in the hearts of the main experiments, known by their acronyms: ALICE (designed to see if it is possible to make and detect a quark-gluon plasma), ATLAS (whose purpose is to find the Higgs), CMS (which will look for evidence of supersymmetry), and LHCb (whose purpose is to make precise measurements of $\mathbb{CP}$-violation in the b-quark sector). A conceptual diagram of the LHC at complex appears in fig. 7.11, and the tables 7.1, 7.2, and 7.3 should help you to keep all the acronyms straight.

TABLE 7.1
Accelerator Acronyms

AAC	Antiproton Accumulator Complex (LHC)
AGS	Alternating Gradient Synchrotron (Brookhaven)
CESR	Cornell Electron Storage Ring
CLIC	Compact LInear Collider (proposed CERN)
DAFNE/DAPHNE	Double Annular Factory for Nice Experiments
EPA	CERN's Electron Positron Accumulator
FMI	Fermilab Main Injector
FNAL	Fermi National Accelerator Laboratory
HERA	Hadron-Electron Ring Accelerator (DESY)
KEK B-Factory	$\mathbb{CP}$-violation in the B meson (KEK)
LEAR/LEIR	Low Energy Ion Ring
LEP	Large Electron Positron collider
LHC	Large Hadron Collider
LIL	Lep Injector Linac
PEP	Positron Electron Project (SLAC)
PSB	Proton Synchrotron Booster (CERN)
PS	Proton Synchrotron (CERN)
RHIC	Relativistic Heavy Ion Collider (Brookhaven)
SLAC	Stanford Linear Acclerator
SPEAR	Stanford Positron Electron Accelerating Ring
SPS	Super Proton Synchrotron
Tevatron	Fermilab's 2-TeV proton-antiproton accelerator

TABLE 7.2

Laboratory Acronyms

ANL	Argonne National Laboratory, (Argonne, Illinois, USA)
BNL	Brookhaven National Laboratory (Upton, Long Island, USA)
CERN	Originally "Conseil Europenne pour Recherches Nuclaires" now European Laboratory for Particle Physics (Geneva, Switzerland)
CLS	Canadian Light Source (Saskatoon, Saskatchewan, Canada)
DESY	Deutches Elektronen SYnchrotron laboratory (Hamburg, Germany)
FNAL	Fermi National Accelerator Laboratory (Batavia, Illinois, USA)
KEK	Koo Energy Ken (Tsukuba, Japan)
LNF	Laboratori Nazionali di Frascati (Rome, Italy)
SLAC	Stanford Linear Accelerator Center (Palo Alto, California)
SNO	Sudbury Neutrino Observatory (Sudbury, Ontario, Canada)
TRIUMF	TRI-University Meson Facility (Vancouver, British Columbia, Canada)

Table 7.3: Experiment/Detector Acronyms

ALEPH	Apparatus for LEP PHysics (CERN)
ALICE	A Large Ion Collider Experiment (LHC, CERN)
AMANDA	Antarctic Muon and Neutrino Detector Array (Antarctica)
	International collaboration to detect high-energy cosmic neutrinos at the South Pole
AMS	Alpha Magnetic Spectrometer (A detector in space to search for antimatter)
APEX	AntiProton Experiment (Fermilab)
ATLAS	A Toroidal LHC ApparatuS (LHC, CERN)
BaBar	$B - \bar{B}$ detector (B-Factory, SLAC)
BELLE	B-meson detector (KEK)
BOREX/Borexino	Underground solar neutrino experiment (Gran Sasso, Italy)
CDF	Collider Detector at Fermilab (Tevatron, Fermilab)
CDMS	Cryogenic Dark Matter Search (Soudan Mine, Minnesota, USA)
CHAOS	Canadian High Acceptance Orbit Spectrometer (TRIUMF)
CHOOZ	Long-baseline reactor neutrino experiment (CHOOZ power station, les Ardennes, France)
CHORUS	CERN Hybrid Oscillation Research Apparatus (CERN)
CLEO	Detector at Cornell's CESR accelerator (CESR)
CMS	Compact Muon Solenoid (LHC, CERN)
COMPASS	CERN's Common Muon and Proton Apparatus for Structure and Spectroscopy (CERN)
CRESST	Cryogenic Rare Event Search with Superconducting Thermometers
	to search WIMPs (Weakly Interacting Massive Particles) (Gran Sasso, Italy)
DELPHI	DEtector with Lepton Photon and Hadron Identification (LEP, CERN)
DONUT	Direct Observation of the Nu Tau (Fermilab)

DZero	Collider detector studies proton-antiproton collisions (Tevatron, Fermilab) (named for location on the Tevatron Ring)
E787	Rare Kaon Decay experiment (AGS, Brookhaven)
FOCUS	FOtoproduction of Charm: Upgraded Spectrometer (Fermilab)
GALLEX	Gallium EXperiment at Gran Sasso (Gran Sasso, Italy)
H1	Collider experiment (DESY)
HEAT	High- Energy Antimatter Telescope (NASA) Balloon-borne experiments to study antimatter in primary cosmic radiation
HERA-B	Fixed-target experiment to investigate CP violation in the B meson (DESY)
HERMES	Fixed-target experiment to explore spin (DESY)
Hi Res Fly's Eye	High-energy cosmic ray experiment (Dugway, Utah, USA)
HOMESTAKE	Solar neutrino experiment i(Homestake Gold Mine, South Dakota, USA)
HYPER-CP	Search for Direct $\mathbb{CP}$-violation in Hyperon decays (Fermilab)
ISOLDE	Isotype On-Line separator (CERN)
K2K& KEK	Long-baseline neutrino experiment (KEK/Kamiokande, Japan)
KAMIOKANDE	Solar neutrino experiment (Kamioka Observatory, Japan)
KARMEN	Karlsruhe-Rutherford Medium-Energy Neutrino Experiment (Rutherford- Appleton Laboratory, United Kingdom)
KLOE	K LOng Experiment (LNF DAFNE)
KTeV	Kaons at the Tevatron (Fermilab)
L3	LEP Detector (LEP, CERN)
LHC-B	Large Hadron Collider B Experiment (LHC, CERN)
LSND	Liquid Scintillator Neutrino Detector (Los Alamos, USA)
MILAGRO	Study of cosmic ray air showers (Los Alamos, USA)
MINOS	Main Injector Neutrino Oscillation Search (Fermilab)

NOMAD	Neutrino Oscillation MAgnetic Detector (CERN)
NuMI	Neutrinos at the Main Injector (Fermilab)
NuSEA	NUcleon SEA (Fermilab)
	Fixed target experiment to measure asymmetry of down and up anti-quarks in the nucleon
NuTeV	Neutrinos at the Tevatron (Fermilab)
	Fixed-target neutrino beam experiment for precision measurement of W mass
OPAL	Omni Purpose Apparatus for LEP (LEP, CERN)
Pierre Auger Project	International experiment to find origin of ultra-high-energy cosmic rays
SAGE	Soviet-American Gallium Experiment (Baksan Mountains, Russia)
SELEX	SEgmented Large X baryon spectrometer EXperiment (Fermilab)
	Study of charm baryons
SLD	SLAC Large Detector (SLAC)
SOUDAN II	Soudan II (Tower-Soudan Iron Mine, Minnesota, USA)
	Search for nucleon decay and study atmospheric neutrino physics
Super-K	Super-Kamiokande (Kamiokande, Japan)
ZEUS	(Not an acronym, but goes with HERA) Collider experiment (HERA, DESY)

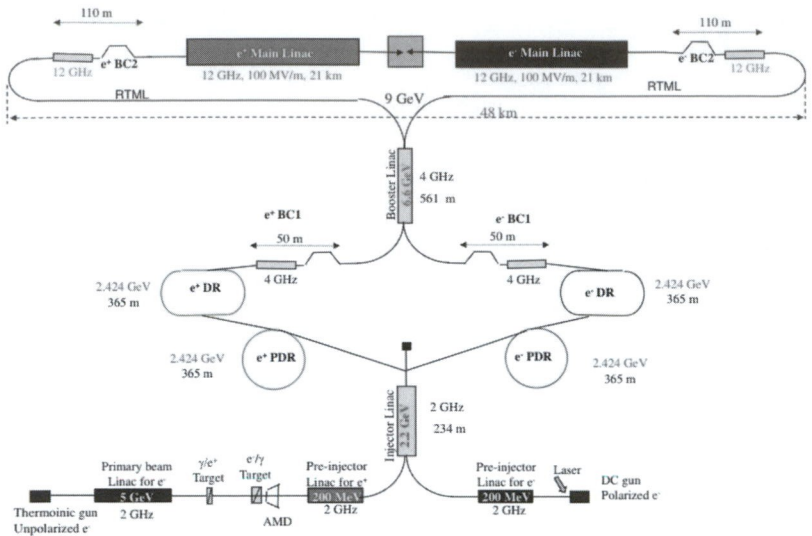

FIGURE 7.12

The proposed CLIC facility (copyright CERN; used with permission).

7.5   The Future of Accelerators

The most important parameters of any accelerator are its final energy and its beam intensity. The energy is proportional to the magnetic field for a given radius of curvature, and so our ability to generate strong magnetic fields is a limiting factor here. With a typical iron magnet the field achieved can be about 20 kG (or 2 Tesla), whereas a superconducting magnet can achieve fields of 50kG (5 Tesla). These are now being used at Fermilab and the LHC. As far as bean intensity is concerned, stochastic cooling techniques can be used to ensure that this is under control.

However, neither is enough to go much beyond the energies of the LHC. The key reason is cost, which runs into the 10s of billions of dollars. New technologies must therefore be explored that will raise the energy threshold, and this is currently under study. One example is the **C**ompact **LI**near **C**ollider, or **CLIC** facility (see fig. 7.12). The basic idea of CLIC is to point two linacs toward each other, and to increase the beam energy to 1-3 TeV in each beam using an RF source of 30 GHz, for a total CMS energy of 2-5 TeV. If this could be implemented, it would be the highest-energy particle acclerator in the world [75].

7.6 Questions

1. Plans are made to upgrade TRIUMF in Vancouver so that it will be a 60GeV synchrotron with a high intensity (200 μA) beam of protons that can be used to produce several new kinds of subatomic particle beams (e.g. K's, π's, η's etc.). However, these beams are difficult to focus: the number of betatron oscillations $\Delta \nu$ varies as $N/\left(\beta^2 \gamma^3\right)$ where β and γ are the relativistic speed and energy parameters and N is the accelerable charge per pulse. A typical 60 GeV synchrotron has an injection energy of 300 MeV in its first stages.

   (a) How much greater must the injection energy be to increase the useable beam current by a factor of 10 without defocusing?

   (b) How does this answer change if a 5% increase in defocusing can be tolerated?

2. In the LHC two beams of protons will collide head on, each with energies of 7 TeV. $E_{tot} = 14$ TeV.

   (a) How does this energy compare to that of a typical cosmic-ray proton?   0.3 GeV

   (b) How much energy must a single proton beam have to yield the same CM energy on a fixed target of hydrogen?   *Lorentz transformation.*

3. (a) A proton beam of kinetic energy 20 MeV enters a dipole magnet 2m in length. How strong must the field be to deflect the beam by 10°?

   (b) Suppose now the beam has 200 GeV of kinetic energy. How much deflection will be induced by a magnetic field of 25 kG?

4. What is the minimum energy needed to produce antiprotons from the collision of two protons? Remember that conservation laws must be respected in the production process.

5. (a) Two proton storage rings with a beam currents of 20 A each are directed to collide into each other. The interaction region has an area of 1 cm^2 and is 5 cm long. The relative velocity of the two beams is approximately the speed of light. How many collisions take place per second?

   (b) Now consider a proton beam colliding with a fixed target of liquid hydrogen that is 5 cm long and 1 cm^2 in area. How much current would have to be in the beam for it to have the same collision rate as in part (a)?

6. The largest magnetic field possible in a superconducting magnet is about 30 Tesla. What is the largest energy that protons can be accelerated to, in an accelerator that circles the Earth's equator?

7. What would the energy loss per turn be for the acclerator in question #6?

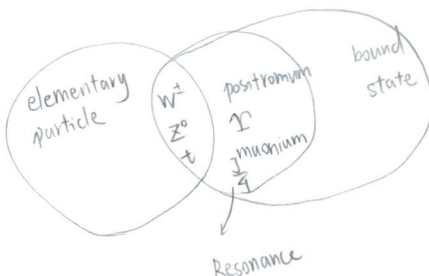

8

Detectors

In a number of ways detectors are more fundamental in expanding our knowledge of particle physics than accelerators are [76]. Without particle acclerators we would still be able to learn much about particle physics by making detectors sensitive enough to probe naturally occuring cosmic rays, neutrinos, photons, and other subatomic particles. Of course an accelerator gives us crucial control over the energy of a given experiment, and its importance to particle physics cannot be understated. Yet without the proper accompanying detectors, accelerators are useless.

After the collision of a particle beam with some target (or another beam) detection of what happens becomes the key task. In order to do this, the particle must leave some imprint of its presence, which is made possible by the fact that particles ultimately transfer energy to the medium they are traversing – if not, we'd never observe them! The construction and design of detectors depends on exploiting this property of energy transfer. There are many ways that this can happen, and we will begin by looking at the various possibilities.

8.1 Energy Transfer and Deposition

dominant for low energy particles

ionization loss
Coulomb scattering
radiation

dominant for high energy particles

8.1.1 Charged Particles

When a charged particle moves through a medium it will interact with the fundamental constituents of that medium: its nuclei and electrons. It can lose energy via three basic means: ionization, coulomb scattering, and radiation. Its primary interactions are with the atomic electrons of the medium, and this forms the dominant mode of energy loss*. The reason for this is that scattering from nuclei causes large changes in the momentum of the incoming particle but relatively small changes in its energy, whereas scattering from atomic electrons (or ionization of nuclei) is an inelastic process that entails

---

*The incoming particle can also have a direct collision with a nucleus, but this is an extremely rare process, and so can be neglected.

substantively more energy transfer. At very high energies, the dominant mode of energy loss is radiation if the incoming particle is an electron or a positron; for more massive particles this process is negligible.

As a charged particle moves through a medium, we expect it lose energy through its interaction with the medium. The key quantity of interest is the rate $\left(\frac{dE}{dx}\right)$ of energy loss per unit path length that the particle traverses whilst in the medium, a quantity called the *stopping power* of the medium. Once we know what it is, we can compute the range R, that any particle will have in the medium:

$$R = \int_0^R dx = \int_0^E \left(\frac{dE}{dx}\right)^{-1} dE \qquad (8.1)$$

Can be used to determine the size of the device (detector)

The stopping power can be calculated for the different ways that the particle can lose energy. Let's look at these.

8.1.1.1 Ionization Loss (low)

At relativistic speeds $v = \beta c$, with Lorentz factor $\gamma = \left(1 - \beta^2\right)^{-1/2}$, the energy loss of a charged particle as a function of distance can be reliably calculated from our knowledge of electromagnetism, and the formula[†]

$$\left(\frac{dE}{dx}\right)_{ionization} = -\frac{4\pi N_A (ze)^2 e^2}{m_e c^2 \beta^2} \left(\frac{Z}{A}\right) \left[\ln\left(\frac{2m_e c^2 \beta^2}{\bar{I}} \gamma^2\right) - \beta^2\right] \qquad (8.2)$$

has been verified experimentally over a wide range of energies for different kinds of particles and media. This formula was derived by Hans Bethe and Felix Bloch [77], and describes the mean rate of energy loss of a particle of charge $q = ze$ due to ionization of a medium with ionization potential $\bar{I}$, atomic number Z and nucleon number A. Here e is the charge on the electron (mass m_e) and $N_A = 10^{23} \text{mol}^{-1}$ is Avagadro's number. Most of the energy loss is due to formation of ion pairs, either by the particle itself, or by the electrons in the ion pairs causing further ionization. The total number of ion pairs is proportional to the energy loss of the incident particle.

The energy loss most strongly depends on the incident velocity β and so can be used to evaluate this quantity. For small β we have

$$\left(\frac{dE}{dx}\right)_{ionization} \to -\frac{4\pi N_A (ze)^2 e^2}{m_e c^2 \beta^2} \left(\frac{Z}{A}\right) \left[\ln\left(\frac{2m_e c^2 \beta^2}{\bar{I}}\right)\right] \qquad (8.3)$$

always negative   *Bethe – Bloch*

and we see that

$$\frac{dE}{dx} \propto \beta^{-2} \ln \beta \simeq \beta^{-2} \simeq (M/p)^2 \qquad (8.4)$$

where M is the mass of the incident particle (to see this, recall that $p = M\gamma v = Mc\beta\gamma$); particles of the same momenta but different mass will have

---

[†]Derivation of this formula is beyond the scope of this text [78]. I have included it here to impress upon you that this energy loss can be quantitatively computed.

different energy loss. The energy loss decreases with increasing particle velocity, reaching a minimum value before growing as $\ln(\gamma^2)$. Note that at very low energies the stopping power becomes negative. This unphysical value indicates a breakdown in the above formula, since a low-energy incident particle can now capture electrons from the medium and form its own atomic systems. Other approximations can be made at lower and higher energies that provide a good description of the stopping power over a very wide range of momentum, as illustrated in figure 8.1.

At high energies eventually long-range interatomic screening effects (ignored in (8.2)) make $\frac{dE}{dx}$ approach a constant value, and it becomes impossible to distinguish particle types based on ionization loss. For solid media the increase is very slight, but for gaseous media there is a rapid increase due to relativistic effects before the plateau is reached. For about 2 orders of magnitude above the ionization minimum, $\frac{dE}{dx}$ is about the same for all materials, having a value

$$\left(\frac{dE}{dx}\right)_{\text{ionization}} \simeq -2\rho\,\text{MeV/g/cm}^2 \qquad (8.5)$$

(constant due to interatomic screening).

where ρ is the density of the absorber.

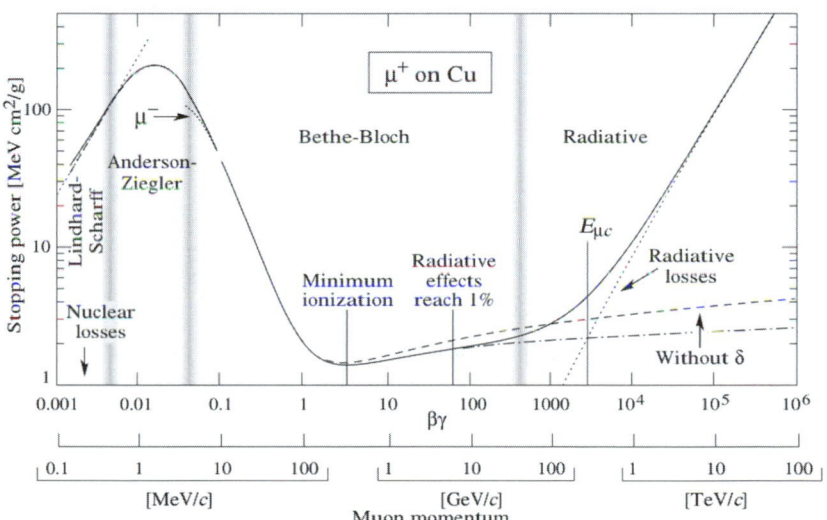

FIGURE 8.1

Diagram of the stopping power for $-dE/dx$ antimuons in copper as a function of momentum, plotted in terms of $\beta\gamma = p/Mc$. The solid curves indicate the total stopping power, and vertical bands indicate boundaries between different approximations. Image courtesy of the Particle Data Group [1].

8.1.1.2   Straggling and Scattering   *(medium)* energy loss $\sim \frac{1}{M}$

Electromagnetism, being long range, will tend to scatter incident charged particles from those in the medium (electrons and nuclei). Since $\frac{dE}{dx} \sim \frac{1}{M}$, only for $M = m_e$ (i.e., electrons and positrons) is this energy loss appreciable. Energy loss of incident nuclei due to electromagnetic interactions is negligible. This kind of energy loss is called *straggling* and occurs because the energy that the incident particles transfer to the medium (i.e., the detector) is variable. Sometimes a lot will be transferred to the particle, and other times only a little energy will be transferred. If we know the functional form that describes the energy transfer, we can compute the mean energy loss and its dispersion.

For example, suppose that the cross-section describing energy transfer to a given detector is

$$\frac{d\sigma}{dq^2} = e^{-8q^2/\lambda^2} \tag{8.6}$$

where q^2 is the square of the momentum transferred to the detector from the beam and λ is a constant characterizing the detector. This formula approximately describes the scattering of nucleons off of nuclei of radius λ^{-1} at low energies. If the detector is made of particles of mass M and the incident (beam) particles are of mass m, then conservation of 4-momentum implies

$$p^\mu_{\text{beam}} + p^\mu_{\text{detector}} = p'^\mu_{\text{beam}} + p'^\mu_{\text{detector}} \tag{8.7}$$

and the transfer of momentum (as we shall see in Chapter 13) is defined to be $q^2 = -\left(p'^\mu_{\text{beam}} - p^\mu_{\text{beam}}\right)^2 = -\left(p'^\mu_{\text{detector}} - p^\mu_{\text{detector}}\right)^2$. If the detector is initially at rest, then we have

$$\begin{aligned}
q^2 &= -\left[\left(E'/c - Mc\right)^2 - |\vec{p}\,'|^2\right] \\
&= -\left[\left(E'/c\right)^2 - |\vec{p}\,'|^2 - 2E'M + M^2c^2\right] \\
&= 2M\left(E' - Mc^2\right)
\end{aligned} \tag{8.8}$$

where $p'^\mu_{\text{detector}} = (E'/c, \vec{p}\,')$.   *mass of the detector*   Hence the kinetic energy transferred to the detector is

$$\text{T} = E' - Mc^2 = \frac{q^2}{2M} \tag{8.9}$$

and we can compute its mean and its dispersion. We find

$$\langle \text{T} \rangle = \frac{\int d\sigma\, \text{T}}{\sigma} = \frac{\int_0^\infty dq^2\, \frac{q^2}{2M}\, e^{-8q^2/\lambda^2}}{\int_0^\infty dq^2 e^{-8q^2/\lambda^2}} = \frac{\lambda^2}{16M} \tag{8.10}$$

$$\langle \text{T}^2 \rangle = \frac{\int_0^\infty dq^2 \left(\frac{q^2}{2M}\right)^2 e^{-8q^2/\lambda^2}}{\int_0^\infty dq^2 e^{-8q^2/\lambda^2}} = \frac{\lambda^4}{128M^2} \tag{8.11}$$

using $\int_0^\infty dx x^n e^{-kx} = n!/k^{n+1}$, and so the dispersion is

dispersion

$$\Delta T = \sqrt{\left\langle (T-\langle T\rangle)^2\right\rangle} = \sqrt{\langle T^2\rangle - \langle T\rangle^2} = \frac{\lambda^2}{16M} \tag{8.12}$$

which in this example happens to equal the mean energy loss.

The detector, predominantly made of nuclei, will have λ proportional to the inverse mean radius of the nucleus, which goes like $1.2A^{1/3}$, where A is the atomic weight. This is almost proportional to the mass and so

$$\Delta T \simeq \frac{1}{16A\left(1.2A^{1/3}\right)^2} = \left(20A^{5/3}\right)^{-1} \text{GeV} \tag{8.13}$$

which is about 0.05 GeV = 50 MeV if the target is protons, but about 7 KeV if the target is lead!

Scattering due to the coulomb field of the nucleus is another effect that needs to be taken into account. This effect limits the precision with which the direction of a particle can be measured. The Rutherford formula[‡]

$$\frac{d\sigma}{d\Omega} = \left(\frac{Qq}{pv}\right)^2 \frac{1}{\sin^4(\theta/2)} \tag{8.14}$$

indicates that the cross section $\sigma(\theta)$ is large for small θ, and so the probability for small deflections is high. Hence in traversing a material, the net scattering is the sum total of a large number of small deviations.

So multiple Coulomb scattering yields a distribution in the net scattering angle. This distribution is approximately Gaussian. The root mean square of the angle is approximately

$$\langle\theta\rangle_{rms} = 20\text{MeV}\,\frac{q}{\beta pc}\sqrt{\frac{L}{X_0}} \tag{8.15}$$

where the incident particles have charge q, velocity βc, and momentum p as they pass through a substance of thickness L and radiation length X_0. This last quantity is one we shall now consider.

8.1.1.3 Radiation Loss (high)

Charged particles radiate electromagnetic energy (photons) when they accelerate or decelerate. Electrons, being so light, are particularly susceptible to this form of radiation loss or *bremsstrahlung*, or braking radiation as it is sometimes called. We saw that the power loss in synchrotron radiation varies

---

[‡]Beginning with Chapter 9 we will learn how to calculate this formula and others like it. For now you can regard the integral of the right-hand-side of eq. (8.14) as giving the effective area a given nucleus presents to the incoming charged particle.

like $(E/m)^4$, so only for small mass particles is this effect appreciable. The bremsstrahlung energy loss per unit length in a given medium is

$$\left(\frac{dE}{dx}\right)_{\text{brem}} = -\frac{E}{X_0} \Rightarrow E = E_0 e^{-x/X_0} \tag{8.16}$$

where X_0 is a constant characteristic of the medium called the *radiation length*. For fast moving electrons this form of energy loss dominates over ionization loss, since as $\beta \to 1$ *Dominates for high energy particles.*

$$\left(\frac{dE}{dx}\right)_{\text{brem}} \to \frac{1}{\sqrt{1-\beta^2}} \quad \text{but} \quad \left(\frac{dE}{dx}\right)_{\text{ionization}} \to \ln\left(\frac{1}{\sqrt{1-\beta^2}}\right) \tag{8.17}$$

Empirically the ratio of these two types of energy loss is approximately

$$\frac{\left(\frac{dE}{dx}\right)_{\text{brem}}}{\left(\frac{dE}{dx}\right)_{\text{ionization}}} = \frac{ZE}{1200Mc^2} \tag{8.18}$$

where Z is the atomic number of the medium and M the rest mass of the particle entering the medium. The critical energy E_c is that energy for which $\left(\frac{dE}{dx}\right)_{\text{brem}} \simeq \left(\frac{dE}{dx}\right)_{\text{ionization}}$. From the preceding equation we deduce

$$E_c \simeq \frac{600}{Z}\text{MeV} \tag{8.19}$$

if electrons are the particles entering the medium.

Table 8.1 lists values of X_0 for various materials.

TABLE 8.1
Radiation Length for Various Materials

Material	Z	Density (g/cm^3)	Critical Energy (Mev)	X_0 (cm)
Liquid H$_2$	1	0.071	340	887
Liquid He	2	0.125	220	745
C	6	1.5	103	28
Al	13	2.7	47	9.0
Fe	26	7.87	24	1.77
Pb	82	11.35	6.9	0.56
Air		0.0012	83	30870
Water		1.0	93	36.4

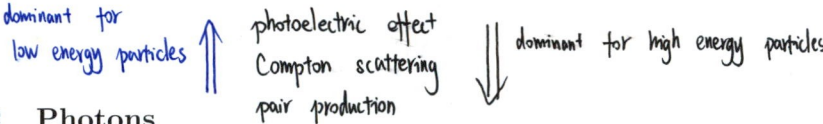

dominant for low energy particles ↑ photoelectric effect, Compton scattering, pair production ⇓ dominant for high energy particles

8.1.2 Photons

Even though they are electromagnetically neutral, photons are carriers of the electromagnetic force, and so can experience interactions with any medium.

In particular, high energy photons (X-rays or γ-rays) traveling through a medium will interact with the atoms of that medium and so lose energy [79]. The change of photon intensity $\mathcal{I}$ per unit thickness of material is proportional to the intensity and so

$$\left(\frac{d\mathcal{I}}{dx}\right) = -\mu\mathcal{I} \Rightarrow \mathcal{I} = \mathcal{I}_0 e^{-\mu x} \tag{8.20}$$

where μ is called the *effective absorption coefficient* of the medium. The value of μ^{-1} is the mean free path for absorption of a photon, or the average distance through which a beam of photons will pass before its intensity is $1/e$ of its original value.

The total absorption coefficient μ is the sum of the absorption coefficients for each process:

$$\mu = \mu_{\text{photo}} + \mu_{\text{Compton}} + \mu_{\text{pair}} \tag{8.21}$$

which signify the three ways that photons can lose energy in a medium. It is proportional to the scattering cross-section, as we shall see in chapter 9. Fig. 8.2 illustrates the relative importance of these three processes (plus a couple of other sub-dominant ones) as a function of energy. Let's briefly consider each.

8.1.2.1 Photoelectric Effect   (low)

This phenomenon refers to the absorption of a photon by a bound electron, which subsequently is ejected from its atom. If the bound electron is in an inner shell of the atom then one of the outer electrons will move to occupy this lower and more stable energy level by emitting an X-ray photon of the appropriate frequency for this transition. Hence the emitted electron can be accompanied by an X-ray photon.

Since the energy for ejection is a chemical energy, this process dominates at low energies. Computation of the cross sections involves details associated with quantum electrodynamics and atomic physics. Setting $M = m_e$ to be the mass of the electron, experimentally the behavior is observed to be

$$\sigma_{\text{photoelec}} \simeq \begin{cases} \frac{Z^5}{E^{7/2}} & \text{for} \quad E < m_e c^2 \\ \frac{Z^5}{E} & \text{for} \quad E > m_e c^2 \end{cases} \tag{8.22}$$

additionally illustrating the importance of this process for high-Z atoms.

8.1.2.2 Compton Scattering   (medium)

At intermediate energies the incoming photon scatters off of electrons. It is not energetic enough to produce pairs, but is too energetic to eject electrons from atoms. Instead it scatters off on an electron, analogous to the manner in which two billiard balls scatter off one another classically. This process is called the Compton Effect. We have already calculated the kinematics in Chapter 2, where we found in eq. (2.39) that

$$E' = \frac{E}{1 + \frac{E}{m_e c^2}(1 - \cos\theta)} \tag{8.23}$$

where E is the energy of the incoming photon that is scattered into a photon of energy E' at an angle θ relative to its original direction. The electron is initially assumed to be free, a good approximation if the incoming photons have enough energy so that atomic binding energies can be neglected. It is clear that $E' < E$ and so the scattered photon loses energy. The cross-section scales as

$$\sigma_{\text{Compton}} \simeq \frac{Z}{E} \tag{8.24}$$

and dominates from photon energies in the range 0.1 to 10 MeV.

8.1.2.3 Pair Production (high)

In pair production a photon is converted into an electron-positron pair. In free-space this is impossible because energy and momentum can't be conserved. The reason for this is as follows. Suppose the photon has a small but nonzero mass. If a single photon could produce a pair of particles, then the total (rest) mass of the particles would have to be smaller than the photon mass, since we can always go to a frame in which the photon is at rest and for which the total energy is its rest mass. This would mean that the photon could only produce pairs of oppositely electromagnetically charged particles that were lighter than half of its mass. Empirically we know that electrons and positrons (the lightest electromagnetically charged particles) are much heavier than the photon; in fact we will later see that we have very good reasons to expect the photon mass to be zero!

So a free photon cannot decay into (or rather produce) pairs. However, a photon traversing the Coulomb field of a nucleus can, because the recoil of the nucleus can balance energy and momentum. The threshold energy for pair production is twice the electron (or positron) mass, which is 1.022 MeV. The cross-section behaves as

$$\frac{d\sigma}{d\Omega} \sim Z^2 \tag{8.25}$$

and we see that it will dominate for high energy photons. Dominance of energy loss due to pair-production sets in for photon energies greater than 10 MeV.

What happens to the positrons that are produced? Along with the produced electrons, they will traverse the medium and lose energy via the mechanisms discussed above for charged particles. At sufficiently low energies they will capture an electron in the medium and bind to it to form positronium. As we saw in Chapter 6, positronium is highly unstable, decaying into a pair of

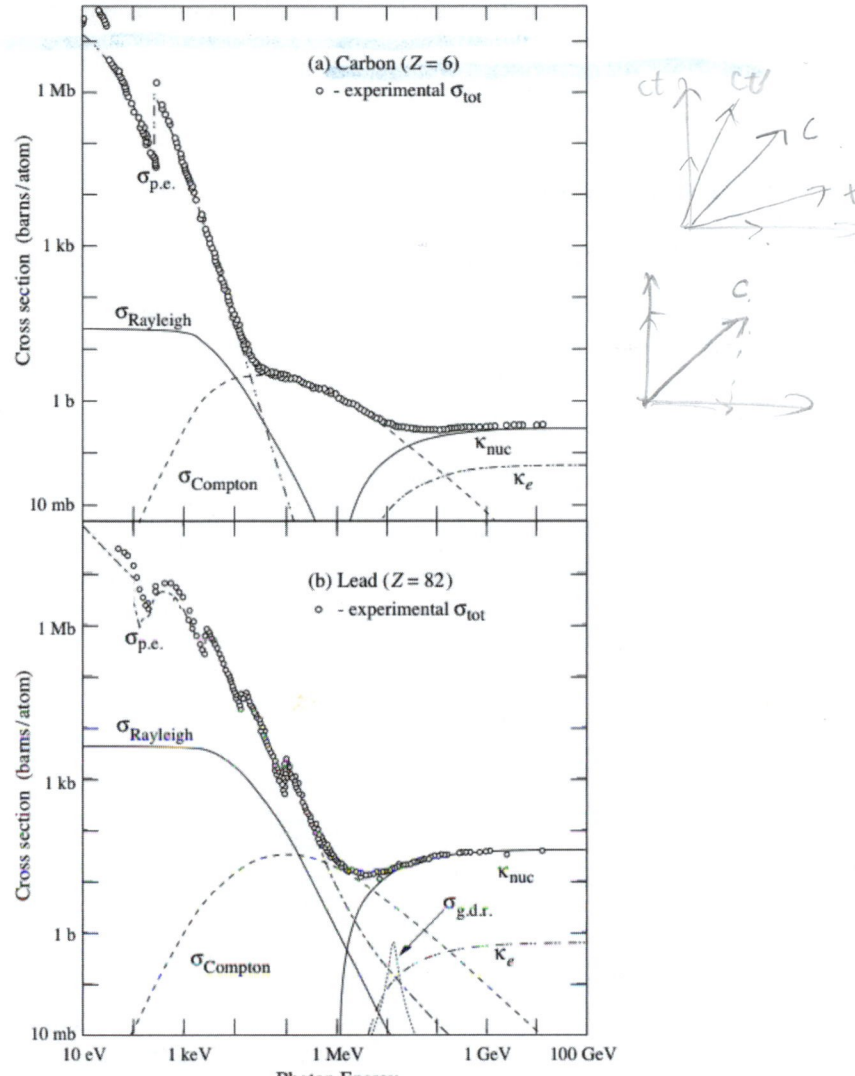

FIGURE 8.2

Absorption cross-sections of photons as a function of energy in carbon and lead. The different contributions are of the photoelectric effect ($\sigma_{p.e.}$), Compton scattering σ_{Compton}, and pair production (κ_{nuc}, due to the nuclear electromagnetic field, and κ_e, the electron field) are shown. Two other relatively small effects not considered in the text – Rayleigh scattering (σ_{Rayleigh} in which the atom is neither excited nor ionized), and Photonuclear interactions ($\sigma_{\text{g.d.r}}$ in which the target nucleus is broken up) – are also illustrated. Image courtesy of the Particle Data Group [1].

photons in about 10^{-10} seconds, with each photon from the produced pair having 0.511 MeV. This provides a way of both detecting positrons and of calibrating detectors at low energies.

8.2 Detector Types

Detection requirements are exacting in particle physics. The key quantities of interest that one would like to measure are the position (or rather trajectory), momentum, energy, arrival time, charge, spin, and any other relevant quantum numbers associated with the identity of the particle. In addition, it is important to know which particles are associated with each other in a given reaction. No one detector can do all this, so in practice combinations of detectors are used to extract the information required.

In a fixed target machine the detector is typically placed behind the target, and the geometrical arrangement is shown in fig. 8.3. However, in a collider the detector must be placed around the target (fig. 8.4) , in order to maximize the amount of information obtained from the collisions. In general, each detector consists of a layered array of materials, with each layer sensitive to detection of a particular kind of particle.

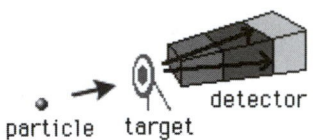

FIGURE 8.3
Geometry of detectors in a fixed-target experiment. The different layers correspond to different kinds of detection materials. Image courtesy of the Particle Data Group [1].

8.2.1 Scintillation Counters

Scintillation counters were the earliest counters invented: called spinthariscopes, the first one was made in 1903 by Crookes [80] and consisted of a ZnS screen that would flash whenever an alpha-particle hit it. The detector was the human eye and the recorder was pen and paper. Geiger and Marsden [81] used two screens to perform the first coincidence experiment in 1910, checking

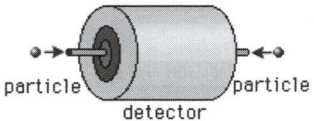

FIGURE 8.4

Geometry of detectors in a collider experiment. The different layers correspond to different kinds of detection materials. Image courtesy of the Particle Data Group [1].

to see if two alpha particles were emitted by a radioactive gas within a given "short" time. The unreliability of the human eye as a detector led to the abandonment of these devices.

In 1944 these detectors were reintroduced, with a photomultiplier replacing the eye. The basic arrangement for such a detector appears in fig. 8.5. When a charged particle passes through the scintillator it excites the medium inside; as the medium de-excites photons are emitted, providing information about the particle's trajectory. It was subsequently discovered that mixtures of organic liquids and aromatic molecules yielded high numbers of photons per unit loss of energy (about 10,000 photons per MeV) [82]. So the most common materials used for the medium in today's scintillators are organic liquids and plastics (which emit UV light, which is then converted to blue light via dye molecules) and inorganic solids (such as sodium iodide, whose dopants capture electron-hole pairs and then emit visible light).

Photomultiplier tubes use the photoelectric effect to record the photons emitted from the medium. These photons liberate electrons at a photocathode (whose conversion efficiency is about 25%), which then travel down a channel of increasing potential electrodes (the dynodes) which emit more electrons causing amplification. There are typically 6-14 dynodes yielding an amplification factor of about 10^4 to 10^9. Sodium iodide doped with thallium has a high detection efficiency but a slow decay time (about 250 nanoseconds) for each pulse. Plastic scintillators (anthrancence or napthalene) can shorten the pulse decay time to 10 nanoseconds (with a resolution of about 200 picoseconds) and so such detectors are ideal triggering devices – in other words their signal is used to decide whether or not to activate the rest of the detector and/or to record the event. However, their efficiency is low.

The signal that emerges is passed through a discriminator, whose function is to eliminate random noise pulses emitted through thermal electron emissions from the cathode and the dynodes. The final signal is not sharp, but instead has a shape that is a consequence of the experimental resolution of the detector. This signal is sent to a pulse-height analyzer, which digitizes and displays the pulses.

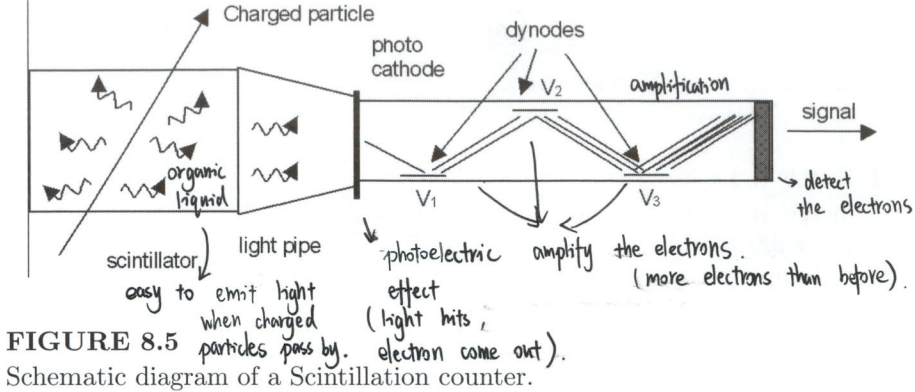

FIGURE 8.5
Schematic diagram of a Scintillation counter.

8.2.2   Cloud Chambers

Cloud chambers were first developed by Charles T.R. Wilson around 1911 for experiments on the formation of rain clouds [83]. The cloud chamber is a sealed environment containing a supercooled, supersaturated water vapor. When a charged particle interacts with the mixture, it ionizes it. The resulting ions act as condensation nuclei, around which a mist forms because the mixture is at the point of condensation. The high energies of the incoming particles mean that a trail is left, due to many ions being produced along the path of the charged particle. These tracks have distinctive shapes. For example an alpha particle's track is broad and straight, while an electron's is thinner and shows more evidence of deflection. Anderson made use of a cloud chamber in 1933 in his discovery of the positron [84].

8.2.3   Bubble Chambers

These descendants of the cloud chamber were invented by Glaser in 1952 [85]. Unlike scintillation counters, which yield only information about the energy of the particle passing through, bubble chambers look at all possible processes/particles passing through them.

They consist of a superheated liquid (one whose pressure is lower than its equilibrium vapor pressure, i.e. "hotter than its boiling point") placed in some (relatively thin) container. As an ionizing particle passes through, the liquid will start boiling by forming bubbles at nucleation centers. The superheated liquid is prepared by starting with the very cold liquid under pressure (about 5 atmospheres and 3K) and then, just before the particle beam arrives, the pressure is reduced suddenly by using a piston to expand the volume by about 1%.

Ions form good nucleation centers, and so a charged particle moving through this detector will leave a trail of bubbles. These bubbles are then photographed, allowing one to directly trace the actual track(s) of the ionizing

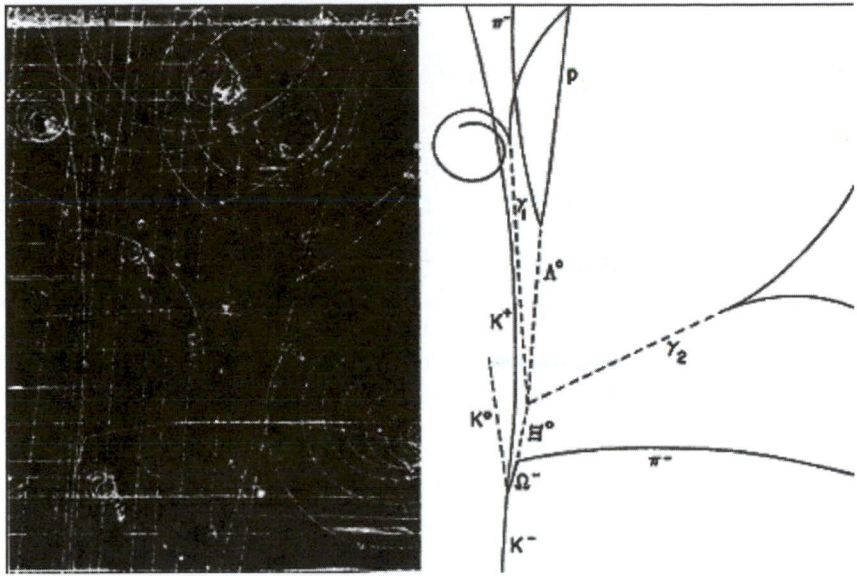

FIGURE 8.6

An important bubble chamber photograph. A K^- impacts upon a proton in the reaction $K^- + p \to \Omega^- + K^+ + K^0$. This photograph (and others like it) provided evidence for the Ω^- particle, whose discovery gave strong support to the quark model. Photo courtesy of Brookhaven National Laboratory.

particle(s). In conjunction with a strong magnetic field (which curves the tracks provided the field is orthogonal to the plane of the thin container), the momentum can be deduced from the track curvature. An example from an actual experiment is shown in fig. 8.6.

Bubble chambers are very useful in studying complicated interactions of many particles. The first bubble chambers contained only a few cc's of liquid, but the volume rapidly increased to be about a million times larger as new ones were built. The best ones produced 35 million photographs per year. Unfortunately they cannot be used in colliders because of the interaction geometry of the beams. They are also not selective because they cannot be triggered – just as a surveillance camera at a bank photographs every visitor, a bubble chamber records everything that goes through it, and so every picture must be developed and scanned to see if there is any useful information present.

8.2.4 Spark Chambers

Short high voltage pulses (10-50 MV/cm) between parallel plate electrodes enclosing a gas will yield short electrical breakdowns in the structure of the gas. An ion trail will then leave "flashes" or "sparks" of light due to such discharges, as Greinacher found in 1931: the passage of an alpha particle through such a system generated a spark between the wire and the plate, allowing for a direct track image to be photographed with good spatial resolution. Like a bubble chamber, events must be recorded by some means and evaluated later. However, spark chambers operate much more rapidly and can be made highly selective by using auxiliary detectors to screen out unwanted events. Because of its selectivity, the spark chamber is most useful in searching for very rare events. They can be highly automated, with data collected and stored electronically instead of photographically, as is necessary with the bubble chamber. A high-speed computer, which may operate simultaneously with the experiment, can then be used to analyze the data. This provides immediate evaluation of the quality of the data and affords optimum operating conditions to be continuously maintained. Very large spark chambers have helped to detect neutrinos. A picture of cosmic rays flying through a spark chamber at CERN in 1999 appears in figure 8.7.

8.2.5 Wire Chambers

Wire chambers were developed by Georges Charpak in 1968 [87] and have excellent time resolution, very good position accuracy and are self-triggered. A schematic diagram outlining the basic features of a wire chamber appears in figure 8.8. These consist of a plane of positively charged wires precisely separated (typically by about 2 mm) from each other. The plane is in between two cathodes (often made of aluminum foil) that are about 1 cm away. This whole structure is encased and then filled with a gas, which ionizes when a charged particle passes through it. Upon ionization of the gas by a charged particle, electrons drift toward the nearest wire, gaining energy. If this energy exceeds the ionization energy of the gas the process will repeat, leading to an avalanche of ion pairs. The amplification factor is typically $\sim 10^5$. Sandwiching a set of these counters in concert (see fig. 8.9) allows for accurate measurement of position.

These multiwire proportional counters (MWPC's), when modified by an electric field, cause the ionized electrons to drift before getting to the high amplification region near the anode. By timing the arrival of the pulse relative to some external fast signal the drift distance can be obtained, yielding the position of the particle. The pulse height can be used to measure the mass of the particle causing the ionization. Hence the track of the incident particle can be reconstructed – with a precision of up to 100 microns!

If a set of MWPC planes is placed on either side of a region that has an applied magnetic field, there will be a change in the trajectory of the particle

FIGURE 8.7

Cosmic rays – created when subatomic particles from outer space collide with the atmosphere – can be seen in this photograph of a large spark chamber at CERN. Their tracks are visible as a series of sparks between metal plates (or planes of parallel wires) several millimeters apart, induced by the passage of each charged particle as it moves through the spark chamber (copyright CERN; used with permission).

as it passes from the first set of planes to the second. This change provides information about the momentum of the particle, using the formula (7.5) $pc = qBr$, where r is the radius of the curved arc traced out by the changing trajectory of the particle of charge q, and B is the magnitude of the applied magnetic field.

8.2.6 Time Projection Chambers

Time projection chambers (TPCs) were invented in 1974 by David Nygren to give spatial resolution of particle tracks in all 3 dimensions over as large a solid angle as possible [88]. A drift chamber is filled with gas (usually some inexpensive mixture of Argon and CH_4) and then uniform electric and magnetic fields are applied parallel to the beam axis. Ion pairs are produced

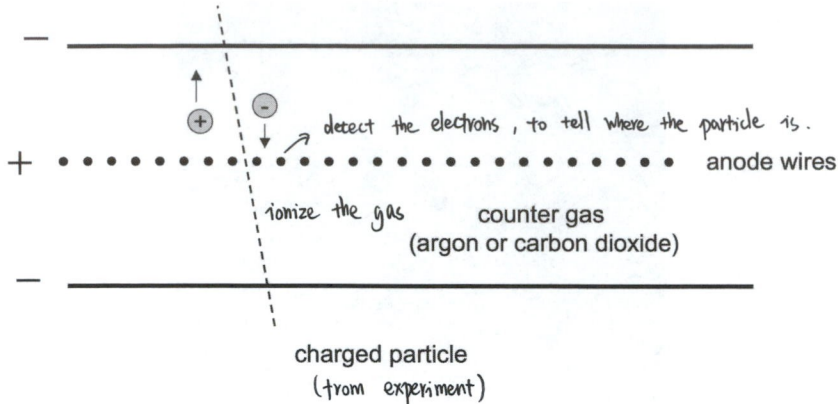

FIGURE 8.8
Schematic diagram of a wire counter.

as the charged particle passes through the TPC. The electric field causes the electrons of the ion pairs to acclerate towards one end of the chamber, and the magnetic field causes the trajectories to become tiny spirals along the magnetic field axis, i.e., parallel to the beam axis. The point of impact of the electrons on the end caps yields the projection of the trajectory, giving two spatial coordinates. The arrival time of the electrons can be used to determine the third spatial coordinate. The total charge deposited at the ends gives the total ionization and hence the total energy lost by the original charged particle. Since the detectors surround the beam pipe completely, a large solid angle is covered. The large number of sensitive elements at each end allows observation of many particles simultaneously, thereby providing efficient pattern recognition of the event.

8.2.7   Cerenkov Counters

Whenever a charged particle moves at a speed faster than c/n (where $n =$ index of refraction of the medium), a coherent wavefront forms as a cone of angle θ about the trajectory as in fig. 8.12. Named after Pavel Cerenkov, who discovered this effect in 1934 [89], this radiation is called Cerenkov radiation and it appears as a continuous spectrum. Measurement of the angle θ of the emitted light, as shown in fig. 8.12, provides a direct measurement of $\beta = v/c$. Note that if $\beta n < 1$ no signal is observed [90].

This effect can also used to identify different particles of the same momenta. Consider protons, pions, and kaons, each moving with a momentum of 1 GeV/c. From the relationship $p = m\gamma v = mv \left(1 - \frac{v^2}{c^2}\right)^{-1/2}$ between velocity and momentum, the respective velocities of these particles are $0.73c, 0.99c,$ and $0.89c$. Observation of each of these particles using a Cerenkov detector

FIGURE 8.9

Georges Charpak (left) works with students on the multiwire proportional counter. Densely packed anode wires and cathode planes are enclosed in a gas-tight chamber. Each wire acts as an individual detector similar to a proportional counter. A particle leaves a trace of ions and electrons that drift toward the nearest wire, which then emits a pulse of current. Several planes of wires with different orientations are used to track the trajectory of the particle very accurately (copyright CERN; used with permission).

requires media of differing refractive indices – in this case n must be larger than 1.37, 1.01, and 1.12 respectively. If we put two Cerenkov detectors in sequence – say one with water (where $n = 1.33$) and one with a gas where $n = 1.03$, then the pion will register a signal in both detectors, the kaon only in the water detector, and the proton will not register at all.

Velocity resolutions of 10^{-7} have been obtained using these kinds of detectors. This was the chief detection method used at the Sudbury Neutrino Observatory [91], where the medium is heavy water, with $n = 1.32828$. Neutrinos in these detectors scatter electrons as they enter the water. As these electrons travel through the heavy water they do so at velocities larger than c/n, and so register a signal which is picked up by photomultiplier tubes.

The main limitation of Cerenkov detectors is that a very feeble signal is produced. In general the number of photons radiated per unit path length ℓ

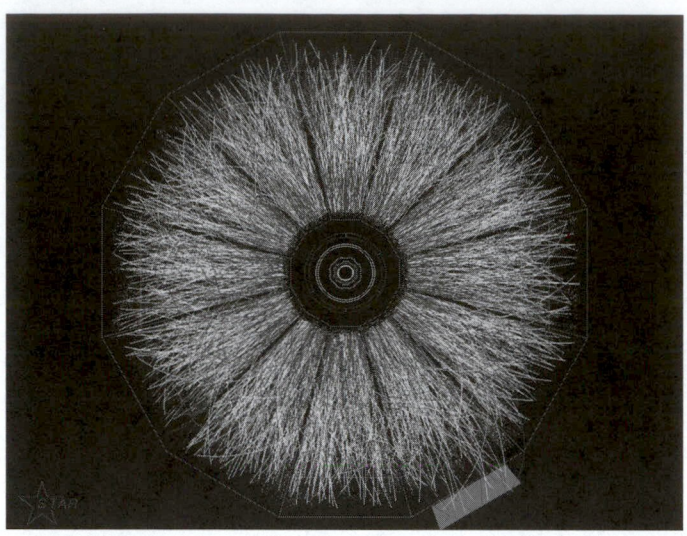

FIGURE 8.10

A view of one of the first full-energy collisions between gold ions at Brookhaven Lab's RHIC (Relativistic Heavy Ion Collider). The image was captured by the Solenoidal Tracker At RHIC (STAR) detector. Each track indicates a path taken by one of the thousands of subatomic particles produced in the collision. Photo courtesy of Brookhaven National Laboratory.

for photons of wavelength λ to $\lambda + d\lambda$ is

$$\frac{dN}{d\ell} d\lambda = 2\pi\alpha \left(1 - \frac{1}{\beta^2 n^2} \right) \frac{d\lambda}{\lambda^2} \tag{8.26}$$

where α is the fine structure constant. The maximum number of photons per cm emitted is in the range of 200 to 300 photons. This means that the detectors typically need to be several meters long to ensure detection of a signal.

8.2.8 Solid State Detectors

The newest technology in detecting charged particles was developed in the 1980s at SLAC [92]. The basic idea is to make use of diodes to detect the movement of charged particles. Wafers of very lightly doped silicon are very sensitive to the passage of charged particles – for example the formation of an electron-hole pair requires only about 3 eV of energy. A diode (i.e. a p-n junction) will be able to detect charged particles that move through its

FIGURE 8.11
The STAR detector at RHIC, a Time Projection Chamber, that tracks and
identifies particles emerging from heavy ion collisions, just before its comple-
tion. Photo courtesy of Brookhaven National Laboratory.

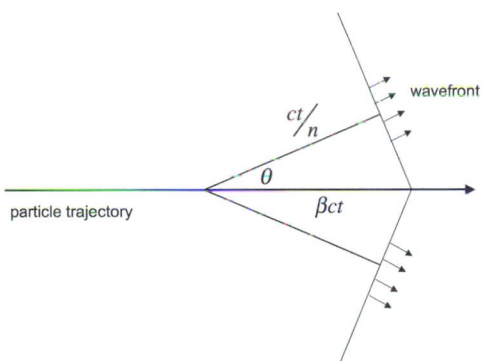

FIGURE 8.12
Cerenkov radiation. The angle $\cos\theta = 1/\beta n$ provides a measurement of the
speed of the particle.

depletion region (the interface between the p and n materials). As the charged
particle moves through the interface it interacts electrically with the lattice,

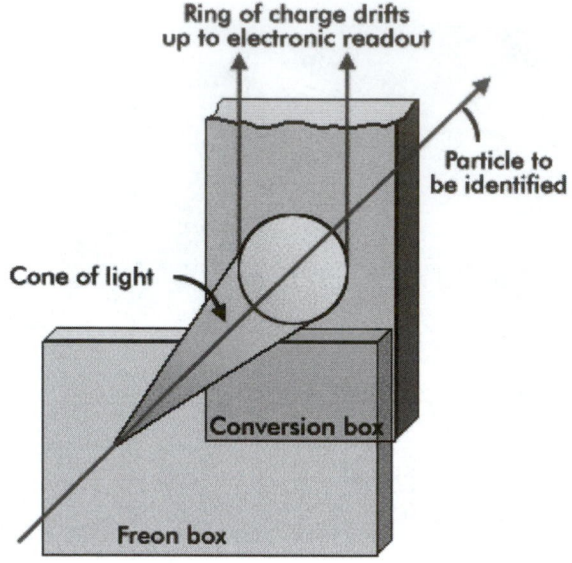

FIGURE 8.13

Schematic diagram of a Ring-Imaging Cerenkov (or RICH) Radiation Detector. The electrons produced by these photons will form a ring pattern. Image courtesy of SLAC National Accelerator Laboratory.

creating electron-hole pairs in a narrow column along its trajectory. The electrons and holes move in opposite directions under the influence of the electric field that has been established by the ionized impurity atoms in the wafer. This creates an electrical signal whose location corresponds to the trajectory of the particle.

The depletion region is made as thick as possible (typically 300 microns) so that the path of the charged particle is as long as possible, thereby yielding a stronger signal. By fabricating the diodes in parallel strips with a spacing of 25 microns (see fig. 8.14), a resolution of about 5 to 10 microns in the trajectory of the particle can be achieved. By gluing two such microstrip wafers together at an angle, the passage of a charged particle gives two independent signals as it moves through the detector, thereby yielding two independent coordinates which allows for the specification of a unique set of points in space corresponding to its path. Solid-state microstrip detectors are now being used in all the major high energy facilities today. They are what permitted the detection of the top quark, and (it is hoped) will assist in the search for the Higgs boson.

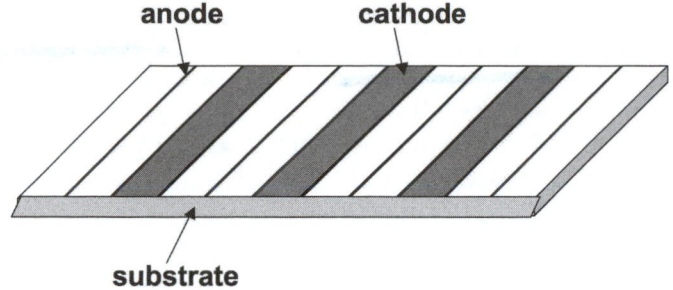

anode   cathode

substrate

FIGURE 8.14
Section of a typical microstrip plane.

8.2.9 Calorimeters

These devices work on the principle that an incident particle generates secondary ones, which generate tertiary ones, ... so that all of the incident energy is absorbed into the medium. A calorimeter absorbs all of the kinetic energy of a particle, yielding a signal that is proportional to this energy. Such detectors have energy resolution like $\sim 1/\sqrt{E}$, allowing for high precision. They also permit quick decisions on event selection.

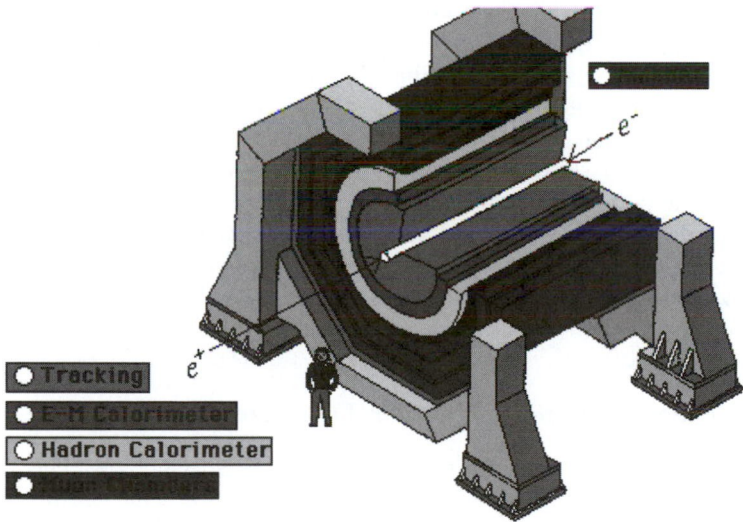

e^-

e^+

Tracking
E-M Calorimeter
Hadron Calorimeter
Muon Chambers

FIGURE 8.15
Schematic diagram of a modern calorimeter. Image courtesy of the Particle Data Group [1].

There are several kinds of calorimeters, each designed for detection of particular kinds of particles. They are often made of an absorbing material that is interspersed with sampling devices that determine the energy of the developing shower of particles. However, they can also be made of a single homogeneous material (for example lead glass) – these have better resolution but are considerably more expensive to construct.

8.3 Modern Collider Detectors

A modern detector in a collider is layered, with each layer performing a different function. Fig. 8.16 illustrates the general structure of a such a device. Let's look in a bit more detail at each layer.

8.3.1 Tracking Chambers

The inner region of any collider detector is called a tracking chamber. Its main function is to detect as accurately as possible the trajectories of the particles that leave the collision. It is generally a segmented combination of multi-wire proportional counters, streamer chambers, and solid-state detectors. Typically the innermost part of this chamber consists of a set of silicon microstrip detectors whose purpose is to provide precise spatial information about the trajectories of the charged particles that emerge from the collision point. They must be sufficiently thin so as to minimize errors due to multiple scattering. Outside of this are drift chambers, typically in an axial magnetic field, that provide measurement of the momentum of any emitted charged particles. The outermost stage is a set of pre-shower counters, which are about 3 radiation lengths of absorbing material followed by scintillation counters. They provide the first evidence of particles that are electromagnetically interacting because high energy photons traveling through the absorbing material will convert into electron/positron pairs that in turn produce a shower in front of the scintillators.

8.3.2 Electromagnetic Shower Detectors

High energy photons and electrons can produce cascade showers. The incident electron produces photons, which produce e^+e^- pairs, which produce more photons, which produce more e^+e^- pairs .... The result is a cascade of photons and electrons increasing exponentially with depth. Coulomb scattering causes the shower to spread laterally. Almost all of the energy of the shower appears as ionization loss of charged particles in the medium. These detectors are built from high-Z materials with small X_0 (typically about 25 radiation lengths, which is one mean free path for hadronic interactions) so the shower is contained in a small volume. They are placed after the tracking chamber, and absorb all of the energy of electrons and photons that pass through them.

8.3.3 Hadron Shower Calorimeters

These work on the same principle as above, except that an incoming hadron produces secondary hadrons, which produce tertiary hadrons ... . These detectors are much larger than the EM ones because the length scale for development of the shower is much longer than the radiation length of the medium. Another complication is that $\sim 30\%$ of the incident energy is lost by either exciting or breaking up nuclei. This can be compensated for by using ^{238}U as the medium; in ^{238}U extra energy is released by fast neutron and proton fission of the nucleus, which compensates for the losses from breakup. These detectors absorb the energy of all of the remaining emitted particles except for muons and neutrinos.

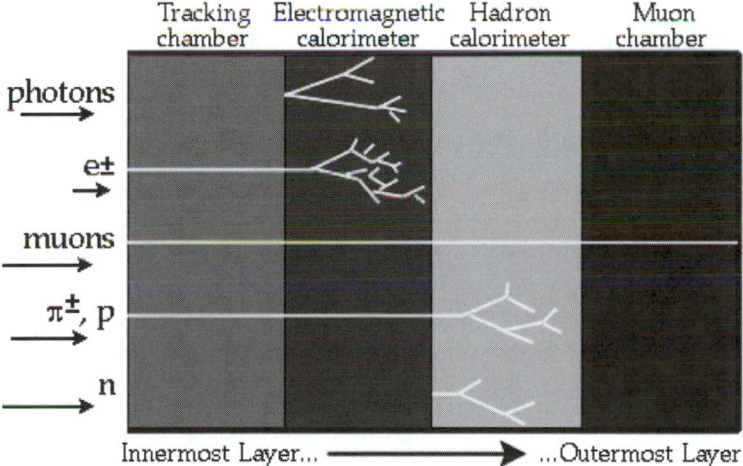

FIGURE 8.16
Schematic diagram of how particles deposit energy in calorimeter detectors. Note that the photons and neutrons, being neutral particles, are not detected in certain layers. Image courtesy of the Particle Data Group [1].

A detector cross-section, showing particle paths

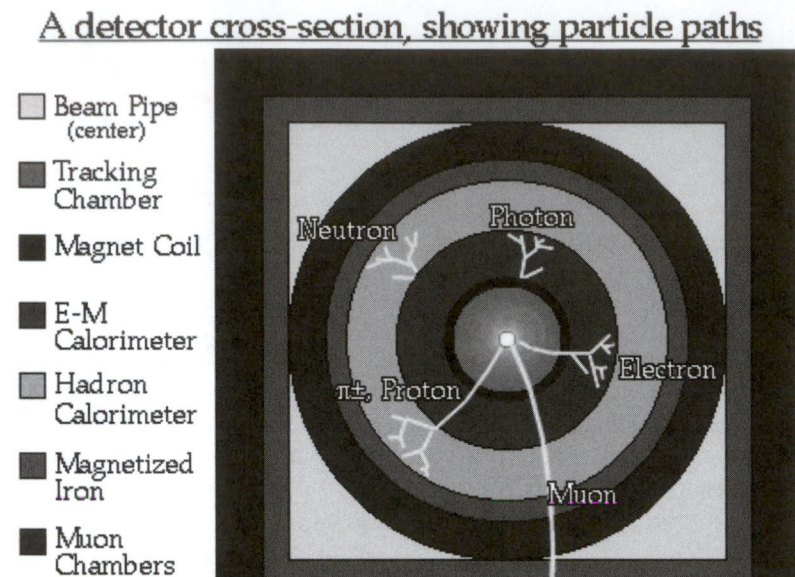

FIGURE 8.17

A cross-section of a calorimeter, showing the various tracks of different particle decays. Image courtesy of the Particle Data Group [1].

8.3.4 Muon Chambers

The only (known) particles that can make it beyond the hadronic calorimeters are high-energy muons and neutrinos. The muons are detected in the outermost layer, which measures their momentum. Their trajectories can be traced back to the inner region by checking consistency with the data from the previous three layers. Only neutrinos escape undetected. Their presence is inferred from conservation of 4-momentum – a lack of momentum and energy balance amongst the remaining particles suggests the presence of one or more particles that do not experience either the strong or electromagnetic interactions.

How do we know the undetected particle is a neutrino? The best way to be sure is to check the overall consistency of the event with a prediction of the Standard model. If the event is not consistent with such predictions, then it may very well signal the existence of a new particle or a new type of interaction. In order to be sure that all of the missing energy is accounted for, the detectors must cover as much of the 4π solid angle around the interaction point as possible and the layered detectors must minimize energy loss due to their structure. This is a great technical challenge. It is a remarkable achievement of modern physics that such detectors can be constructed, providing us with reliable empirical information about the subatomic world.

8.4 Questions

1. Antiprotons were first discovered in an experiment by Chamberlain et al. that produced them at a momentum of 1.2 GeV. In order for them to be detected by a Cerenkov counter, what would its minimum refractive index have to be?

2. A 400 GeV muon passes through a block of iron 500 cm thick. To what accuracy can its incident angle be measured?

3. Two particles of the same momentum but different mass travel between two scintillation counters separated by a distance L.

   (a) What is the difference in their time of arrival?

   (b) What is the minimum flight path necessary to distinguish Kaons from pions with momentum 4 GeV/c if the time of flight can be measured to an accuracy of 200 picoseconds?

4. How long would a gas Cerenkov counter of refractive index n have to be in order to distinguish Kaons from pions of identical momentum p? Assume that at least 100 photons in the visible range are needed to ensure a detection.

5. The rate of energy loss of protons traveling a material of density ρ typically follows a power law

$$\frac{dE}{dx} = -\frac{K}{\rho} \left(\frac{E}{E_0} \right)^{-p}$$

where the constants K and p are characteristic of the material and E_0 is a reference calibration energy that we can take to be 1 MeV.

   (a) Find an expression for the thickness of material that will reduce the energy of a proton by half the value it has upon entering the material.

   (b) For Carbon, $K = 212$ g-MeV/cm^2, and $p = 0.757$; for Lead, $K = 89.5$ g-MeV/cm^2, and $p = 0.694$. How much thicker must a slab of Carbon be than a slab of Lead in order to reduce a proton of incident energy 150 MeV by half?

6. Alpha particles of energy 300 MeV pass through a copper foil 100 microns thick. How much energy do they lose?

7. In a photomultiplier tube, each incident photon produces n electrons. Over N repetitions of this process the average number of electrons produced is

$$\bar{n} = \frac{1}{N} \sum_{i=1}^{N} n_i$$

and the probability of observing a particular number n of electrons for a typical measurement is

$$P(n) = \frac{\bar{n}^n}{n!} e^{-\bar{n}}$$

a distribution called a *Poisson distribution*.

(a) Show that $\sum_n P(n) = 1$.

(b) Compute the average $\langle n \rangle = \sum_n nP(n)$

(c) Find the standard deviation $\sqrt{\left\langle (n - \langle n \rangle)^2 \right\rangle}$ of the number of electrons observed.

8. An underground scintillator at SNOlab counts five muons per hour on average. Suppose this experiment is run for 1000 hours. How often will counts of 2, 4, 6, 8 and 10 muons be recorded?

9

Scattering

We obtain empirical information about particle physics from a variety of sources: from studying cosmic rays, the cosmic microwave background, and bound states – the latter is particularly important when it comes to quark physics. However, most of what we know experimentally in particle physics comes from data on the decays of unstable particles and on the scattering of one particle from another. The accelerators and detectors that were discussed in chapters 7 and 8 are the primary tools used to measure these phenomena. We want to make use of the data that emerges from these experiments to modify, corroborate, and falsify our theories of the subatomic world, the ultimate goal being that of obtaining the deepest understanding possible of the laws of nature.

In order to do this – that is, in order for theory and experiment to make contact – we need to deal with quantities that the theorist can calculate and that the experimentalist can measure. There are a number of such quantities, but the most important two are *lifetimes* and *cross-sections*.

9.1 Lifetimes

The lifetime τ of an elementary particle is the average amount of time it takes for the particle to decay into something else. If it doesn't decay into anything then we expect the lifetime to be infinite; we say that such a particle is stable. In practice we can experimentally only set lower bounds on the lifetime of stable particles. For example the lifetime of the proton [93] has been measured to be $\tau_p > 2.1 \times 10^{29}$ years; for the electron [94] it is $\tau_{e^-} > 6.4 \times 10^{24}$ years*. We will assume that particles that have only lower bounds on their lifetimes are indeed stable, never decaying into anything. There are only a few particles in nature of this type. The vast majority of elementary particles are unstable and will generally decay into some other particles.

---

*These lifetimes are for the proton and electron to decay into anything – known particles or not. If we add the additional requirement that the decays be into known particles, then these lifetimes increase by a factor of 100-1000 [95].

In principle, measurement of the lifetime of a particle is simple: just watch it to see how long it takes to decay! However, in practice this is complicated by relativistic effects (faster moving particles take longer to decay than slower moving ones) and quantum effects (particles moving at the same speed will decay at different rates due to the uncertainty principle – we can't arbitrarily reduce the measurement error of the time it takes to decay).

Both of these problems may be circumvented by considering a group of identical particles in their rest frame. In this case relativistic effects average out and although we cannot predict the lifetime of a given particle in the group, we can predict (and measure!) the rate of decay of the group as a whole. So we are interested in how many particles per unit time, on average, will decay in a large group. This is the decay rate Γ: the probability per unit time that a particle will decay (as measured in its rest frame).

So if we have, say, $N(t)$ unstable particles at time t, the probability that any one of them will decay in time Δt is $\Gamma \Delta t$. Hence the entire sample consisting of $N(t)$ particles will decrease by an amount ΔN where

$$\Delta N = -N\Gamma\Delta t \Longrightarrow N(t) = N_0 e^{-\Gamma t} \tag{9.1}$$

and so the number of particles decreases exponentially with time. Here N_0 is the number of particles in the sample at $t = 0$. It's easy to show that the mean lifetime is just $\tau = 1/\Gamma$.

Most particles decay by several routes: for example, the τ-lepton can decay into $\mu^- + \bar{\nu}_\mu + \nu_\tau$, or into $e^- + \bar{\nu}_e + \nu_\tau$, or into $\pi^- + \nu_\tau$, or to a number of other types of elementary particles of smaller mass. The total decay rate is the sum of the individual rates for each process

$$\Gamma_{\text{tot}} = \sum_{i=1}^{n} \Gamma_i \tag{9.2}$$

and the mean lifetime is

$$\tau = \frac{1}{\Gamma_{\text{tot}}} \tag{9.3}$$

The *branching ratio* is the fraction of all decays into a given mode:

$$B_i = \frac{\Gamma_i}{\Gamma_{\text{tot}}} \tag{9.4}$$

and in practice this is the quantity of interest.

The task of the theorist is then to compute as accurately as possible the decay rates Γ_i for each particle described by his or her proposed theory. The task of the experimentalist is to measure each Γ_i as precisely as possible to test the theory.

9.2 Resonances

Unstable particles are observed as *resonances*. These are local maxima in the cross-section as a function of energy, or as a maximum in the invariant mass distribution of the particles in the final state of a reaction.

To understand better the character of a resonance, let's consider at atom modeled as a massive pointlike nucleus with a charge at the origin surrounded by a cloud of equal but opposite charge. If the nucleus is displaced from the origin relative to the cloud then the system will undergo oscillations. In any given spatial direction there will be an infinite set of damped modes, each with a characteristic frequency and width. Since there are three independent spatial directions there is a triply-infinite set of modes. Treating the system classically, suppose we let X be the coordinate measuring the distance of the system from equilibrium, where $X = X_0$. To a very good approximation we will have

$$X(t) = X_0 \exp\left(-\frac{\Gamma}{2}t\right) \cos(\omega t) \qquad (9.5)$$

lifetime

for some characteristic frequency ω and width Γ.

Suppose now that the system is subjected to a periodic force with frequency Ω and amplitude proportional to F_0 in the direction of motion. This could be done by directly shining monochromatic light onto the atom. The behavior now becomes

$$X(t) = X_0 \exp\left(-\frac{\Gamma}{2}t\right) \cos(\omega t) + \frac{kF_0}{\sqrt{(\omega^2 - \Omega^2)^2 + \Omega^2\Gamma^2}} \cos(\Omega t + \varphi) \qquad (9.6)$$

and we see that for late times, the first term becomes negligible and the system undergoes periodic motion. The average (kinetic) energy of the atom is easily found to be

$$\langle E \rangle = \frac{1}{4} \frac{m^2 k^2 F_0^2 \Omega^2}{(\omega^2 - \Omega^2)^2 + \Omega^2\Gamma^2} = R(\omega) \frac{m^2 k^2 F_0^2}{4\Gamma^2} \qquad (9.7)$$

where m is the mass of the atom, and I have written this average in terms of the response function

$$R(\omega) = \frac{\Omega^2\Gamma^2}{(\omega^2 - \Omega^2)^2 + \Omega^2\Gamma^2} \qquad (9.8)$$

that tells us the general dependence of the average energy of the atom independently of its mass, and the amplitude of the driving force. The response function is a maximum when $\omega = \Omega$, and we can approximate its behavior near the maximum via

$$R(\omega) \simeq \frac{\left(\frac{\Gamma}{2}\right)^2}{(\omega - \Omega)^2 + \left(\frac{\Gamma}{2}\right)^2} \qquad (9.9)$$

particle energy   *particle mass*

an expression known as the *Breit-Wigner shape function* [96].

Applying these ideas to particle physics, we can regard the ground state of the atom as one kind of "particle" and an excited level as another kind of "particle". This may sound a bit strange at first, but each of these states of the atom has their own characteristic energy, frequency and lifetime, where the latter is infinite for the ground state and finite for all of the excited states. The search for unstable particles in particle physics is analogous to the search for unstable atomic states of an atom. To find such unstable atomic states we could shine a monochromatic light beam whose frequency we could vary and then search for resonant behavior of the atom (initially in its ground state) as the frequency is varied. Peaks in the intensity of the scattered light will correspond to the excited atomic states (the unstable "particles"). The resonance will occur at a definite energy (the value ω, which is the "mass" of the "particle") and will have a definite width proportional to Γ. The shape of the peak will be given by the Breit-Wigner shape function. Similarly, to find unstable subatomic particles, we collide two beams together and search for increases in the number of particles reaching the detector – that is, for peaks in the cross-section (described in detail below) – as a function of energy.

We can also find resonances by producing them as intermediate states. For example, consider the reaction

$$A + B \to R + E \to C + D + E \tag{9.10}$$

which means that particles A and B collide to temporarily form a particle E and a metastable particle R (the resonance), that in turn decays into particles C and D. We expect in such cases that the mass M_{CD} of the final state of C and D will be described by a Breit-Wigner function peaked at the mass of the resonance with width Γ. If there is no resonance then the mass M_{CD} can have any value consistent with energy, momentum, and angular momentum conservation and will be a smooth function of collision energy that is not peaked.

9.3 Cross Sections

If we have two particles at our disposal then we can have them collide to see what happens. Again this is a conceptually simple idea – just throw the two particles against each other (perhaps by placing one as a fixed target, or perhaps by throwing both) and measure what comes out. However, this simple idea is fraught with complications: due to the uncertainty principle it's impossible to aim the particles with arbitrarily precise accuracy. As with the previous case, we deal with this inherent randomness by considering groups of particles scattering off of one another.

Let's consider what happens in a frame where one group of particles (the target) is at rest. For example we might consider a beam of neutrons impacting upon hydrogen gas (essentially protons). We then have a situation of the type described in fig. 9.1. The beam consists of a flux of particles with luminosity $\mathcal{L}$: the number of particles per unit time incident on a given area that is perpendicular to the axis of the beam. As a particle in the beam approaches the target it will experience some kind of interaction with the target (or rather, with the potential generated by the target) and will scatter off at some angle θ relative to its initial trajectory. This angle depends upon the impact parameter b, the distance by which the incoming particle would have missed the scattering center had it not been scattered. The particles with impact parameters between b and $b + db$ will scatter at angles between θ and $\theta - d\theta$, where the minus sign expresses the fact that as the impact parameter gets larger the scattering angle gets smaller. In general the dependence of θ on b will depend upon the particular type of interaction between the particle and the target. In other words, it will depend upon our underlying fundamental physical theory.

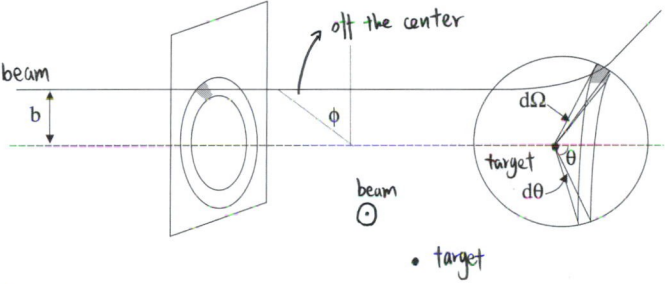

FIGURE 9.1

Particle incident in an area $d\sigma$ (the shaded region on the plane) scatters into a solid angle $d\Omega$ (the shaded region on the sphere).

The beam of particles between b and $b+db$ will form an annular ring centered about the beam axis. The part of the beam between angles ϕ and $d\phi$ (where ϕ is the angle around the annular ring relative to some axis perpendicular to the beam axis) will have $d\mathcal{N} = \mathcal{L}d\sigma$ particles passing through the infinitesimal area $d\sigma$ of the annular ring per unit time. This part of the beam will be scattered into some solid angle $d\Omega$ between Ω and $\Omega + d\Omega$.

The *differential cross-section* $\mathcal{D}(\theta, \phi)$ is defined as the ratio between these two quantities:

$$\mathcal{D}(\theta, \phi) = \frac{d\sigma}{d\Omega} = \frac{1}{\mathcal{L}} \frac{d\mathcal{N}}{d\Omega} \tag{9.11}$$

Hence an experimentalist can measure the differential cross section[†] by count- ing the number of particles scattered into a given solid angle per unit time and dividing by the luminosity, which is controlled by the apparatus. Most of the time $\mathcal{D}(\theta, \phi)$ is independent of ϕ since we deal with spherically symmetric potentials.

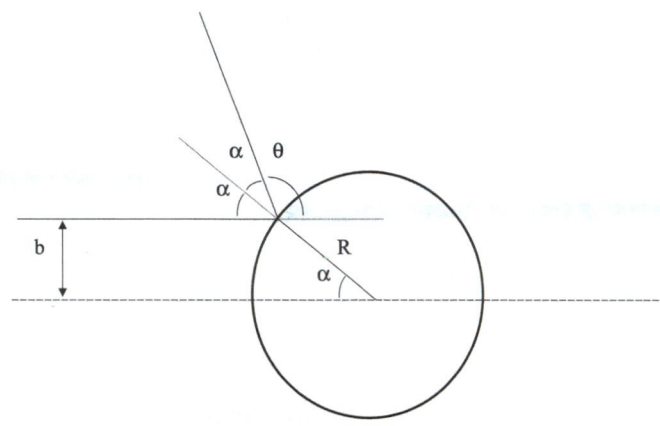

FIGURE 9.2
Hard sphere scattering

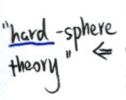

As an example, suppose we scatter a particle off of a hard sphere. Our "theory" is that the scattering center consists of a potential that is zero ev- erywhere except at the edge of the sphere of fixed radius R, where it becomes infinite. Hence any particle encountering the sphere will elastically bounce off of it, with its angle of incidence α equalling its angle of reflection. We can see from fig. 9.2 that the scattering angle $\theta = \pi - 2\alpha$. If the impact parameter is b and the radius of the sphere is R we have $b = R \sin \alpha = R \cos \frac{\theta}{2}$, yielding (indirectly) the dependence of θ on b. Geometrically $d\sigma = |b \, db \, d\phi|$ and the scattering solid angle $d\Omega = |\sin \theta \, d\theta \, d\phi|$. Consequently

$$\frac{d\sigma}{d\Omega} = \frac{|b \, db \, d\phi|}{|\sin \theta \, d\theta \, d\phi|} = \frac{\left| R \cos \frac{\theta}{2} \, d \left(R \cos \frac{\theta}{2} \right) \right|}{|\sin \theta \, d\theta|} = \frac{R^2 \cos \frac{\theta}{2} \sin \frac{\theta}{2}}{2 \sin \theta} = \frac{R^2}{4} \quad (9.12)$$

and integrating over solid angle we have $\sigma = \int d\Omega \frac{R^2}{4} = \pi R^2$. So the effective

---

[†]Note that in mathematical terms the quantity $\mathcal{D}(\theta, \phi)$ is not really a differential at all. In practice people refer to $\frac{d\sigma}{d\Omega}$ as the differential cross-section, and sometimes even abbreviate this to just $d\sigma$, with the implicit understanding that an integral over the solid angle must be carried out to find σ.

area the sphere presents to the beam is the projected area of its forward hemisphere: any particles within πR^2 will scatter; all others will not!

This example illustrates the character of a cross-section: it is the effective area that the target presents to the beam or, more generally, it is the effective area that one particle presents to another. However, it is important to be aware of the limitations of this example – in general the target is not one that you either hit or miss (as was the case here) but rather is one for which the closer you come the greater the deflection. The dependence of the differential cross-section as a function of angle (in turn determined by underlying theory) in general terms encodes this "soft" behavior.

For cross-sections, we often are interested only in the case where the final state consists of a particular type of particle (or set of particles). As with the decay rate we have many possible outcomes for a given set of incident particles, each of which will have their own cross-section σ_i. The total (or inclusive) cross-section is obtained by summing over all possible channels (or modes) of scattering:

$$\sigma_{\text{tot}} = \sum_{i=1}^{n} \sigma_i \qquad \text{for } n \text{ possible modes of scattering} \qquad (9.13)$$

In general σ is inversely proportional to velocity, since the cross section should be proportional to the amount of time the incident particle spends near the target. Hence if we graph σ vs. velocity (or, more commonly σ vs. Energy), we expect to see it montonically decrease. However, this behavior is not universal: a bump in this graph means that the particle "likes" to be near the target at that energy – the target presents a much larger area to the incident beam. This is another example of a resonance, which in this case is a short-lived semibound state of particle and target. Hunting for resonances in the plot of σ vs E is the chief way in which new (unstable) particles are discovered.

The absorption coefficient described in Chapter 8 can be related to the scattering cross section by recognizing that any particle scattered out of a beam produces an increase in the counting rate for scattering or equivalently a drop in the beam intensity. Suppose that there are n scattering centers per unit volume of material, each scattering center having a cross-section σ. Then number of particles scattered through a material of thickness dx is $n\sigma dx$, and this must correspond to the fraction of the beam $d\mathcal{I}/\mathcal{I}$ that is attenuated. Hence

$$n\sigma dx = \left|\frac{d\mathcal{I}}{\mathcal{I}}\right| = \mu dx \qquad (9.14)$$

using the definition of absorption coefficient from eq. (8.20), which implies

$$\mu = n\sigma \qquad (9.15)$$

9.4 Matrix Elements

The essential job of the theorist is to take his or her favorite theory and use it to compute Γ and $\frac{d\sigma}{d\Omega}$ for various processes of interest. These quantities are computed from something called the *scattering amplitude* [97].

The scattering amplitude (or S-matrix element) is defined to be

distant future → distant past

$$S_{if} =_{T\to\infty} \langle \Phi |S| \Psi \rangle_{T\to-\infty} \equiv \text{S-matrix element} \tag{9.16}$$

i.e. it is the amplitude for some state $|\Psi\rangle$ in the distant past (where we idealize all particles emitted by the source to be initially non-interacting) to evolve into some other state $\langle\Phi|$ in the future (where the detector is again idealized to absorb only free non-interacting particles). This action is just a time translation, so S is unitary.

Most of the time in a scattering reaction, particles just zip by each other and nothing happens. To account for this we write

miss each other

$$S_{if} = \delta_{if} + iM_{if} = 1 + i \langle p'_1 \cdots p'_m |M| p_1 \cdots p_n \rangle \tag{9.17}$$

where the "1" is the identity matrix in the Hilbert space spanning initial and final states. It is the matrix element M – the part of the S-matrix that is not the identity – that a theorist computes from a given theory.

Quantum-mechanically, the probability for scattering is given by the square of the magnitude of the amplitude. Hence we expect for decay rates that

$$d\Gamma \propto |\langle p'_1 \cdots p'_m |M| p_1 \rangle|^2 \tag{9.18}$$

since there is just one initial particle, and for cross sections

$$d\sigma \propto |\langle p'_1 \cdots p'_m |M| p_1 p_2 \rangle|^2 \tag{9.19}$$

since there are two particles in the initial state.

You might think that we could begin calculating once we had M, but this isn't quite right for several reasons:

(a) we need to conserve momenta between initial and final states

(b) because of the uncertainty principle we can't prepare our initial states with exactly the momenta $\vec{p}_i$

(c) again, because of the uncertainty principle, we can't precisely measure the final momenta $\vec{p}'_i$ of the outgoing particles

The first problem (a) is easy to correct. We just multiply by a δ-function that ensures the sum of the incoming 4-momenta equals the sum of the outgoing 4-momenta

phase space P ↑ *uncertainty.*

$$(2\pi)^4 \, \delta^{(4)} \left(\sum_{i=1}^{m} p_i' - p_2 - p_1 \right)$$

$$= (2\pi)^4 \, \delta^{(3)} \left(\sum_{i=1}^{m} \vec{p}_i' - \vec{p}_2 - \vec{p}_1 \right) \delta \left(\sum_{i=1}^{m} E_i' - E_2 - E_1 \right) \tag{9.20}$$

where the factor of $(2\pi)^4$ is a convention. To address (b) and (c) we should multiply by the uncertainty $(\Delta p)^3$ in momentum for each $\vec{p}_i'$ and for $\vec{p}_2$ and $\vec{p}_1$. However, this quantity isn't Lorentz-invariant. Instead, we multiply by $\frac{(\Delta p)^3}{E}$. This is because in a frame moving with velocity $\beta = v/c$ along some axis we have

$$\frac{(\Delta p')^3}{E'} = \frac{\left(\Delta p_\parallel' \right) (\Delta \vec{p}_\perp')}{E'} = \frac{\gamma \left(\Delta \left(p_\parallel - \beta E \right) \right) (\Delta \vec{p}_\perp)}{\gamma \left(E - \beta p_\parallel \right)}$$

Lorentz transformed ↙

$$= \frac{\left(\left(1 - \beta \frac{\Delta E}{\Delta p_\parallel} \right) \right) (\Delta \vec{p}_\perp)(\Delta p_\parallel)}{E \left(1 - \beta \frac{p_\parallel}{E} \right)} = \frac{\left(1 - \beta \frac{p_\parallel}{E} \right) (\Delta p)^3}{E \left(1 - \beta \frac{p_\parallel}{E} \right)}$$

$$= \frac{(\Delta p)^3}{E} \qquad \begin{array}{l} E^2 = p_\perp^2 + p_\parallel^2 + m^2 \\[4pt] E \, \Delta E = p_\parallel \, \Delta p_\parallel \implies \frac{\Delta E}{\Delta p_\parallel} = \frac{p_\parallel}{E} \end{array} \tag{9.21}$$

where $\vec{p}_\perp$ is the component of momentum orthogonal to the direction of the moving frame and $p_\parallel$ is the component parallel to the moving frame[‡]. So it is the quantity $\frac{(\Delta p)^3}{E}$ that is Lorentz invariant, and we multiply by this.

Putting it all together, for cross-sections we get

↗ *final state*

$$d\sigma \propto W = W_0 \prod_{i=1}^{m} \frac{(\Delta p_i')^3}{2E_i' \, (2\pi\hbar)^3} \, |\langle p_1' \cdots p_m' \, |\mathsf{M}| \, p_1 p_2 \rangle|^2 \, \frac{(\Delta p_1)^3}{2E_1 \, (2\pi\hbar)^3} \frac{(\Delta p_2)^3}{2E_2 \, (2\pi\hbar)^3}$$

$$\times (2\pi)^4 \, \delta^{(4)} \left(\sum_{i=1}^{m} p_i' - p_2 - p_1 \right)$$

$$= W_0 \left[\left(\frac{(\Delta p_1')^3}{2E_1' \, (2\pi\hbar)^3} \right) \left(\frac{(\Delta p_2')^3}{2E_2' \, (2\pi\hbar)^3} \right) \cdots \left(\frac{(\Delta p_m')^3}{2E_m' \, (2\pi\hbar)^3} \right) \right] \tag{9.22}$$

$$\times |\langle p_1' \cdots p_m' \, |\mathsf{M}| \, p_1 p_2 \rangle|^2 \, \frac{(\Delta p_1)^3}{2E_1 \, (2\pi\hbar)^3} \frac{(\Delta p_2)^3}{2E_2 \, (2\pi\hbar)^3}$$

$$\times (2\pi)^4 \, \delta^{(4)} \left(\sum_{i=1}^{m} p_i' - p_2 - p_1 \right)$$

[‡]The prime in eq. (9.21) refers to the moving frame (and not the final state).

where $(2\pi\hbar)^3$ is the quantum phase space volume and W_0 is a normalization factor. In order to turn the proportionality into an equality, note that

$$d\sigma = \begin{pmatrix} \text{Probability of scattering} \\ \text{to final state} \end{pmatrix} \times \text{area}$$

$$= \mathcal{S}W \div \begin{pmatrix} \text{incident} \\ \text{(target) density} \end{pmatrix} \div \begin{pmatrix} \text{incident} \\ \text{(beam) flux} \end{pmatrix}$$

statistical factor

$$= \mathcal{S}W \div \mathcal{N}_1 \div \left[\mathcal{N}_2 \times \begin{pmatrix} \text{relative} \\ \text{velocity} \end{pmatrix} \right]$$

$$= \frac{\mathcal{S}}{\mathcal{N}_1 \mathcal{N}_2 \left| \vec{v}_2 - \vec{v}_1 \right|} W \tag{9.23}$$

where $\mathcal{N}_1$ and $\mathcal{N}_2$ are the incident particle densities and $\mathcal{S}$ is a statistical factor that corrects for overcounting of identical particles (we'll look at it in a bit more detail below). Now recall [98] that the density of states for a particle with quantum numbers n_x, n_y and n_z is

$$(\Delta p)^3 = \left(\frac{2\pi\hbar}{L} \right)^3 (\Delta n)^3 = \frac{(\Delta n)^3}{L^3} (2\pi\hbar)^3 \propto (2\pi\hbar)^3 \mathcal{N} \tag{9.24}$$

where the particle is in a volume $V = L^3$. Applying this to the incident particles, we have

$$d\sigma = \frac{\mathcal{S}W_0}{\mathcal{N}_1 \mathcal{N}_2 \left| \vec{v}_2 - \vec{v}_1 \right|} \prod_{i=1}^{m} \frac{(\Delta p_i')^3}{2E_i' (2\pi\hbar)^3} \left| \langle p_1' \cdots p_m' \left| \mathsf{M} \right| p_1 p_2 \rangle \right|^2$$

$$\times \frac{\mathcal{N}_1 \mathcal{N}_2}{4E_1 E_2} (2\pi)^4 \delta^{(4)} \left(\sum_{i=1}^{m} p_i' - p_2 - p_1 \right) \tag{9.25}$$

or

Lorentz - covariant factor *phase space*

$$d\sigma = \frac{\hbar^2 \mathcal{S}}{4\sqrt{(p_1 \cdot p_2)^2 - p_1^2 p_2^2}} \left[\prod_{i=1}^{m} \frac{c (\Delta p_i')^3}{2E_i' (2\pi)^3} \right] \left| \langle p_1' \cdots p_m' \left| \mathsf{M} \right| p_1 p_2 \rangle \right|^2$$

matrix element

$$\times (2\pi)^4 \delta^{(4)} \left(\sum_{i=1}^{m} p_i' - p_2 - p_1 \right) \tag{9.26}$$

choosing the normalization W_0 to obtain equality and the correct units. This is the cross section for a process in which particle $1'$ has a 3-momentum that lies in the range $\Delta p_1'$ around the value $\vec{p}_1'$, particle $2'$ has a 3-momentum that lies in the range $\Delta p_2'$ around the value $\vec{p}_2'$, etc. Typically the momenta of all but one of the particles (say particle $1'$) in the final state are integrated over; what remains after integration is the differential cross-section for particle $1'$ to emerge into the solid angle $d\Omega$.

If we only had one particle in the initial state, we'd be interested in the decay rate, and a similar analysis would give

$$d\Gamma = \frac{\mathcal{S}c^2}{2\hbar E} \left[\prod_{i=1}^{m} \frac{c\left(\Delta p_i'\right)^3}{2E_i'\left(2\pi\right)^3} \right] |\langle p_1' \cdots p_m' \,|\mathbf{M}|\, p \rangle|^2 \left(2\pi\right)^4 \delta^{(4)} \left(\sum_{i=1}^{m} p_i' - p \right) \quad (9.27)$$

where the decaying particle initially has momentum $\vec{p}$.

9.4.1 General Features of Decay Rates and Cross-Sections

Some comments on these formulae are appropriate.

1. The factors in front of the cross-section and decay rate come from Lorentz covariance. It is not hard to show that $d\Gamma$ has inverse time-dilation properties (i.e. it transforms as an inverse time). The factor in front of the expression for $d\sigma$ can be written as

$$c^3 \sqrt{\left(p_1 \cdot p_2\right)^2 - p_1^2 p_2^2} = E_1 E_2 \,|\vec{v}_2 - \vec{v}_1| \quad (9.28)$$

   for particles in a head-on collision, where $\vec{p}_2 = -\vec{p}_1$.

2. The factor $\mathcal{S}$ is a statistical factor – it equals $1/n!$ for each group of n identical particles in the final state. For example, if a particle decays into four identical particles of one kind, and three of another kind, then $\mathcal{S} = \frac{1}{4!}\frac{1}{3!} = \frac{1}{144}$.

3. The factor $\left[\prod_{i=1}^{m} \frac{c\left(\Delta p_i'\right)^3}{2E_i'\left(2\pi\right)^3} \right] \left(2\pi\right)^4 \delta^{(4)} \left(\sum_{i=1}^{m} p_i' - p \right)$ is called the *phase space of the final state*, where $E_i' = \sqrt{|\vec{p}_i'|^2 + m_i^2}$. Physically it is the kinematic information in the scattering process: it tells us how many different ways the available energy and momentum can be partitioned into the final state. If a heavy particle decays into very light ones, then lots of phase space is available because only a little bit of the initial energy will go into the rest masses of the final particles. But if a particle decays into one close to its own rest mass (such as a neutron decaying into a proton, an electron, and an antineutrino), then very little phase space is available, reducing the decay rate. The phase space factor is Lorentz invariant. If we want to sum over all possible final states, we must integrate over this quantity:

$$\sigma = \frac{\hbar^2 \mathcal{S}}{4\sqrt{\left(p_1 \cdot p_2\right)^2 - p_1^2 p_2^2}} \left[\int \prod_{i=1}^{m} \frac{c d^3 p_i'}{2E_i'\left(2\pi\right)^3} \right] |\mathcal{M}|^2$$

$$\times \left(2\pi\right)^4 \delta^{(4)} \left(\sum_{i=1}^{m} p_i' - p_2 - p_1 \right) \quad (9.29)$$

$$\Gamma = \frac{Sc^2}{2\hbar E} \left[\int \prod_{i=1}^{m} \frac{d^3 p_i'}{2E_i'(2\pi)^3} \right] |\mathcal{M}|^2 (2\pi)^4 \delta^{(4)} \left(\sum_{i=1}^{m} p_i' - p \right) \quad (9.30)$$

which gives the total cross section (or decay rate) for the process under consideration. Here I have written $|\mathcal{M}|^2 = |\langle f |M| i \rangle|^2$ as the square of the matrix element between initial and final states.

4. The quantity $|\mathcal{M}|$ is a scalar function of the momenta. If the particles emitted have spin, then $|\mathcal{M}|$ is still a scalar function, but it can now depend on things like $\vec{p}_1' \cdot \vec{s}_1$, $\vec{p}_1' \cdot \vec{s}_2$, and $\vec{s}_1 \cdot \vec{s}_2$. However, we will always work with spin-averaged quantities, in which case these terms average out to zero.

One other comment: The factors of (2π) can seem like a nuisance, but they are essential to get the numerical values of the answers correct. And they are not too hard to remember: just put a (2π) in the numerator for every δ (so $\delta^{(4)}$ gets a $(2\pi)^4$ and put a (2π) in the denominator for every d (so each $d^3 p_i$ is divided by $(2\pi)^3$). Once you've done this, you can of course cancel them out accordingly.

Enrico Fermi called the formulae (9.26) and (9.27) for computing cross-sections and decay rates the *Golden Rule* [99], though most of the work leading up to it was done by Dirac [100]. Basically the rule says that to find the transition rate for any process, take the modulus of the amplitude, square it, and multiply by the phase space. The Golden Rule not only occurs in particle physics, but also in non-relativistic time-dependent perturbation theory in quantum mechanics [101], where you may have already encountered it.

9.5 2-Body Formulae

The general formulae for decays and cross-sections are rather formidable. Fortunately they really simplify if – as is very common – there are only two particles in the final state. This will prove to be very useful in understanding much of particle physics, so we will list these formulae here.

9.5.1 2-Body Decay Rate

In the rest frame of a decaying particle the 2-Body decay rate becomes

$$\Gamma_{\text{2-Body}} = \frac{S |\vec{p}| c}{8\pi\hbar (Mc)^2} |\mathcal{M}|^2 \quad (9.31)$$

where $\vec{p}$ is the momentum of either of the outgoing momenta in the final state (it's the same for each final particle since momentum is conserved), and M is the rest mass of the decaying particle.

9.5.2   2-Body CM Cross-Section

The cross section can have two useful forms if the final state has only two particles. In the center-of-momentum frame (CM frame), when both particles collide head-on

$$\left(\frac{d\sigma}{d\Omega}\right)_{\text{2-Body CM}} = \left(\frac{\hbar c}{8\pi}\right)^2 \frac{\mathcal{S}\,|\mathcal{M}|^2}{(E_1 + E_2)^2} \frac{|\vec{p}\,'|}{|\vec{p}|} \quad \text{CM frame:} \quad \begin{cases} \vec{p}_1 = -\vec{p}_2 = \vec{p} \\ \vec{p}_1{}' = -\vec{p}_2{}' = \vec{p}\,' \end{cases}$$
$$(9.32)$$

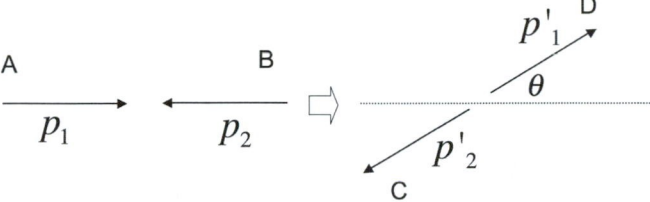

FIGURE 9.3
Two body scattering in the center-of-momentum frame.

9.5.3   2-Body Lab Cross-Section

In the lab frame $\vec{p}_2 = 0$. The differential cross section becomes somewhat messy in this frame unless we have elastic scattering ($A + B \longrightarrow A + B$), in which case

$$\left(\frac{d\sigma}{d\Omega}\right)_{\text{2-Body elastic-LAB}} = \left(\frac{\hbar}{8\pi}\right)^2 \frac{\mathcal{S}\,|\mathcal{M}|^2\,|\vec{p}_1{}'|^2}{M_2\,|\vec{p}_1|\,[\,|\vec{p}_1{}'|\,(E_1 + M_2 c^2) - |\vec{p}_1|\,E_1' \cos\theta]}$$
$$\text{Lab frame:} \quad \begin{cases} \vec{p}_2 = 0 \\ E_2 = M_2 c^2 \end{cases} \quad (9.33)$$

Note that these formulae are valid no matter what $|\mathcal{M}|$ is. This is why they are so useful – we don't have to do the delta-function integration every time. I've left the derivation of these formulae for you to do in the problems at the end of this chapter.

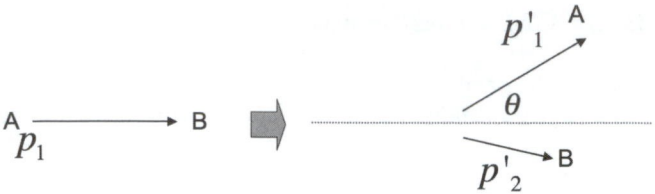

FIGURE 9.4
Elastic scattering in the Lab frame.

9.6   Detailed Balance Revisited

The factor

$$|\mathcal{M}| = |\langle f\,|\mathsf{M}|\,i\rangle| = |\langle p'_1 \cdots p'_m\,|\mathsf{M}|\,p_1 p_2\rangle| = |\mathcal{M}_{fi}| \tag{9.34}$$

is what we compute from a given physical theory such as quantum electro-dynamics (QED), or quantum chromodynamics (QCD). For a given physical process, different competing theories will give different values for this factor: this is where theory and experiment "meet." For particles with spin we have

$$|\mathcal{M}| = |\langle f\,|\mathsf{M}|\,i\rangle| = |\langle p'_1, s'_1; \ldots; p'_m, s'_m\,|\mathsf{M}|\,p_1, s_1; p_2, s_2\rangle| = |\mathcal{M}_{fi}| \tag{9.35}$$

The spins can really complicate the formulae for Γ or σ. Fortunately we can avoid this complication in most cases because experiments use beams of unpolarized particles for the initial states and detectors are insensitive to the final state spins[§]. Hence we define

$$\overline{|\mathcal{M}|}^2 = \begin{pmatrix} \text{average over initial spins} \\ \text{and sum over final spins} \end{pmatrix} |\mathcal{M}|^2 \tag{9.36}$$

$$= \frac{1}{(2s_1+1)(2s_2+1)} \sum_{s_1, s_2, s'} |\langle p'_1, s'_1; \ldots; p'_m, s'_m\,|\mathsf{M}|\,p_1 s_1; p_2 s_2\rangle|^2$$

whenever the incoming or outgoing particles have spin.

Note under time-reversal and parity

$$\mathbb{P}\mathbb{T}\,\langle p'_1, s'_1; \ldots; p'_m, s'_m\,|\mathsf{M}|\,p_1, s_1\,; p_2, s_2\rangle$$

$$= \mathbb{P}\,\langle -p_1, -s_1; -p_2, -s_2\,|\mathsf{M}|-p'_1, -s'_1; \ldots; -p'_m, -s'_m\rangle$$

$$= \langle p_1, -s_1; p_2 - s_2\,|\mathsf{M}|\,p'_1, -s'_1; \ldots; p'_m, -s'_m\rangle \tag{9.37}$$

[§]This is true most of the time and it is the only situation we will consider. However, keep in mind that for polarized beam experiments, such as those carried out sometimes at the Stanford Linear Collider this is NOT true, and a more detailed analysis must be carried out.

and if these are both symmetries of a given interaction then

$$\sum_{s_1 s_2} \sum_{s'} |\langle p_1', s_1'; \cdots; p_m', s_m' \, |\mathsf{M}| \, p_1 s_1; p_2 s_2 \rangle|^2$$

$$= \sum_{s_1 s_2} \sum_{s'} |\langle p_1, -s_1; p_2 - s_2 \, |\mathsf{M}| \, p_1', -s_1'; \cdots; p_m', -s_m' \rangle|^2 \quad (9.38)$$

since summing over a given spin from $-s$ to $+s$ is the same as summing from $+s$ to $-s$. Inserting factors of $(2s_i + 1)$ to take care of spin-averaging gives

$$\prod_{i=1}^{2} (2s_i + 1) \overline{|\mathcal{M}_{fi}|}^2 = \prod_{i=1}^{m} (2s_i' + 1) \overline{|\mathcal{M}_{if}|}^2 \quad (9.39)$$

which is the *principle of detailed balance*!

9.6.1 Pion Spin

Let's apply this to a general 2-Body scattering situation. The 2-Body cross-section is

$$\sigma (AB \to CD) = \left(\frac{\hbar c}{8\pi} \right)^2 \frac{\mathcal{S}_{CD}}{E_0^2} \frac{|\vec{p}\,'|}{|\vec{p}|} \int d\Omega \overline{|\mathcal{M}_{fi}|}^2 \quad (9.40)$$

in the CM frame, with $E_A + E_B = E_0$. For the reverse reaction carried out at the same total energy we find

$$\sigma (CD \to AB) = \left(\frac{\hbar c}{8\pi} \right)^2 \frac{\mathcal{S}_{AB}}{E_0^2} \frac{|\vec{p}|}{|\vec{p}\,'|} \int d\Omega \overline{|\mathcal{M}_{if}|}^2 \quad (9.41)$$

and so detailed balance implies the relation

$$\frac{\sigma (AB \to CD)}{\sigma (CD \to AB)} = \frac{(2s_C + 1)(2s_D + 1)}{(2s_A + 1)(2s_B + 1)} \frac{|\vec{p}\,'|^2}{|\vec{p}|^2} \frac{\mathcal{S}_{CD}}{\mathcal{S}_{AB}} \quad (9.42)$$

which can be tested in experiments.

For example, consider the reaction

$$p + p \longrightarrow \pi^+ + \mathrm{D} \quad (9.43)$$

which is the collision of two protons forming a deuteron D and a pion. We have from the above

$$\frac{\sigma (p + p \longrightarrow \pi^+ + \mathrm{D})}{\sigma (\pi^+ + \mathrm{D} \longrightarrow p + p)} = \frac{(2s_\pi + 1)(2s_d + 1)}{(2s_p + 1)(2s_p + 1)} \frac{|\vec{p}_\pi|^2}{|\vec{p}_p|^2} \frac{1}{(1/2)} \quad (9.44)$$

where we have a symmetry factor of $\frac{1}{2}$ because the protons are identical. Independent measurements indicate that the proton spin is $\frac{1}{2}$ and the deuteron spin is 0, so for $|\vec{p}_\pi|^2 = |\vec{p}_p|^2$ ⇝ experimental choice

$$\frac{\sigma (p + p \longrightarrow \pi^+ + \mathrm{D})}{\sigma (\pi^+ + \mathrm{D} \longrightarrow p + p)} = \frac{(2s_\pi + 1)}{2} \quad (9.45)$$

Measurements by Cartwright, Clark and Durbin [102] showed that

$$\frac{\sigma\left(\pi^{+} + D \longrightarrow p + p\right)}{\sigma\left(p + p \longrightarrow \pi^{+} + D\right)} = 2.0 \pm 0.4 \Longrightarrow s_{\pi} = 0 \tag{9.46}$$

from which we empirically determine that the pion is a spin-0 particle!

As a final comment, for the weak interactions, parity and time-reversal are not symmetries, and so detailed balance does not hold in general. However, first order perturbation theory in the weak interactions indicates that $\mathcal{M}_{if} = \mathcal{M}_{fi}$, and so detailed balance will hold to lowest order in the weak coupling.

9.7 Questions

1. Show that the 2-Body decay rate is

$$\Gamma = \frac{S\,|\vec{p}|\,c}{8\pi\hbar\,(Mc)^2}\,|\mathcal{M}|^2$$

2. Show that the formula for the 2-Body cross-section in the CM frame is

$$\left(\frac{d\sigma}{d\Omega}\right)_{\text{CM}} = \left(\frac{\hbar c}{8\pi}\right)^2 \frac{S\,|\mathcal{M}|^2\,|\vec{p}_f|}{(E_1 + E_2)^2\,|\vec{p}_i|}$$

3. Show that the resonance function

$$R\left(\omega\right) = \frac{\Omega^2\Gamma^2}{\left(\omega^2 - \Omega^2\right)^2 + \Omega^2\Gamma^2}$$

    can be approximated by the Breit-Wigner function near $\omega = \Omega$.

4. Consider a non-relativistic particle of mass M scattering off of a fixed repulsive potential $V = \frac{k}{r^2}$, where k is constant.

    (a) Find the scattering angle θ as a function of the impact parameter b.

    (b) Find the differential cross-section for this process.

    (c) Compute the total cross-section.

5. (a) Show that

$$\sqrt{\left(p_1 \cdot p_2\right)^2 - p_1^2 p_2^2} = \left(E_1 + E_2\right)|\vec{p}_1| = E_1 E_2 |\vec{v}_2 - \vec{v}_1|$$

    in the CM frame.

    (b) Compute this quantity in the lab frame.

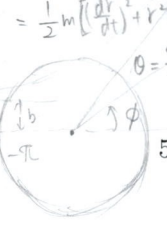

6. For elastic scattering $A + B \rightarrow A + B$ in the lab frame, show that the differential cross-section is

$$\left(\frac{d\sigma}{d\Omega}\right)_{\text{LAB}} = \left(\frac{\hbar}{8\pi}\right)^2 \frac{\mathcal{S}\,|\mathcal{M}|^2\,|\vec{p}_1'|^2}{M_2\,|\vec{p}_1|\,[|\vec{p}_1'|\,(E_1 + M_2 c^2) - |\vec{p}_1|\,E_1'\cos\theta]}$$

and simplify it in the limit that the recoil energy of the target B is negligible, i.e. $m_B c^2 \gg E_A$.

7. Find the expression for the differential cross-section in the lab frame when the final state particles are massless

8. For elastic scattering with the incident particle massless, find the expression for the differential cross-section in the lab frame.

9. Find the general expression for the decay rate of a massive particle into two massless ones.

$$\frac{d\Gamma}{d\Omega} = \frac{\hbar^2 S}{4\sqrt{(p_1 \cdot p_2)^2 - p_1^2 p_2^2}} \left[\int \frac{d^3 p_1'}{2E_1'(2\pi)^3} \cdot \int \frac{d^3 p_2'}{2E_2'(2\pi)^3}\right] |M|^2 \, (2\pi)^4 \, \delta^3\left(\Sigma \vec{p}_i' - \vec{p}_i - p_2\right)$$

$$\delta\left(E_1' + E_2' - E_1 - E_2\right)$$

$$d^3 p_1' = \int d|\vec{p}_1'| \, |\vec{p}_1'|^2 \, d\Omega$$
$$= \int d|\vec{p}_1'| \, |\vec{p}_1'|^2 \sin\theta \, d\theta \, d\phi$$

$$\vec{p}_2' = \vec{p}_1 - \vec{p}_1'$$

$$|\vec{p}_2'|^2 = |\vec{p}_1|^2 + |\vec{p}_1'|^2 - 2|\vec{p}_1||\vec{p}_1'| \cos\theta$$

$$\int_{-\infty}^{+\infty} dx \, \delta(f(x)) = \int_{-\infty}^{+\infty} dy \, \frac{\delta(y)}{f'(x)} = \sum_{\text{roots of } f} \frac{1}{f'(x)}$$

10

A Toy Theory

We have now established all of the necessary background for confronting theory with experiment. The key task from a theoretical viewpoint is to compute the matrix element $\mathcal{M}$ for a given process. However, before trying to evaluate "real-world" processes such as those that occur in Quantum Electrodynamics (QED) or Quantum Chromodynamics (QCD), it's useful to look at what physicists call a "toy"-theory: a theory that is not necessarily intended to model the real world but whose purpose is to illustrate the essential features of the method.

We are interested in computing the matrix element, i.e., the quantity

$$|\mathcal{M}| = |\langle f\,|\mathsf{M}|\,i\rangle| = |\langle p_1', s_1'; \cdots; p_m', s_m'\,|\mathsf{M}|\,p_1, s_1; p_2, s_2\rangle| = |\mathcal{M}_{fi}| \qquad (10.1)$$

for a given theory. The key method for doing this is to use a set of diagrams called Feynman diagrams*. Each diagram corresponds to a set of mathematical rules (called Feynman rules) that are in turn used to compute the matrix elements of interest. This is the method that we want to illustrate with our toy theory.

In order to properly understand where the rules come from we need to make use of more advanced methods from quantum field theory – an extension of quantum mechanics that allows for the creation and annihilation of particles in physical processes. This requires a rather formidable mathematical background that is beyond the scope of this book. Fortunately we don't need to have all of this background in order to actually employ the rules to compute matrix elements. So the approach that I will take here is to show you what the rules are and how to use them. If you want to know for a given theory how to derive the rules (in other words find out why the rules are what they are) you will have to study quantum field theory [106].

We can see from the above that in general a matrix element depends on both the momenta and spins of the initial and final states. Spin is an essential feature of the real world (all matter has spin-1/2, and photons have spin-1) but unfortunately its inclusion leads to a fair amount of messy algebra when calculating cross sections, decay rates, etc. Fortunately spin has nothing to do with how to calculate a Feynman diagram as such. Hence in our first attempt

---

*These were developed by Richard Feynmann in 1949 [103] and were concurrently derived from quantum field theory by Freeman Dyson [104, 105].

to learn the rules we would like to get rid of the complications introduced by spin.

This means that our toy theory will contain only spin-0 (or scalar) particles. In order to calculate $\mathcal{M}$ we will need to know what the Feynman rules are for such particles. These rules are of two types – rules (and conventions) that are valid for any physical theory, and then rules that are specific to our toy theory.

10.1 Feynman Rules

Let's first look at the rules that hold for any physical theory of spinless particles. For simplicity I will set the speed of light c and Planck's constant $\hbar$ equal to one, restoring them in the final expressions for the cross-sections and decay rates.

1. **NOTATION.** Label the incoming (outgoing) four-momenta by $p_1, p_2, \ldots, p_n$ $(p'_1, p'_2, \ldots, p'_m)$ and label the internal four-momenta $q_1, q_2, \ldots, q_j$, as shown in figure 10.1

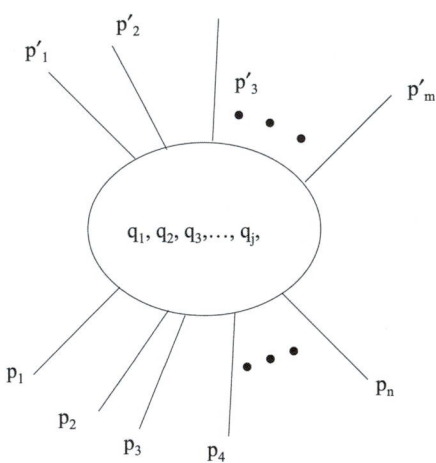

FIGURE 10.1
The general form of a Feynman diagram in a theory of spinless particles.

Although fig. 10.1 is a diagram in momentum space[†], you can think of time as flowing from bottom to top and of space as being in the horizontal direction. In diagram 10.1 we have n different (spin-0) particles coming in from distant regions, idealized as free plane-wave states in the distant past from where they are produced by some source. They then collide together, interacting in a region that is symbolized by the "blob" in the middle, which refers to all possible processes that can cause the initial state to become the final state. We will see that these interactions can be understood in terms of particle exchange – in fig. 10.1 we see that there j particles exchanged during the interaction. After all interactions have taken place there are m particles produced that fly off to distant regions, again idealized as plane-wave states, where they may be detected. Note that n, m and j are not in general equal.

These processes are dictated by the laws of physics; the rules encode these laws into the diagram. The labeling (p's for external particles, q's for internal ones) is a convention – you can of course choose any labels you want for the momenta – but in first learning how to use Feynman diagrams it is helpful to stick to a common set of conventions. All physical theories have diagrams of this basic form.

2. **INTERNAL LINES.** As noted above, inside the blob there will be various particles interacting with one another. The trajectory of each particle is represented by a line. In our toy theory with only scalar particles, each internal line contributes a factor as follows:

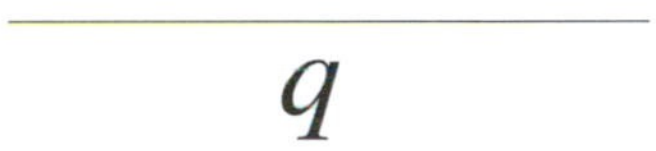

$$\frac{}{q} = \frac{i}{q^2 - m^2}$$

FIGURE 10.2
An internal line in ABB theory.

where m is the mass of the particle of momentum q. The factor $\frac{i}{q^2-m^2}$ is called the *propagator*.

You might think that the propagator should be infinite – after all, isn't the square of the 4-momentum of a particle equal to the square of its mass, as in eq. (2.34)? However, for internal particles, which are called *virtual particles*, $q^2 \neq m^2$. Only for free particles in plane-wave states does $q^2 = m^2$. Particle physicists say that a virtual particle is not on its mass shell, which

---

[†]This is because it corresponds to the Fourier transformation of the amplitude in physical space, another result from quantum field theory.

is another way of saying that $q^2 \neq m^2$. Virtual particles do not obey their free equations of motion (considered in Chapter 11). Rather they are regarded as being in intermediate states (hence the term "virtual") – they are not themselves directly detectable but instead mediate physical processes that change initial states into final states.

Note that since there are two particles in our toy theory, we will have two propagators: one for particle A ($\frac{i}{q^2-m_A^2}$) and one for particle B ($\frac{i}{q^2-m_B^2}$).

Rule 2 depends on the spin of the particle, as we will see later on when we look at realistic physical theories. All spin-0 particles contribute this factor to any Feynman diagram.

3. CONSERVATION OF ENERGY AND MOMENTUM In order for interactions to take place we will need to define something called a vertex, which is a point at which interactions take place. We shall do this later, since the rules for a vertex are specific to a given theory. However, regardless of the theory there is a general rule stating that for each vertex, insert a delta function factor of the form

$$(2\pi)^4 \delta^{(4)} (k_1 + k_2 + k_3 + \cdots + k_N)$$

where the k's are the four-momenta coming into the vertex (i.e. each k^μ will be either a q^μ, a p^μ, or a p'^μ). If the momentum leads outward, then k^μ is *minus* the four-momentum of that line. This rule imposes conservation of energy and momentum at each vertex (and hence for the diagram as a whole) because the delta function vanishes unless the sum of the incoming momenta at the vertex equals the sum of the outgoing momenta. All physical theories respect this rule.

4. INTEGRATE OVER INTERNAL MOMENTA For each internal momentum q, insert a factor

$$\frac{d^4 q}{(2\pi)^4}$$

and integrate. The idea here is that the virtual particles will – for any given process – have any momenta possible that is consistent with conservation of energy and momentum. On probabilistic grounds, all such possibilities will occur and so we have to add them all up – this is what this integration does. The rule holds for all physical theories.

5. CANCEL THE DELTA FUNCTION

After performing the integrations the result will include a factor

$$(2\pi)^4 \delta^{(4)} (p_1' + p_2' \cdots + p_m' - p_1 - p_2 - \cdots - p_n)$$

corresponding to overall energy-momentum conservation. Cancel this factor (because we have already included it in our expressions for decay rates and cross-sections), and what remains is $-i\mathcal{M}$. The rule holds for all physical theories.

After a bit of practice you can compress rules 3,4, and 5 into a single rule: *integrate over all undetermined internal momenta*. This way is faster, but in first learning the method it is easier to break this up into these 3 distinct rules. The factors of (2π) are necessary – their origin is rooted in Fourier transformations of wavefunctions of the various states[‡], and they must be included if the correct numerical answer is to be obtained. Most of them cancel out against each other, but you must keep careful track of them.

Except for rule 2, all of the above rules hold for any physical theory, and rule 2 holds for any spin-0 particle. However, we don't yet have a theory because we don't know how the particles can interact. This is where the next rule comes in. We will suppose that we have 2 spinless particles – A and B (with masses m_A and m_B respectively) – and then propose a rule for how they can interact. This kind of rule – dependent on which particles exist and how they interact – is what distinguishes one theory from another. The rule is the "vertex factor" rule. For our A, B theory we will propose the simplest rule possible [107].

6. **VERTEX FACTORS** For every interaction between particles A and B, draw a point with three lines coming out signifying one A particle and two B particles as shown in fig. 10.3, and include a factor of $-ig$ for every vertex that

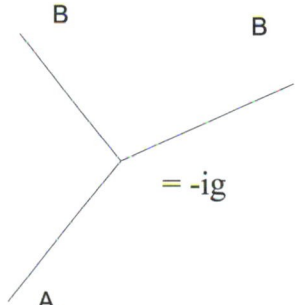

FIGURE 10.3
The vertex of ABB theory.

appears in a given diagram. Due to the choice of interaction, I will call this theory[§] ABB-theory. The quantity g is the *coupling constant* of ABB-theory:

[‡]Again, a proper derviation of this follows from the advanced methods of quantum field theory.

[§]Many particle physicists would prefer to call this a *model* rather than a *theory*, reserving the term *theory* for the more general formalism that describes interacting quantum scalar

it measures the strength of our proposed interaction.

Of course I could have proposed other interactions with still other particles. For example either of the two vertices in fig. 10.4 would be valid choices for an interaction between three scalar particles (A, B, and C). The theory described by the vertex on the right allows for only one interaction where each kind of particle meets together at a single point in spacetime. The vertex on the left corresponds to a theory having what is called a 4-point interaction: in this theory, the only interaction that can take place is one in which four particles meet at a single point in spacetime – two must be distinct and the B particle must appear twice. I could also draw other vertices, and I could use more

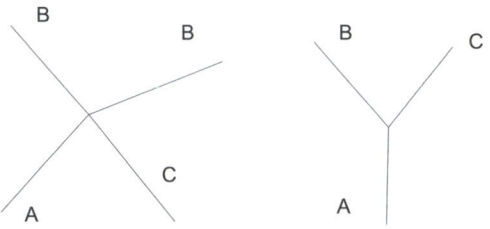

FIGURE 10.4

Other possible vertices for theories of spinless particles.

than one vertex in my theory.

So which vertices should I use? This depends on my mood, which is why we call this a "toy" theory: the interaction is something I invent for fun. Less fancifully, theorists develop toy theories to illustrate ideas and/or methods that they believe have something to do with reality. In the real world, the vertices model interactions that we hypothesize to be true in nature. Guided by experiment, in a real physical theory, we want these interactions to accurately describe the world we observe. For now we'll stick to ABB theory, that has only the single 3-point vertex in figure 10.3.

An important difference to note here between the toy theory and a real-world theory is that the coupling constant g has units of momentum. In real-world theories (like QED), the coupling constant is dimensionless. The dimensionality of g will have physical consequences, as we will see when we consider particle decay in ABB theory.

There is one more rule that we need in order to be able to start calculating. This rule is the one that tells us how to construct the diagrams from the rules

---

fields. I shall stick to the term *theory*, but you should keep this subtle distinction in mind.

and put them all together. It's called the topology rule, and holds for all physical theories.

7. **TOPOLOGY** To get all contributions for a given process, draw diagrams by joining up all internal vertex points either to the external lines or to each other by internal lines in all possible arrangments that are topologically inequivalent, consistent with rule 6. The number of ways a given diagram can be drawn is the topological weight of the diagram. The sum of all diagrams is equal to $-i\mathcal{M}$.

Given the above set of rules, there are infinitely many possible diagrams that one can draw for any physical process. So how could we possibly add them all up? The answer is hidden in rule 6: each diagram has a factor of a coupling constant. If the coupling constant is small, then diagrams with more factors of the coupling (more vertices or more interactions) are less important (are numerically smaller) than diagrams with fewer couplings. So in practice we group diagrams by the number of powers of the coupling, and truncate the sum at whatever given order of the coupling we desire. In other words, the diagrams are to be understood in a perturbative sense: the matrix element will have the general form

$$-i\mathcal{M} = -i\mathcal{M}_1 g + -i\mathcal{M}_2 g^2 + -i\mathcal{M}_3 g^3 + \cdots \tag{10.2}$$

where $\mathcal{M}_1$ is computed from a finite number of diagrams of order g, $\mathcal{M}_2$ from a finite number of order g^2, etc. In understanding the basic physics in a scattering process, it is typically sufficient to retain only the terms of leading order.

Now we're ready to calculate some processes in ABB theory.

10.2 A-Decay

The simplest diagram we can draw is given in figure 10.5. It has no internal lines and only one vertex. Applying rule 3 we get

$$-i\mathcal{M} = -ig(2\pi)^4 \delta^{(4)} \left(p'_{B_1} + p'_{B_2} - p_A \right) \tag{10.3}$$

and then discarding the overall δ-function (rule 5), we get

$$-i\mathcal{M} = -ig \Rightarrow \mathcal{M} = g \tag{10.4}$$

The two-body decay rate is, in the rest-frame of the A:

$$\Gamma = \frac{S \, |\vec{p}| \, c}{8\pi\hbar \, (Mc)^2} \, |\mathcal{M}|^2 = \frac{|\vec{p}| \, g^2}{16\pi\hbar m_A^2 c} \tag{10.5}$$

FIGURE 10.5
Decay of the A particle.

where $\vec{p}$ is the magnitude of the outgoing momentum of either of the two B-particles and $\mathcal{S} = \frac{1}{2!}$ since the final particles are indistinguishable. Momentum conservation due to the δ-function implies

$$m_A = \sqrt{|\vec{p}|^2 + m_B^2} + \sqrt{|\vec{p}|^2 + m_B^2}$$

$$\Rightarrow p = \sqrt{\frac{m_A^2}{4} - m_B^2} \tag{10.6}$$

and so we get

$$\Gamma = \frac{g^2}{32\pi\hbar c m_A}\sqrt{1 - 4\frac{m_B^2}{m_A^2}} \tag{10.7}$$

for the decay rate of the A particle.

For any given theory the masses m_A and m_B are fixed, reflecting the fact that in the real world an elementary particle cannot change its rest mass. However, we can mathematically adjust the mass to be whatever we like – in so doing we are really comparing a whole continuum of theories dependent on the choice of the particle masses. We can also do the same for the couplings. Such comparsions afford important physical insight.

For the decay process above, we see that the smaller the value of g, the smaller the decay rate, in accord with our intuition that a weaker coupling makes for a less probable interaction or a less likely decay. If $m_B > \frac{m_A}{2}$ (i.e. if the B-particle has a rest-mass more than half that of the A), then the decay rate becomes imaginary – in other words there is no decay! This corresponds to a stable A particle as we expect. But note that for large m_A, the decay rate $\Gamma \sim m_A^{-1}$: it is small, yielding a long lifetime for a heavy A particle. This leads to the counter-intuitive result that a heavier A particle (i.e. heavy relative to the B) would decay less rapidly than a lighter one! We actually find a maximum for the decay rate when $m_A = 2\sqrt{2}m_B$, and so a toy theory with this mass value would have the shortest-lived unstable A particle. We will see later on that a similar situation exists in the real world for pion decay, in which it is less probable for the pion to decay to relatively light products than relatively heavy ones.

The reason for this strange result – contrary to what we saw in Chapter 9 from general phase-space considerations – is that the coupling constant g has units of momentum (or alternatively of (mass)×(speed of light)) in the toy theory, whereas the analog of the coupling g for a real world theory is dimensionless. In the real world, it is the matrix element that provides the proper dimensionality to ensure that Γ has the correct units. This typically means that it introduces additional powers of the mass, forcing $\Gamma \sim m$ and leading to the result that in the real world heavy particles typically decay with more rapidity than light ones. However, there are exceptions to this as the toy theory illustrates, and we will see in Chapter 21 that in the real world the pion is one of them.

10.3 Scattering in the Toy Theory

Let's compute the cross section for the elastic scattering process $B + B \longrightarrow A + A$. The diagrams are (to lowest order in g – rule 7) given in figure 10.6, where we note that we combine the diagrams in all possible ways consistent with rule 6. The diagram on the left gives

$$-i\mathcal{M}_I = \tag{10.8}$$

$$\underbrace{(-ig)^2}_{\text{rule 6}} \underbrace{\int \frac{d^4q}{(2\pi)^4}}_{\text{rule 4}} \underbrace{(2\pi)^4\delta^{(4)}\left(p_1' + q - p_1\right)}_{\text{rule 3}} \underbrace{(2\pi)^4\delta^{(4)}\left(p_2' - q - p_2\right)}_{\text{rule 3}} \underbrace{\frac{i}{q^2 - m_B^2}}_{\text{rule 2}}$$

and the one on the right gives

$$-i\mathcal{M}_{II} = (-ig)^2 \int \frac{d^4q}{(2\pi)^4}(2\pi)^4\delta^{(4)}\left(p_2' + q - p_1\right)$$

$$\times (2\pi)^4\delta^{(4)}\left(p_1' - q - p_2\right)\frac{i}{q^2 - m_B^2} \tag{10.9}$$

The integrations are easy, since[¶]

$$\int \frac{d^4q}{(2\pi)^4}(2\pi)^4\delta^{(4)}\left(q - k\right)F(q) = F(k) \tag{10.10}$$

[¶]See the appendix on Dirac delta functions if you are rusty on this.

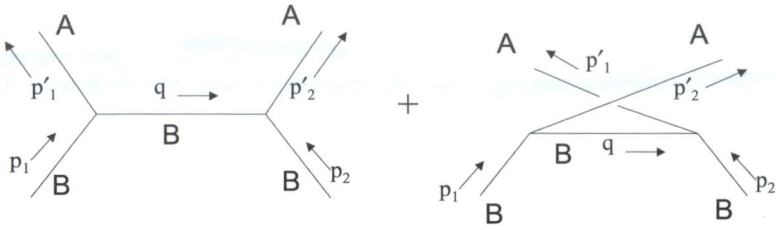

FIGURE 10.6
Lowest order diagrams for $B + B \longrightarrow A + A$.

Integrating over q we find that the first delta-function gives for $\mathcal{M}_I$

$$\mathcal{M}_I = (2\pi)^4 \delta^{(4)} \left(p'_1 + p'_2 - p_1 - p_2 \right) \frac{g^2}{\left(p'_1 - p_1 \right)^2 - m_B^2} \tag{10.11}$$

and for $\mathcal{M}_{II}$

$$\mathcal{M}_{II} = (2\pi)^4 \delta^{(4)} \left(p'_1 + p'_2 - p_1 - p_2 \right) \frac{g^2}{\left(p'_2 - p_1 \right)^2 - m_B^2} \tag{10.12}$$

Adding them together and using rule 5 gives

$$\mathcal{M} = g^2 \left[\frac{1}{\left(p'_1 - p_1 \right)^2 - m_B^2} + \frac{1}{\left(p'_2 - p_1 \right)^2 - m_B^2} \right] \tag{10.13}$$

for the lowest-order matrix element which describes this process.

Of course we could have integrated over the second delta-function in eqs. (10.9) and (10.10). Had we done so, the intermediate steps would have differed, but the end result (10.13) would have been the same.

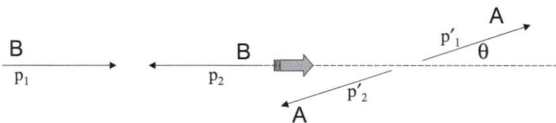

FIGURE 10.7
$B + B \longrightarrow A + A$ scattering picture.

Now we can take $|\mathcal{M}|^2$ and put this into the 2-Body cross-section we computed last time. In the CM system

$$E_1 = E_2 = E, \quad \vec{p}_1 = -\vec{p}_2 = \vec{p}$$
$$E_1' = E_2' = E', \quad \vec{p}_1' = -\vec{p}_2' = \vec{p}\,'$$

and energy conservation implies that $E = E'$. For simplicity, let's take $m_A = 0$ and $m_B = m$. Then

$$(p_1' - p_1)^2 - m_B^2 = (p_1')^2 + (p_1)^2 - 2p_1' \cdot p_1 - m^2$$
$$= 0 + m^2 - 2\left(E'E - \vec{p} \cdot \vec{p}\,' \cos\theta\right) - m^2$$
$$= -2E\left(E - |\vec{p}| \cos\theta\right) \qquad (10.14)$$

and by similar arithmetic

$$(p_2' - p_1)^2 - m_B^2 = -2E\left(E + |\vec{p}| \cos\theta\right) \qquad (10.15)$$

Putting this into the formula for $\frac{d\sigma}{d\Omega}$ in the CM system gives

$$\frac{d\sigma}{d\Omega} = \left(\frac{\hbar c}{8\pi}\right)^2 \frac{S\,|\mathcal{M}|^2}{(E_1 + E_2)^2} \frac{|\vec{p}\,'|}{|\vec{p}|}$$
$$= \left(\frac{\hbar c}{8\pi}\right)^2 \frac{(1/2)E'}{4\,|\vec{p}|\,E^2} \left|g^2 \left[-\frac{1}{2E\left(E - |\vec{p}| \cos\theta\right)} - \frac{1}{2E\left(E + |\vec{p}| \cos\theta\right)}\right]\right|^2$$
$$= \frac{1}{2}\left(\frac{\hbar c g^2}{16\pi}\right)^2 \frac{c^3}{E\,|\vec{p}|\left(E^2 - |\vec{p}c|^2 \cos^2\theta\right)^2} \qquad (10.16)$$

where $S = 1/2$ due to identical A's in the final state and I have inserted the correct factors of c in the last line.

Let's pause to note some things about this formula. At high energies $|\vec{p}c| \simeq E$, and $\frac{d\sigma}{d\Omega} \to \frac{(\hbar c)^2(cg)^4}{E^6(1-\cos^2\theta)^2} \sim \frac{(\hbar c)^2(cg)^4}{E^6 \sin^4\theta/2}$. The cross-section falls off rapidly with increasing energy (high-energy particles are less likely to interact) and with increasing angle (it's more likely to scatter in a forward (or backward) direction than in a perpendicular one). Note also that the units are correct: gc has units of energy and $\hbar c$ has units of (energy)×(length), so the cross-section does indeed have units of area.

For a massless A particle, ABB theory is a toy version of QED, with A playing the role of a spinless photon, and B playing the role of a spinless electron. The scattering process in this example is pair-annihilation, where two B particles destroy each other and become a pair of "photons." We will encounter the real-world version of this process a bit later on, and we will see that the basic physics described above will be preserved.

10.4   Higher-Order Diagrams

All we've looked at so far are "tree" diagrams that are lowest-order in g. For example each diagram in figure 10.6 is of order g^2. Higher order diagrams

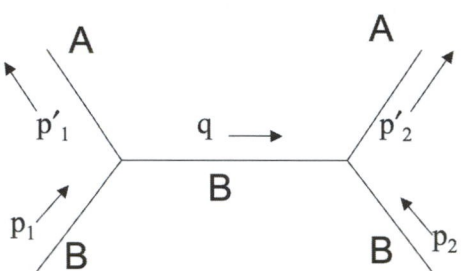

FIGURE 10.8
A lowest order diagram for BB annihilation.

will contribute to this process; for example the diagrams in figure 10.9 all

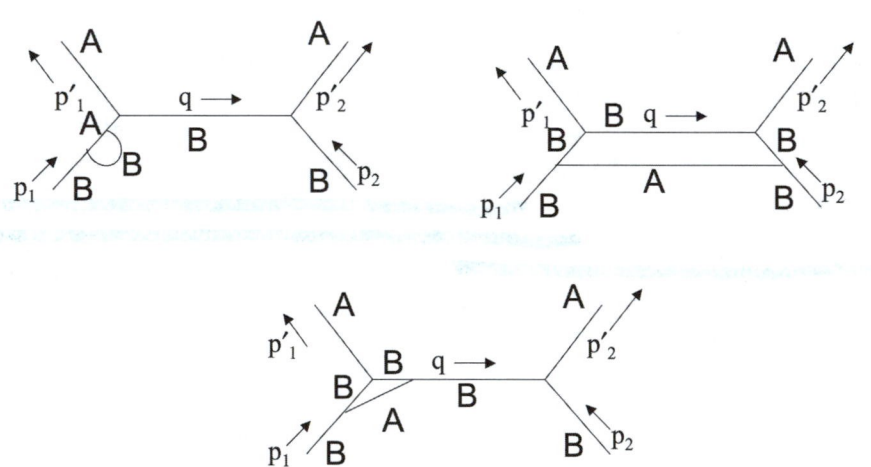

FIGURE 10.9
Some higher order diagrams for BB annihilation.

contribute to order g^4. I've drawn the three distinct diagrams in which the added line starts on the lower-left B-line and connects to another line according to the rules. There are another three for the lower-right B-line, but we've already counted one of these (the diagram on the upper-right in 10.9), so we get two more. Each of the final A-lines can have a B-bubble, and there is one more diagram in which an A-particle leaves and returns to the internal B-line (an A-bubble). So in all there are eight diagrams, plus another eight for the crossed version (where the final state A's have switched places – the diagram at the right in figure 10.6). Note that disconnected diagrams, those that can be separated into two distinct parts without cutting any line, don't count.

So there are 16 diagrams to compute at the next order. Let's compute the diagram in figure 10.10 for which the internal B line has a bubble. There

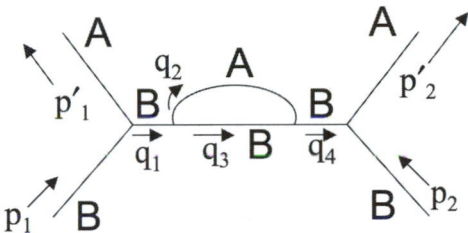

FIGURE 10.10
An A-bubble diagram.

are four internal lines, and so rule 4 gives four integrations. There are four propagators (rule 2) and four delta functions (rule 3). Hence we get

$$(-ig)^4 \int \frac{d^4 q_1}{(2\pi)^4} \int \frac{d^4 q_2}{(2\pi)^4} \int \frac{d^4 q_3}{(2\pi)^4} \int \frac{d^4 q_4}{(2\pi)^4} \frac{i}{q_1^2 - m_B^2} \frac{i}{q_2^2 - m_A^2}$$

$$\times \frac{i}{q_3^2 - m_B^2} \frac{i}{q_4^2 - m_B^2} (2\pi)^4 \delta^{(4)} \left(p_1' + q_1 - p_1\right) (2\pi)^4 \delta^{(4)} \left(q_2 + q_3 - q_1\right)$$

$$\times (2\pi)^4 \delta^{(4)} \left(q_4 - q_2 - q_3\right) (2\pi)^4 \delta^{(4)} \left(p_2' - q_4 - p_2\right) \qquad (10.17)$$

The delta-functions make the integrations fairly easy to do. Integrating over q_1 means setting $q_1 = p_1 - p_1'$, and integrating over q_4 means setting $q_4 = p_2' - p_2$. This gives

$$(-ig)^4 \frac{1}{\left[(p_1' - p_1)^2 - m_B^2\right]} \frac{1}{\left[(p_2' - p_2)^2 - m_B^2\right]} \int \frac{d^4 q_2}{(2\pi)^4} \frac{i}{q_2^2 - m_A^2}$$

$$\times \int \frac{d^4 q_3}{(2\pi)^4} \frac{i}{q_3^2 - m_B^2} (2\pi)^4 \delta^{(4)} (q_2 + q_3 + p_1' - p_1)$$
$$\times (2\pi)^4 \delta^{(4)} (p_2' - p_2 - q_2 - q_3) \tag{10.18}$$

Integrating over q_2 gives

$$(2\pi)^4 \delta^{(4)} (p_1' + p_2' - p_1 - p_2) \frac{g^4}{\left[(p_1' - p_1)^2 - m_B^2 \right]^2}$$
$$\times \int \frac{d^4 q_3}{(2\pi)^4} \frac{1}{(p_1' + q_3 - p_1)^2 - m_A^2} \frac{1}{q_3^2 - m_B^2} \tag{10.19}$$

where I have used the delta-function to set $p_1' - p_1 = p_2 - p_2'$ to simplify things. Using rule 5 gives (dropping the redundant "3" index on q_3):

$$-i\mathcal{M} = \frac{g^4}{\left((p_1' - p_1)^2 - m_B^2 \right)^2} \int \frac{d^4 q}{(2\pi)^4} \frac{1}{(p_1' + q - p_1)^2 - m_A^2} \frac{1}{q^2 - m_B^2} + \cdots \tag{10.20}$$

where the dots refer to contributions from other diagrams.

We see that we have one integration left to do. At this point we run into big trouble: the remaining integral is infinite. This is easy to see: there are four integrations to do but only four powers of q in the denominator. For large q we get

$$\int \frac{d^4 q}{(q^2)^2} = [\text{angular integration}] \times \int_0^\infty \frac{dq}{q} \sim \int_0^\infty \frac{dq}{q} \sim \ln(q)|^\infty = \infty \tag{10.21}$$

and we see that the integral diverges logarithmically.

This situation is endemic to all quantum field theories: matrix elements are power series in small couplings (here g) with infinite coefficients. This was first realized in 1930 and remained an unsolved problem until Tomonaga (in 1946) [108] and Feynman and Schwinger (in 1947) [103, 109] proposed a solution called *renormalization*.

How does renormalization work? Let's parametrize the infinity by "cutting off" the integral for large q, i.e., let's take

$$\int d^4 q \sim -i \int q^3 dq \longrightarrow -i \int_0^\Lambda q^3 dq$$

where Λ is some large momentum and the factor of $-i$ arises from analytic continuation$^\|$. Here q is a radial type of variable in a 4-dimensional space;

$^\|$This is a technical point – it is easiest to do the integral if we set $q_0 \to -iq_4$ and then used methods from complex analysis to calculate the result. This procedure is called Wick rotation.

there will also be an angular integration, but it will just multiply the integral by a constant factor of $2\pi^2$. The rest of the integral gives

$$\mathcal{M} = \left(\frac{g}{2\pi}\right)^4 \frac{2\pi^2}{\left[(p_1' - p_1)^2 - m_B^2\right]^2} \left[\ln\left(\frac{\Lambda}{m^2}\right) + \cdots\right] \qquad (10.22)$$

where for simplicity I've taken $m_B = m_A = m$. Of course if $\Lambda \to \infty$ this is still infinite.

But now a miracle happens: by adding this to the first diagram of $\mathcal{O}(g)$ in the $B + B \longrightarrow A + A$ process we computed above, we find that this $\ln(\Lambda)$ term corrects the mass:

$$\mathcal{M} = \frac{g^2}{(p_1' - p_1)^2 - m_B^2} + \left(\frac{g^2}{2\pi}\right)^2 \frac{1}{\left[(p_1' - p_1)^2 - m_B^2\right]^2} \left[\ln\left(\frac{\Lambda}{m^2}\right) + \cdots\right]$$

$$\simeq \frac{g^2}{(p_1' - p_1)^2 - \left[m_B^2 + \left(\frac{g}{2\pi}\right)^2 \ln\left(\frac{\Lambda}{m^2}\right)\right]} + \cdots + \mathcal{O}(g^6) \qquad (10.23)$$

where the dots refer to finite terms. The second diagram in $B + B \longrightarrow A + A$ is similarly corrected by the crossed-version of the A-bubble diagram. So we interpret

$$m_B^{\text{PHYSICAL}} = m_B + \delta m_B$$

$$= \sqrt{m_B^2 + \left(\frac{g}{2\pi}\right)^2 \ln\left(\frac{\Lambda}{m^2}\right)}$$

$$\simeq m_B + \frac{1}{2m_B}\left(\frac{g}{2\pi}\right)^2 \ln\left(\frac{\Lambda}{m^2}\right) \qquad (10.24)$$

Other diagrams with infinite contributions *all* either correct the masses of the particles or the coupling g. So when we add *all* such infinities together we find

$$m_{\text{PHYSICAL}} = m + \delta m \qquad g_{\text{PHYSICAL}} = g + \delta g \qquad (10.25)$$

The quantities δm and δg are infinite as $\Lambda \to \infty$, but that's okay: we can't measure them anyway! All we *can* measure are m_{PHYSICAL} (for A and B) and g_{PHYSICAL}. These are finite by definition. So we can take account of the infinities by using m_{PHYSICAL} and g_{PHYSICAL} in the diagrams instead of the original m's and g we started with (which themselves are assumed infinite to cancel the infinities in δm and δg).

Of course this trickery doesn't come without a price. The price we pay is that the masses and coupling in our theory can *never* be predicted. They must be input from experiment. Perhaps this isn't too bad a price to pay: there are only 2 masses and 1 coupling in ABB-theory, so all we need are 3 inputs. A *renormalizable* theory is one in which all infinites can be absorbed by these kinds of redefinitions of a finite number of parameters. *ABB* theory is an example of a renormalizable theory. QED and QCD – indeed all gauge theories – have also been shown to be renormalizable. Renormalizable theories are very important since they have predictive power – they yield answers for decay rates and cross-sections that do not depend upon the cutoff.

Non-renormalizable theories do not have this feature – they have an infinite number of infinities, and so yield meaningless results. It would be nice if we could just ignore these kinds of theories, but we can't – it turns out that our best theory of gravity – general relativity – is a non-renormalizable theory. This is one of the major problems of quantum gravity – the methods that yield predictive quantum theories for the non-gravitational forces of nature (described by gauge theories like QED or QCD) yield meaningless results when applied to gravity.

What do the finite contributions do? They also correct m_A, m_B, and g by finite calculable amounts that are functions of the four-momenta of the external particles. This means the effective masses and couplings depend on these four momenta, i.e. on the energies of the particles involved. We call them "running" masses and "running" couplings. This dependence is slight, but it does have observable consequences in both QED (in the form of the Lamb shift) and in QCD (in the form of asymptotic freedom), as we shall see in subsequent chapters.

10.5   Appendix: n-Dimensional Integration

It is common in particle physics to compute integrals over spaces of different dimensionality. Most of the time these integrals come from loops that appear in Feynman diagrams. These integrals are generalizations of the switch from Cartesian coordinates to polar coordinates in two dimensions, or from Cartesian coordinates to spherical coordinates in three dimensions. In general computing the integral involves transforming from Cartesian coordinates in n dimensions to generalized spherical coordinates in n dimensions.

Recall that in two dimensions the transformation is

$$\left. \begin{array}{l} x^1 = r\cos\theta \\ x^2 = r\sin\theta \end{array} \right\} \Rightarrow d^2x = dx^1 dx^2 = r\,dr\,d\theta \tag{10.26}$$

where I have written $(x, y) = (x^1, x^2)$. The factor of r comes from computing the Jacobian of the transformation:

$$J = \left| \det \begin{bmatrix} \frac{\partial x^1}{\partial r} & \frac{\partial x^2}{\partial r} \\ \frac{\partial x^1}{\partial \theta} & \frac{\partial x^2}{\partial \theta} \end{bmatrix} \right| = \left| \det \begin{bmatrix} \cos\theta & \sin\theta \\ -r\sin\theta & r\cos\theta \end{bmatrix} \right| = r\cos^2\theta + r\sin^2\theta = r \tag{10.27}$$

The procedure applies to any number of dimensions. We write $(x^1, x^2, \ldots, x^n)$ for the Cartesian coordinates and $(r, \theta^1, \ldots, \theta^{n-1})$ for the angular coordinates. The transformation between them is

$$
\begin{aligned}
x^1 &= r \cos \theta^1 \\
x^2 &= r \sin \theta^1 \cos \theta^2 \\
&\vdots \\
x^{n-1} &= r \sin \theta^1 \sin \theta^2 \cdots \cos \theta^{n-1} \\
x^n &= r \sin \theta^1 \sin \theta^2 \cdots \sin \theta^{n-1}
\end{aligned}
\tag{10.28}
$$

and every angle has a range from 0 to π, except for θ^{n-1} which has a range from 0 to 2π. The Jacobian is

$$
\det \begin{bmatrix}
\frac{\partial x^1}{\partial r} & \frac{\partial x^2}{\partial r} & \cdots & \frac{\partial x^n}{\partial r} \\
\frac{\partial x^1}{\partial \theta^1} & \frac{\partial x^2}{\partial \theta^1} & \cdots & \frac{\partial x^n}{\partial \theta^1} \\
\vdots & & \ddots & \vdots \\
\frac{\partial x^1}{\partial \theta^{n-1}} & \cdots & \cdots & \frac{\partial x^n}{\partial \theta^{n-1}}
\end{bmatrix}
= r^{n-1} \left(\sin \theta^1 \right)^{n-2} \left(\sin \theta^2 \right)^{n-3} \cdots \sin \theta^{n-1}
\tag{10.29}
$$

and can be obtained by induction. So we have

$$
\begin{aligned}
d^n x &= dx^1 dx^2 \cdots dx^n \\
&= r^{n-1} \left(\sin \theta^1 \right)^{n-2} \left(\sin \theta^2 \right)^{n-3} \cdots \left(\sin \theta^{n-2} \right) dr d\theta^1 \cdots d\theta^{n-1} \\
&= r^{n-1} dr d\Omega^{n-1}
\end{aligned}
\tag{10.30}
$$

where $d\Omega^{n-1}$ is shorthand notation for the angular part of the integral.

For example, the three dimensional volume element is

$$
d^3 x = dx^1 dx^2 dx^3 = r^2 \left(\sin \theta^1 \right) dr d\theta^1 d\theta^2 = r^2 \sin \theta d\theta d\phi
\tag{10.31}
$$

where I have written the more familiar notation $\theta^1 = \theta$ and $\theta^2 = \phi$ in the last term.

How can we do the angular integration? There is a trick to this that makes use of Gaussian integrals and the Gamma function. Recall that a Gaussian integral gives

$$
\int_{-\infty}^{\infty} e^{-z^2} dz = \sqrt{\pi}
\tag{10.32}
$$

and that the Gamma function (the factorial function) is defined as

$$
\Gamma(n) = \int_0^{\infty} z^{n-1} e^{-z} dz
\tag{10.33}
$$

Let's consider the integral

$$
\int \int_0^{\infty} e^{-r^2} r^{n-1} dr d\Omega^{n-1} = \int_0^{\infty} e^{-z} \left(\sqrt{z} \right)^{n-1} \frac{dz}{2\sqrt{z}} \int d\Omega^{n-1}
$$

$$= \frac{1}{2} \int_0^\infty e^{-z} z^{n/2-1} dz \int d\Omega^{n-1}$$

$$= \frac{1}{2}\Gamma\left(\frac{n}{2}\right) \int d\Omega^{n-1} \tag{10.34}$$

wher I have set $r = \sqrt{z}$ in the first line on the right-hand side; the final line easily follows from the definition of $\Gamma(n)$. Since $r^2 = (x^1)^2 + (x^2)^2 + \cdots + (x^n)^2$, we have $e^{-r^2} = e^{-\left[(x^1)^2 + (x^2)^2 + \cdots + (x^n)^2\right]} = e^{-(x^1)^2} e^{-(x^2)^2} \cdots e^{-(x^n)^2}$ — the exponential of the sum is the product of the exponentials. Hence we can write

$$\int\int_0^\infty e^{-r^2} r^{n-1} dr d\Omega^{n-1} = \int_{-\infty}^\infty dx^1 \cdots \int_{-\infty}^\infty dx^n e^{-(x^1)^2} e^{-(x^2)^2} \cdots e^{-(x^n)^2}$$

$$= \int\int_0^\infty e^{-r^2} dx^1 dx^2 \cdots dx^n = \int_{-\infty}^\infty dx^1 e^{-(x^1)^2} \int_{-\infty}^\infty dx^2 e^{-(x^2)^2} \cdots \int_{-\infty}^\infty dx^n e^{-(x^n)^2}$$

(transform back to "Cartesian"). $= (\sqrt{\pi})^n \tag{10.35}$

where the last line follows since each integral gives the same factor of $\sqrt{\pi}$. We can equate the two expressions to obtain

$$\int d\Omega^{n-1} = \frac{2\pi^{\frac{n}{2}}}{\Gamma\left(\frac{n}{2}\right)} \tag{10.36}$$

for the $(n-1)$-dimensional angular integral. For $n = 2$ this gives $\int d\theta = \frac{2\pi}{\Gamma(1)} = 2\pi$, which is the circumference of a circle (the area of a 1-dimensional "sphere") and for $n = 3$ this gives $\int d\theta = \frac{2\pi^{3/2}}{\Gamma(\frac{3}{2})} = \frac{4}{3}\pi$, which is the area of a sphere of unit radius.

Our 4-dimensional momentum integrals are typically of the form

$$\prod_A \int d^4 q F_A (\underbrace{(q_0 - p_{0A})^2 - |\vec{q} - \vec{p_A}|^2}_{\text{momentum}})$$

or in other words, a product of functions of relativistic invariants. The easiest way to do these integrals is to analytically continue the time components, so that $q_0 \to iq_4$, $p_{0A} \to ip_{4A}$, a procedure called Wick rotation. One then has a 4-dimensional integral that can be carried out, and then final answer is obtained by Wick rotating back, so that $p_{4A} \to -ip_{0A}$. There are some mathematical properties the integrand has to obey in order for this procedure to be valid, but they will be satisfied for all the integrals we consider in this book. In the simplest case we have, after Wick rotating

$$\int d^4 q \to i \int dq^1 \int dq^2 \int dq^3 \int dq^4$$

$$= i \int d\tilde{q}\tilde{q}^3 \int d\Omega^3$$

$$= i\frac{2\pi^2}{\Gamma(2)} \int d\tilde{q}\tilde{q}^3 \qquad \int d\,\tilde{q}^4 \ ?$$

$$= 2i\pi^2 \int d\tilde{q}\tilde{q}^3 \tag{10.37}$$

where $(\tilde{q})^2 = \left(q^1\right)^2 + \left(q^2\right)^2 + \left(q^3\right)^2 + \left(q^4\right)^2$ is the "radial" momentum variable.

10.6 Questions

1. Consider ABB theory in which the A is massless.

(a) Find the differential cross-section in the lab-frame for the reaction

$$B + A \longrightarrow B + A$$

What is the angular dependence of the cross-section at high energies?

(b) Find the differential cross-section in the CMS for the reaction

$$A + A \longrightarrow B + B$$

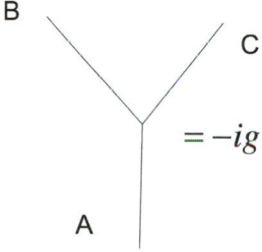

FIGURE 10.11
Vertex for ABC theory.

2. Consider a generalization of ABB theory to ABC theory, whose vertex rule is given in figure 10.11 and in which the C is massless and the A and B have equal mass m.

(a) Draw all allowed lowest-order diagrams for the reaction

$$C + A \longrightarrow C + A$$

and find the matrix element and differential cross-section in the lab-frame to this order.

(b) Draw all allowed lowest-order diagrams for the reaction

$$C + C \longrightarrow 2B$$

and find the matrix element and differential cross-section in the CMS to this order.

3. For the ABC theory of question #2:

(a) Find the differential cross-section in the lab-frame for the reaction

$$C + B \longrightarrow C + B$$

(b) Find the differential cross-section in the CMS for the reaction

$$C + C \longrightarrow A + A$$

4. (a) Consider a process in the ABC theory of question #2 that has n_A external A-lines, n_B external B-lines and n_C external C-lines. Find a simple rule that shows which processes are allowed and which are forbidden in ABC theory.

(b) Test your rule on the following processes. Draw a diagram for each allowed process.

$$B \longrightarrow C + C + C \qquad\qquad B + B \longrightarrow A + C + C$$
$$A + C \longrightarrow B + B + B \qquad\qquad A + B + C \longrightarrow A + B + C$$

5. Consider a variant of the ABC theory in question #2, in which both B and C are massless and the A has mass m.

(a) Find the differential cross-section in the lab-frame for the reaction

$$A + C \longrightarrow A + C$$

(b) Find the differential cross-section in the CMS for the reaction

$$C + C \longrightarrow B + B$$

6. Consider the process $A \to B + B$ in ABB theory. Draw all diagrams relevant for this process to order g^3.

7. Consider a theory of three spinless particles A, B and C. Each particle is its own antiparticle. The C is massless and $m_A > m_B$. The Feynman rules for the theory are in figure 10.12

(a) What are the possible parities for each particle if parity is conserved?

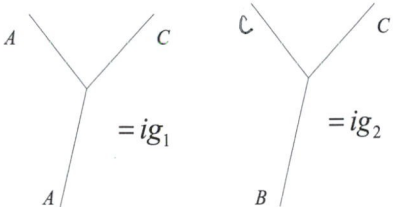

FIGURE 10.12
Vertices for question 7.

(b) Under what circumstances can $\mathbb{C} = -1$ for the C-particle?

(c) Which particles can be unstable in this theory? Under what conditions?

(d) Compute the decay rate in terms of the masses and the coupling constant(s) for any one of the unstable particles that satisfy the conditions you find in part (c) to lowest order in the coupling(s).

(e) Draw and label all diagrams which correct the process in part (c) to next-lowest order in the couplings.

8. For the theory in question #7:

(a) Consider scattering of a B particle with an A particle. What particles can appear in the final state (to lowest order in the couplings)? Draw and label the diagram(s) associated with this process.

(b) Find the matrix element for this process (i.e., for $B + A \rightarrow$ your answer) to lowest order in g_1 and g_2. Simplify your answer in the lab system, where the A particle is at rest. What does it become in the limit that the incident energy of the B particle is large?

(c) Draw two diagrams that are the next-order corrections to the process in part (b). Label only the names of the particles in each diagram.

11

Wave Equations for Elementary Particles

In order to compute matrix elements for real world theories, we need a better understanding of how to write down the incoming and outgoing states for actual physical particles. We saw in Chapter 5 that the group of spacetime symmetries – space-and-time translations, rotations, parity – have allowed us to classify particles according to some basic properties, which I have recapitulated in table 11.1.

TABLE 11.1
Particle Classification

Space and time Translations	Mass-energy (the rest mass, m)
Rotations	Intrinsic spin (spin $s = 0, \frac{\hbar}{2}, \hbar, ...$)
Parity	Intrinsic parity ($\mathbb{P} = \pm 1$)
Charge Conjugation	Intrinsic charge conjugation ($\mathbb{C} = \pm 1$)

However, we have not yet seen how relativity (i.e. boost covariance) aids in the classification. To do this, we'll need wave equations for the irreps we've developed so far (i.e. for wavefunctions of definite mass and spin) so that we can do quantum mechanics, compute matrix elements, and make predictions for decay rates and cross sections.

Clearly we need to go beyond the free-particle Schroedinger equation

$$i\hbar\frac{\partial}{\partial t}\Psi = H\Psi = -\frac{\hbar^2\nabla^2}{2m}\Psi \tag{11.1}$$

because it is not relativistic: space and time are treated differently. However, we can use it to get a hint of how to find relativistic wave equations by noting that its solution is a plane wave

$$\Psi = \Psi_0 \exp\left[-i\left(Et - \vec{p}\cdot\vec{x}\right)/\hbar\right] \qquad \text{where} \qquad E = \frac{\vec{p}\cdot\vec{p}}{2m} = \frac{|\vec{p}|^2}{2m} \tag{11.2}$$

where we interpret E as the energy and $\vec{p}$ as the momentum of the particle. The only thing non-relativistic about the solution Ψ is the relationship between E and $\vec{p}$, namely that $E = \frac{|\vec{p}|^2}{2m}$. In general, for stationary states in which $H\Psi = E\Psi$, we have

$$i\hbar\frac{\partial}{\partial t}\Psi = E\Psi \Rightarrow \Psi = \Psi(\vec{x})\exp\left[-iEt/\hbar\right] \tag{11.3}$$

in other words the time derivative of the wave function gives us a constant that we interpret as the energy. Note that the minus sign in the exponent of the plane-wave solution is crucial because it ensures that the energy of the particle is positive. Had we written $\Psi = \Psi_0\exp\left[+i\left(Et - \vec{p}\cdot\vec{x}\right)/\hbar\right]$, we would have obtained $E = -\frac{|\vec{p}|^2}{2m}$, i.e., this would correspond to a particle of negative kinetic energy, which does not solve the free-particle Schroedinger equation (11.1) above.

11.1 Klein-Gordon Equation

From now on let's set $\hbar = c = 1$ for simplicity. The preceding approach suggests a guess that a relativistic free particle has the wavefunction

$$\phi = \phi_0\exp\left[-i\left(Et - \vec{p}\cdot\vec{x}\right)\right] = \phi_0\exp\left[-ip_\mu x^\mu\right] \tag{11.4}$$

$$\text{where now we require} \quad E = \sqrt{\vec{p}\cdot\vec{p} + m^2} \quad \text{and} \quad p^\mu = (E, \vec{p})$$

and the sign is chosen in accord with the non-relativistic case. We can find the equation that ϕ obeys by differentiating it twice with respect to $\vec{x}$:

$$\nabla^2\phi = i\vec{p}\cdot\vec{\nabla}\phi = (-\vec{p}\cdot\vec{p})\phi = (m^2 - E^2)\phi = m^2\phi + \frac{\partial^2\phi}{\partial t^2}$$

$$\Longrightarrow \left(\frac{\partial^2}{\partial t^2} - \nabla^2\right)\phi + m^2\phi = 0 \qquad \text{using non-relativistic SE.} \tag{11.5}$$

which we can also write as

$$\boxed{\left(\partial_\mu\partial^\mu + m^2\right)\phi = 0} \tag{11.6}$$

an equation called the *Klein-Gordon equation.* Here $\partial_\mu \equiv \left(\frac{\partial}{\partial t}, \vec{\nabla}\right) = \frac{\partial}{\partial x^\mu}$ is the relativistic gradient, with $t = x^0$. Note that

$$\partial_\mu\partial^\mu = g^{\mu\nu}\frac{\partial}{\partial x^\mu}\frac{\partial}{\partial x^\nu} = (+1)\frac{\partial}{\partial t}\frac{\partial}{\partial t} - \frac{\partial}{\partial x}\frac{\partial}{\partial x} - \frac{\partial}{\partial y}\frac{\partial}{\partial y} - \frac{\partial}{\partial z}\frac{\partial}{\partial z} = \frac{\partial^2}{\partial t^2} - \nabla^2 \tag{11.7}$$

Since the operator ∇^2 is invariant under rotations, so is $\partial_\mu\partial^\mu$, and so the rotation properties of ϕ must be trivial – in other words ϕ is a scalar (i.e. it has spin-0). We interpret $p^\mu = (E, \vec{p})$ as the 4-momentum of the particle whose wavefunction is ϕ. Note that for small $\vec{p}$, we have $E = \sqrt{\vec{p} \cdot \vec{p} + m^2} \simeq m + \frac{|\vec{p}|^2}{2m}$ and the solution reduces (up to a phase) to the solution for the free-particle Schroedinger equation.

The Klein-Gordon equation (or KG equation) was the first attempt* at developing a relativistic version of Schroedinger's equation (and was actually first proposed by Schroedinger [110]!). It is relativistically invariant, since

$$x'^{\mu} = \Lambda^{\mu}{}_{\nu} x^{\nu} \Rightarrow \partial'_{\nu} = \frac{\partial}{\partial x'^{\mu}} = \frac{\partial x^{\mu}}{\partial x'^{\nu}} \frac{\partial}{\partial x^{\mu}} = \Lambda^{\mu}{}_{\nu} \partial_{\mu} \Rightarrow \partial'_{\mu} \partial'^{\mu} = \partial_{\mu} \partial^{\mu} \quad (11.8)$$

and so

$$\left(\partial'_{\mu} \partial'^{\mu} + m^2\right) \phi'(x') = \left(\partial_{\mu} \partial^{\mu} + m^2\right) \phi(x) = 0 \quad (11.9)$$

This is nice. However, there is an awkward snag – unlike the Schroedinger equation above, the KG equation has another solution

$$\hat{\phi} = \tilde{\phi}_0 \exp\left[+ip_{\mu} x^{\mu}\right] = \tilde{\phi}_0 \exp\left[+i\left(Et - \vec{p} \cdot \vec{x}\right)\right] = \tilde{\phi}_0 \exp\left[-i\left(-Et + \vec{p} \cdot \vec{x}\right)\right] \quad (11.10)$$

where the 4-momentum of the particle is now $p^{\mu} = (-E, -\vec{p})$. The time-derivative of $\hat{\phi}$ is easily seen to be $i\frac{\partial \hat{\phi}}{\partial t} = -E\hat{\phi} = -\sqrt{\vec{p} \cdot \vec{p} + m^2}\hat{\phi}-$ in other words the particle has negative energy! Of course this is what we expect from a 2nd order differential equation, namely that there are two independent solutions. Inserting either solution into the KG equation gives

$$\left(\partial_{\mu} \partial^{\mu} + m^2\right) \phi(x) = -\left(p^2 - m^2\right) \phi(x) = 0$$
$$\left(\partial_{\mu} \partial^{\mu} + m^2\right) \hat{\phi}(x) = -\left(p^2 - m^2\right) \hat{\phi}(x) = 0$$

This other solution was quite puzzling to theorists in the 1920s. Not only were there negative energy solutions, but the probability density associated with the equation could be shown to be positive for positive-energy states and negative for negative energy states! Even worse, the equation was not able to incorporate the newly discovered property of electron spin. It turned out that these problems were not occupying too much attention in the physics community at that time because shortly afterward Dirac came out with a different equation that incorporated relativity into quantum mechanics.

11.2 Dirac Equation

The problem with the negative energy solution appears to arise because there are two time derivatives in the KG equation (this is what gives $p^2 - m^2 = 0$,

---

*In 1926 Pauli wrote in a letter to Schroedinger that he didn't "believe that the relativisitic equation of 2nd order with the many fathers corresponds to reality." The Klein-Gordon equation has a rather interesting history [111], involving many people in its early development.

which has two solutions for the energy). If we wanted an equation that had only one time derivative, then it must have only one space derivative as well. So we might guess

$$(a^\mu \partial_\mu + m) \Psi = 0 \tag{11.11}$$

Note that if a^μ is just a 4-vector, then for a plane wave this equation gives $ia^\mu p_\mu + m = 0$, which is the wrong relation between 4-momentum and mass. This is not what we want. However, if Ψ were a multi-component wavefunction (with components ψ_b), then maybe we could get the right relationship if a^μ were a *matrix*. Let's call this matrix γ^μ – kind of like a 4-vector, but with each entry a matrix.

So let's guess (summing over repeated indices!)

$$\left(i \left(\gamma^\mu\right)_a{}^b \partial_\mu - m\delta_a{}^b\right) \psi_b = 0$$

or　　　　$(i\gamma^\mu \partial_\mu - m) \psi = 0$　　$\begin{array}{l}\text{suppressing}\\ \text{matrix indices}\end{array}$ 　$\tag{11.12}$

Note that we write $mI = m$, where I is the identity matrix (whose components are $\delta_a{}^b$). At this point we don't know the dimensionality of either the identity or of the $(\gamma^\mu)_a{}^b$ (i.e., of the range of the indices a, b), but we will soon find out what this must be.

We need to be sure that we can recover the usual relationship $p^2 - m^2 = 0$ between the 4-momentum p^μ and the mass of the multicomponent wavefunction ψ. If ψ had no spin, it would behave like a KG field, i.e. like $e^{-ip \cdot x}$. Let's guess that something similar happens here, and set

$$\psi_a = u_a(p)e^{-ip \cdot x} \tag{11.13}$$

where $u_a(p)$ is some N-component "column matrix" (recall that we don't yet know the value of N; all we know so far is that $a = 1, 2, \ldots, N$) that is a function of the 4-momentum. Our proposed equation becomes

$$0 = (i\gamma^\mu \partial_\mu - m) \psi = e^{-ip \cdot x} (\gamma^\mu p_\mu - m) u(p) \Rightarrow (\gamma^\mu p_\mu - m) u(p) = 0 \tag{11.14}$$

and so we have a constraint on the N-component object $u(p)$. Let's multiply eq. (11.14) by the matrix $(\gamma^\mu p_\mu + m)$:

$$\begin{aligned} 0 &= (\gamma^\nu p_\nu + m) (\gamma^\mu p_\mu - m) u(p) \\ &= \left((\gamma^\nu p_\nu) (\gamma^\mu p_\mu) + m (\gamma^\mu p_\mu) - m (\gamma^\nu p_\nu) - m^2\right) u(p) \\ &= \left(\gamma^\mu \gamma^\nu p_\mu p_\nu - m^2\right) u(p) \end{aligned} \tag{11.15}$$

where in the second line I have put brackets around the objects that are matrices. The set of manipulations involved in getting the last line is

$$\begin{aligned} (\gamma^\nu p_\nu) (\gamma^\mu p_\mu) &= (\gamma^\nu p_\nu)_a{}^c (\gamma^\mu p_\mu)_c{}^b = (\gamma^\nu)_a{}^c p_\nu (\gamma^\mu)_c{}^b p_\mu \\ &= (\gamma^\nu)_a{}^c (\gamma^\mu)_c{}^b p_\mu p_\nu = (\gamma^\mu)_a{}^c (\gamma^\nu)_c{}^b p_\nu p_\mu \\ &= (\gamma^\mu)_a{}^c (\gamma^\nu)_c{}^b p_\mu p_\nu \end{aligned} \tag{11.16}$$

which follows from a recognition that the γ^μ's are matrices (and so don't commute) but the p_μ's aren't (and so do commute).

This looks like an extra constraint on the column matrix $u(p)$. If we don't want this to be an extra constraint then we must require it to hold because $p^2 - m^2 = 0$ and not for any other reason. Hence we must constrain the γ-matrices to obey

$$(\gamma^\mu)_a^{\ c} (\gamma^\nu)_c^{\ b} p_\mu p_\nu - m^2 \delta_a^{\ b} = (p^2 - m^2) \delta_a^{\ b} = 0$$

$$\text{or alternatively} \quad \gamma^\mu \gamma^\nu p_\mu p_\nu = p^2 = g^{\mu\nu} p_\mu p_\nu \tag{11.17}$$

As noted previously the γ^μ's are matrices and the p_μ's are not, so we can write

$$\gamma^\mu \gamma^\nu p_\mu p_\nu = \frac{1}{2} \gamma^\mu \gamma^\nu (p_\mu p_\nu + p_\nu p_\mu)$$

$$= \frac{1}{2} \gamma^\mu \gamma^\nu p_\mu p_\nu + \frac{1}{2} \gamma^\mu \gamma^\nu p_\nu p_\mu$$

$$= \frac{1}{2} (\gamma^\mu \gamma^\nu + \gamma^\nu \gamma^\mu) p_\mu p_\nu \tag{11.18}$$

where the last line follows from relabeling the indices. In order to ensure that $\gamma^\mu \gamma^\nu p_\mu p_\nu = p^2$ we must have

$$\frac{1}{2} (\gamma^\mu \gamma^\nu + \gamma^\nu \gamma^\mu) p_\mu p_\nu = g^{\mu\nu} p_\mu p_\nu \tag{11.19}$$

or alternatively, since this must hold for any p_μ :

$$(\gamma^\mu \gamma^\nu + \gamma^\nu \gamma^\mu) = 2g^{\mu\nu} I \tag{11.20}$$

where I is the identity matrix.

The preceding relation provides a very important constraint on the γ^μ matrices. We can write it as

$$\boxed{\{\gamma^\mu, \gamma^\nu\} = 2g^{\mu\nu}} \tag{11.21}$$

where $\{\gamma^\mu, \gamma^\nu\} \equiv (\gamma^\mu \gamma^\nu + \gamma^\nu \gamma^\mu)$ is the anticommutator. For any two operators or matrices, the anticommutator is defined as

$$\{A, B\} \equiv AB + BA \tag{11.22}$$

So any set of 4 γ-matrices that obey eq. (11.21) will guarantee that $p^2 - m^2 = 0$ for the ψ wavefunction without imposing any further constraints on ψ. The simplest solution[†] to (11.21) is when $N = 4$, i.e., the γ-matrices are 4×4:

[†]How do we know that this is the simplest solution? It's done by brute force. The matrices must be at least 2-dimensional, and must obey $(\gamma^0)^2 = I = -(\gamma^1)^2 = -(\gamma^2)^2 = -(\gamma^3)^2$ as well as anticommute with each other. The three 2×2 Pauli matrices do this, but we need a 4th matrix (the γ^0 matrix) that also does this, so 2×2 ($N = 2$) is too small. You can show that there are no 3×3 solutions. The next simplest thing is the 4×4 case shown here.

$$\gamma^0 = \begin{pmatrix} I & 0 \\ 0 & -I \end{pmatrix} \qquad \gamma^i = \begin{pmatrix} 0 & \sigma^i \\ -\sigma^i & 0 \end{pmatrix} \tag{11.23}$$

where $I = \begin{pmatrix} 1 & 0 \\ 0 & 1 \end{pmatrix}$ is the 2×2 identity matrix and the σ^i are the 3 Pauli matrices. Note that we can obtain many other 4×4 solutions to (11.21) by writing $\gamma^{\mu'} = U^\dagger \gamma^\mu U$ where U is a 4×4 unitary matrix. It doesn't matter which γ-matrices you use as long as they satisfy (11.21).

So we see that since the γ-matrices are 4×4, the wavefunction ψ must be a 4-component wavefunction ψ, and that it obeys the equation

$$(i\gamma^\mu \partial_\mu - m)\,\psi = 0 \tag{11.24}$$

which we now call the *Dirac equation*, proposed by Dirac [112] in 1927 as a relativistic generalization of the Schroedinger equation. Any wavefunction that obeys this equation automatically satisfies the KG equation (but not the converse!) since all its solutions will have $p^2 - m^2 = 0$. In this sense the Dirac equation is like the "square-root" of the Klein-Gordon equation.

Note that, although ψ has four components it is *not* a 4-vector, in other words it does not transform under a Lorentz transformation the way the a 4-vector does. That's why I used Latin indices a, b to label the components of ψ.

11.3   Physical Interpretation

So what is ψ? Let's look at the solutions to the Dirac equation to see if we can find out. The block-structure of the γ-matrices suggests that it might be useful to write

$$\psi = u(p)e^{-ip \cdot x} = \begin{pmatrix} \xi \\ \chi \end{pmatrix} e^{-ip \cdot x} \tag{11.25}$$

where ξ and χ are each 2-component objects. Recalling that $p_\mu = (E, -\vec{p})$, and using $\gamma^\mu \partial_\mu = \gamma^0 \frac{\partial}{\partial t} + \vec{\gamma} \cdot \vec{\nabla}$, the Dirac equation (11.24) then can be decomposed as follows:

$$i\left(\gamma^0 \frac{\partial}{\partial t} + \vec{\gamma} \cdot \vec{\nabla}\right)\left(u(p)e^{-ip \cdot x}\right) - m\left(u(p)e^{-ip \cdot x}\right) = 0$$

$$\Rightarrow e^{-ip \cdot x}\left[i\left(-i\gamma^0 E + i\vec{\gamma} \cdot \vec{p}\right)(u(p)) - m\,(u(p))\right] = 0$$

$$\Rightarrow e^{-ip \cdot x} \begin{pmatrix} E - m & -\vec{p} \cdot \vec{\sigma} \\ \vec{p} \cdot \vec{\sigma} & -E - m \end{pmatrix} \begin{pmatrix} \xi \\ \chi \end{pmatrix} = 0 \tag{11.26}$$

This equation breaks up into two parts

$$\left.\begin{array}{l} (E - m)\,\xi - (\vec{p}\cdot\vec{\sigma})\,\chi = 0 \\ -(E + m)\,\chi + (\vec{p}\cdot\vec{\sigma})\,\xi = 0 \end{array}\right\} \Rightarrow \chi = \frac{\vec{p}\cdot\vec{\sigma}}{(E+m)}\xi = \frac{|\vec{p}|}{(E+m)}\,(\hat{p}\cdot\vec{\sigma})\,\xi \quad (11.27)$$

where I used the lower equation to determine χ in terms of ξ. Inserting this solution into the first equation just gives

$$(E - m)\,\xi - (\vec{p}\cdot\vec{\sigma})\,\chi = (E - m)\,\xi - \frac{(\vec{p}\cdot\vec{\sigma})^2}{(E+m)}\xi = \frac{E^2 - m^2 - |\vec{p}|^2}{E + m}\xi = 0$$
$$(11.28)$$

since $(\vec{p}\cdot\vec{\sigma})^2 = \vec{p}\cdot\vec{p}$, and $E^2 = m^2 + |\vec{p}|^2$. Consequently the components of ξ are not determined, and we have

$$\psi = \mathfrak{N}(p) \begin{pmatrix} \xi \\ \frac{|\vec{p}|}{(E+m)}\,(\hat{p}\cdot\vec{\sigma})\,\xi \end{pmatrix} e^{-ip\cdot x} = \mathfrak{N}(p) \begin{pmatrix} \xi \\ \sqrt{\frac{E-m}{E+m}}\,(\hat{p}\cdot\vec{\sigma})\,\xi \end{pmatrix} e^{-ip\cdot x}$$
$$(11.29)$$

where $\mathfrak{N}(p)$ is an undetermined normalization. Let's choose it to eliminate the energy term in the denominator of ψ, so that $\mathfrak{N}(p) = \psi_0\sqrt{E + m}$, giving

$$\psi = \psi_0\sqrt{2m} \begin{pmatrix} \sqrt{\frac{E+m}{2m}}\,\xi \\ \sqrt{\frac{E-m}{2m}}\,(\hat{p}\cdot\vec{\sigma})\,\xi \end{pmatrix} e^{-ip\cdot x} \quad (11.30)$$

where ψ_0 is just an arbitrary constant.

Now notice something. Suppose we are in the rest frame of ψ, where $\vec{p} = 0$. In this case $E = m$ and we have

$$\psi\,(\vec{p} = 0) = \psi_0\sqrt{2m} \begin{pmatrix} \xi \\ 0 \end{pmatrix} e^{-imt} \quad (11.31)$$

and so there are really only two independent components of ψ, namely the two components of ξ! We can normalize these so that

$$\xi^{(\uparrow)} = \begin{pmatrix} 1 \\ 0 \end{pmatrix} \quad \text{and} \quad \xi^{(\downarrow)} = \begin{pmatrix} 0 \\ 1 \end{pmatrix} \quad (11.32)$$

which means that we actually have two independent solutions to the Dirac equation

$$\psi^{(i)}(x) = u^{(i)}(x) = \psi_0\sqrt{2m} \begin{pmatrix} \sqrt{\frac{E+m}{2m}}\,\xi^{(i)} \\ \sqrt{\frac{E-m}{2m}}\,(\hat{p}\cdot\vec{\sigma})\,\xi^{(i)} \end{pmatrix} e^{-ip\cdot x} \quad (11.33)$$

corresponding to each of these possibilities, with $i = \uparrow, \downarrow$.

Dirac interpreted these solutions to correspond to the spin-up electron and the spin-down electron – something I've anticipated in the notation above.

The solution with $\xi^{(\uparrow)}$ corresponds to a plane-wave solution of a particle whose spin is parallel with respect to the $\hat{p}$-axis, and the solution with $\xi^{(\downarrow)}$ corresponds to a plane-wave solution of a particle whose spin is anti-parallel with respect to the $\hat{p}$-axis. Non-relativistically we can ignore the lower two components of ψ, as the above derivation shows. Indeed, since non-relativistically there are only two independent solutions, the wavefunction ψ must correspond to a spin-1/2 particle, because this is the only representation of the rotation group that has two components!

11.4 Antiparticles

Hence the solutions to the Dirac equation are those of a relativistic particle of spin-1/2: a relativistic spinor! Superficially it appears that the lower two components have no physical content, but Dirac soon realized this was not correct. Since the Dirac equation is a 4-component coupled first-order differential equation, it must have 4 independent solutions. The natural thing to do to find these other solutions is to write

$$\psi = v(p)e^{+ip \cdot x} = \begin{pmatrix} \xi \\ \chi \end{pmatrix} e^{+ip \cdot x} \tag{11.34}$$

and see what happens. Using the same calculational approach as before, we find

$$(\gamma^\mu p_\mu + m)\, v(p) = 0$$

$$\Rightarrow e^{+ip \cdot x} \begin{pmatrix} E+m & -\vec{p} \cdot \vec{\sigma} \\ \vec{p} \cdot \vec{\sigma} & -E+m \end{pmatrix} \begin{pmatrix} \xi \\ \chi \end{pmatrix} = 0 \tag{11.35}$$

$$\Rightarrow \left. \begin{array}{r} (E+m)\,\xi - (\vec{p} \cdot \vec{\sigma})\,\chi = 0 \\ (E-m)\,\chi - (\vec{p} \cdot \vec{\sigma})\,\xi = 0 \end{array} \right\} \Rightarrow \xi = \frac{\vec{p} \cdot \vec{\sigma}}{(E+m)} \chi = \frac{|\vec{p}|}{(E+m)} (\hat{p} \cdot \vec{\sigma})\,\chi$$

where we now write ξ in terms of χ so as to have a well-defined non-relativistic limit. Hence

$$\hat{\psi}^{(i)} = v^{(i)}(x) = \tilde{\psi}_0 \sqrt{2m} \begin{pmatrix} \sqrt{\frac{E-m}{2m}}\,(\hat{p} \cdot \vec{\sigma})\,\xi^{(i)} \\ \sqrt{\frac{E+m}{2m}}\,\xi^{(i)} \end{pmatrix} e^{ip \cdot x} \tag{11.36}$$

are two other solutions to the Dirac equation. Now the non-relativistic limit $(E \to m)$ removes the top two components instead of the bottom two!

Note that the $e^{ip \cdot x}$ factor is just like the negative-energy solution to the KG equation – and so we expect that the $v(p)$ solution is a negative-energy spin-1/2 plane wave. It looks like we have not cured the problem that we encountered in the spin-0 case after all.

Dirac thought that perhaps the $v(p)$ solutions could be eliminated from the theory[‡]. Since the solutions to the Dirac equation have spin-1/2, they should obey the Pauli principle: there can only be one particle of a given spin per state. So Dirac hypothesized that all states corresponding to the $v(p)$ solutions were filled: there was a completely filled sea of negative energy electron states!

Weird as this might sound, it offered a reason as to why we don't observe the negative energy solutions – they are all occupied, and so do not physically appear in any experiment ever carried out[§]. However, Dirac also realized that if one of these states ever received an energy larger than $2mc^2$, the electron in that state would "appear" as a positive energy particle, since it was excited out of the occupied sea. It would also be accompanied by a hole in the sea that had the same energy but opposite spin, momentum, and charge as the electron. The hole was interpreted as a positively charged electron: the *positron* [113]. Hence the relativistic generalization of a spinor led to a remarkable prediction:

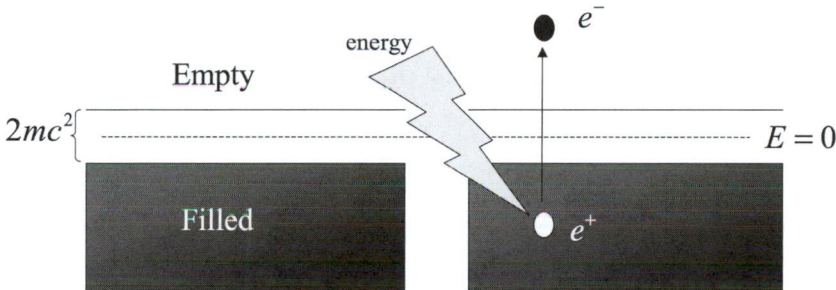

FIGURE 11.1

Schematic Diagram of the Dirac sea. Energy (represented by the lightning bolt) can excite a state in the filled sea up into the empty region. We observe this state as an electron; the hole left in the sea is observed as a positron.

the existence of antimatter! Positrons – the antiparticles of electrons – were first observed in 1933 by Anderson [84], confirming Dirac's prediction.

[‡]Dirac originally wanted to interpret these other two solutions as corresponding to the two spin states of the proton. However, because of the $e^{ip\cdot x}$ factor, these solutions appear to have negative energy (as did the $e^{ip\cdot x}$ solution to the Klein-Gordon equation) and so this can't be right. Furthermore, this would force the mass of the proton to equal that of the electron, which is also experimentally unacceptable.

[§]Note that we can't use the same reasoning for the KG equation because its solutions don't obey the Pauli principle.

Nowadays we don't think in terms of a filled sea – we regard the $e^{-ip\cdot x}$ solutions as corresponding to particles and the $e^{+ip\cdot x}$ solutions as corresponding to antiparticles. This approach means that we can also make this interpretation for the KG solutions as well. Indeed the KG equation is valid for all spin-0 particles and their antiparticles; the Dirac equation is valid for all spin-1/2 particles and their antiparticles[¶].

Antiparticles also have another interpretation. Notice under time reversal that

$$\mathbb{T}\left(\exp\left[+ip_\mu x^\mu\right]\right) = \mathbb{T}\left(\exp\left[-i\left(-Et + \vec{p}\cdot\vec{x}\right)\right]\right)$$
$$= \exp\left[-i\left(Et - \vec{p}\cdot\vec{x}\right)\right]$$
$$= \exp\left[-ip_\mu x^\mu\right] \qquad (11.37)$$

which is like a positive energy solution! This was Feynman's interpretation of the negative energy states: an antiparticle moving forward in time is equivalent to a negative energy solution moving backward in time!

The Lorentz-transformation properties of the relativistic scalar are quite simple, and we looked at them above. However, the Lorentz-transformation properties of the relativistic spinor are quite complicated. The details don't matter much for the remainder of the text, and so I've included these properties in an Appendix.

11.5   Appendix: The Lorentz Group and Its Representations

In order to see how spinors – or any other relativistic wavefunctions – transform, we'll need to look at the underlying symmetry group. This group is the Lorentz group, the group of all matrices $\Lambda^\mu{}_\nu$ which obey

$$g_{\mu\nu}\Lambda^\mu{}_\alpha\Lambda^\mu{}_\beta = g_{\alpha\beta} \qquad (11.38)$$

Since this is similar to rotations $(R^T R = I)$, let's proceed the way we did in Chapter 3. In that case we wrote $R(\theta) = \exp\left[i\vec{\theta}\cdot\vec{J}\right]$, and then used closure under small rotations to get the commutation relations for the rotation group, whose generators are the 3 $\vec{J}$'s. In a similar manner, Lorentz transformations are either of the form

$$\Lambda_{\text{rot}} \qquad (11.39)$$

[¶]This interpretation of particle and antiparticle wavefunctions properly requires the conceptual and calculational foundations of quantum field theory [97].

$$= \left\{ \begin{pmatrix} 1 & 0 & 0 & 0 \\ 0 & 1 & 0 & 0 \\ 0 & 0 & \cos\theta_x & \sin\theta_x \\ 0 & 0 & -\sin\theta_x & \cos\theta_z \end{pmatrix}, \begin{pmatrix} 1 & 0 & 0 & 0 \\ 0 & \cos\theta_y & 0 & -\sin\theta_y \\ 0 & 0 & 1 & 0 \\ 0 & \sin\theta_y & 0 & \cos\theta_y \end{pmatrix}, \begin{pmatrix} 1 & 0 & 0 & 0 \\ 0 & \cos\theta_z & \sin\theta_z & 0 \\ 0 & -\sin\theta_z & \cos\theta_z & 0 \\ 0 & 0 & 0 & 1 \end{pmatrix} \right\}$$

or, for boosts

$$\Lambda_{\text{boost}} \tag{11.40}$$

$$= \left\{ \begin{pmatrix} \gamma_x & -\beta_x\gamma_x & 0 & 0 \\ -\beta_x\gamma_x & \gamma_x & 0 & 0 \\ 0 & 0 & 1 & 0 \\ 0 & 0 & 0 & 1 \end{pmatrix}, \begin{pmatrix} \gamma_y & 0 & -\beta_y\gamma_y & 0 \\ 0 & 1 & 0 & 0 \\ -\beta_y\gamma_y & 0 & \gamma_y & 0 \\ 0 & 0 & 0 & 1 \end{pmatrix}, \begin{pmatrix} \gamma_z & 0 & 0 & -\beta_z\gamma_z \\ 0 & 1 & 0 & 0 \\ 0 & 0 & 1 & 0 \\ -\beta_z\gamma_z & 0 & 0 & \gamma_z \end{pmatrix} \right\}$$

where $\gamma_x = 1/\sqrt{1-\beta_x^2}$ etc.

We write $\vec{\theta} = \theta\hat{n}$ to denote a rotation of angle θ about an axis pointing in the direction of the unit vector $\hat{n}$, and $\vec{\beta} = \beta\hat{n}$ to denote a boost to velocity βc along the $\hat{n}$ direction. For small $|\vec{\theta}|$ and $|\vec{\beta}|$ we obtain the generators

$$\left\{ \hat{\mathcal{J}}^x, \hat{\mathcal{J}}^y, \hat{\mathcal{J}}^z \right\} \tag{11.41}$$

$$= \left\{ -i\begin{pmatrix} 0 & 0 & 0 & 0 \\ 0 & 0 & 0 & 0 \\ 0 & 0 & 0 & 1 \\ 0 & 0 & -1 & 0 \end{pmatrix}, -i\begin{pmatrix} 0 & 0 & 0 & 0 \\ 0 & 0 & 0 & -1 \\ 0 & 0 & 0 & 0 \\ 0 & 1 & 0 & 0 \end{pmatrix}, -i\begin{pmatrix} 0 & 0 & 0 & 0 \\ 0 & 0 & 1 & 0 \\ 0 & -1 & 0 & 0 \\ 0 & 0 & 0 & 0 \end{pmatrix} \right\}$$

$$\left\{ \hat{\mathcal{K}}^x, \hat{\mathcal{K}}^y, \hat{\mathcal{K}}^z \right\} \tag{11.42}$$

$$= \left\{ -i\begin{pmatrix} 0 & -1 & 0 & 0 \\ -1 & 0 & 0 & 0 \\ 0 & 0 & 0 & 0 \\ 0 & 0 & 0 & 0 \end{pmatrix}, -i\begin{pmatrix} 0 & 0 & -1 & 0 \\ 0 & 0 & 0 & 0 \\ -1 & 0 & 0 & 0 \\ 0 & 0 & 0 & 0 \end{pmatrix}, -i\begin{pmatrix} 0 & 0 & 0 & -1 \\ 0 & 0 & 0 & 0 \\ 0 & 0 & 0 & 0 \\ -1 & 0 & 0 & 0 \end{pmatrix} \right\}$$

which can easily be shown to give the commutation relations

$$\begin{aligned} \left[\hat{\mathcal{J}}^x, \hat{\mathcal{J}}^y \right] &= i\hat{\mathcal{J}}^z \quad + \text{cyclic} \\ \left[\hat{\mathcal{K}}^x, \hat{\mathcal{K}}^y \right] &= -i\hat{\mathcal{J}}^z \quad + \text{cyclic} \\ \left[\hat{\mathcal{J}}^x, \hat{\mathcal{K}}^y \right] &= i\hat{\mathcal{K}}^z \quad + \text{cyclic} \end{aligned} \tag{11.43}$$

The first part is the familiar set of commutation relations from the rotation group **SO**(3). The rest of the commutation relations indicate how the boost generators $\hat{\mathcal{K}}^i$ interact with the rest of the group.

TABLE 11.2
Rotations compared to Lorentz transformations

Spin	Irrep	Terminology	Full Transformation
$s=0$	$S^{\mu\nu}=0$	Scalar	$\phi'(x)=\phi\left(\Lambda^{-1}x\right)$
$s=\frac{1}{2}$	$(S^{\mu\nu})_a^{\ b}=\frac{i}{4}\left([\gamma^\mu,\gamma^\nu]\right)_a^{\ b}\equiv(\Sigma^{\mu\nu})_a^{\ b}$	Spinor	$\psi'_a(x)=\exp\left[\frac{i}{2}\omega_{\mu\nu}\Sigma^{\mu\nu}\right]_a^{\ b}\psi_b\left(\Lambda^{-1}x\right)$
$s=1$	$(S^{\mu\nu})_\alpha^{\ \beta}=i\left(g^{\mu\beta}\delta^\nu_\alpha-g^{\nu\beta}\delta^\mu_\alpha\right)\equiv(J^{\mu\nu})_\alpha^{\ \beta}$	Vector	$A'_\alpha(x)=\exp\left[\frac{i}{2}\omega_{\mu\nu}J^{\mu\nu}\right]_\alpha^{\ \beta}A_\beta\left(\Lambda^{-1}x\right)$

$a,b=1,2,3,4$: Dirac indices $\qquad$ $\alpha,\beta=0,1,2,3$: Lorentz indices

The algebra of the Lorentz group is a bit awkward written in this way. We can clean things up by writing

$$\mathfrak{J}^{0i} = -\mathfrak{J}^{i0} = i\widehat{\mathcal{K}}^i \quad \mathfrak{J}^{ij} = \epsilon^{ijk}\widehat{\mathcal{J}}^k \tag{11.44}$$

so that

$$[\mathfrak{J}^{\mu\nu}, \mathfrak{J}^{\lambda\sigma}] = -i\left(g^{\mu\lambda}\mathfrak{J}^{\nu\sigma} - g^{\mu\sigma}\mathfrak{J}^{\nu\lambda} + g^{\nu\sigma}\mathfrak{J}^{\mu\lambda} - g^{\nu\lambda}\mathfrak{J}^{\mu\sigma}\right) \tag{11.45}$$

which embodies all the relations between the $\widehat{\mathcal{J}}$'s and $\widehat{\mathcal{K}}$'s above.

We now regard any solution to (11.45) as a valid way to represent a Lorentz transformation. The most general solution is

$$(\mathfrak{J}^{\mu\nu})^M_N = \left[\mathfrak{L}^{\mu\nu}\delta^M_N + (\mathfrak{S}^{\mu\nu})^M_N\right] \tag{11.46}$$

where $\mathfrak{L}^{\mu\nu}$ is the orbital part and $\mathfrak{S}^{\mu\nu}$ the spin part of the Lorentz generators. Analogous to the rotation group we have

$$\mathfrak{L}^{\mu\nu} = i\left(x^\mu\partial^\nu - x^\nu\partial^\mu\right) \tag{11.47}$$

and it is not too hard to show that $\mathfrak{L}^{\mu\nu}$ is a solution to (11.45). In the rest-frame of a particle, this will vanish when acting on any wavefunction; all that will remain is the spin-operator $\mathfrak{S}^{\mu\nu}$ acting on the wavefunction.

Hence we classify particles by their spin-values – i.e. by the possible $\mathfrak{S}^{\mu\nu}$'s that are irreducible solutions to (11.45). Finding all of these is an exercise in group theory beyond the scope of this text. The answer turns out to be analogous to what happened with rotations, as shown in table 11.2.

The matrix $\Sigma^{\mu\nu} = \frac{i}{4}\left([\gamma^\mu, \gamma^\nu]\right)$, and the associated tensor $\omega_{\mu\nu}$ is a parameter matrix; it's analogous to the $\vec{\theta}$ parameters in rotations. In fact, by comparing the Lorentz rotations to the standard rotations, we find

$$\omega_{ij} = \epsilon_{ijk}\theta_k \quad \text{and} \quad \sinh\omega_{0j} \equiv \sinh\eta_j = \beta_j\gamma_j = \frac{\beta_j}{\sqrt{1-\beta_j^2}} \tag{11.48}$$

The quantities θ_k are called the rotation angles and the quantities η_j are called the *rapidity parameters*. Note also that the Lorentz and Dirac indices are quite distinct from each other, even though both take on 4 different values. This is a coincidence that holds only in 4 spacetime dimensions$^\|$. And don't confuse the relativistic $\gamma_j = \left(1-\beta_j^2\right)^{-1/2}$ with the Dirac gamma matrices – they are mathematically distinct quantities that for reasons of historical accident make use of the same symbol.

$^\|$The range of a Lorentz index is from 0 to $d-1$ in d spacetime dimensions, whereas the range of a Dirac index is from 0 to $2^{[d/2]}$ where $[d/2]$ is the integer just less than $d/2$. Hence a spinor in 11 dimensions has 32 components, but a vector has only 11.

Let's look a little more closely at the spinor transformation matrices. We have

$$\Sigma^{0i} = \frac{i}{4}\left([\gamma^0, \gamma^i]\right) = -\frac{i}{2}\begin{pmatrix} 0 & \sigma^i \\ \sigma^i & 0 \end{pmatrix} \tag{11.49}$$

$$\Sigma^{ij} = \frac{i}{4}\left([\gamma^i, \gamma^j]\right) = -\frac{i}{4}\begin{pmatrix} [\sigma^i, \sigma^j] & 0 \\ 0 & [\sigma^i, \sigma^j] \end{pmatrix} = \frac{1}{2}\epsilon_{ijk}\begin{pmatrix} \sigma^k & 0 \\ 0 & \sigma^k \end{pmatrix} \tag{11.50}$$

So for rotations

$$\frac{i}{2}\omega_{\mu\nu}\Sigma^{\mu\nu} = \frac{i}{2}\omega_{ij}\Sigma^{ij} = \frac{i}{2}\epsilon_{ijk}\theta_k\Sigma^{ij} = \frac{i}{2}\begin{pmatrix} \vec{\theta}\cdot\vec{\sigma} & 0 \\ 0 & \vec{\theta}\cdot\vec{\sigma} \end{pmatrix}$$

$$\Rightarrow \exp\left[\frac{i}{2}\epsilon_{ijk}\theta_k\Sigma^{ij}\right] = \begin{pmatrix} \exp\left(\frac{i}{2}\vec{\theta}\cdot\vec{\sigma}\right) & 0 \\ 0 & \exp\left(\frac{i}{2}\vec{\theta}\cdot\vec{\sigma}\right) \end{pmatrix} \tag{11.51}$$

and for boosts

$$\frac{i}{2}\omega_{\mu\nu}\Sigma^{\mu\nu} = i\omega_{0i}\Sigma^{0i} = i\eta_k\Sigma^{0k} = \frac{1}{2}\begin{pmatrix} 0 & \vec{\eta}\cdot\vec{\sigma} \\ \vec{\eta}\cdot\vec{\sigma} & 0 \end{pmatrix}$$

$$\Rightarrow \exp\left(i\eta_k\Sigma^{0k}\right) = \exp\left[\frac{1}{2}\begin{pmatrix} 0 & \vec{\eta}\cdot\vec{\sigma} \\ \vec{\eta}\cdot\vec{\sigma} & 0 \end{pmatrix}\right]$$

$$= \cosh\left(\frac{|\vec{\eta}|}{2}\right) + \hat{\eta}\cdot\begin{pmatrix} 0 & \vec{\sigma} \\ \vec{\sigma} & 0 \end{pmatrix}\sinh\left(\frac{|\vec{\eta}|}{2}\right) \tag{11.52}$$

where the last line follows by taking the power series of the exponential function and noting that

$$\begin{pmatrix} 0 & \vec{\eta}\cdot\vec{\sigma} \\ \vec{\eta}\cdot\vec{\sigma} & 0 \end{pmatrix}^2 = \begin{pmatrix} (\vec{\eta}\cdot\vec{\sigma})^2 & 0 \\ 0 & (\vec{\eta}\cdot\vec{\sigma})^2 \end{pmatrix}$$

$$= \begin{pmatrix} \vec{\eta}\cdot\vec{\eta} & 0 \\ 0 & \vec{\eta}\cdot\vec{\eta} \end{pmatrix}$$

$$= |\vec{\eta}|^2\begin{pmatrix} I & 0 \\ 0 & I \end{pmatrix}$$

$$= |\vec{\eta}|^2 I_{4\times 4} \tag{11.53}$$

where $I_{4\times 4}$ is the 4×4 identity matrix.

Consider a spinor of the form

$$\psi = \begin{pmatrix} \varsigma \\ \varrho \end{pmatrix}$$

Under rotations it transforms as

$$\psi' = \exp\left[\frac{i}{2}\begin{pmatrix} \vec{\theta}\cdot\vec{\sigma} & 0 \\ 0 & \vec{\theta}\cdot\vec{\sigma} \end{pmatrix}\right]\psi = \begin{pmatrix} \exp\left(\frac{i}{2}\vec{\theta}\cdot\vec{\sigma}\right)\varsigma \\ \exp\left(\frac{i}{2}\vec{\theta}\cdot\vec{\sigma}\right)\varrho \end{pmatrix} \tag{11.54}$$

and so we see that rotations do not mix up the upper and lower components. This is consistent with our non-relativistic experience that an electron wavefunction (which would have only the upper two components) can always be rotated into a form that is fully spin-up or fully spin-down.

If $\vec{\theta} = 2\pi\hat{z}$, then we can easily show that $\psi' = -\psi$, which is what a spinor must do under a 2π rotation as we saw in Chapter 5 (see eq. (5.18)). Under a boost it transforms as

$$
\psi' = \exp\left[\frac{1}{2}\begin{pmatrix} 0 & \vec{\eta}\cdot\vec{\sigma} \\ \vec{\eta}\cdot\vec{\sigma} & 0 \end{pmatrix}\right]\psi
$$

$$
= \begin{pmatrix} \cosh\left(\frac{|\vec{\eta}|}{2}\right)\zeta + \hat{\eta}\cdot\vec{\sigma}\sinh\left(\frac{|\vec{\eta}|}{2}\right)\varrho \\ \cosh\left(\frac{|\vec{\eta}|}{2}\right)\varrho + \hat{\eta}\cdot\vec{\sigma}\sinh\left(\frac{|\vec{\eta}|}{2}\right)\zeta \end{pmatrix} \tag{11.55}
$$

and so we see that boosts do mix up the upper and lower components. Hence a moving spinor is a linear combination of electron-like and positron-like wavefunctions.

One other issue – what exactly is γ^μ? Is it a 4-vector or a matrix? Actually it's a bit of both. Using the relationships above it is possible to show that

$$
\mathbf{S}^{-1}\gamma^\mu\mathbf{S} = \Lambda^\mu_\nu\gamma^\nu \tag{11.56}
$$

where $\exp\left[\frac{i}{2}\omega_{\mu\nu}\Sigma^{\mu\nu}\right] = \mathbf{S}$ is the Lorentz-transformation matrix for a spinor under a Lorentz transformation $x'^\mu = \Lambda^\mu_\nu x^\nu$. Using this it is not too hard to show the following relationships

$$
\overline{\psi}'\psi' = \overline{\psi}\psi \tag{11.57}
$$

$$
\overline{\psi}'\gamma^5\psi' = \det(\Lambda)\left(\overline{\psi}\gamma^5\psi\right) \tag{11.58}
$$

$$
\overline{\psi}'\gamma^\mu\psi' = \Lambda^\mu_\nu\left(\overline{\psi}\gamma^\nu\psi\right) \tag{11.59}
$$

$$
\overline{\psi}'\gamma^\mu\gamma^5\psi' = \det(\Lambda)\Lambda^\mu_\nu\left(\overline{\psi}\gamma^\nu\gamma^5\psi\right) \tag{11.60}
$$

where $\gamma^5 = i\gamma^0\gamma^1\gamma^2\gamma^3$ is the product of all the γ-matrices, and $\overline{\psi} \equiv \psi^\dagger\gamma^0$. We see that $\overline{\psi}\psi$ transforms as a scalar, but $\overline{\psi}\gamma^5\psi$ transforms as a pseudoscalar, since $\det(\Lambda) = -1$ under a reflection. Similarly, $\overline{\psi}\gamma^\nu\psi$ has the properties of a 4-vector and $\overline{\psi}\gamma^\nu\gamma^5\psi$ the properties of a 4-pseudovector.

11.6   Questions

1. Show for Dirac spinors $u(p)$ and $v(p)$ that

$$
\sum_{i=\uparrow,\downarrow} u_a^{(i)}(p)\overline{u}_b^{(i)}(p) = (\not{p} + m)_{ab} \quad \text{and} \quad \sum_{i=\uparrow,\downarrow} v_a^{(i)}(p)\overline{v}_b^{(i)}(p) = (\not{p} - m)_{ab}
$$

2. Find spinors u that satisfy $(\gamma^\mu p_\mu - m)\, u(p) = 0$ that have positive energy and that are eigenstates of the operator $\hat{p} \cdot \vec{S}$ where is the unit vector of the 3-momentum of the spinor and

$$\vec{S} = \frac{\hbar}{2}\vec{\Sigma} = \frac{\hbar}{2}\begin{pmatrix} \vec{\sigma} & 0 \\ 0 & \vec{\sigma} \end{pmatrix}$$

is the spin angular momentum operator.

3. (a) Write the Dirac equation in Hamiltonian form by isolating the time-derivative of the spinor ψ on the left-hand side of the equation. The Hamiltonian H will be the operator on the right-hand side of the equation. What is H?

(b) Find the commutator of H with the orbital angular momentum operator $\vec{L}$.

(c) Find the commutator of H with the spin angular momentum operator $\vec{S}$.

(d) Find the commutator of H with the total angular momentum operator $\vec{J} = \vec{L} + \vec{S}$.

(e) Show that all spinors are eigenstates of $\vec{S} \cdot \vec{S} = S^2$. Since $S^2 = \hbar^2 s\,(s+1)$, what is the value of s for a Dirac spinor?

4. Prove the following

$$\mathrm{Tr}\,[I] = 4 \quad \text{and} \quad \mathrm{Tr}\,[\gamma^\mu \gamma^\nu] = 4g^{\mu\nu}$$
$$\mathrm{Tr}\,[\text{odd \# of } \gamma\text{-matrices}] = 0$$
$$\mathrm{Tr}\,[\gamma^\mu \gamma^\nu \gamma^\alpha \gamma^\beta] = 4\left(g^{\mu\nu}g^{\alpha\beta} - g^{\mu\alpha}g^{\nu\beta} + g^{\mu\beta}g^{\nu\alpha}\right)$$
$$\mathrm{Tr}\,[\gamma^5 \gamma^\mu \gamma^\nu \gamma^\alpha \gamma^\beta] = -4i\varepsilon^{\mu\nu\alpha\beta} \quad \text{with} \quad \varepsilon^{0123} = -1$$
$$\gamma_\alpha \gamma^\nu \gamma^\alpha = -2\gamma^\nu$$
$$\gamma_\alpha \gamma^\mu \gamma^\nu \gamma^\alpha = 4g^{\mu\nu}$$
$$\mathrm{Tr}\,[\gamma^5] = 0$$

where $\varepsilon^{\mu\nu\alpha\beta}$ is the 4-dimensional Levi-Civita symbol (or epsilon-tensor) which obeys

$$\varepsilon^{0123} = -1 \qquad \varepsilon_{0123} = +1$$
$$\varepsilon^{\mu\nu\alpha\beta} = \begin{cases} -1 & (\text{if } \mu\nu\alpha\beta \text{ is an even permutation of 0123}) \\ +1 & (\text{if } \mu\nu\alpha\beta \text{ is an odd permutation of 0123}) \end{cases}$$
$$\varepsilon_{\mu\nu\alpha\beta} = \begin{cases} +1 & (\text{if } \mu\nu\alpha\beta \text{ is an even permutation of 0123}) \\ -1 & (\text{if } \mu\nu\alpha\beta \text{ is an odd permutation of 0123}) \end{cases}$$

5. Writing $\exp\left[\frac{i}{2}\omega_{\mu\nu}\Sigma^{\mu\nu}\right] = S$, show

$$\Sigma^{\mu\nu} = \frac{i}{4}[\gamma^\mu, \gamma^\nu]$$
$$S^{-1}\gamma^\mu S = \Lambda^\mu_{\ \nu}\gamma^\nu$$
$$\Lambda^\mu_{\ \nu} = \left(e^{-\omega}\right)^\mu_{\ \nu}$$
$$= \delta^\mu_{\ \nu} - \omega^\mu_{\ \nu} + \frac{1}{2!}\omega^\mu_{\ \alpha}\omega^\alpha_{\ \nu} - \cdots$$

where $x'^\mu = \Lambda^\mu_{\ \nu} x^\nu$ is a Lorentz-transformation. (Hint: use $\exp(z) = \lim_{N \to \infty} \left(1 + \frac{z}{N}\right)^N$).

6. Verify the following Lorentz transformation properties:

$$\overline{\psi}' \psi' = \overline{\psi} \psi$$
$$\overline{\psi}' \gamma^5 \psi' = \det(\Lambda) \, \overline{\psi} \gamma^5 \psi$$
$$\overline{\psi}' \gamma^\mu \psi' = \Lambda^\mu_{\ \nu} \overline{\psi} \gamma^\nu \psi$$
$$\overline{\psi}' \gamma^\mu \gamma^5 \psi' = \det(\Lambda) \, \Lambda^\mu_{\ \nu} \overline{\psi} \gamma^\nu \gamma^5 \psi$$

for a Dirac spinor ψ.

7. Charge-conjugation transforms a spinor ψ according to the relation

$$\mathbb{C}\psi = i\gamma^2 \psi^*$$

where the $*$ denotes the complex-conjugate. Compute the charge-conjugates of $v^{(\uparrow)}(x)$ and $v^{(\downarrow)}(x)$. How do they compare to $u^{(\uparrow)}(x)$ and $u^{(\downarrow)}(x)$?

$\mathrm{ad}_A \equiv [A, B]$ 　　　　BCH identity

$e^A B e^{-A} = \exp[\mathrm{ad}_A] B$

$= (1 + A + \cdots) B (1 - A + \cdots)$

$= B + AB - BA + \cdots$

$= B + [A, B] = B + \mathrm{ad}_A B = \mathrm{ad}_A^\circ B + \mathrm{ad}_A B + \cdots$

12

Gauge Invariance

Perhaps the most powerful symmetry principle in physics associated with the non-gravitational interactions is gauge invariance. It forms the foundation of our understanding of the Standard Model and all of its generalizations. Its origins emerged in the 1820s with the discovery of electromagnetism and the first theory of electrodynamics. Over a period of several decades of experimental study and theoretical refinement, physicists realized that different forms of the vector potential result in the same observable forces. James Clerk Maxwell [114] formulated the equations of electromagnetism that embody the first known gauge principle, though the nomenclature "gauge" was not used then and the equations were written in a rather obscure way that many people found hard to understand. The quest to understand relativistic quantum mechanics in 1926 led Klein [115] to formulate the Klein-Gordon equation in such a way that Fock [116] discovered it was invariant with respect to multiplication of the wave function by a phase factor that depended on position and location, provided one incorporated the vector potential in a suitable way [117].

Hermann Weyl [118] declared this invariance as a general principle and called it Eichinvarianz in German and *gauge invariance* in English. In this chapter I will introduce you to this principle in its simplest context.

12.1 Solutions to the Dirac Equation

We have seen that the Dirac equation

$$(i\gamma^\mu \partial_\mu - m)\,\psi = 0 \tag{12.1}$$

has the following complete set of solutions

$$\left\{ u^{(i)}(p)e^{-ip\cdot x}, v^{(i)}(p)e^{ip\cdot x} \right\} \qquad i = \uparrow, \downarrow$$

$$(11.14) \quad (\not{p} - m)\,u(p) = 0 \qquad (\not{p} + m)\,v(p) = 0 \tag{12.2}$$

$$u^{(i)}(p) = \sqrt{2m} \begin{pmatrix} \sqrt{\frac{E+m}{2m}}\,\xi^{(i)} \\ \sqrt{\frac{E-m}{2m}}\,(\hat{p}\cdot\vec{\sigma})\,\xi^{(i)} \end{pmatrix} \qquad v^{(i)}(p) = \sqrt{2m} \begin{pmatrix} \sqrt{\frac{E-m}{2m}}\,(\hat{p}\cdot\vec{\sigma})\,\xi^{(i)} \\ \sqrt{\frac{E+m}{2m}}\,\xi^{(i)} \end{pmatrix}$$

where for convenience I have employed the "slash" notation, in which any 4-vector that is multiplied by a γ-matrix is written with a slash through it: for example, $\not{p} = \gamma^\mu p_\mu$ and $\not{\partial} = \gamma^\mu \partial_\mu$.

If we wanted to obtain quantum probabilities from these solutions, we might expect that the relevant quantity to compute is $\psi^\dagger \psi$. Let's try this for a specific u-type solution that is, say, spin-up. We find

$$\psi^{\dagger(\uparrow)}\psi^{(\uparrow)}$$

$$= u^{\dagger(\uparrow)}(p)u^{(\uparrow)}(p)$$

$$= \left(\psi_0\sqrt{2m}\right)^2 \left(\sqrt{\tfrac{E+m}{2m}}\,\xi^{(\uparrow)\dagger} \quad \sqrt{\tfrac{E-m}{2m}}\,\xi^{(\uparrow)\dagger}\,(\hat{p}\cdot\vec{\sigma}^\dagger)\right) \begin{pmatrix} \sqrt{\tfrac{E+m}{2m}}\,\xi^{(\uparrow)} \\ \sqrt{\tfrac{E-m}{2m}}\,(\hat{p}\cdot\vec{\sigma})\,\xi^{(\uparrow)} \end{pmatrix}$$

$$= 2m\psi_0^2 \left[\left(\frac{E+m}{2m}\right)\xi^{(\uparrow)\dagger}\xi^{(\uparrow)} + \left(\frac{E-m}{2m}\right)\xi^{(\uparrow)\dagger}(\hat{p}\cdot\vec{\sigma}^\dagger)(\hat{p}\cdot\vec{\sigma})\,\xi^{(\uparrow)}\right]$$

$$= 2m\psi_0^2 \left[\left(\frac{E+m}{2m}\right) + \left(\frac{E-m}{2m}\right)\right]$$

$$= 2E\psi_0^2 \tag{12.3}$$

which is not Lorentz invariant! The final answer depends on the energy, whose value depends upon the (boosted) frame of reference, and so the probability will depend on the frame as well.

Clearly $\psi^\dagger \psi$ is not the correct thing to compute. But what is? How can we fix this problem?

Notice that if the two factors in the second-last line were subtracted, then we would have a Lorentz-invariant quantity. This means that we need the lower two components to subtract instead of add. We can arrange for this to happen by defining the conjugate of ψ to be $\overline{\psi} \equiv \psi^\dagger\gamma^0$. In this case we find

$$\overline{\psi}^{(\uparrow)}\psi^{(\uparrow)}$$

$$= \overline{u}^{(\uparrow)}(p)u^{(\uparrow)}(p)$$

$$= \left(\psi_0\sqrt{2m}\right)^2 \left(\sqrt{\tfrac{E+m}{2m}}\,\xi^{(\uparrow)\dagger} \quad \sqrt{\tfrac{E-m}{2m}}\,\xi^{(\uparrow)\dagger}\,(\hat{p}\cdot\vec{\sigma}^\dagger)\right)\begin{pmatrix} I & 0 \\ 0 & -I \end{pmatrix} \begin{pmatrix} \sqrt{\tfrac{E+m}{2m}}\,\xi^{(\uparrow)} \\ \sqrt{\tfrac{E-m}{2m}}\,(\hat{p}\cdot\vec{\sigma})\,\xi^{(\uparrow)} \end{pmatrix}$$

$$= \left(\psi_0\sqrt{2m}\right)^2 \left(\sqrt{\tfrac{E+m}{2m}}\,\xi^{(\uparrow)\dagger} \quad \sqrt{\tfrac{E-m}{2m}}\,\xi^{(\uparrow)\dagger}\,(\hat{p}\cdot\vec{\sigma}^\dagger)\right)\begin{pmatrix} \sqrt{\tfrac{E+m}{2m}}\,\xi^{(\uparrow)} \\ -\sqrt{\tfrac{E-m}{2m}}\,(\hat{p}\cdot\vec{\sigma})\,\xi^{(\uparrow)} \end{pmatrix}$$

$$= 2m\psi_0^2 \left[\left(\frac{E+m}{2m}\right)\xi^{(\uparrow)\dagger}\xi^{(\uparrow)} - \left(\frac{E-m}{2m}\right)\xi^{(\uparrow)\dagger}(\hat{p}\cdot\vec{\sigma}^\dagger)(\hat{p}\cdot\vec{\sigma})\,\xi^{(\uparrow)}\right]$$

$$= 2m\psi_0^2 \left[\left(\frac{E+m}{2m}\right) - \left(\frac{E-m}{2m}\right)\right]$$

$$= 2m\psi_0^2 \tag{12.4}$$

which is Lorentz-invariant!

The idea that the conjugate of ψ is not just its complex-conjugate transpose (i.e. the dagger) may seem strange, but it actually is what we need to make sense of the conjugate of the Dirac equation. Taking its dagger, we find

$$0 = ((i\gamma^\mu \partial_\mu - m)\psi)^\dagger = -i\partial_\mu \psi^\dagger (\gamma^\mu)^\dagger - m\psi^\dagger \tag{12.5}$$

which doesn't look much like the conjugate of the Dirac equation. However, using the relation* $(\gamma^\mu)^\dagger = \gamma^0\gamma^\mu\gamma^0$ (and the fact that $(\gamma^0)^2 = 1$) we get

$$\psi^\dagger = \overline{\psi}\mathbb{1}$$

$$0 = ((i\gamma^\mu \partial_\mu - m)\psi)^\dagger = -i\partial_\mu \psi^\dagger (\gamma^\mu)^\dagger - m\psi^\dagger = -\left(i\partial_\mu \overline{\psi}\gamma^\mu + m\overline{\psi}\right)\gamma^0$$

$$\Rightarrow \overline{\psi}\left(i\overleftarrow{\partial}_\mu \gamma^\mu + m\right) = 0 \tag{12.6}$$

where the symbol $\overleftarrow{\partial}_\mu$ means that the derivative operator acts on objects to its left.

So it is $\overline{\psi}$ that obeys the conjugate Dirac equation, and we say that $\overline{\psi}$ is the adjoint of ψ.

12.2 Conserved Current

How then do we interpret $\psi^\dagger\psi$? Well, we can write $\psi^\dagger\psi = \overline{\psi}\gamma^0\psi$, which looks like the 0-th component of a 4-vector. Indeed, from eq. (11.59) in Chapter 11, we know that $\overline{\psi}\gamma^\mu\psi$ has the transformation properties of a 4-vector. Furthermore, the Dirac equation implies that

$$\partial_\mu\left(\overline{\psi}\gamma^\mu\psi\right) = \left(\partial_\mu\overline{\psi}\gamma^\mu\right)(\psi) + \overline{\psi}\gamma^\mu\partial_\mu\psi = im\overline{\psi}\psi - im\overline{\psi}\psi = 0 \tag{12.7}$$

which means that $\overline{\psi}\gamma^\mu\psi$ is conserved! Suppose we set $\psi^\dagger\psi \propto \rho$ and $\psi^\dagger\gamma^0\vec{\gamma}\psi \propto \vec{\mathcal{J}}$. We then find that

$$0 = \partial_\mu\left(\overline{\psi}\gamma^\mu\psi\right) = \partial_0\left(\psi^\dagger\psi\right) + \vec{\nabla}\cdot\left(\psi^\dagger\gamma^0\vec{\gamma}\psi\right) \propto \frac{\partial\rho}{\partial t} + \vec{\nabla}\cdot\vec{\mathcal{J}} \tag{12.8}$$

which is just like the conservation of an electric current $\mathcal{J}^\mu = (\rho, \vec{\mathcal{J}})$:

$$\frac{\partial\rho}{\partial t} + \vec{\nabla}\cdot\vec{\mathcal{J}} = 0 \Leftrightarrow \partial_\mu\mathcal{J}^\mu = 0 \tag{12.9}$$

In other words, the Dirac equation automatically has a conserved current $\overline{\psi}\gamma^\mu\psi$!

Indeed, if we assume that the wavefunction ψ is that of an electron, it should have a charge $(-e)$, and it is tempting to interpret $\rho = -e\psi^\dagger\psi$ as its

---

*This is easily proved by brute force. Just check it for each value of μ.

charge density and $\vec{\mathcal{J}} = -e\left(\psi^\dagger\gamma^0\vec{\gamma}\psi\right) = -e\overline{\psi}\vec{\gamma}\psi$ as its electric current. Dirac thought that this was correct, and suggested using it in Maxwell's equations

$$\vec{\nabla}\cdot\vec{E} = \rho = -e\psi^\dagger\psi \tag{12.10}$$

$$-\frac{\partial\vec{E}}{\partial t} + \vec{\nabla}\times\vec{B} = \vec{\mathcal{J}} = -e\overline{\psi}\vec{\gamma}\psi \tag{12.11}$$

as the current and charge density.

12.3   The Gauge Principle

The presence of a conserved current indicates the presence of a symmetry. Let's see what that symmetry is.

If we are going to use the 4-current $-e\overline{\psi}\gamma^\mu\psi$ as a source for Maxwell's equations, then we should also modify the Dirac equation; otherwise the Dirac wavefunction will influence the evolution of the electromagnetic field (it provides a source for the field), but the electromagnetic field won't affect the evolution of the electron (which it must, from observation).

How can we do this? Consider the phase of $\psi(x)$. The charge and current densities remain unchanged if we redefine $\psi(x) \to e^{i\alpha}\psi(x)$, where α is some constant. This is as we expect – the diffraction pattern of an electron beam is insensitive to phase changes. However, we actually have more: we would find that ρ and $\vec{\mathcal{J}}$ would remain the same even if $\alpha = \alpha(x)$ – even if we changed the phase differently at every point in space and time! Can we impose this much more powerful symmetry?

At first, it looks like it might be hard to do this. Under the transformation $\psi(x) \to \psi'(x) = e^{i\alpha(x)}\psi(x)$ we find that the Dirac equation changes

$$(i\gamma^\mu\partial_\mu - m)\,\psi' = e^{i\alpha(x)}\left[(i\gamma^\mu\partial_\mu - m)\,\psi - \gamma^\mu\,(\partial_\mu\alpha)\,\psi\right] \tag{12.12}$$

because we pick up derivatives of α. Since we want this more powerful invariance to hold (i.e., we want $\alpha = \alpha(x)$), we must invent a new kind of derivative operator D_μ such that $(D_\mu\psi)' = e^{i\alpha(x)}D_\mu\psi$. For this to work, we will need to introduce another 4-vector – let's call it $A_\mu(x)$ – that ensures this transformation property. This 4-vector – referred to as a vector field, since it depends on space and time – has to compensate for the $\partial_\mu\alpha$ term. So let's define

$$D_\mu\Psi = \partial_\mu\Psi + ieA_\mu\Psi \tag{12.13}$$

which can be valid for any wavefunction Ψ (and not just a Dirac wavefunction). This new derivative operator will cancel out the α-derivative provided we require the transformation to be

$$\Psi'(x) = e^{i\alpha(x)}\Psi(x) \quad \text{and} \quad A'_\mu = A_\mu - \frac{1}{e}\partial_\mu\alpha \tag{12.14}$$

so that

$$(D_\mu \Psi)' = e^{i\alpha(x)} \left[\partial_\mu \Psi + i \left(\partial_\mu \alpha \right) \Psi + ie \left(A_\mu - \frac{1}{e} \partial_\mu \alpha \right) \Psi \right] = e^{i\alpha(x)} D_\mu \Psi$$

$$(12.15)$$

This transformation is called a *gauge* transformation. If α is constant then we call it a *global* gauge transformation, whereas if α is a function we call it a *local* gauge transformation[†]. Note that the group of transformations depends upon one continuous parameter α. From our discussion on Lie Groups in Chapter 3 (recall table 3.3) we know that this group of transformations is $\mathbf{U}(1)$: the set of transformations of unitary 1×1 matrices (i.e. of complex phases). The derivative operator D_μ is called a gauge covariant derivative because it "co-varies" along with the gauge transformation (i.e. it transforms the same way that Ψ does). The object A_μ must be a 4-vector by Lorentz covariance, and is in general a function of space and time.

So the locally gauge invariant Dirac equation is

$$(i\gamma^\mu D_\mu - m) \psi = 0$$

$$(12.16)$$

which couples the wavefunction ψ to A_μ.

12.4 The Maxwell-Dirac Equations

Now we need an equation for A_μ to obey – one that is gauge invariant, Lorentz-covariant and (we hope) simple. We could try

$$\partial^\mu A_\mu = 0$$

$$(12.17)$$

which is simple, but not gauge invariant. The next-simplest thing to try is something with two derivatives[‡]. So let's try

$$a\partial_\mu \partial^\mu A_\nu + b\partial_\nu \partial^\mu A_\mu = 0$$

$$(12.18)$$

where a and b are constants, and demand that it be gauge-invariant:

$$0 = a\partial_\mu \partial^\mu A'_\nu + b\partial_\nu \partial^\mu A'_\mu = a\partial_\mu \partial^\mu A_\nu + b\partial_\nu \partial^\mu A_\mu - \frac{1}{e}(a+b)\left(\partial_\mu \partial^\mu \partial_\nu \alpha\right) \quad (12.19)$$

[†]The terms local and global can be understood like this. Consider a cubic box filled with billiard balls all of the same color that are closely packed. Rotation of this box about any axis through its center by a $90°$ angle leaves the box looking the same. This is a global transformation because we have done the same thing to every ball in the box. However, if every ball is individually rotated by any amount about any axis (while remaining inside the box of course) then this operation also leaves the box looking the same. This is a local transformation – we have acted differently on every ball in the box.

[‡]Note that we can't use the γ-matrices here because A_μ has no spinor indices.

which forces $a + b = 0$. Scaling out the constant a gives

$$\partial_\mu \partial^\mu A_\nu - \partial_\nu \partial^\mu A_\mu = 0 \tag{12.20}$$

which may be rewritten as

$$\partial^\mu F_{\mu\nu} = 0 \quad \text{where} \quad F_{\mu\nu} = \partial_\mu A_\nu - \partial_\nu A_\mu \tag{12.21}$$

The quantity $F_{\mu\nu}$ is obviously gauge invariant:

$$F'_{\mu\nu} = \partial_\mu A'_\nu - \partial_\nu A'_\mu = \partial_\mu \left(A_\nu - \frac{1}{e}\partial_\nu \alpha \right) - \partial_\nu \left(A_\mu - \frac{1}{e}\partial_\mu \alpha \right)$$
$$= \partial_\mu A_\nu - \partial_\nu A_\mu = F_{\mu\nu} \tag{12.22}$$

It also obeys

$$\partial_\lambda F_{\mu\nu} + \partial_\mu F_{\nu\lambda} + \partial_\nu F_{\lambda\mu} = 0 \tag{12.23}$$

Now notice if we write

$$F_{i0} = \frac{\partial A_0}{\partial x^i} - \frac{\partial A_i}{\partial t} = E_i \quad \text{and} \quad F_{ij} = \frac{\partial A_j}{\partial x^i} - \frac{\partial A_i}{\partial x^j} = \varepsilon_{ijk} B_k \tag{12.24}$$

then the previous two equations become

$$\partial^\mu F_{\mu\nu} = 0 \Rightarrow \begin{cases} \vec{\nabla} \cdot \vec{E} = 0 \\ -\frac{\partial \vec{E}}{\partial t} + \vec{\nabla} \times \vec{B} = 0 \end{cases} \tag{12.25}$$

$$\partial_\lambda F_{\mu\nu} + \partial_\mu F_{\nu\lambda} + \partial_\nu F_{\lambda\mu} = 0 \Rightarrow \begin{cases} \vec{\nabla} \cdot \vec{B} = 0 \\ \frac{\partial \vec{B}}{\partial t} + \vec{\nabla} \times \vec{E} = 0 \end{cases} \tag{12.26}$$

which are eqs. (12.10,12.11) with $\rho = \mathcal{J} = 0$: the source-free Maxwell-equations!

We therefore interpret $A_\mu = (\varphi, \vec{A})$ to be the electromagnetic 4-vector potential, with $\vec{A}$ the usual vector potential and $\varphi = A_0$ the Coulomb potential. Since A_μ has a single 4-vector index it transforms like a spin-1 wavefunction under Lorentz-transformations: $A'_\mu = \Lambda_\mu{}^\nu A_\nu$. The quantity $F_{\mu\nu}$ is the field strength of this potential (sometimes referred to as the Faraday tensor) since it encodes both the electric and magnetic fields that are determined from A_μ.

Based on Dirac's idea above, we now see how to couple A_μ to ψ. We write

$$\partial^\mu F_{\mu\nu} = \mathcal{J}_\nu \tag{12.27}$$

and this will modify the first two Maxwell equations as already noted. The current $\mathcal{J}_\nu$ can actually be anything provided it is conserved since

$$\partial^\nu \mathcal{J}_\nu = \partial^\nu \partial^\mu F_{\mu\nu} = 0 \tag{12.28}$$

where the latter equality holds because $F_{\mu\nu} = -F_{\nu\mu}$ and $\partial^\nu\partial^\mu = \frac{\partial}{\partial x_\nu}\frac{\partial}{\partial x_\mu} = \frac{\partial}{\partial x_\mu}\frac{\partial}{\partial x_\nu} = \partial^\mu\partial^\nu$. However, we want to couple A_μ to ψ and vice versa, so we set $\mathcal{J}^\mu = -e\overline{\psi}\gamma^\mu\psi$ and write the following closed system of equations:

$$\begin{aligned} \partial^\mu F_{\mu\nu} &= -e\overline{\psi}\gamma_\nu\psi \\ (i\gamma^\mu D_\mu - m)\,\psi &= 0 \end{aligned} \tag{12.29}$$

which are called the *Maxwell-Dirac Equations.*

The Maxwell-Dirac equations, when quantized, are the theory of *Quantum Electrodynamics*, or QED. Formulated by Tomonaga [108], Feynman [103], and Schwinger [109], it was the first quantum field theory constructed, and the one with the most spectacular agreement between theory and experiment of any non-gravitational theory known – up to parts in a trillion!

12.4.1 Physical Features of the Maxwell-Dirac Equations

The Maxwell-Dirac equations have a number of important features that I shall summarize here.

1. All non-gravitational interactions are founded on the principles we used to construct QED, namely Lorentz-covariance and local gauge invariance. For QED the local gauge invariance is just a local phase invariance. For the weak and strong interactions there is also a local gauge invariance that has a conceptually similar character but becomes more complicated in detail, as we'll see in subsequent chapters.

2. The gauge principle – plus the simplicity of minimal coupling (the fewest number of derviatives and terms possible in each equation) gave us a set of four wave equations for the spin-1 wavefunction A_μ. We know these equations as Maxwell's equations. As a consequence of the gauge principle the A_μ wavefunction (or gauge field as it is more commonly called) is massless. A mass term m would modify eq. (12.27) to be

$$\partial^\mu F_{\mu\nu} + m^2 A_\nu = \mathcal{J}_\nu$$

but this would not be invariant under the gauge transformation (12.14), and so we must set the mass to $m = 0$. This is a general consequence of gauge invariance – the gauge fields are massless. In QED we call the particle whose wavefunction is A_μ the *photon*.

3. The group of gauge transformations depends on one parameter – the phase function α – and so the symmetry group of the theory is $\mathbf{U}(1)$.

4. The local gauge invariance implies that the charge $Q = -e \int d^3x\, \overline{\psi}\gamma^0 \psi$ is conserved[§].

5. Gauge-invariant theories are renormalizable: they have predictive power.

12.5   The Wavefunction of the Photon

We can solve the source-free Maxwell equations for A_μ quite easily. Suppose we have a solution A_μ for the source-free Maxwell equations. Then $A_\mu + \partial_\mu \lambda$ also solves this system, and we can use the gauge freedom in the function λ to impose a constraint on A_μ:

$$\partial^\mu A_\mu = 0 \tag{12.30}$$

a constraint known as the Lorentz condition[¶]. In this case Maxwell's equations become

$$\partial_\mu \partial^\mu A_\nu = 0 \tag{12.31}$$

which is like a set of four massless Klein-Gordon equations – one for each component of A_μ. Hence the solution is

$$A_\nu(x) = a_0 \varepsilon_\nu(p) e^{-ip\cdot x} \tag{12.32}$$

where a_0 is a constant and the Maxwell equations and the Lorentz condition imply

$$p^2 = 0 \quad \text{and} \quad \varepsilon \cdot p = 0 \tag{12.33}$$

[§] This can be shown as follows.

$$\frac{d}{dt}Q = -e \int d^3x\, \frac{\partial}{\partial t}\left(\overline{\psi}\gamma^0\psi\right)$$

$$= +e \int d^3x\, \vec{\nabla}\cdot\left(\overline{\psi}\vec{\gamma}\psi\right)$$

$$= e \oint d\vec{S}\cdot\left(\overline{\psi}\vec{\gamma}\psi\right)$$

$$= 0$$

where the 3rd line follows from Gauss' theorem and the last line follows by taking the surface intergral to be at large distance where ψ vanishes.

[¶] Two years after Maxwell the Danish physicist Ludvig Lorenz [119] also formulated the basic equations for electromagnetism, and reached the same conclusions Maxwell did about the relationships between light, charge, and current. He also formulated eq. (12.30), so we really should call this the Lorenz condition. However, more than 40 years later the Dutch physicist H. A. Lorentz wrote extensive encyclopedia articles and a book on electromagnetism [120] establishing him as an authority in classical electrodynamics and leading the community to refer to eq. (12.30) as the Lorentz condition. So don't feel too guilty if you accidentally drop the 't' every now and then!

The quantity $\varepsilon_\mu(p)$ characterizes the spin of the photon. It looks like it has three independent components, since $\varepsilon \cdot p = 0$ is only one condition. However, this is not right: the Lorentz condition (12.30) does not allow us to uniquely specify A_μ because we could perform a further gauge transformation with functions ζ that obey $\partial^\mu \partial_\mu \zeta = 0$. To eliminate this last bit of indeterminacy we can impose

$$A_0 = 0 \qquad \vec{\nabla} \cdot \vec{A} = 0 \tag{12.34}$$

a set of constraints known as the Coulomb gauge, in which case

$$\varepsilon^\mu = (0, \hat{e}) \quad \text{where } \hat{e} \cdot \hat{e} = 1 \ \text{ and } \hat{e} \cdot \vec{p} = 0 \tag{12.35}$$

and so $\varepsilon_\nu(p)$ actually has only two independent components. These correspond to the two possible polarization states – or spin states – of the photon.

For a photon moving in the $\hat{z}$-direction, we can write

$$\hat{\mathcal{E}}_z = 0 \quad \text{since } \hat{\mathcal{E}} \cdot \vec{P} = 0$$

$$\varepsilon^{\mu(x)} = (0, 1, 0, 0) \qquad \varepsilon^{\mu(y)} = (0, 0, 1, 0) \tag{12.36}$$

and any photon wavefunction will consist of some complex linear combinations of these two polarizations. For example a circularly polarized photon would have

$$\varepsilon^{\mu(+)} = -\frac{1}{\sqrt{2}}(0, 1, i, 0) \qquad \varepsilon^{\mu(-)} = \frac{1}{\sqrt{2}}(0, 1, -i, 0) \tag{12.37}$$

and these correspond to the two spin states $s_z = \pm 1$ of the photon.

So instead of four independent components for the photon wavefunction A_μ we have only two. But a massive spin-s particle has $2s + 1$ spin states as we saw in Chapter 5. Since the photon has spin $s = 1$, shouldn't it have $2s + 1 = 3$ independent spin states instead of 2? The reason it has only two independent spin states is because the photon is massless. This means that there is no rest frame for the photon (or any massless particle), which in turn means that is not possible for an observer traveling in the same direction as the particle to move faster than the particle and observe a reversal of the component of its spin along its direction of motion. Hence the component of spin along the direction of motion for any massless particle must be fixed, which means that it must be either aligned or antialigned with its direction of motion. Hence a massless spin-s particle has only two spin states regardless of the value of s (unless $s = 0$ in which case it has only one spin state).

We are now in a position to be able to write down the Feynman rules for QED, the subject of the next chapter.

12.6 Questions

1. Apply the local gauge invariance principle to the Schroedinger equation.

$$\vec{\nabla} \rightarrow \vec{\nabla} - ie\vec{A} \ .$$

2. Find the Maxwell-Klein-Gordon equations, applying the same procedure that was used to find the Maxwell-Dirac equations. Note that you will have to consider a complex scalar field φ in order to make this work.

3. Show that the current

$$ie\left[\varphi^*\left(D_\mu\varphi\right)-\left(D_\mu\varphi\right)^*\varphi\right]$$

is conserved for solutions to the Maxwell-Klein-Gordon equations.

4. (a) Find the charge density ρ and current density $\vec{J}$ from the current in question #3.

   (b) Consider a gauge transformation that makes φ purely a real scalar. How does your answer to part (a) change?

5. Write the equation
$$\left(i\gamma^\mu D_\mu - m\right)\psi = 0$$

in Hamiltonian form by isolating the time-derivative of the spinor ψ on the left-hand side of the equation. The Hamiltonian H will be the operator on the right-hand side of the equation. What is H?

6. (a) For a wavefunction Ψ with charge e show that

$$\left[D_\mu, D_\nu\right]\Psi = ieF_{\mu\nu}\Psi$$

   (b) Use the preceding relation to show that $F_{\mu\nu}$ is gauge-invariant. This provides another way to obtain the field strength of the electromagnetic potential.

7. (a) Consider a modification to the Maxwell-Dirac equation

$$\left(i\gamma^\mu D_\mu - m\right)\psi = g\varphi\psi$$

where φ is a scalar wavefunction and g is a constant. How must φ transform if this equation is to remain gauge-covariant?

   (b) Find the corresponding equation obeyed by $\bar{\psi}$ that is gauge-covariant.

8. Consider a theory of one complex scalar particle with wavefunction φ and two spin-$\frac{1}{2}$ particles ψ and χ, each of which couples to the photon via the equations

$$\left(i\gamma^\mu D_\mu - m_\psi\right)\psi = g\varphi\chi$$
$$\left(i\gamma^\mu D_\mu - m_\chi\right)\psi = g\varphi^*\psi$$
$$D^\mu D_\mu\varphi + m_\varphi^2\varphi = g\bar{\psi}\chi$$

where m_φ^2, m_ψ, and m_χ are the respective masses of the φ, ψ and χ particles.

(a) Find the most general local phase transformation of φ, ψ and χ that leaves this system gauge-covariant.

(b) What is the relationship between the charges of φ, ψ and χ ?

(c) In Maxwell's equations

$$\partial^\mu F_{\mu\nu} = J_\nu$$

what is the current J_ν for this theory?

$$\psi' = e^{?}\psi$$
$$\varphi = e^{??}\varphi$$
$$\chi = e^{???}\chi$$

$A_\mu \to$ changes in the same manner for all three particles

$$\psi' = e^{ied}\psi$$

13

Quantum Electrodynamics

Quantum Electrodynamics – or QED – is the quantum theory of electromagnetism that is founded on the Maxwell-Dirac Equations:

$$\begin{matrix} \partial^\mu F_{\mu\nu} = \mathcal{J}_\nu \\ (i\gamma^\mu D_\mu - m)\psi = 0 \end{matrix} \quad \text{where} \quad \begin{matrix} D_\mu = \partial_\mu + ieA_\mu \\ F_{\mu\nu} = \partial_\mu A_\nu - \partial_\nu A_\mu \end{matrix} \tag{13.1}$$

The current $\mathcal{J}_\nu$ can be anything that obeys the conservation law $\partial^\nu \mathcal{J}_\nu = 0$. For fermions it has the form

$$\mathcal{J}_\nu = e_1 \overline{\psi}_1 \gamma_\nu \psi_1 + e_2 \overline{\psi}_2 \gamma_\nu \psi_2 + \cdots + \mathcal{J}_\nu^{\text{ext}} \tag{13.2}$$

where e_{J} is the charge of the "J-th" particle (e.g. $-e$ for electrons, $+\frac{2}{3}e$ for up-quarks, etc.), and ψ_{J} is its wavefunction. Each ψ_{J} obeys its own Dirac equation

$$(i\gamma^\mu (\partial_\mu + ie_{\text{J}} A_\mu) - m_{\text{J}})\psi_{\text{J}} = 0 \tag{13.3}$$

where the repeated index "J" is NOT summed over, since it denotes which kind of particle we are considering. The $\mathcal{J}_\nu^{\text{ext}}$ part is called an *external* current; we include it if we want to consider particles moving in some background electromagnetic field*.

QED is one of three parts of the Standard Model of Particle Physics (usually referred to as the "Standard Model"), the other two parts being the theories of the strong and weak interactions. The basic structure and foundations of QED [121] are incorporated into the other parts of the Standard Model, and so what we learn from QED will be useful when it comes to learning about the other parts of the Standard Model. We will also see beginning in Chapter 23 that QED is actually subsumed into a more comprehensive theory called Electroweak theory, and in this sense the Standard Model has only two parts. For now we will treat QED separately.

Table 13.1 recapitulates the solutions to these equations that we obtained in previous chapters (or that, as for the completeness relations, you can deduce from calculation as per the questions at the end of this chapter).

---

*A background field, by definition, is assumed to be purely classical. Of course as far as we know quantum mechanics is pervasive and there is no physical system that is *purely* classical. So what we mean by a background quantity is one whose quantum effects are negligible relative to its classical ones. For example, if we act on an electron using a large magnet, we can generally treat the magnet and its magnetic field using classical physics, in which case we say that the magnet generates a background magnetic field.

TABLE 13.1
Free-Particle Solutions to the Maxwell and Dirac Equations

$(i\gamma^\mu \partial_\mu - m)\psi = 0$

$\Rightarrow (i\gamma^\mu \partial_\mu - m) u^{(i)}(p) e^{-ipx} = 0$

$\Rightarrow (i\gamma^\mu \partial_\mu u^{(i)}(p)) e^{-ipx}$

$+ i\gamma^\mu u^{(i)}(p)(\partial_\mu e^{-ipx})$

$- m u^{(i)}(p) e^{-ipx} = 0$

$\Rightarrow (i\gamma^\mu \partial_\mu u^{(i)}(p)) e^{-ipx}$

$+ \gamma^\mu u^{(i)}(p) e^{-ipx} \frac{\partial(p \cdot x)}{\partial x^\mu}$

$- m u^{(i)}(p) e^{-ipx} = 0.$

$\Rightarrow (\gamma^\mu p_\mu - m) u^{(i)}(p) e^{-ipx} = 0.$

$\Rightarrow (\not{p} - m) u^{(i)}(p) = 0$

		Fermions $(i = \uparrow, \downarrow)$	Antifermions $(i = \uparrow, \downarrow)$	Photons $(\lambda = +, -)$
	Free-particle Wavefunction	$\psi(x) = u^{(i)}(p)e^{-ip\cdot x}$	$\psi(x) = v^{(i)}(p)e^{ip\cdot x}$	$A_\mu(x) = \varepsilon_\mu^{(i)}(p)e^{-ip\cdot x}$
	Spin/Momentum constraints	$(\not{p} - m)\, u^{(i)}(p) = 0$ $p^2 = m^2$	$(\not{p} + m)\, v^{(i)}(p) = 0$ $p^2 = m^2$	$\varepsilon^{(\lambda)} \cdot p = 0$ $p^2 = 0$
	Adjoint Condition	$\overline{u}^{(i)}(p)\,(\not{p} - m) = 0$	$\overline{v}^{(i)}(p)\,(\not{p} + m) = 0$	$\varepsilon^{(\lambda)} \cdot p = 0$
	Normalization	$\overline{u}^{(i)}(p)u^{(i)}(p) = 2m$	$\overline{v}^{(i)}(p)v^{(i)}(p) = -2m$	$\varepsilon^{\mu(+)*}\varepsilon_\mu^{(+)} = \varepsilon^{\mu(-)*}\varepsilon_\mu^{(-)} = -1$
	Orthogonality	$\overline{u}^{(\uparrow)}(p)u^{(\downarrow)}(p) = 0$	$\overline{v}^{(\uparrow)}(p)v^{(\downarrow)}(p) = 0$	$\varepsilon^{\mu(+)*}\varepsilon_\mu^{(-)} = 0$
	Completeness	$\sum_{i=\uparrow,\downarrow} u_a^{(i)}(p)\overline{u}_b^{(i)}(p)$ $= (\not{p} + m)_{ab}$	$\sum_{i=\uparrow,\downarrow} v_a^{(i)}(p)\overline{v}_b^{(i)}(p)$ $= (\not{p} - m)_{ab}$	$\sum_{\lambda=+,-} \varepsilon_i^{(\lambda)*}\varepsilon_j^{(\lambda)} = \delta_{ij} - \widehat{p}_i\widehat{p}_j$

We can test QED by determining its predictions. This is done in two ways.

1. **Bound States**: Electromagnetism is the force that binds atoms together. By considering the Maxwell-Dirac equations for bound systems we can search for phenomena peculiar to QED. We'll look at this in Chapter 14.

2. **Scattering**: As with the toy theory, we wish to compute the matrix element $\mathcal{M}$

$$-i\mathcal{M} = \langle p_1' \cdots p_m' \,|\mathrm{M}|\, p_1 \cdots p_n \rangle = \begin{array}{l} \text{sum of all relevant} \\ \text{Feynman Diagrams} \end{array} \quad (13.4)$$

$$= \alpha f(p', p) + \alpha^2 g(p', p) + \cdots$$

where α is the fine-structure constant that we encountered in Chapter 6, experimentally given by

$$\alpha = \frac{e^2}{4\pi \hbar c} = \frac{1}{137.035989} \simeq \frac{1}{137} \quad (13.5)$$

It is a dimensionless number that characterizes the strength of the electromagnetic coupling. Since it is small, we compute all matrix elements as a perturbation series, using α as the expansion parameter.

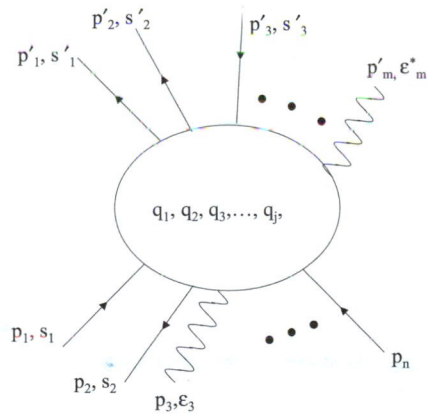

FIGURE 13.1
A general QED diagram.

13.1 Feynman Rules for QED

Now let's write down the Feynman rules for QED. For comparison I will include the rules for scalars where appropriate.

1. NOTATION. Label the incoming (outgoing) four-momenta as $p_1, p_2, \ldots, p_n$ $(p'_1, p'_2, \ldots, p'_m)$, the incoming (outgoing) spins as $s_1, s_2, \ldots, s_n$ $(s'_1, s'_2, \ldots, s'_m)$, the incoming (outgoing) photon polarizations as $\varepsilon_1^\mu, \varepsilon_2^\mu, \ldots$ $(\varepsilon_1^{\mu'}, \varepsilon_2^{\mu'}, \ldots)$, and label the internal four-momenta $q_1, q_2, \ldots, q_j$. Assign arrows to the lines as shown in fig. 13.1.

Note that in fig. 13.1 time flows from bottom to top, wiggly lines are photons, and lines with arrows are fermions (if they point upward with time) or antifermions (if they point downward against time).

2. EXTERNAL LINES. Each external line contributes a factor as follows where I've included scalars for comparison. Note that the factors associated

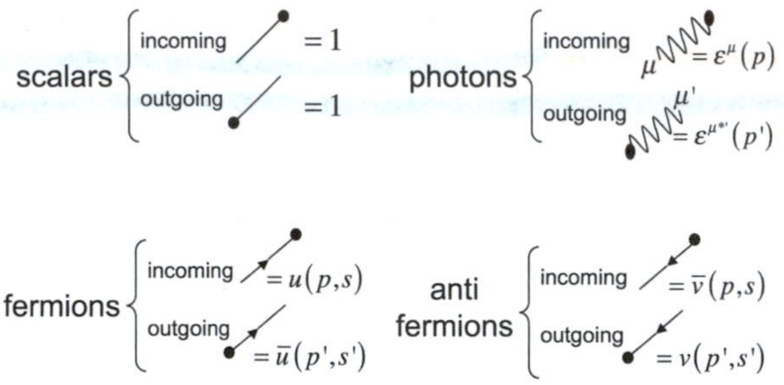

FIGURE 13.2
External lines in QED.

with external lines correspond to the incoming/outgoing plane-wave states; this is consistent with our scattering assumption that all particles are free particles in the past/future asymptotic limits. For the scalars in the toy theory this rule was not required because the factor was unity.

3. INTERNAL LINES. Each internal line contributes a factor shown in figure 13.3, where m is the mass of the particle. As before $q^2 \neq m^2$ because the particle flowing through the line is virtual (i.e. it does not obey its equations of motion). These internal lines are called *propagators*.

The next two rules are the same as for ABB theory.

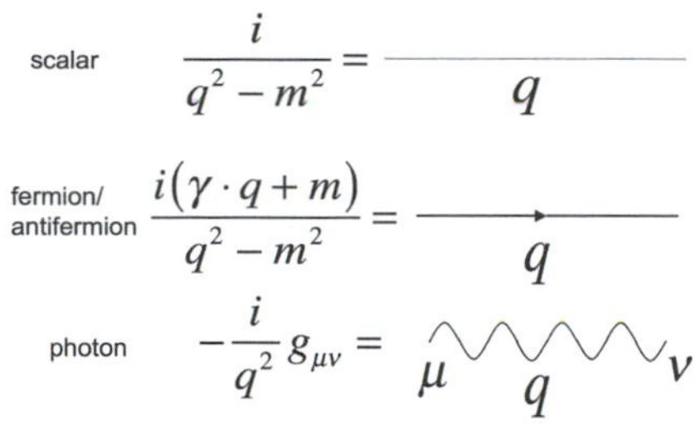

$$\text{scalar} \qquad \frac{i}{q^2 - m^2} = \underline{}_{q}$$

$$\begin{array}{l}\text{fermion/}\\ \text{antifermion}\end{array} \qquad \frac{i(\gamma \cdot q + m)}{q^2 - m^2} = \xrightarrow{}_{q}$$

$$\text{photon} \qquad -\frac{i}{q^2} g_{\mu\nu} = \overset{}{\underset{\mu \quad q}{\wwww}} \nu$$

FIGURE 13.3

Internal lines in QCD.

4. CONSERVATION OF ENERGY AND MOMENTUM For each vertex, write a delta function of the form

$$(2\pi)^4 \delta^{(4)} \left(k_1 + k_2 + k_3 + \cdots + k_N \right)$$

where the k's are the four-momenta coming into the vertex (i.e. each k^μ will be either a q^μ or a p^μ). If the momentum leads outward, then k^μ is *minus* the four-momentum of that line). This factor imposes conservation of energy and momentum at each vertex (and hence for the diagram as a whole) because the delta function vanishes unless the sum of the incoming momenta equals the sum of the outgoing momenta.

5. INTEGRATE OVER INTERNAL MOMENTA For each internal momentum q, insert a factor

$$\frac{d^4 q}{(2\pi)^4}$$

and integrate.

6. VERTEX FACTOR This is the rule that characterizes QED. For every interaction between charged particles and photons draw a point with three lines coming out signifying one photon and two fermions as shown in figure 13.4. and insert a factor of $-ig_e(\gamma^\mu)_{ab}$, where a is the spinor index of the fermion pointing away from the vertex (the "barred" fermion) and b that of the fermion pointing toward the vertex (the "unbarred" one), and where $g_e = e = \sqrt{4\pi\hbar c\alpha}$ is the dimensionless coupling for electrons/positrons. In general it is the magnitude of the fermion charge entering/leaving the vertex in units of the positron charge.

7. TOPOLOGY To get all contributions for a given process, draw diagrams by joining up all internal vertex points either to the external lines or to each other by internal lines in all possible arrangments that are topologically in-equivalent, consistent with rule 6. The number of ways a given diagram can

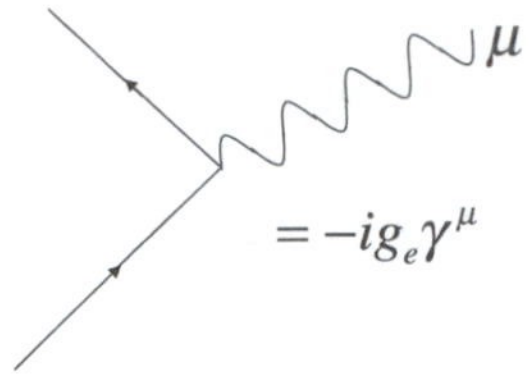

FIGURE 13.4
The QED vertex, where ε^{μ} is the photon polarization.

be drawn is the topological weight of the diagram. The result is equal to $-i\mathcal{M}$.

The next two rules apply to all theories that have fermions.

8. ANTISYMMETRIZATION Because fermion wavefunctions anticommute, we must include a minus sign between diagrams that differ

  (a) only in the interchange of two incoming (or outgoing) fermions/antifermions of the same kind

or

  (b) only in the interchange of an incoming fermion with an outgoing antifermion of the same kind (or vice versa).

9. LOOPS Every fermion loop gets a factor of (-1).

10. CANCEL THE DELTA FUNCTION The result will include a factor

$$(2\pi)^4 \delta^{(4)} \left(p_1' + p_2' \cdots + p_m' - p_1 - p_2 - \cdots - p_n\right)$$

corresponding to overall energy-momentum conservation. Cancel this factor, and what remains is $-i\mathcal{M}$.

13.2 Examples

The best way to learn the rules is to use them. We will consider here a variety of physical processes to illustrate how the rules are used in QED. Unless otherwise stated, all computations will be to lowest order in α.

13.2.1 Electron-Muon Scattering

Let's begin by looking at the simplest scattering process that occurs in QED – namely the scattering of one charged particle from another of a different kind. Electron-muon scattering is a realistic physical example:

$$e^- + \mu^- \longrightarrow e^- + \mu^-$$

To lowest order in the coupling, there is only one diagram since the particles are all distinct[†]. Applying the rules we get

$$-i\mathcal{M} = \int \underbrace{\frac{d^4q}{(2\pi)^4}}_{\text{rule 5}} \left[\underbrace{\overline{u}^{(i'_1)}(p'_1)}_{\text{rule 2}} \underbrace{(-ig_e\gamma^\mu)}_{\text{rule 6}} \underbrace{u^{(i_1)}(p_1)}_{\text{rule 2}}\right] \underbrace{\left(\frac{-ig_{\mu\nu}}{q^2}\right)}_{\text{rule 3}}$$

$$\times \left[\underbrace{\overline{u}^{(i'_2)}(p'_2)}_{\text{rule 2}} \underbrace{(-ig_e\gamma^\nu)}_{\text{rule 6}} \underbrace{u^{(i_2)}(p_2)}_{\text{rule 2}}\right] \qquad \text{pay attention to } \mu, \nu \text{ and } g_{\mu\nu}$$

$$\times \underbrace{(2\pi)^4\delta\,(p'_2 - q - p_2)\,(2\pi)^4\delta\,(p'_1 + q - p_1)}_{\text{rule 4}} \qquad (13.6)$$

It's easiest to see this by taking an outgoing fermion line (e.g. the row-matrix

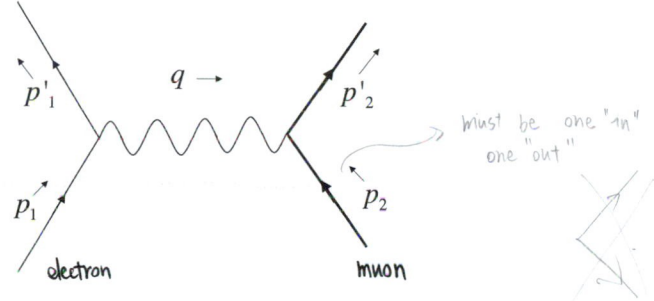

annotations: $q \rightarrow$; must be one "in" one "out"

FIGURE 13.5

Electron-muon Scattering Diagram. The muon is drawn with a thicker line to distinguish it from the electron.

$\overline{u}_a^{(i'_1)}(p'_1)$), following it backward to attach it to the vertex (e.g., $(-ig_e\gamma^\mu)_{ab}$, where the index μ is the same as that of the "internal polarization" ε^μ of the virtual photon), and then following it back to the incoming line (e.g. the column-matrix $u_b^{(i_1)}(p_1)$). The result is the first term in the above expression[‡].

[†]Note that we will get the same kind of answer for any two distinct charged fermions scattering off of each other (e.g., electron-quark, muon-tau, etc). All that will change are what we put in for the charges and the masses.

[‡]I've put this expression in square brackets to emphasize that this is the product [(row-matrix)×(square-matrix)×(column-matrix)].

Doing this for the muon line gives the second term, where we must attach a different polarization ε^ν to the photon at that vertex. The indices on the internal photon line are chosen so as to join these together with a metric, whose indices are the same as the two γ's.

Doing the integral in (13.6) is easy, since it is just an integral over delta-functions, just like what we did in eq. (10.9); carrying out the same procedure and simplifying a bit gives

$$-i\mathcal{M} = \left[\overline{u}^{(i'_1)}(p'_1)\gamma^\mu u^{(i_1)}(p_1) \right] \frac{ie^2}{(p'_1 - p_1)^2} \left[\overline{u}^{(i'_2)}(p'_2)\gamma_\mu u^{(i_2)}(p_2) \right] \quad (13.7)$$

Notice the structure of $\mathcal{M}$ in eq. (13.7): it is of the form

$$(\text{current})^\mu \times (\text{propagator}) \times (\text{current})_\mu$$

This is the general form of all fermion-fermion interactions.

To proceed further we would have to put in the initial conditions for the electron spin-state $u^{(i_1)}(p_1)$ and the muon spin-state $u^{(i_2)}(p_2)$, as well as the final states for each. The resultant matrix element $\mathcal{M}$ would then be squared and put into a cross-section formula to yield a predicted scattering rate. This is very tedious and we won't do this here – we will see later that there is an alternative method that avoids resorting to such initial and final conditions.

13.2.2 Bhabha Scattering (electron-positron scattering)

The next simplest process is that of the scattering of a particle off of its own antiparticle. Let's consider electrons scattering off of positrons, since these kinds of beams can be easily produced in a laboratory setting:

$$e^- + e^+ \longrightarrow e^- + e^+$$

To lowest order in g_e we have two diagrams, shown in fig. 13.6 where the one on the right is similar to the one we computed in eq. (13.6):

$$-i\mathcal{M}_{\text{right}} = \int \frac{d^4q}{(2\pi)^4} \left[\overline{u}^{(i'_1)}(p'_1) \left(-ig_e\gamma^\mu \right) u^{(i_1)}(p_1) \right] \left(\frac{-ig_{\mu\nu}}{q^2} \right)$$
$$\times \left[\overline{v}^{(i_2)}(p_2) \left(-ig_e\gamma^\nu \right) v^{(i'_2)}(p'_2) \right]$$
$$\times (2\pi)^4 \delta\left(p'_2 - q - p_2\right) (2\pi)^4 \delta\left(p'_1 + q - p_1\right) \quad (13.8)$$

except that we have positron-wavefunctions $v^{(i_1)}(p)$ instead of the muon wavefunctions[§].

[§]The easiest way to construct the $\left[\overline{v}^{(i_2)}(p_2) \left(-ig_e\gamma^\nu \right) v^{(i'_2)}(p'_2) \right]$ expression for the positron is to follow the arrow backwards from bottom to top. The incoming antifermion yields a factor of $\left(\overline{v}^{(i_2)}(p_2) \right)_b$; the vertex yields the term $\left(-ig_e\gamma^\nu \right)_{ba}$; and the outgoing antifermion yields a factor of $\left(v^{(i'_2)}(p'_2) \right)_a$, yielding the product $\left(\overline{v}^{(i_2)}(p_2) \right)_b \left(-ig_e\gamma^\nu \right)_{ba} \left(v^{(i'_2)}(p'_2) \right)_a$. Summing over the repeated $\{a, b\}$ indices gives $\left[\overline{v}^{(i_2)}(p_2) \left(-ig_e\gamma^\nu \right) v^{(i'_2)}(p'_2) \right]$

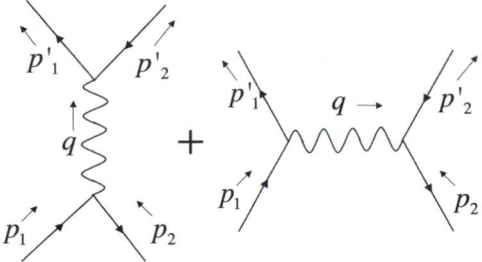

FIGURE 13.6
Bhabha scattering to lowest order.

The diagram on the left represents annihilation of the electron and positron into a single virtual photon which then produces an electron-positron pair. Here the rules give

$$-i\mathcal{M}_{\text{left}} = \int \frac{d^4 q}{(2\pi)^4} \left[\overline{u}^{(i'_1)}(p'_1) \left(-ig_e\gamma^{\mu}\right) v^{(i'_2)}(p'_2) \right] \left(\frac{-ig_{\mu\nu}}{q^2} \right)$$
$$\times \left[\overline{v}^{(i_2)}(p_2) \left(-ig_e\gamma^{\nu}\right) u^{(i_1)}(p_1) \right]$$
$$\times (2\pi)^4 \delta\left(p'_1 + p'_2 - q\right) (2\pi)^4 \delta\left(q - p_1 - p_2\right) \quad (13.9)$$

Now we have to apply rule 8, which tells us whether or not one diagram gets a minus sign (i.e. whether or not we add or subtract these two diagrams). This means that we need to see if these diagrams are the same or not if we switch, say, the incoming electron with the outgoing positron. From figure 13.7 it's easy to see that they are, and so rule 8 says that these diagrams subtract.

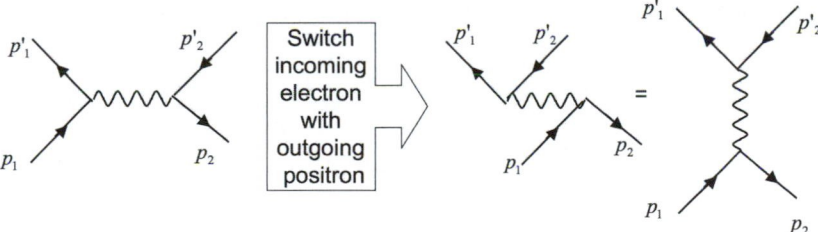

FIGURE 13.7
Equivalence of the two diagrams under particle interchange.

Hence we obtain

$$-i\mathcal{M} = -i\mathcal{M}_{\text{left}} - \left(-i\mathcal{M}_{\text{right}}\right)$$

$$= (-ig_e)^2 \left[\bar{u}^{(i_1')}(p_1')\gamma^\mu v^{(i_2')}(p_2')\right] \frac{i g^2}{(p_1+p_2)^2} \left[\bar{v}^{(i_2)}(p_2)\gamma^\nu u^{(i_1)}(p_1)\right]$$

$$- (-ig_e)^2 \left[\bar{u}^{(i_1')}(p_1')\gamma^\mu u^{(i_1)}(p_1)\right] \frac{i g^2}{(p_1'-p_1)^2} \left[\bar{v}^{(i_2)}(p_2)\gamma^\nu v^{(i_2')}(p_2')\right] (13.10)$$

for the matrix element for this process.

13.2.3 Compton Scattering

This process involves a photon scattering off of an electron:

$$e^- + \gamma \longrightarrow e^- + \gamma$$

Again we have two diagrams to lowest order as shown in figure 13.8 and now

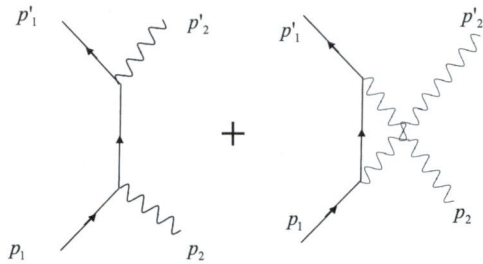

FIGURE 13.8
Compton scattering to lowest order.

we add them since fermion interchange is irrelevant. Applying the rules now gives

$$-i\mathcal{M}_{\text{left}} = \varepsilon_\mu^{*\prime}(p_2') \int \frac{d^4q}{(2\pi)^4} \left[\bar{u}^{(i_1')}(p_1')\left(-ig_e\gamma^\mu\right) \frac{i\left(\gamma\cdot q + m\right)}{q^2 - m^2}\left(-ig_e\gamma^\nu\right)u^{(i_1)}(p_1)\right]$$

$$\times \varepsilon_\nu(p_2)(2\pi)^4\delta\left(p_1' + p_2' - q\right)(2\pi)^4\delta\left(q - p_1 - p_2\right) \qquad (13.11)$$

and

$$-i\mathcal{M}_{\text{right}} = \varepsilon_\nu^{*\prime}(p_2') \int \frac{d^4q}{(2\pi)^4} \left[\bar{u}^{(i_1')}(p_1')\left(-ig_e\gamma^\mu\right) \frac{i\left(\gamma\cdot q + m\right)}{q^2 - m^2}\left(-ig_e\gamma^\nu\right)u^{(i_1)}(p_1)\right]$$

$$\times \varepsilon_\mu(p_2)(2\pi)^4\delta\left(p_2' + q - p_1\right)(2\pi)^4\delta\left(p_1' - q - p_2\right) \qquad (13.12)$$

Note that the only differences have to do with where the polarizations contract and with how the delta functions conserve momenta.

13.3 Obtaining Cross-Sections

In order to finish any of these calculations we would have to specify all of the initial and final spins in the problem, insert the corresponding free-particle solutions into the expression for $\mathcal{M}$ and then integrate over the solid angle to get the cross-section. This is enormously tedious and fortunately unnecessary – as we discussed in Chapter 9, most particle beams are unpolarized, with spins randomly distributed along any axis, and most particle detectors count particles regardless of spin. We therefore apply the spin-summing/averaging trick of eq. (9.36):

$$\left|\overline{\mathcal{M}}\right|^2 = \left(\frac{1}{2}\right)^2 \sum_{i_1,i_2=\uparrow,\downarrow} \sum_{i'_1,i'_2=\uparrow,\downarrow} |\mathcal{M}|^2 \tag{13.13}$$

where the factor of $\left(\frac{1}{2}\right)^2$ arises because we have two initial spins to average over. The expressions then simplify due to the following identities

$$\sum_{i=\uparrow,\downarrow} u_a^{(i)}(p)\overline{u}_b^{(i)}(p) = (\not{p} + m)_{ab} \qquad \sum_{i=\uparrow,\downarrow} v_a^{(i)}(p)\overline{v}_b^{(i)}(p) = (\not{p} - m)_{ab} \tag{13.14}$$

which you can prove yourself. These identities are used in what are called the "Casimir tricks" [122]:

$$\sum_{i_A,i_B=\uparrow,\downarrow} \left[\overline{u}^{(i_A)}(p_A)\Gamma_{\mathrm{I}}u^{(i_B)}(p_B)\right]^\dagger \left[\overline{u}^{(i_A)}(p_A)\Gamma_{\mathrm{II}}u^{(i_B)}(p_B)\right]$$

$$= \mathrm{Tr}\left[\overline{\Gamma}_{\mathrm{I}}\left(\not{p}_A + m_A\right)\Gamma_{\mathrm{II}}\left(\not{p}_B + m_B\right)\right] \tag{13.15}$$

$$\sum_{i_A,i_B=\uparrow,\downarrow} \left[\overline{v}^{(i_A)}(p_A)\Gamma_{\mathrm{I}}v^{(i_B)}(p_B)\right]^\dagger \left[\overline{v}^{(i_A)}(p_A)\Gamma_{\mathrm{II}}v^{(i_B)}(p_B)\right]$$

$$= \mathrm{Tr}\left[\overline{\Gamma}_{\mathrm{I}}\left(\not{p}_A - m_A\right)\Gamma_{\mathrm{II}}\left(\not{p}_B - m_B\right)\right] \tag{13.16}$$

$$\sum_{i_A,i_B=\uparrow,\downarrow} \left[\overline{v}^{(i_A)}(p_A)\Gamma_{\mathrm{I}}u^{(i_B)}(p_B)\right]^\dagger \left[\overline{v}^{(i_A)}(p_A)\Gamma_{\mathrm{II}}u^{(i_B)}(p_B)\right]$$

$$= \mathrm{Tr}\left[\overline{\Gamma}_{\mathrm{I}}\left(\not{p}_A - m_A\right)\Gamma_{\mathrm{II}}\left(\not{p}_B + m_B\right)\right] \tag{13.17}$$

for any two 4×4 matrices Γ_{I} and Γ_{II}, where $\overline{\Gamma}_{\mathrm{I}} \equiv \gamma^0 \Gamma_{\mathrm{I}}^\dagger \gamma^0$. I'll relegate the proof of these relations to an Appendix. For now, let's use them to finish the calculations.

Let's consider electron-muon scattering. First we need $|\mathcal{M}|^2$:

$$|\mathcal{M}|^2 = \mathcal{M}^\dagger \mathcal{M}$$

$$= \frac{e^4}{(p_1' - p_1)^4} \left(\left[\overline{u}^{(i_1')}(p_1') \gamma^\mu u^{(i_1)}(p_1) \right] \left[\overline{u}^{(i_2')}(p_2') \gamma_\mu u^{(i_2)}(p_2) \right] \right)^\dagger$$

$$\times \left(\left[\overline{u}^{(i_1')}(p_1') \gamma^\nu u^{(i_1)}(p_1) \right] \left[\overline{u}^{(i_2')}(p_2') \gamma_\nu u^{(i_2)}(p_2) \right] \right)$$

$$= \frac{e^4}{(p_1' - p_1)^4} \left(\left[\overline{u}^{(i_2')}(p_2') \gamma_\mu u^{(i_2)}(p_2) \right]^\dagger \left[\overline{u}^{(i_1')}(p_1') \gamma^\mu u^{(i_1)}(p_1) \right]^\dagger \right)$$

$$\times \left(\left[\overline{u}^{(i_1')}(p_1') \gamma^\nu u^{(i_1)}(p_1) \right] \left[\overline{u}^{(i_2')}(p_2') \gamma_\nu u^{(i_2)}(p_2) \right] \right)$$

$$= \frac{e^4}{(p_1' - p_1)^4} \left(\left[\overline{u}^{(i_1')}(p_1') \gamma^\mu u^{(i_1)}(p_1) \right]^\dagger \left[\overline{u}^{(i_1')}(p_1') \gamma^\nu u^{(i_1)}(p_1) \right] \right)$$

$$\times \left(\left[\overline{u}^{(i_2')}(p_2') \gamma_\mu u^{(i_2)}(p_2) \right]^\dagger \left[\overline{u}^{(i_2')}(p_2') \gamma_\nu u^{(i_2)}(p_2) \right] \right) \tag{13.18}$$

Note that I've done two things here. In the first line, I relabeled the Greek index μ (in the $\mathcal{M}^\dagger$ part) to ν (in the $\mathcal{M}$ part) so I won't get confused between the two. Then, in going from the 2nd line to the 3rd line, I regrouped the square-bracketed pieces with common factors of momenta so that I can use the Casimir trick. For this particular case, the Γ_I and Γ_II matrices are going to be given by γ^μ and γ^ν. The trick implies that

$$\sum_{i_1, i_1' = \uparrow, \downarrow} \left[\overline{u}^{(i_1')}(p_1') \gamma^\mu u^{(i_1)}(p_1) \right]^\dagger \left[\overline{u}^{(i_1')}(p_1') \gamma^\nu u^{(i_1)}(p_1) \right]$$

$$= \mathrm{Tr} \left[\overline{\gamma}^\mu \left(\not{p}_1' + m \right) \gamma^\nu \left(\not{p}_1 + m \right) \right] \tag{13.19}$$

$$\sum_{i_2, i_2' = \uparrow, \downarrow} \left[\overline{u}^{(i_2')}(p_2') \gamma_\mu u^{(i_2)}(p_2) \right]^\dagger \left[\overline{u}^{(i_2')}(p_2') \gamma_\nu u^{(i_2)}(p_2) \right]$$

$$= \mathrm{Tr} \left[\overline{\gamma}_\mu \left(\not{p}_2' + M \right) \gamma_\nu \left(\not{p}_2 + M \right) \right] \tag{13.20}$$

where m is the electron mass and M is the muon mass. So we get

$$\overline{|\mathcal{M}|}^2 = \left(\frac{1}{2} \right)^2 \sum_{i_1, i_2 = \uparrow, \downarrow} \sum_{i_1', i_2' = \uparrow, \downarrow} |\mathcal{M}|^2$$

$$= \frac{1}{4} \left(\frac{e^2}{(p_1' - p_1)^2} \right)^2 \mathrm{Tr} \left[\overline{\gamma}^\mu \left(\not{p}_1' + m \right) \gamma^\nu \left(\not{p}_1 + m \right) \right]$$

$$\times \mathrm{Tr} \left[\overline{\gamma}_\mu \left(\not{p}_2' + M \right) \gamma_\nu \left(\not{p}_2 + M \right) \right] \tag{13.21}$$

and so we see that spin-summing/averaging has turned the calculation into a problem of computing products of traces over γ-matrices!

How do we do these? The first thing to note is that

$$\overline{\gamma}^\mu \equiv \gamma^0 \left(\gamma^\mu \right)^\dagger \gamma^0 = \gamma^\mu \tag{13.22}$$

which you can show by brute force from the definition of the γ-matrices. The next step is to note that, since the γ-matrices are 4×4 matrices, we get the following two simple identities

$$\text{Tr}\,[I] = 4 \quad \text{and} \quad \text{Tr}\,[\gamma^\mu \gamma^\nu] = 4g^{\mu\nu} \tag{13.23}$$

proved in problem 2 from Chapter 11, along with other results for traces and products. These are listed in section 13.4.2 of the appendix for convenience.

Using these results we find

$$\text{Tr}\left[\overline{\gamma}^\mu \left(\slashed{p}'_1 + m\right)\gamma^\nu\left(\slashed{p}_1 + m\right)\right]$$

Tr [odd # of γ's] = 0.

$$= \text{Tr}\left[\overline{\gamma}^\mu \slashed{p}'_1 \gamma^\nu \slashed{p}_1\right] + m^2\text{Tr}\left[\overline{\gamma}^\mu \gamma^\nu\right]$$

$\overline{\gamma}^\mu = \gamma^\mu$ (13.22)

$$= \text{Tr}\left[\gamma^\mu \slashed{p}'_1 \gamma^\nu \slashed{p}_1\right] + m^2\text{Tr}\left[\gamma^\mu \gamma^\nu\right]$$

$$= \text{Tr}\left[\gamma^\mu \left(p'_{1\lambda}\gamma^\lambda\right)\gamma^\nu \left(p_{1\eta}\gamma^\eta\right)\right] + 4m^2 g^{\mu\nu}$$

definition of $\slashed{p}'_1$ and $\slashed{p}_1$.

$$= p'_{1\lambda}p_{1\eta}\text{Tr}\left[\gamma^\mu \gamma^\lambda \gamma^\nu \gamma^\eta\right] + 4m^2 g^{\mu\nu}$$

$p'_{1\lambda}$ and $p_{1\eta}$ are just numbers.

$$= 4p'_{1\lambda}p_{1\eta}\left(g^{\mu\lambda}g^{\nu\eta} - g^{\mu\nu}g^{\lambda\eta} + g^{\mu\eta}g^{\nu\lambda}\right) + 4m^2 g^{\mu\nu}$$

$p'_{1\lambda}p_{1\eta}\,g^{\mu\nu}g^{\lambda\eta}$
$= g^{\mu\nu}\,p'^{\,\eta}_1\,p_{1\eta}$

$$= 4\left(p'^\mu_1 p^\nu_1 + p'^\nu_1 p^\mu_1 + g^{\mu\nu}\left[m^2 - p'_1 \cdot p_1\right]\right)$$

$= g^{\mu\nu}\,p'_1\cdot p_1$

$$\equiv 4L^{\mu\nu}(p'_1, p_1; m^2) \tag{13.24}$$

(definition of scalar products).

where note that in going from the 3rd to the 4th line the p's come out of the trace because they are not matrices. In going from the 4th to the 5th line I used the relation

$$\text{Tr}\left[\gamma^\mu \gamma^\lambda \gamma^\nu \gamma^\eta\right] = 4\left(g^{\mu\lambda}g^{\nu\eta} - g^{\mu\nu}g^{\lambda\eta} + g^{\mu\eta}g^{\nu\lambda}\right) \tag{13.25}$$

listed in the appendix. Similarly

$$\text{Tr}\left[\overline{\gamma}_\mu \left(\slashed{p}'_2 + M\right)\gamma_\nu\left(\slashed{p}_2 + M\right)\right] = 4\left(p'_{2\mu}p_{2\nu} + p'_{2\nu}p_{2\mu} + g_{\mu\nu}\left[M^2 - p'_2 \cdot p_2\right]\right)$$

$$= 4L_{\mu\nu}(p'_2, p_2; M^2) \tag{13.26}$$

So the matrix element is

(definition of $L_{\mu\nu}$ corrected to be consistent with Chapter 17.

$$|\mathcal{M}|^2 = \frac{16}{4}\left(\frac{e^2}{(p'_1 - p_1)^2}\right)^2 L^{\mu\nu}(p'_1, p_1; m^2)L_{\mu\nu}(p'_2, p_2; M^2)$$

$$= 4\left(\frac{e^2}{(p'_1 - p_1)^2}\right)^2 \left(p'^\mu_1 p^\nu_1 + p'^\nu_1 p^\mu_1 + g^{\mu\nu}\left[m^2 - p'_1 \cdot p_1\right]\right)$$

$$\times \left(p'_{2\mu}p_{2\nu} + p'_{2\nu}p_{2\mu} + g_{\mu\nu}\left[M^2 - p'_2 \cdot p_2\right]\right)$$

expands the brackets.

$$= 4\left(\frac{e^2}{(p'_1 - p_1)^2}\right)^2 \left[2\left(p'_1 \cdot p'_2\right)\left(p_1 \cdot p_2\right) + 2\left(p'_1 \cdot p_2\right)\left(p_1 \cdot p'_2\right)\right]$$

$$+2\left(p'_2 \cdot p_2\right)\left[m^2 - p'_1 \cdot p_1\right] + 2\left(p'_1 \cdot p_1\right)\left[M^2 - p'_2 \cdot p_2\right]$$
$$+4\left[m^2 - p'_1 \cdot p_1\right]\left[M^2 - p'_2 \cdot p_2\right]]$$
$$= 8\left(\frac{e^2}{\left(p'_1 - p_1\right)^2}\right)^2 \left[\left(p'_1 \cdot p'_2\right)\left(p_1 \cdot p_2\right) + \left(p'_1 \cdot p_2\right)\left(p_1 \cdot p'_2\right)\right.$$
$$\left. - \left(p'_1 \cdot p_1\right)M^2 - \left(p'_2 \cdot p_2\right)m^2 + 2m^2 M^2\right] \tag{13.27}$$

which is a Lorentz-invariant scalar!

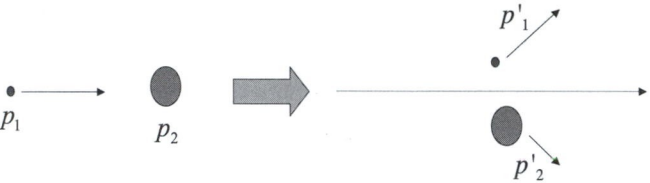

FIGURE 13.9

We could now evaluate this quantity in some frame of reference and compute a cross-section from it! For example, in the lab frame, with the muon initially at rest as illustrated in figure 13.9:

$$p_1^\mu = (E, \vec{p}) \qquad p_2^\mu = (M, 0) \qquad p_1'^\mu = (E', \vec{p'}) \tag{13.28}$$

and we need to compute all of the "dot products" in the expression. These are:

$$p_1 \cdot p_2 = EM \qquad p'_1 \cdot p_2 = E'M \qquad p'_1 \cdot p_1 = EE' - |\vec{p}||\vec{p'}|\cos\theta$$
$$p_1 \cdot p'_2 = p_1 \cdot (p_1 + p_2 - p'_1) = m^2 + EM - (EE' - |\vec{p}||\vec{p'}|\cos\theta)$$
$$p'_1 \cdot p'_2 = p'_1 \cdot (p_1 + p_2 - p'_1) = (EE' - |\vec{p}||\vec{p'}|\cos\theta) + E'M - m^2$$
$$p'_2 \cdot p_2 = p_2 \cdot (p_1 + p_2 - p'_1) = EM + M^2 - E'M \tag{13.29}$$
$$\left(p'_1 - p_1\right)^2 = m^2 + m^2 - 2\left(EE' - |\vec{p}||\vec{p'}|\cos\theta\right)$$

and so

$$\left(p'_1 \cdot p'_2\right)\left(p_1 \cdot p_2\right) + \left(p'_1 \cdot p_2\right)\left(p_1 \cdot p'_2\right) - \left(p'_1 \cdot p_1\right)M^2 - \left(p'_2 \cdot p_2\right)m^2 + 2m^2 M^2$$
$$= \left[\left(EE' - |\vec{p}||\vec{p'}|\cos\theta\right) + E'M - m^2\right]EM$$
$$+ E'M\left[m^2 + EM - \left(EE' - |\vec{p}||\vec{p'}|\cos\theta\right)\right] - \left[EE' - |\vec{p}||\vec{p'}|\cos\theta\right]M^2$$
$$- \left[EM + M^2 - E'M\right]m^2 + 2m^2 M^2$$
$$= M\left(E' - E + M\right)|\vec{p}||\vec{p'}|\cos\theta + M^2\left(EE' + m^2\right)$$
$$+ M\left(EE' - 2m^2\right)\left(E - E'\right) \tag{13.30}$$

giving

$$|\overline{\mathcal{M}}|^2 = \frac{8e^4}{4\left[m^2 - (EE' - |\vec{p}|\,|\vec{p}\,'|\cos\theta)\right]^2}$$
$$\times \left[M(E' - E + M)|\vec{p}|\,|\vec{p}\,'|\cos\theta\right.$$
$$\left. + M^2\left(EE' + m^2\right) + M\left(EE' - 2m^2\right)(E - E')\right] \quad (13.31)$$

which can be put into the lab-frame cross-section and evaluated:

$$\left(\frac{d\sigma}{d\Omega}\right)_{\text{LAB}} = \left(\frac{\hbar}{8\pi}\right)^2 \frac{|\overline{\mathcal{M}}|^2\,|\vec{p}\,'|^2}{M\,|\vec{p}|\left[|\vec{p}\,'|\left(E + Mc^2\right) - |\vec{p}|\,E'\cos\theta\right]} \quad (13.32)$$

The preceding expression is quite cumbersome, but we can understand its physics using a simple approximation. Suppose that the muon is so heavy that we can neglect its recoil. This implies that $M >> E, E'$ and $|\vec{p}| \simeq |\vec{p}\,'|$, $E \simeq E' \simeq \sqrt{|\vec{p}|^2 + m^2}$ so eq. (13.31) reduces in this limit to

$$|\overline{\mathcal{M}}|^2 \simeq \frac{8e^4 M^2 \left[|\vec{p}|^2(1 + \cos\theta) + 2m^2\right]}{4\,|\vec{p}|^4\,[1 - \cos\theta]^2} = \frac{e^4 M^2 \left[|\vec{p}|^2\cos^2\frac{\theta}{2} + m^2\right]}{|\vec{p}|^4\sin^4\frac{\theta}{2}} \quad (13.33)$$

and eq.(13.32) becomes (putting into $\mathcal{M}$ the correct factors of c and remembering that $e = \sqrt{4\pi\alpha}$)

$$\left(\frac{d\sigma}{d\Omega}\right)_{\text{Mott}} \simeq \left(\frac{\hbar}{8\pi M}\right)^2 \frac{M^2 e^4 \left[|\vec{p}|^2\cos^2\frac{\theta}{2} + (mc)^2\right]}{|\vec{p}|^4\sin^4\frac{\theta}{2}}$$
$$= \left(\frac{\hbar\alpha}{2\,|\vec{p}|^2\sin^2\frac{\theta}{2}}\right)^2 \left[|\vec{p}|^2\cos^2\frac{\theta}{2} + (mc)^2\right] \quad (13.34)$$

which is called the Mott formula [123].

The Mott formula is a good approximation for electron-muon scattering or for that matter electron-proton scattering if we take M to be the proton mass. Note that if the incident electron is non-relativistic then $|\vec{p}| \sim m\,|\vec{v}| << (mc)$ and eq. (13.34) reduces to

$$\left(\frac{d\sigma}{d\Omega}\right)_{\text{Mott}} \simeq \left(\frac{\hbar c\alpha}{2mv^2\sin^2\frac{\theta}{2}}\right)^2 = \left(\frac{d\sigma}{d\Omega}\right)_{\text{Rutherford}} \quad (13.35)$$

which is the original Rutherford scattering formula [18], first developed by Ernest Rutherford to explain the data from his experiments scattering alpha particles off of a gold foil. Mott's formula was developed by Neville Mott to describe the analogous process when alpha particles are replaced by electrons which, being much lighter, have relativistic momenta necessitating use of the more detailed formula (13.34).

13.4 Appendix: Mathematical Tools for QED

13.4.1 The Casimir Trick

It is very common to encounter terms of the form $\left[\overline{u}^{(i_A)}(p_A)\Gamma u^{(i_B)}(p_B)\right]$ when evaluating Feynman diagrams, where Γ will be some product of matrices (typically composed of γ matrices). The Casimir trick helps us deal with these quantities when we employ spin-averaging and spin-summing to our expressions.

Consider the following expression:

$$\overline{u}^{\dagger(i_A)}(p_A) = \left(u\gamma^0\right)^{\dagger}$$
$$= \left(\gamma^0\right)^{\dagger}\left(u^{\dagger}\right)^{\dagger}$$
$$= \gamma^0 u$$

$$\left[\overline{u}^{(i_A)}(p_A)\Gamma_{\mathrm{I}}u^{(i_B)}(p_B)\right]^{\dagger} = \left[u^{\dagger(i_B)}(p_B)\Gamma_{\mathrm{I}}^{\dagger}\overline{u}^{\dagger(i_A)}(p_A)\right]$$
$$= \left[u^{\dagger(i_B)}(p_B)\gamma^0\gamma^0\Gamma_{\mathrm{I}}^{\dagger}\gamma^0 u^{(i_A)}(p_A)\right]$$
$$= \left[\overline{u}^{(i_B)}(p_B)\gamma^0\Gamma_{\mathrm{I}}^{\dagger}\gamma^0 u^{(i_A)}(p_A)\right]$$
$$= \left[\overline{u}^{(i_B)}(p_B)\overline{\Gamma}_{\mathrm{I}}u^{(i_A)}(p_A)\right] \tag{13.36}$$

The second line follows because $\left(\gamma^0\right)^2 = I$ and because $\overline{u}^{\dagger} = \left(\left(u^{\dagger}\right)\gamma^0\right)^{\dagger} = \gamma^0 u$ (since $\left(\gamma^0\right)^{\dagger} = \gamma^0$). The last line follows from the definition $\overline{\Gamma}_{\mathrm{I}} = \gamma^0\Gamma_{\mathrm{I}}^{\dagger}\gamma^0$. Now consider

$$\sum_{i_A,i_B=\uparrow,\downarrow}\left[\overline{u}^{(i_A)}(p_A)\Gamma_{\mathrm{I}}u^{(i_B)}(p_B)\right]^{\dagger}\left[\overline{u}^{(i_A)}(p_A)\Gamma_{\mathrm{II}}u^{(i_B)}(p_B)\right]$$
$$= \sum_{i_A,i_B=\uparrow,\downarrow}\left[\overline{u}^{(i_B)}(p_B)\overline{\Gamma}_{\mathrm{I}}u^{(i_A)}(p_A)\right]\left[\overline{u}^{(i_A)}(p_A)\Gamma_{\mathrm{II}}u^{(i_B)}(p_B)\right]$$
$$= \sum_{i_A,i_B=\uparrow,\downarrow}\overline{u}_a^{(i_B)}(p_B)\left(\overline{\Gamma}_{\mathrm{I}}\right)_{ab}u_b^{(i_A)}(p_A)\overline{u}_c^{(i_A)}(p_A)\left(\Gamma_{\mathrm{II}}\right)_{cd}u_d^{(i_B)}(p_B) \tag{13.37}$$

where in the last line I have explicitly written out all of the Dirac indices that are summed over, so that for example $\Gamma_{\mathrm{II}}u^{(i_B)}(p_B) = \sum_{d=1}^{4}\left(\Gamma_{\mathrm{II}}\right)_{cd}u_d^{(i_B)}(p_B) = \left(\Gamma_{\mathrm{II}}\right)_{cd}u_d^{(i_B)}(p_B)$ using the summation convention. We can now sum over the spin-indices:

$$\sum_{i_A,i_B=\uparrow,\downarrow}\overline{u}_a^{(i_B)}(p_B)\left(\overline{\Gamma}_{\mathrm{I}}\right)_{ab}u_b^{(i_A)}(p_A)\overline{u}_c^{(i_A)}(p_A)\left(\Gamma_{\mathrm{II}}\right)_{cd}u_d^{(i_B)}(p_B)$$

$$= \sum_{i_B=\uparrow,\downarrow}\overline{u}_a^{(i_B)}(p_B)\left(\overline{\Gamma}_{\mathrm{I}}\right)_{ab}\left(\sum_{i_A=\uparrow,\downarrow}u_b^{(i_A)}(p_A)\overline{u}_c^{(i_A)}(p_A)\right)\left(\Gamma_{\mathrm{II}}\right)_{cd}u_d^{(i_B)}(p_B)$$

$$= \sum_{i_B=\uparrow,\downarrow}\overline{u}_a^{(i_B)}(p_B)\left(\overline{\Gamma}_{\mathrm{I}}\right)_{ab}\left(\not{p}_A + m_A\right)_{bc}\left(\Gamma_{\mathrm{II}}\right)_{cd}u_d^{(i_B)}(p_B) \tag{13.38}$$

where the last line follows from the identity $\sum_{i=\uparrow,\downarrow} u_a^{(i)}(p)\bar{u}_b^{(i)}(p) = (\not{p}+m)_{ab}$. Now I can do the other spin-sum by moving the components of $u_d^{(i_B)}(p_B)$ all the way to the left (since they're just functions, this is allowed)

$$\sum_{i_B=\uparrow,\downarrow} \bar{u}_a^{(i_B)}(p_B)\left(\overline{\Gamma}_{\mathrm{I}}\right)_{ab}\left(\not{p}_A+m_A\right)_{bc}\left(\Gamma_{\mathrm{II}}\right)_{cd} u_d^{(i_B)}(p_B)$$

$$= \sum_{i_B=\uparrow,\downarrow} u_d^{(i_B)}(p_B)\bar{u}_a^{(i_B)}(p_B)\left(\overline{\Gamma}_{\mathrm{I}}\right)_{ab}\left(\not{p}_A+m_A\right)_{bc}\left(\Gamma_{\mathrm{II}}\right)_{cd}$$

$$= \left(\not{p}_B+m_B\right)_{da}\left(\overline{\Gamma}_{\mathrm{I}}\right)_{ab}\left(\not{p}_A+m_A\right)_{bc}\left(\Gamma_{\mathrm{II}}\right)_{cd} \qquad (13.39)$$

using the same identities as before. Note now that the last line is just the trace of the product of the 4 matrices $\left(\not{p}_B+m_B\right), \overline{\Gamma}_{\mathrm{I}}, \left(\not{p}_A+m_A\right)$, and Γ_{II}.

Hence we find

$$\sum_{i_A,i_B=\uparrow,\downarrow} \left[\bar{u}^{(i_A)}(p_A)\Gamma_{\mathrm{I}} u^{(i_B)}(p_B)\right]^\dagger\left[\bar{u}^{(i_A)}(p_A)\Gamma_{\mathrm{II}} u^{(i_B)}(p_B)\right]$$

$$= \left(\not{p}_B+m_B\right)_{da}\left(\overline{\Gamma}_{\mathrm{I}}\right)_{ab}\left(\not{p}_A+m_A\right)_{bc}\left(\Gamma_{\mathrm{II}}\right)_{cd}$$

$$= \mathrm{Tr}\left[\left(\not{p}_B+m_B\right)\overline{\Gamma}_{\mathrm{I}}\left(\not{p}_A+m_A\right)\Gamma_{\mathrm{II}}\right]$$

$$= \mathrm{Tr}\left[\overline{\Gamma}_{\mathrm{I}}\left(\not{p}_A+m_A\right)\Gamma_{\mathrm{II}}\left(\not{p}_B+m_B\right)\right] \qquad (13.40)$$

where the last line follows from trace cyclicity ($\mathrm{Tr}(AB) = \mathrm{Tr}(BA)$). This is the Casimir trick. By similar reasoning (try it!) you can show

$$\sum_{i_A,i_B=\uparrow,\downarrow} \left[\bar{v}^{(i_A)}(p_A)\Gamma_{\mathrm{I}} v^{(i_B)}(p_B)\right]^\dagger\left[\bar{v}^{(i_A)}(p_A)\Gamma_{\mathrm{II}} v^{(i_B)}(p_B)\right]$$

$$= \mathrm{Tr}\left[\overline{\Gamma}_{\mathrm{I}}\left(\not{p}_A-m_A\right)\Gamma_{\mathrm{II}}\left(\not{p}_B-m_B\right)\right] \qquad (13.41)$$

$$\sum_{i_A,i_B=\uparrow,\downarrow} \left[\bar{v}^{(i_A)}(p_A)\Gamma_{\mathrm{I}} u^{(i_B)}(p_B)\right]^\dagger\left[\bar{v}^{(i_A)}(p_A)\Gamma_{\mathrm{II}} u^{(i_B)}(p_B)\right]$$

$$= \mathrm{Tr}\left[\overline{\Gamma}_{\mathrm{I}}\left(\not{p}_A-m_A\right)\Gamma_{\mathrm{II}}\left(\not{p}_B+m_B\right)\right] \qquad (13.42)$$

13.4.2 Dirac γ−Matrices and Their Traces

The γ−matrices are

$$\gamma^0 = \begin{pmatrix} I & 0 \\ 0 & -I \end{pmatrix} \qquad \gamma^i = \begin{pmatrix} 0 & \sigma^i \\ -\sigma^i & 0 \end{pmatrix} \qquad (13.43)$$

and obey the relation

$$\{\gamma^\mu, \gamma^\nu\} = 2g^{\mu\nu} \qquad (13.44)$$

We define some auxiliary matrices via the relations

$$\gamma^5 = i\gamma^0\gamma^1\gamma^2\gamma^3 \qquad \Sigma^{\mu\nu} = \frac{i}{4}[\gamma^\mu, \gamma^\nu] \tag{13.45}$$

$$\Sigma^{0i} = -\frac{i}{2}\begin{pmatrix} 0 & \sigma^i \\ \sigma^i & 0 \end{pmatrix} \qquad \Sigma^{ij} = \frac{1}{2}\epsilon_{ijk}\begin{pmatrix} \sigma^k & 0 \\ 0 & \sigma^k \end{pmatrix} \tag{13.46}$$

Note that we now must distinguish between raised and lowered indices: $\gamma^0 = \gamma_0$ and $\gamma^i = -\gamma_i$. The following identities can be directly verified:

$$
\begin{array}{lll}
\left(\gamma^0\right)^\dagger = \gamma^0 & \left(\gamma^i\right)^\dagger = -\gamma^i & \overline{\gamma}^\mu = \gamma^\mu \\
\left(\gamma^0\right)^2 = I & \left(\gamma^i\right)^2 = -I & \gamma_\alpha\gamma^\alpha = 4 \\
\gamma_\alpha\gamma^\nu\gamma^\alpha = -2\gamma^\nu & & \gamma_\alpha\slashed{a}\gamma^\alpha = -2\slashed{a} \\
\gamma_\alpha\gamma^\mu\gamma^\nu\gamma^\alpha = 4g^{\mu\nu} & & \gamma_\alpha\slashed{a}\slashed{b}\gamma^\alpha = 4a\cdot b \\
\gamma_\alpha\gamma^\mu\gamma^\nu\gamma^\lambda\gamma^\alpha = -2\gamma^\lambda\gamma^\mu\gamma^\nu & & \gamma_\alpha\slashed{a}\slashed{b}\slashed{c}\gamma^\alpha = -2\slashed{c}\slashed{b}\slashed{a}
\end{array}
$$

Using these relations, we get the following trace theorems

$$
\begin{array}{ll}
\text{Tr}\,[I] = 4 & \text{Tr}\,[\text{odd \# of } \gamma\text{-matrices}] = 0 \\
\text{Tr}\,[\gamma^\mu\gamma^\nu] = 4g^{\mu\nu} & \text{Tr}\,[\slashed{a}\slashed{b}] = 4a\cdot b \\
\text{Tr}\,[\gamma^\mu\gamma^\nu\gamma^\alpha\gamma^\beta] = 4\left(g^{\mu\nu}g^{\alpha\beta} - g^{\mu\alpha}g^{\nu\beta} + g^{\mu\beta}g^{\nu\alpha}\right) & \\
\text{Tr}\,[\slashed{a}\slashed{b}\slashed{c}\slashed{d}] = 4\left(a\cdot b\, c\cdot d - a\cdot c\, b\cdot d + a\cdot d\, b\cdot c\right) & \\
\text{Tr}\,[\gamma^5] = 0 & \text{Tr}\,[\gamma^5\gamma^\mu\gamma^\nu] = 0 \\
\text{Tr}\,[\gamma^5\gamma^\mu\gamma^\nu\gamma^\alpha\gamma^\beta] = 4i\varepsilon^{\mu\nu\alpha\beta} & \text{Tr}\,[\gamma^5\slashed{a}\slashed{b}\slashed{c}\slashed{d}] = 4i\varepsilon^{\mu\nu\alpha\beta}a_\mu b_\nu c_\alpha d_\beta
\end{array}
$$

where $\varepsilon^{\mu\nu\alpha\beta}$ is the 4-dimensional Levi-Civita symbol (or epsilon-tensor) which obeys

$$\varepsilon^{0123} = -1 \qquad \varepsilon_{0123} = +1$$

$$\varepsilon^{\mu\nu\alpha\beta} = \begin{cases} -1 & \text{(if } \mu\nu\alpha\beta \text{ is an even permutation of 0123)} \\ +1 & \text{(if } \mu\nu\alpha\beta \text{ is an odd permutation of 0123)} \end{cases} \tag{13.47}$$

$$\varepsilon_{\mu\nu\alpha\beta} = \begin{cases} +1 & \text{(if } \mu\nu\alpha\beta \text{ is an even permutation of 0123)} \\ -1 & \text{(if } \mu\nu\alpha\beta \text{ is an odd permutation of 0123)} \end{cases}$$

Using this it is possible to prove the following

$$\varepsilon^{\mu\nu\alpha\beta}\varepsilon_{\mu\rho\tau\varkappa} = -\delta^\nu_\rho\delta^\alpha_\tau\delta^\beta_\varkappa - \delta^\alpha_\rho\delta^\beta_\tau\delta^\nu_\varkappa - \delta^\beta_\rho\delta^\nu_\tau\delta^\alpha_\varkappa$$
$$+\delta^\nu_\rho\delta^\beta_\tau\delta^\alpha_\varkappa + \delta^\beta_\rho\delta^\alpha_\tau\delta^\nu_\varkappa + \delta^\alpha_\rho\delta^\nu_\tau\delta^\beta_\varkappa \tag{13.48}$$

$$\varepsilon^{\mu\nu\alpha\beta}\varepsilon_{\mu\nu\tau\varkappa} = -2\left(\delta^\alpha_\tau\delta^\beta_\varkappa - \delta^\alpha_\varkappa\delta^\beta_\tau\right) \tag{13.49}$$

$$\varepsilon^{\mu\nu\alpha\beta}\varepsilon_{\mu\nu\alpha\varkappa} = -6\delta^\beta_\varkappa \tag{13.50}$$

$$\varepsilon^{\mu\nu\alpha\beta}\varepsilon_{\mu\nu\alpha\beta} = -24 \tag{13.51}$$

as I will later demonstrate in the appendix of Chapter 20.

13.5 Questions

1. Consider a theory in which the electron couples to a massless pseudoscalar particle (the ϕ) with the vertex

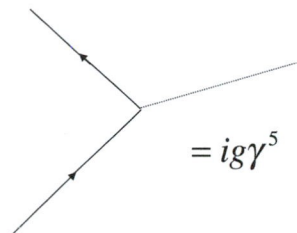

$$= ig\gamma^5$$

FIGURE 13.10

(a) Draw the lowest-order Feynman diagrams for the process $e^+ + e^- \longrightarrow 2\phi$.

(b) Compute to lowest order the differential cross-section for the process in (a) in the CMS frame.

2. For the theory in question #1

(a) Draw the lowest-order Feynman diagrams for the process $\phi + \phi \longrightarrow e^+ + e^-$.

(b) Compute to lowest order the differential cross-section for the process in (a) in the CMS frame.

3. For the theory in question #1:

(a) Draw the lowest-order Feynman diagrams for the process $\phi + e^- \longrightarrow \phi + e^-$

(b) Compute to lowest order the differential cross-section for the process in (a) in the lab frame.

4. The lowest-order diagram for the process $e^- + \mu^- \longrightarrow e^- + \mu^-$ is shown in fig. 13.11. Draw (but don't calculate) and label all relevant diagrams to next-lowest order in the coupling constant.

5. The lowest-order diagrams for the process $e^+ + e^- \longrightarrow e^+ + e^-$ are shown in fig. 13.12. Draw (but don't calculate) and label all relevant diagrams to next-lowest order in the coupling constant.

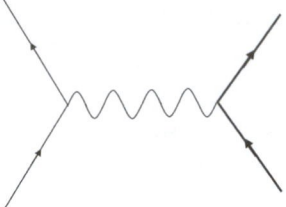

FIGURE 13.11
Electron-muon Scattering Diagram. The muon is drawn with a thicker line.

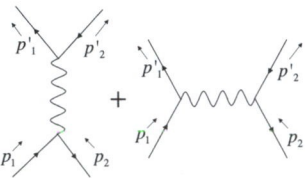

FIGURE 13.12
Electron-positron Scattering Diagrams to lowest order.

6. Consider the Dirac equation in $(1 + 1)$ dimensions (1 spatial + 1 temporal):
$$(i\gamma^\mu \partial_\mu - m)\, \psi = 0$$
where the index $\mu = (0, 1)$, $\partial_\mu = \left(\frac{\partial}{\partial t}, \frac{\partial}{\partial x}\right)$ and the 2x2 γ-matrices are:
$$\gamma^0 = \begin{pmatrix} 1 & 0 \\ 0 & -1 \end{pmatrix} \qquad \gamma^1 = \begin{pmatrix} 0 & 1 \\ -1 & 0 \end{pmatrix}$$

(a) Compute $\{\gamma^\mu, \gamma^\nu\}$. What is the physical interpretation of this result?

(b) Find the solutions to the Dirac equation in $(1 + 1)$ dimensions. Physically interpret your results. Do the particles have spin? Are there antiparticles?

(c) Apply the principle of local gauge invariance to this equation and write down the results.

(d) Show that the quantity $\mathcal{J}_\nu = e\bar{\psi}\gamma_\nu\psi$ is conserved. What is the physical interpretation of the components?

7. Find all diagrams of order e^4 for the processes below. Label them but do not calculate them.

(a) $e^+ + \gamma \longrightarrow e^+ + \gamma$

(b) $\gamma + \gamma \longrightarrow e^+ + e^-$

14

Testing QED

In Chapter 13 we saw that the matrix element for electromagnetically scattering an electron (mass m, initial momentum p_1^μ) off of a muon (mass M, initial momentum p_2^μ) is, from figure 14.1

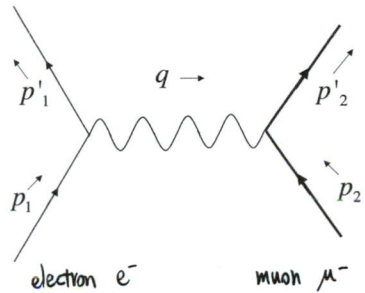

FIGURE 14.1
Electron-muon Scattering Diagram. The muon is drawn with a thicker line to distinguish it from the electron.

$$\mathcal{M} = i \left[\overline{u}^{(i_1')}(p_1')\gamma^\mu u^{(i_1)}(p_1) \right] \frac{e^2}{(p_1' - p_1)^2} \left[\overline{u}^{(i_2')}(p_2')\gamma_\mu u^{(i_2)}(p_2) \right] \qquad (14.1)$$

which after spin-summing/averaging boiled down to

$$|\overline{\mathcal{M}}|^2 = \frac{8e^4 \mathcal{F}(E, M, m, \theta)}{4 \left[m^2 - (EE' - |\vec{p}| \, |\vec{p}\,'| \cos\theta) \right]^2}$$

with

$$\mathcal{F}(E, M, m, \theta) = [M \, (E' - E + M) \, |\vec{p}| \, |\vec{p}\,'| \cos\theta \qquad (14.2)$$
$$+ M^2 \left(EE' + m^2 \right) + M \left(EE' - 2m^2 \right) (E - E')]$$

in the lab-frame. This is an elastic scattering process ($A + B \longrightarrow A + B$), so we obtain in the Lab frame

$$p_1^\mu = (E, \vec{p}) \qquad p_1'^\mu = (E', \vec{p}\,') \qquad \vec{p}_2 = 0 \quad E_2 = M$$

(where the muon is at rest) the expression

$$\frac{d\sigma}{d\Omega} = \left(\frac{\hbar}{8\pi}\right)^2 \frac{\mathcal{S}\left|\overline{\mathcal{M}}\right|^2 |\vec{p}\,'|^2}{M\,|\vec{p}|\,[|\vec{p}\,'|\,(E + Mc^2) - |\vec{p}|\,E'\cos\theta]} \tag{14.3}$$

where the statistical factor $\mathcal{S} = 1$.

The final energy E' of the muon is determined in terms of the initial conditions $(E, \vec{p},$ and $M)$ and the scattering angle θ, which is the only independent variable. We can see this by noting that momentum conservation gives

$$p_2'^2 = (p_2 + p_1 - p_1')^2 \Rightarrow M^2 = M^2 + p_1^2 + p_1'^2 + 2\,(p_2 \cdot (p_1 - p_1')) - p_1 \cdot p_1'$$
$$\Rightarrow 0 = m^2 + M\,(E - E') - (EE' - |\vec{p}|\,|\vec{p}\,'|\cos\theta) \tag{14.4}$$

Proceeding along the lines given in Chapter 2, we solve for E'

$$E' = \frac{(M + E)\,(m^2 + ME) + |\vec{p}|^2\cos\theta\sqrt{M^2 - m^2(1 - \cos\theta)}}{(M + E)^2 - |\vec{p}|^2\cos^2\theta} \tag{14.5}$$

in terms of $(E, \vec{p},$ and $M)$. When the mass m of the electron can be neglected we obtain

$$\begin{aligned}
E' &\simeq \frac{(M + E)\,(0 + ME) + |\vec{p}|^2\cos\theta\sqrt{M^2}}{(M + E)^2 - |\vec{p}|^2\cos^2\theta} \\
&= \frac{ME\,(M + E) + ME^2\cos\theta}{[(M + E) - E\cos\theta]\,[(M + E) + E\cos\theta]} \\
&= \frac{ME}{E(1 - \cos\theta) + M} \tag{14.6}
\end{aligned}$$

which is the result (2.39) that we found in Chapter 2!
(electron acts as a massless particle)

14.1 Basic Features of QED Scattering

Putting all of this together – inserting eqs. (14.2) and (14.5) into eq. (14.3) to get the resultant differential cross-section – gives a very complicated looking formula! Despite this, we can extract some meaningful physical insight from it by looking at some key features it has in common with other 2-fermion QED scattering diagrams.

14.1.1 Coupling

Note that

$$\left|\overline{\mathcal{M}}\right|^2 \propto e^4 \sim \alpha^2 \Rightarrow \frac{d\sigma}{d\Omega} \propto \alpha^2 \tag{14.7}$$

This is because the amplitude has two vertices, each with a factor of $e \sim \sqrt{\alpha}$. The next order correction goes like α^4, and higher orders contribute further corrections proportional to even powers of α. The power of α from any diagram is always $\alpha^{n/2}$ where n is its number of vertices.

14.1.2 Propagator

The denominator of the matrix element comes from the internal line, which gives a factor

$$\frac{1}{(p_1' - p_1)^2} = \frac{1}{2\left[m^2 - (EE' - |\vec{p}|\,|\vec{p}\,'|\cos\theta)\right]^2}$$

$$\simeq \frac{1}{2\left[EE'\,(1 - \cos\theta)\right]^2} + \mathcal{O}\left(\left(\frac{m}{E}\right)^2, \left(\frac{M}{E}\right)^2\right)$$

$$= \frac{1}{8\,(EE')^2 \sin^4\frac{\theta}{2}} \tag{14.8}$$

at high energies, where $|\vec{p}| \simeq E$ and $|\vec{p}\,'| \simeq E'$. If the matrix element had no other structure, then $\frac{d\sigma}{d\Omega} \propto \frac{1}{\sin^4\frac{\theta}{2}}$ which is the salient term in the Rutherford scattering formula (13.35). The exchange of a single virtual particle between two scattering particles will always produce a $\sin^4\frac{\theta}{2}$ factor in the denominator at sufficiently high energy.

$$\approx \left[M\left(E' - E + M\right) EE' \cos\theta + M^2 EE' \right.$$
$$\left. + M\, EE'\left(E - E'\right) \right]$$
$$= MEE'\left(E - E'\right)\left(1 - \cos\theta\right) + M^2 EE'\left(\cos\theta + 1\right).$$
$$\approx 2M^2 EE' \cos^2\frac{\theta}{2}$$

14.1.3 Matrix element

The numerator of the matrix element is

$$\mathcal{F}(E, M, m, \theta) = \left[M\left(E' - E + M\right) |\vec{p}|\,|\vec{p}\,'| \cos\theta \right.$$
$$\left. + M^2\left(EE' + m^2\right) + M\left(EE' - 2m^2\right)\left(E - E'\right) \right]$$

$$\simeq 2M^2 EE' \cos^2\frac{\theta}{2} \tag{14.9}$$

again at high energies, where $|\vec{p}| \simeq E$ and $|\vec{p}\,'| \simeq E'$, indicating that in this limit $\frac{d\sigma}{d\Omega} \propto \cos^2\frac{\theta}{2}$. This feature is due to the spin of the fermions and does not occur in a scalar theory like *ABB* theory. The reason is due to helicity conservation. Suppose that at high energies, the incoming electron is moving along the z-axis with spin up (i.e., spin aligned with its direction of motion). It will be in the state

$$u^{(\uparrow)}(p) = \sqrt{2m}\left(\begin{array}{c} \sqrt{\frac{E+m}{2m}}\,\xi^{(\uparrow)} \\ \sqrt{\frac{E-m}{2m}}\,(\hat{p}\cdot\vec{\sigma})\,\xi^{(\uparrow)} \end{array} \right) \simeq \left(\begin{array}{c} \begin{pmatrix} \sqrt{E} \\ 0 \end{pmatrix} \\ \sigma_3 \begin{pmatrix} \sqrt{E} \\ 0 \end{pmatrix} \end{array} \right) + \mathcal{O}\left(\frac{m}{E}\right) = \sqrt{E}\begin{pmatrix} 1 \\ 0 \\ 1 \\ 0 \end{pmatrix}$$

$$\tag{14.10}$$

whereas the outgoing electron – which is scattered at an angle θ – is effectively in a frame that is rotated by an angle θ relative to the z-axis. Setting, say, $\vec{\theta} = \theta\hat{y}$, it has the form

$$u^{(\uparrow)}(p') = \begin{pmatrix} \exp\left[\frac{i}{2}\vec{\theta}\cdot\vec{\sigma}\right] & 0 \\ 0 & \exp\left[\frac{i}{2}\vec{\theta}\cdot\vec{\sigma}\right] \end{pmatrix} u^{(\uparrow)}(p', \theta = 0)$$

$$= \begin{pmatrix} \left(\cos\frac{\theta}{2} + i\sigma_2\sin\frac{\theta}{2}\right) & 0 \\ 0 & \left(\cos\frac{\theta}{2} + i\sigma_2\sin\frac{\theta}{2}\right) \end{pmatrix} u^{(\uparrow)}(p', \theta = 0)$$

$$\simeq \begin{pmatrix} \cos\theta/2 & \sin\theta/2 & 0 & 0 \\ -\sin\theta/2 & \cos\theta/2 & 0 & 0 \\ 0 & 0 & \cos\theta/2 & \sin\theta/2 \\ 0 & 0 & -\sin\theta/2 & \cos\theta/2 \end{pmatrix} \sqrt{E'} \begin{pmatrix} 1 \\ 0 \\ 1 \\ 0 \end{pmatrix}$$

$$= \sqrt{E'} \begin{pmatrix} \cos\theta/2 \\ -\sin\theta/2 \\ \cos\theta/2 \\ -\sin\theta/2 \end{pmatrix} \tag{14.11}$$

in the same high-energy limit. This gives

$$\bar{u}^{(\uparrow)}(p')\gamma^0 u^{(\uparrow)}(p) \simeq \sqrt{EE'}\left(\cos\frac{\theta}{2}, -\sin\frac{\theta}{2}, \cos\frac{\theta}{2}, -\sin\frac{\theta}{2}\right)(\gamma^0)^2 \begin{pmatrix} 1 \\ 0 \\ 1 \\ 0 \end{pmatrix}$$

$$= 2\sqrt{EE'}\cos\frac{\theta}{2} \tag{14.12}$$

and so $\left|\bar{u}^{(\uparrow)}(p')\gamma^0 u^{(\uparrow)}(p)\right|^2 \simeq EE'\cos^2\frac{\theta}{2}$, consistent with our formula above.

14.1.4 Dimensionality

At high energies, we have

$$|\mathcal{M}|^2 \propto \frac{EE'}{(EE')^2} = \frac{1}{EE'} \quad \text{and} \quad \frac{d\sigma}{d\Omega} \propto |\mathcal{M}|^2 \frac{E'}{E} \Rightarrow \frac{d\sigma}{d\Omega} \propto \frac{1}{E^2} \tag{14.13}$$

because σ is like an area$\sim$(length)$^2 \sim \hbar^2/(\text{momentum})^2 \sim (\hbar c)^2/(\text{energy})^2$. At high energies particle masses are negligible, and dimensionality forces the cross-section to behave like $1/E^2 \sim 1/(\text{CMS energy})^2$ since this is the only energy scale in the problem. This feature is shared by all 2-Body cross-sections in QED.

14.1.5 Antiparticles

We would get exactly the same answer for positron-muon scattering, or electron-antimuon scattering, or positron-antimuon scattering. This is because QED

<mark>is charge-conjugation invariant.</mark>

14.2 Major Tests of QED

There are a number of different ways to test QED. The most important of these have to do with measurements of scattering processes, anomalous magnetic moments, atomic energy level shifts, and changes in the coupling strength. Let's look at a brief sketch of each.

14.2.1 Scattering Processes

There are various scattering processes of interest in QED, which are given in the diagrams in figures 14.2 and 14.3 These can be computed using the

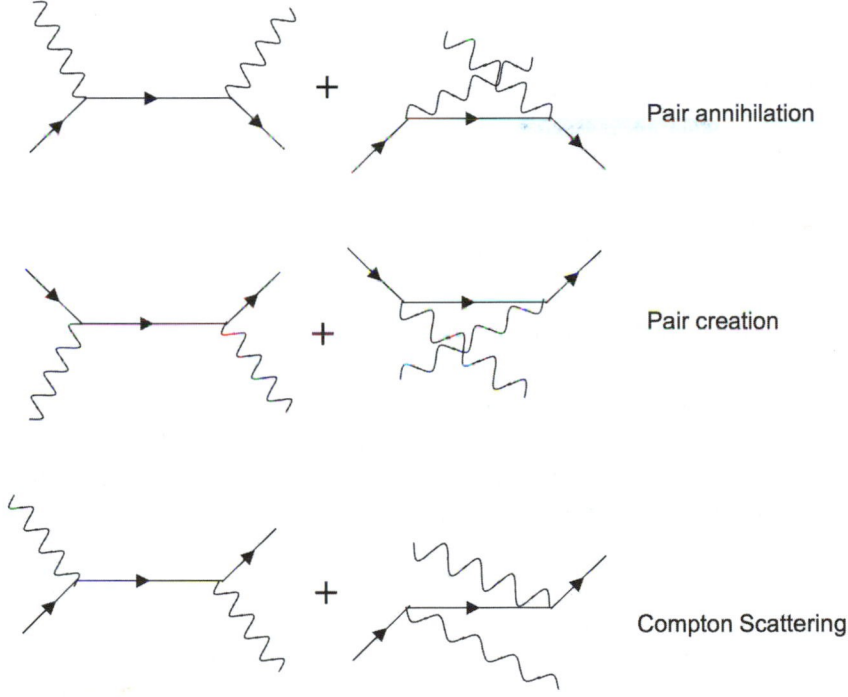

FIGURE 14.2

Lowest order <mark>inelastic scattering</mark> processes in QED.

e-μ　scattering

Electron-electron scattering (Moller scattering)

$+$

Electron-positron scattering (Bhaba scattering)

$+$

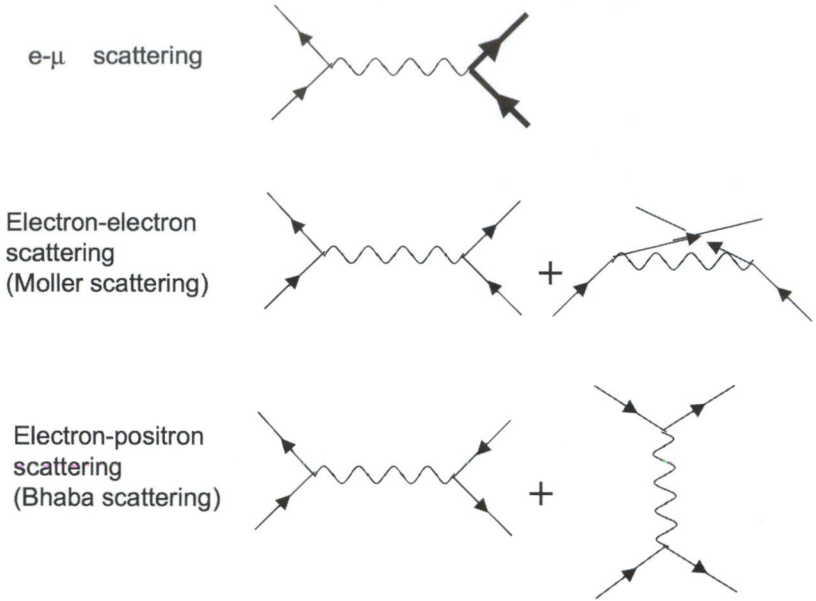

FIGURE 14.3

Lowest order elastic scattering processes in QED.

diagrammatic formalism of Chapter 13, and the resultant cross-sections have been shown to agree with experiment to a high degree of precision.

14.2.2　Anomalous Magnetic Moments

All charged objects with angular momenta have magnetic moments, so we expect electrons (and muons, and taus) to also have magnetic moments. An object with a magnetic moment $\vec{\mu}$ will (non-relativistically) have its interaction with a magnetic field $\vec{B}$ described by a term in the Hamiltonian that is proportional to the magnetic field. This is

$$H_{\text{mag}} = -\vec{\mu} \cdot \vec{B} \quad \text{where } \vec{\mu} = g\mu_B \vec{S} = g\frac{e\hbar}{2mc}\vec{S} \qquad (14.14)$$

where $\vec{S}$ is the spin operator with eigenvalues $\left|\vec{S}\right|^2 = s(s+1)$ for a particle of spin-s, and $\mu_B = \frac{e\hbar}{2mc} = 5.78838263 \times 10^{-5}$ eV T^{-1} is the Bohr magneton in units of electron volts per Tesla. The quantity g is called the *gyromagnetic ratio* of the particle. It is the ratio of the magnetic dipole moment to the mechanical angular momentum of a system – a dimensionless number that indicates the strength of the magnetic moment in Bohr magnetons.

The Dirac equation predicts $g = 2$. We can see this by rewriting the Dirac equation so that it is in non-relativistic Hamiltonian form. Writing $A_\mu = \left(\phi, \vec{A}\right)$, we have

$$\left(i\gamma^\mu D_\mu - m\right)\psi = 0 \Rightarrow i\gamma^0 \frac{\partial\psi}{\partial t} = \left[-i\vec{\gamma}\cdot\vec{\nabla}\psi + e\phi\gamma^0\psi - e\vec{\gamma}\cdot\vec{A}\psi + m\psi\right] \tag{14.15}$$

Anticipating taking the non-relativistic limit, let's write $\psi = e^{-imt}\begin{pmatrix}\varphi\\\chi\end{pmatrix}$ and insert it into (14.15). This gives a pair of 2-component spinor equations

$$i\frac{\partial\varphi}{\partial t} = e\phi\varphi + \vec{\sigma}\cdot\left(\vec{p} - e\vec{A}\right)\chi \tag{14.16}$$

$$i\frac{\partial\chi}{\partial t} = \left(-2m + e\phi\right)\chi + \vec{\sigma}\cdot\left(\vec{p} - e\vec{A}\right)\varphi \tag{14.17}$$

where $\vec{p} = -i\vec{\nabla}$. So far this is an exact result. Now let's take the non-relativistic limit, where we have

$$m >> e\phi, \left|\frac{\partial\chi}{\partial t}\right| \tag{14.18}$$

and so the 2nd equation has the approximate solution

$$\chi \simeq \frac{\vec{\sigma}\cdot\left(\vec{p} - e\vec{A}\right)}{2m}\varphi \tag{14.19}$$

$$\Rightarrow i\frac{\partial\varphi}{\partial t} \simeq e\phi\varphi + \frac{1}{2m}\left[\vec{\sigma}\cdot\left(\vec{p} - e\vec{A}\right)\right]^2\varphi \tag{14.20}$$

where eq. (14.20) is the Pauli equation (named after Wolfgang Pauli who first proposed the concept of electron spin [124]). It is an equation for a 2-component spinor φ that generalizes the non-relativistic Schroedinger equation to include spin!

Now we can use the identities (5.32) $\sigma_i\sigma_j = \delta_{ij} + i\epsilon_{ijk}\sigma_k$ to write

$$\left[\vec{\sigma}\cdot\left(\vec{p} - e\vec{A}\right)\right]^2 = \left|\vec{p} - e\vec{A}\right|^2 - e\vec{\sigma}\cdot\vec{B} \tag{14.21}$$

the proof of which I have left as a problem. This gives

$$i\frac{\partial\varphi}{\partial t} \simeq \left[\frac{\left|\vec{p} - e\vec{A}\right|^2}{2m} + e\phi - \frac{e}{2m}\vec{\sigma}\cdot\vec{B}\right]\varphi = H\varphi \tag{14.22}$$

The first two terms correspond to the Hamiltonian for a particle of charge e moving in an electromagnetic field with vector potential $\vec{A}$ and Coulomb potential ϕ. Since $\vec{S} = \frac{\vec{\sigma}}{2}$ for a spin-$\frac{1}{2}$ particle, the last term on the right-hand-side suggests that we identify

$$H_{\text{mag}} = -\frac{e}{2m}\vec{\sigma} \cdot \vec{B} = -2\frac{e}{2m}\left(\frac{\vec{\sigma}}{2} \cdot \vec{B}\right) = -2\frac{e\hbar}{2mc}\left(\vec{S} \cdot \vec{B}\right) \qquad (14.23)$$

(putting in the correct factors of $\hbar$ and c) which implies that $g = 2$ for a Dirac particle of charge e, as promised*.

Here then is a clear prediction from the Dirac equation, one that can be easily compared to experiment by putting an electron (or muon or tau) in a magnetic field and checking the value of the coupling. The g-values of the e^- and μ^- leptons have indeed been measured – but they *disagree* with the value $g = 2$ by $\sim 0.2\%$! It's a tiny discrepancy – but one that is not within limits of error.

Thus the Dirac picture of a pointlike e^- and μ^- is not exact. Why?

The actual reason has to do with quantum field theory: the quantum formalism in which the wavefunctions of the e^- or μ^- and photons are themselves quantized so that they can create or destroy particles. This feature allows the e^- (or μ^-) to continually emit and reabsorb its own photons. From the diagrams in figure 14.4 we see that the g-factor should be corrected. The actual computation of these diagrams is quite tedious – I've put in the answers that come from computing them. The diagram at the bottom comes from adding up the two diagrams in the middle row of 14.4. Upon comparison with H_{mag} above, we see that

$$g = 2\left(1 + \frac{\alpha}{2\pi}\right) \qquad (14.24)$$

and so the magnetic moment receives a correction of order α with a coefficient of $\frac{1}{2\pi}$, responsible for the 0.2% correction noted above and first obtained by Schwinger [125].

We also see that the charge of the electron (or muon) is corrected by a term of order α

$$e_{\text{electron}} = e\left(1 + \frac{\alpha}{\pi}\int \frac{dq}{q}\right) \qquad (14.25)$$

but with a coefficient of infinite magnitude!

What to do? This is the problem of diverging quantities noted earlier in the ABB theory. The results of quantum field theory indicate that physical quantities are perturbatively corrected in powers of the coupling α. Sometimes these corrections are finite (as with the magnetic moment), but sometimes they are not. For the electron charge this problem is dealt with by regarding the charge e as not being directly measurable – instead, what is measured in the lab is the quantity e_{electron} above. It is the quantity e_{electron} that is defined

---

*Experimentally the magnetic moments of the neutron and proton differ widely from this value. As we will see beginning in Chapter 16, this implies that these particles have internal structure.

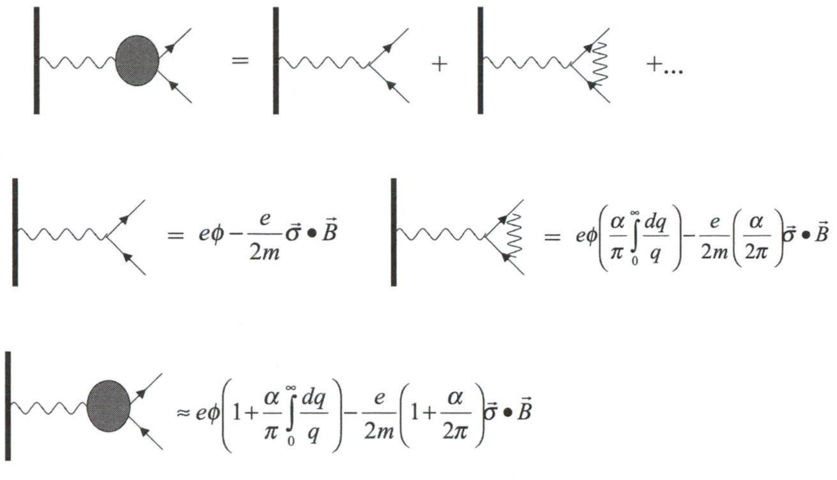

$$\text{diagram} = e\phi - \frac{e}{2m}\vec{\sigma}\cdot\vec{B} \qquad \text{diagram} = e\phi\left(\frac{\alpha}{\pi}\int_0^\infty \frac{dq}{q}\right) - \frac{e}{2m}\left(\frac{\alpha}{2\pi}\right)\vec{\sigma}\cdot\vec{B}$$

$$\approx e\phi\left(1+\frac{\alpha}{\pi}\int_0^\infty \frac{dq}{q}\right) - \frac{e}{2m}\left(1+\frac{\alpha}{2\pi}\right)\vec{\sigma}\cdot\vec{B}$$

FIGURE 14.4

Lowest-order diagrams contributing to the anomalous magnetic moment of the electron. Only the interaction terms the arise from computing the diagram have been written.

to agree with experiment. The quantity e (the bare charge) is assumed to be divergent in such a manner as to render e_{electron} finite.

As with the ABB theory, this crazy idea works because QED is *renormalizable*: once this redefinition of the charge is made (plus a similar redefinition of the mass and the wavefunctions), all other quantities in QED (e.g., cross-sections, decay rates) are finite and uniquely calculable. One of these quantities is the magnetic moment, given above, which to this order is predicted to be $g = 2\left(1+\frac{\alpha}{2\pi}\right) = 2.00232....$

Can we test this? The answer is yes, and the comparison between theory and experiment is nothing short of spectacular [126]. The above procedure for computing the magnetic moment has been carried out to 5th order in α. These computations involve $\sim 1,000$ diagrams. The result for the electron is [127]

$$\left(\frac{g-2}{2}\right)_e^{\text{QED}} = \frac{\alpha}{2\pi} - 0.328478444\left(\frac{\alpha}{\pi}\right)^2 + 1.181234017\left(\frac{\alpha}{\pi}\right)^3$$
$$-1.7283\left(\frac{\alpha}{\pi}\right)^4 + 0.0\left(\frac{\alpha}{\pi}\right)^5 + 1.71173\times 10^{-12}\cdots$$
$$= .0011596521884 \pm .0000000000043$$
$$= 1,159,652,188.4(4.3)\times 10^{-12} \tag{14.26}$$

The first term is represented by a Feynman diagram with one closed loop, and requires no more than a page or two of hand calculation. The second term represents seven diagrams, and originally took seven years to calculate. Seventy-two Feynman diagrams are needed for the third term, and all of them have been evaluated exactly using symbolic manipulation programs on computers, after nearly thirty years of hard work. The fourth term requires the evaluation of 891 four-loop Feynman diagrams, and has been estimated numerically using large-scale computations on supercomputers. The fifth term is even more laborious, and the coefficient 0.0 has an error of ±3.7. The last term (not involving the fine-structure constant) is a small correction caused by particles in the loop other than the electron, and strong and weak interaction corrections.

The particle data book lists the experimental value as [1]

$$\left(\frac{g-2}{2}\right)_e^{\text{QED-expt}} = 1,159,652,181.11 \pm (0.074) \times 10^{-12} \tag{14.27}$$

which is a world average over all experiments. This agreement between the theoretical value above and the experimental value(s) given here is the most precise for any non-gravitational theory and experiment to date.

To properly compare experiment and theory, we need to know the value of the fine-structure constant. The measurements of the value of this constant aren't as accurate as those of $(g-2)$! There are many ways of doing this [127]. For example one can measure α using the quantum Hall effect, get $1/137.0360037(27)$ (an accuracy of 0.020 ppm), and predict

$$\left(\frac{g-2}{2}\right)_e^{\text{QED/qH}} = 1,159,652,156.4(22.9) \times 10^{-12} \tag{14.28}$$

One can measure α using the ac Josephson effect to be $1/137.0359770(77)$ (0.056 ppm), and predict

$$\left(\frac{g-2}{2}\right)_e^{\text{QED/acJ}} = 1,159,652,378.0(65.3) \times 10^{-12} \tag{14.29}$$

Or, one can measure Planck's constant $\hbar$ and the mass of the neutron, and derive α to be $1/137.03601082(524)$ (0.039 ppm) to predict

$$\left(\frac{g-2}{2}\right)_e^{\text{QED/n}} = 1,159,652,092.2(44.4) \times 10^{-12} \tag{14.30}$$

The numbers in parentheses are due to the uncertainty in the experimental value of α; the errors in the computer-measurement of the theoretical formula is much smaller (plus or minus 1.2). If you trust QED, you can work backwards to an even better estimate of α: $1/137.03599993(52)$, with an estimated error of 0.0038 ppm.

For the muon the result is

$$\left(\frac{g-2}{2}\right)_\mu^{\text{QED}} = \frac{\alpha}{2\pi} - 0.765857408 \left(\frac{\alpha}{\pi}\right)^2 + 24.05050959 \left(\frac{\alpha}{\pi}\right)^3$$

$$+ 130.9916 \left(\frac{\alpha}{\pi}\right)^4 + 663 \left(\frac{\alpha}{\pi}\right)^5 + 709.4 \times 10^{-10} \quad (14.31)$$

where the last part is due to strong and weak interaction corrections. The best theoretical calculations yield

$$\left(\frac{g-2}{2}\right)_\mu^{\text{QED}} = \begin{cases} 1,165,847,181.0(1.6) \times 10^{-12} & \begin{array}{c}\text{pure}\\\text{QED}\end{array} \\ 1,165,917,880(20)(460)(350) \times 10^{-12} & \begin{array}{c}\text{including strong}\\ \text{\& weak corrections}\end{array} \end{cases}$$

$$(14.32)$$

where in the top line the quantities in brackets indicate the error due to uncertainties in the fine structure constant, and in the bottom they indicate the errors due to electroweak, lowest-order hadronic, and higher-order hadronic contributions, respectively. The best data [1] indicate that

$$\left(\frac{g-2}{2}\right)_\mu^{\text{expt}} = 1,165,920,800(540)(330) \times 10^{-12} \quad (14.33)$$

where the errors in brackets are due to statistical and systematic errors respectively.

As you can see, strong and weak interactions provide significant corrections that are still not fully understood to 12 decimal place precision. The discrepancy is

$$\Delta\left(\frac{g-2}{2}\right) \equiv \Delta a_\mu = 292(63)(58) \times 10^{-11} \quad (14.34)$$

which is 3.4 standard deviations away from the expected result.

Experiment E821 at Brookhaven [128] have indicated that there is something we don't understand about the anomalous magnetic moment of the muon. The current situation (as of June 2009) is described in figure 14.5. Notice that theory and experiment do *not* agree within limits of error: the present experimental value differs from that predicted by QED, or more properly the Standard Model, since relevant weak and strong corrections have been included. The uncertainty on the Standard Model theoretical values is dominated by the uncertainty on the lowest-order hadronic vacuum polarization. This contribution can be determined directly from the annihilation of e^+e^- to hadrons through a dispersion integral (the lower open circle point). The indirect determination using data from hadronic τ decays, the conserved vector current hypothesis, plus the appropriate isospin corrections, appear to improve the agreement between theory and experiment, but nobody knows why.

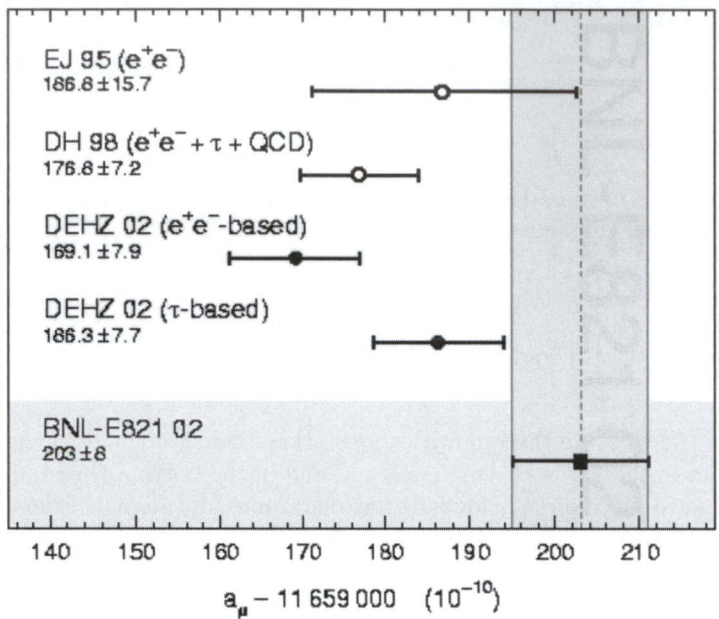

FIGURE 14.5

Results for calculations of the anomalous magnetic moment of the muon, subtracted by the central value of the experimental average, given in by the bottommost point on the right. The shaded band indicates the experimental error. The Standard Model predictions (DEHZ, DH, EJ) are from various groups doing calculations that take into account different ways of incorporating hadronic corrections. Image courtesy of Particle Data Group [209]; used with permission.

A definitive understanding of the muon g factor will have to await further refinements of both the experimental and the theoretical values. Major advances may be some time in coming. On the experimental side, the E821 project has been shut down by the United States Department of Energy, at least for the time being. From the theoretical viewpoint, the next major stage will require evaluation of the five-loop Feynman diagrams. There are 12,672 of those. Anyone eager to do this calculation?

14.2.3   Lamb Shift

The Dirac equation can be solved exactly in the presence of a Coulomb potential, where $A_\mu = (\frac{Ze}{r}, 0)$. The computation is a tedious but analogous extension of the computation of the energy levels for a Hydrogen-like atom using Schroedinger's equation. For the Dirac case, the energy levels are also fully and uniquely calculable and the result is [129]

$$E = \frac{m_r c^2}{\sqrt{1 + \frac{(Z\alpha)^2}{\left[n-(j+1/2)+\sqrt{(j+1/2)^2-(Z\alpha)^2}\right]^2}}} \qquad \begin{array}{l} n = \text{principal quantum number} \\ j = \frac{1}{2}, \frac{3}{2}, \frac{5}{2}, \cdots \\ = \text{total angular momentum} \end{array}$$

$$= m_r c^2 \left(1 - \frac{1}{2}\frac{(Z\alpha)^2}{n^2} + \frac{3}{8}\frac{(Z\alpha)^4}{n^4} - \frac{(Z\alpha)^4}{n^3(2j+1)} + \cdots \right) \qquad (14.35)$$

where m_r is the reduced mass of the electron and Ze is the charge of the spinless nucleus. The expansion is in powers of $Z\alpha$, and the first non-trivial term is the usual part that one gets from the Schroedinger equation.

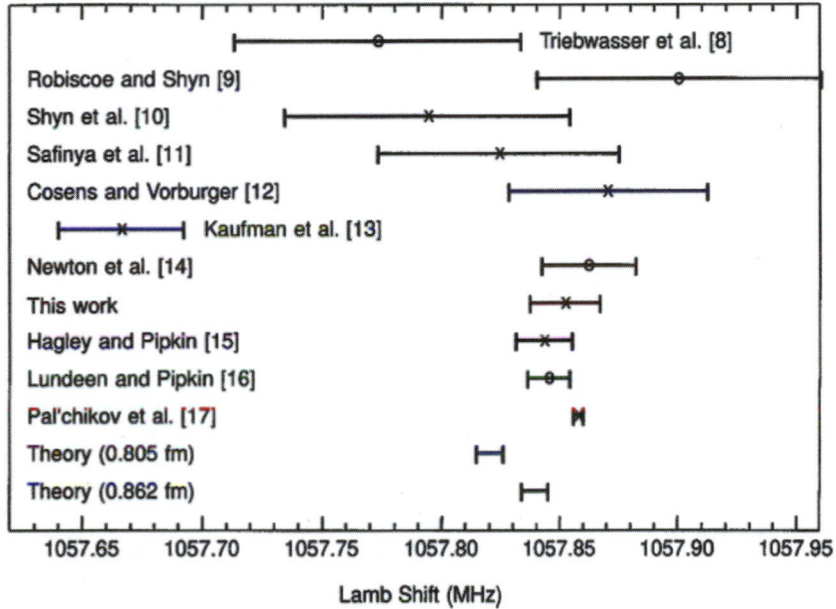

FIGURE 14.6
Comparison of theory and experiment for the Lamb shift of hydrogen. Open circles (o) denote direct measurements, and crosses (x) denote indirect measurements dependent upon the $2P_{1/2}$ lifetime of 1.5961887 (15) $\times 10^{-9}$ s. The measurement "This work" is by van Wijngarten et al. *Lamb-shift measurement in hydrogen by the anisotropy method* Can. J. Phys. **76**, 95 (1998), fig. 3 pg. 102 [131]; other results appear in ref. [133]. Copyright 2008 NRC Canada or its licensors. Reproduced with permission.

Note that there is no explicit dependence on either the orbital angular momentum ℓ or on the electron spin s; instead the energy levels just depend on the total angular momentum $j = \ell + s$. Hence states with differing ℓ but the same j will have the same energy – for example, the $2S_{1/2}$ (i.e. $n = 2$, $\ell = 0$) and the $2P_{1/2}$ (i.e. $n = 2$, $\ell = 1$) are predicted to be degenerate. Hence

$$\left(E_{2S_{1/2}} - E_{2P_{1/2}}\right)^{\text{Dirac eqn}} = 0 \tag{14.36}$$

which upon comparison to experiment [131]

$$\left(E_{2S_{1/2}} - E_{2P_{1/2}}\right)^{\text{expt}} = 1057.852 \pm .015 \; \hbar\text{MHz} \simeq 4.374 \times 10^{-6}\text{eV} \tag{14.37}$$

indicates a small but clear discrepancy.

This discrepancy is explained by quantum-field-theoretic effects in QED: the self-energy of an electron bound to a nucleus depends on the orbital angular momentum ℓ and causes the above shift, first measured in 1947 by Willis Lamb and Robert Retherford [132]. The (rather long) calculation gives to leading order [130]

$$\left(E_{2S_{1/2}} - E_{2P_{1/2}}\right)^{\text{QED}} = \frac{m\alpha^5}{6\pi}\left(\ln\left(\frac{m\langle E_{2P}\rangle}{2\text{Ryd}\langle E_{2S}\rangle}\right) + \frac{91}{120}\right) = 1052.1 \; \hbar\text{MHz} \tag{14.38}$$

an early triumph of QED!

Other tests of QED involve investigation of spin-flip transitions in the ground states of positronium (e^+e^-) and muonium (μ^+e^-). There is an ongoing effort in atomic physics to push for ever-greater precision in confronting QED with experiment to see just how accurate the Standard Model is.

14.2.4 Running Coupling Constant

A test charge inside a polarizable (dielectric) medium is shielded by the dielectric and so has a smaller Coulomb potential than in free space. This is a well-known effect in crystals. In a quantum field theory such as QED, the vacuum itself is a polarizable medium, as illustrated in figure 14.7. A charged particle can emit and reabsorb virtual photons which themselves emit and reabsorb e^+e^- pairs which shield the particle: this is called *vacuum polarization*.

We observe a charged particle from large distances when it is "fully shielded." As we probe closer and closer to the "core" of the particle (i.e. within a Compton wavelength), shielding becomes small and the potential due to the "bare" charge is observed.

How do we actually measure this effect? Recall that if we have 2 charged particles of charges $Z_1 e$ and $Z_2 e$ that the lowest order scattering diagram is given by fig. 14.8 for which the matrix element is

FIGURE 14.7
Schematic diagram of vacuum polarization.

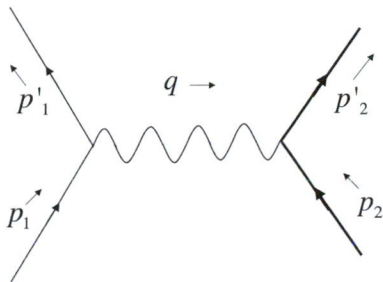

FIGURE 14.8
Lowest order QED scattering.

$$\mathcal{M} = - \left[\overline{u}(p'_1)\gamma^\mu u(p_1)\right] \frac{Z_1 Z_2 e^2}{(p'_1 - p_1)^2} \left[\overline{u}(p'_2)\gamma_\mu u(p_2)\right] \qquad (14.39)$$

However, higher-order corrections modify this. One of the most interesting is shown in figure 14.9, which contributes

$$\delta\mathcal{M} = -i \left[\overline{u}(p'_1)\gamma^\mu u(p_1)\right] \frac{Z_1 Z_2 e^4}{(p'_1 - p_1)^4} \mathcal{I}_{\mu\nu} \left[\overline{u}(p'_2)\gamma^\nu u(p_2)\right] \qquad (14.40)$$

to the matrix element, where the quantity $\mathcal{I}_{\mu\nu}$ is, using the rules for QED,

$$\mathcal{I}_{\mu\nu} = -\int \frac{d^4 k}{(2\pi)^4} \frac{\mathrm{Tr}\left[\gamma_\mu \left(\slashed{k} + m\right)\gamma_\nu \left(\slashed{q} - \slashed{k} + m\right)\right]}{(k^2 - m^2)\left((q-k)^2 - m^2\right)}$$

$$= -ig_{\mu\nu} \frac{q^2}{12\pi^2} \left[\lim_{\Lambda\to\infty} \int_{m^2}^{\Lambda} \frac{dx}{x} - f\left(-\frac{q^2}{m^2}\right)\right] + q_\mu q_\nu K(q^2) \quad (14.41)$$

with $f\left(-\frac{q^2}{m^2}\right)$ and $K(q^2)$ calculable functions, and m is the mass of the particle in the fermion loop in fig. 14.9, here taken to be an electron. The

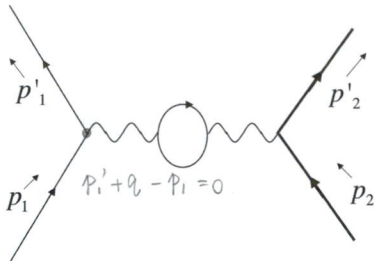

FIGURE 14.9
A higher order correction to the process in fig. 14.8.

$K(q^2)$ contributes nothing to $\mathcal{M}$, since it is proportional to $q_\mu q_\nu$ and so is annihilated by the $[\overline{u}\gamma^\mu u]$ terms. For example

$$q_\mu \overline{u}(p_1')\gamma^\mu u(p_1) = \overline{u}(p_1')\not{q}u(p_1) = \overline{u}(p_1')\left(\not{p}_1' - \not{p}_1\right)u(p_1)$$

$$= \overline{u}(p_1')\left(m - m\right)u(p_1) = 0 \tag{14.42}$$

(recall $(\not{p} - m)u(p) = 0$ for any u-spinor). So the full matrix element is

$$\mathcal{M}_{\text{tot}} = \mathcal{M} + \delta\mathcal{M}$$

$$= -\frac{Z_1 Z_2 e^2}{(p_1' - p_1)^2}\,[\overline{u}(p_1')\gamma^\mu u(p_1)]\,[\overline{u}(p_2')\gamma_\mu u(p_2)]$$

$$\times \left(1 + \frac{e^2}{12\pi^2}\left[\lim_{\Lambda\to\infty}\int_{m^2}^{\Lambda}\frac{dx}{x} - f\left(-\frac{q^2}{m^2}\right)\right]\right) \tag{14.43}$$

If we take the limit $\Lambda \to \infty$ we obtain an infinite answer. However, if we define a renormalized coupling via

$$e_{\text{R}} = e\left(1 - \frac{e^2}{12\pi^2}\lim_{\Lambda\to\infty}\int_{m^2}^{\Lambda}\frac{dx}{x}\right)^{1/2} \tag{14.44}$$

then (since $(p_1' - p_1)^\mu = q^\mu$) we obtain to this order in e^2

$$\mathcal{M}_{\text{tot}} = -\,[\overline{u}(p_1')\gamma^\mu u(p_1)]\,\frac{Z_1 Z_2 e_{\text{R}}^2}{q^2}\left(1 + \frac{e_{\text{R}}^2}{12\pi^2}f\left(-\frac{q^2}{m^2}\right)\right)[\overline{u}(p_2')\gamma_\mu u(p_2)]$$

$$= -\,[\overline{u}(p_1')\gamma^\mu u(p_1)]\,\frac{Z_1 Z_2 e_{\text{R}}^2}{q^2\left(1 - \frac{e_{\text{R}}^2}{12\pi^2}f\left(-\frac{q^2}{m^2}\right)\right)}[\overline{u}(p_2')\gamma_\mu u(p_2)] \tag{14.45}$$

where we can interpret the terms $Z_1 e_R \left[\bar{u}(p'_1) \gamma^\mu u(p_1) \right] = \mathcal{J}_1^\mu$ and $Z_2 e_R \left[\bar{u}(p'_2) \gamma^\mu u(p_2) \right] = \mathcal{J}_2^\mu$ as the currents associated with the 2 charged particles.

Let's see how to interpret this. Suppose that the two charged particles are approximately static, and each is spin-up. Then setting $\xi = \xi^{(\uparrow)}$ and $E \simeq m$ in eq. (11.33) gives $\left[\bar{u}(p'_1) \gamma^\mu u(p_1) \right] \simeq \sqrt{2m_1}$ and $\left[\bar{u}(p'_2) \gamma^\mu u(p_2) \right] \simeq \sqrt{2m_2}$, and $q^\mu = (p'_1 - p_1)^\mu = (E'_1 - E_1, \vec{p}'_1 - \vec{p}_1) \simeq (0, \vec{p}'_1 - \vec{p}_1) = (0, \vec{q})$. In this case the Fourier transform of eq. (14.45) is

$$\mathcal{M}_{\text{tot}} = Z_1 Z_2 \sqrt{2m_1} \sqrt{2m_2} V(r)$$

where $r = |\vec{r}| = |\vec{x}_1 - \vec{x}_2|$ is the separation between the two charged particles, and we can interpret the function

$$V(r) = \int \frac{d^3 q}{(2\pi)^3} \frac{e_R^2 \; e^{-i\vec{q}\cdot\vec{r}}}{-|\vec{q}|^2 \left(1 - \frac{e_R^2}{12\pi^2} f\left(\frac{|\vec{q}|^2}{m^2} \right) \right)}$$

$$\simeq -e^2 \int \frac{d^3 q}{(2\pi)^3} \frac{e^{-i\vec{q}\cdot\vec{r}}}{|\vec{q}|^2} = -\frac{e^2}{4\pi |\vec{r}|} \tag{14.46}$$

as the potential between them, where the last expression is obtained by retaining only the lowest order in e. Including the next-order loop corrections [134] and Fourier transforming gives

$$V(r) = -\frac{\alpha(r)}{r} = -\frac{\alpha}{r} \left[1 - \frac{2\alpha}{3\pi} Q(r) \right]^{-1} \tag{14.47}$$

where $\alpha = \frac{e_R^2}{4\pi\hbar c}$ is the familar fine-structure constant and

$$Q(r) = \int_1^\infty du \frac{2u^2 + 1}{u^4} \sqrt{u^2 + 1} e^{-2mru} + \mathcal{O}(\alpha)$$

$$= \begin{cases} \ln\left(\frac{1}{mr} \right) - \gamma_E - \frac{5}{6} + \cdots & mr \ll 1 \\ \frac{3\sqrt{\pi}}{8(mr)^{3/2}} e^{-2mr} & mr \gg 1 \end{cases} \tag{14.48}$$

with $\gamma_E = .5772...$ which is Euler's constant.

So for large $r \gg 1/m$, the quantum corrections $Q(r)$ to the Coulomb potential fall off exponentially, and $V(r) \approx -\frac{\alpha}{r}$, which is the usual Coulomb potential. However, for $r \ll 1/m$, the quantum corrections $Q(r)$ grow logarithmically – in other words, $V(r)$ *increases* with decreasing distance!

We can express this phenomenon by modifying the electromagnetic coupling $\alpha \to \alpha(r)$. We say that the coupling varies (or runs with decreasing distance, or increasing energy q^2. For QED

$$\alpha(r) = \frac{\alpha}{\left[1 - \frac{2\alpha}{3\pi} Q(r) \right]} \tag{14.49}$$

and so the ==running coupling== increases with decreasing distance. In fact, it becomes singular at $Q(r) = \frac{3\pi}{2\alpha}$, or at

$$r \simeq \frac{\hbar}{mc} \exp\left[-\frac{3\pi}{2\alpha}\right] = 3.86 \times 10^{-291} \text{cm} \tag{14.50}$$

In other words the quantum potential of the electron becomes infinite at a non-zero distance (albeit a very tiny one) from its center!

This singularity is called a ==Landau singularity,== after the Russian physicist Lev Landau who argued that its presence in QED indicates that strong vacuum polarization effects screen the electric charge completely at short distances [135]. Others, including Shirkov [136], have called it the Landau ghost contending that it instead indicates the internal inconsistency of quantum electrodynamics, since this singularity cannot be removed by renormalization. Shirkov's viewpoint has by and large been the more persuasive one, with the singularity regarded as the distance scale in QED at which perturbation theory definitely cannot be trusted, though whether or not it implies that QED is fundamentally inconsistent is unknown. In any event, effects from quantum gravity become relevant at 10^{-33}cm, a much larger distance from the center of the electron than where the Landau singularity is.

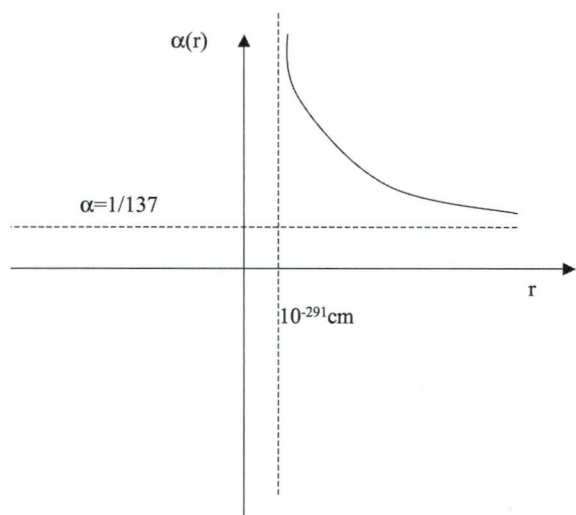

FIGURE 14.10

Qualitative behavior of the running coupling constant in QED

Although I have given the dependence of α as a function of distance, it is actually common practice in particle physics to express the running of α as a function of the momentum transfer q^2. The result is

$$\alpha(q^2) = \frac{\alpha\left(\mu^2\right)}{\left[1 - z_f \frac{2\alpha(\mu^2)}{3\pi} \ln\left(\frac{\sqrt{|q^2|}}{\mu}\right)\right]} \tag{14.51}$$

where $z_f = \sum_j Q_j^2$ is the sum of the squares of the charges (in units of the electron charge) of the fermions that contribute for all energies less than or equal to $\sqrt{|q^2|}$. For example, for energies less than 100 GeV, all leptons and quarks except for the top quark will contribute to the sum, which equals $\frac{20}{3}$ (since we must include each color of quark as a separate charge). The quantity μ is an arbitrary energy scale that cannot be fixed in QED.

At low energies $|q^2| \to \mu^2 \to 0$, the logarithm in the denominator vanishes, and $\alpha(q^2) \to \alpha\left(\mu^2\right) \to \alpha\left(0\right) = \frac{1}{137}$. A high-precision determination of α at the Z-mass yielded [1]

$$\alpha^{-1}(M_Z^2) = 128.936 \pm 0.046 \tag{14.52}$$

where I have expressed the result in terms of α^{-1} for convenience.

14.3 Questions

1. Show that

$$\left[\vec{\sigma} \cdot \left(\vec{p} - e\vec{A}\right)\right]^2 = \left|\vec{p} - e\vec{A}\right|^2 - e\vec{\sigma} \cdot \vec{B}$$

Hint: remember that $\vec{p} = -i\hbar\vec{\nabla}$ and that the vector potential $\vec{A}$ depends on space and time.

2. (a) Calculate to lowest order the spin-averaged spin-summed matrix element for the process $e^- + \gamma \longrightarrow e^- + \gamma$.

(b) Your answer will depend upon the polarizations $\left(\varepsilon_i^{(\lambda)}, \varepsilon_f^{(\lambda)}\right)$ of the respective initial and final photon states. Show that

$$\sum_{\lambda=1}^{2}\left(\varepsilon_i^{(\lambda)} \bullet \varepsilon_f^{(\lambda)}\right)^2 = 1 + \cos^2\theta$$

where θ is the scattering angle.

(c) Use this to obtain a differential cross-section that is averaged over initial photon polarizations and summer over final photon polarizations. Work in the lab-frame.

Note: This is a tedious and somewhat lengthy calculation. I recommend that you make use of the simplifiying tricks given in table 13.1, as well as average over initial fermion spins and summing over final fermion spins. You will also find it helpful to choose the polarizations of the initial and final photons to be orthogonal to the initial electron momentum.

3. (a) Calculate to lowest-order the spin-averaged spin-summed matrix element for the process $\gamma + \gamma \longrightarrow e^- + e^+$.

   (b) Find the differential cross-section that is averaged over initial photon polarizations and summed over the electron/positron spins. Work in the CM frame.

4. Show that

$$\int \frac{d^3 q}{(2\pi)^3} \frac{e^{-i\vec{q}\cdot\vec{r}}}{|\vec{q}|^2 + m^2} = \frac{e^{-m|\vec{r}|}}{4\pi|\vec{r}|}$$

   where m is a constant. What does this become in the limit $m \to 0$?

5. Given that $\alpha^{-1}(M_Z^2) = 129$ where $M_Z = 91$ GeV, find the value of the fine structure constant at $\sqrt{|q^2|} = 1$ TeV, assuming that no other particles exist beyond the known quarks and leptons.

6. Suppose the fine structure constant is found to have a value 10 times its low-energy value at $\sqrt{|q^2|} = 2$ TeV. How many new quarks and leptons must there be at energies between 100 GeV and 2 TeV if the pattern of generations in the Standard Model is upheld?

15

From Nuclei to Quarks

Strong interaction physics first began with Rutherford's discovery of the nucleus in 1910 [18]. The solar-system model of the atom that emerged from this discovery soon gave way to Bohr's theory of the atom, and then to quantum mechanics. Rapid progress was made in understanding atomic physics during this period. However, for 20 years after Rutherford's discovery the structure of the nucleus remained quite mysterious: it had to be positively charged, but why didn't electromagnetic forces cause it to explode apart? In 1932 James Chadwick discovered the neutron [137]: this led to the realization that the nucleus is a composite object, bound together by a nuclear or "strong" force, one much more powerful than the electromagnetic force. The strong force – by definition – is experienced only by particles referred to as hadrons*.

Understanding the nature of the nucleus and its structure was a long and arduous process of calculation, experimentation, and intelligent guesswork. The subject of nuclear physics today is indeed a complicated subject, since it is a many-body problem that is still not fully understood. Our interest here is in understanding the elementary constituents of nuclear matter, which are known as quarks. In order to appreciate the origin of quarks (both in their conception and in their discovery), we'll proceed historically, looking at the development of key concepts as they occurred.

15.1 Range of the Nuclear Force

Shortly after Chadwick's discovery, Wigner realized that nuclear forces must have a very short range (of only a few fm, where 1 fm=10^{-13}cm) and must be very strong [138]. The reason for this came from a consideration of the binding energies of the deuteron (hydrogen with one extra neutron), tritium (hydrogen with two extra neutrons, sometimes called triton), and the alpha particle (helium). As table 15.1 indicates the binding energies are millions of times larger than the electromagnetic energies that bind electrons in atoms.

---

*Actually there are two nuclear forces – a strong force and a weak one – but the distinction between them did not become clear until the 1950s.

TABLE 15.1

Binding Energies of Small Nuclei

	# of Bonds	Binding Energy in MeV		
		Total	per Particle	per Bond
^{2}H	1	2.2	1.1	2.2
^{3}H	3	8.5	2.8	2.8
^{4}He	6	28	7	4.7

Table 15.1 displays the binding energies of these three nuclides, as well as the energies per particle and per bond. The energy per bond is not roughly the same, and so something else is needed to explain the large increase in binding energy. Wigner reasoned that a short-range force can explain this, because as the number of bonds increases the nucleons pull closer together, and thereby experience a deeper potential well. This gives an additional increase to the binding energies per particle and per bond.

The mass and binding energy for ^{3}He were observed to be almost the same as for ^{3}H – this provided evidence that the forces between any two nucleons are the same, apart from small corrections due to electromagnetic forces, an empirical observation for which there is overwhelming evidence today.

15.2 Isospin

At about the same time, Heisenberg [139] suggested that the proton and neutron are really 2 different states of one particle called the *nucleon*, analogous to the way that the spin-↑ e^- and spin-↓ e^- are 2 different states of the electron. A comparison appears in table 15.2. Just as we can write an electron

TABLE 15.2

Comparison between Spin and Isospin

Spin-1/2 particle in ordinary space	Nucleon in in Isospin space
up	proton
down	neutron

wavefunction as a linear combination of spin-↑ and spin-↓, this perspective implies that we can write the wavefunction of a nucleon $|N\rangle$ as

$$|N\rangle = \varsigma |p\rangle + \zeta |n\rangle \tag{15.1}$$

i.e. as a complex linear combination of proton and neutron wavefunctions. This seems strange: the charges of the proton and neutron differ, and so we don't know how to define the charge of the state $|N\rangle$. However, Heisenberg's hypothesis entails that the nuclear force is independent of the electric charge. We therefore can combine the two states, ignoring the effects of electromagnetism, which are much smaller. To every such state there will also be an orthogonal state $|P\rangle$:

$$|P\rangle = \alpha |p\rangle + \beta |n\rangle \tag{15.2}$$

so that

$$\begin{pmatrix} |P\rangle \\ |N\rangle \end{pmatrix} = \begin{pmatrix} \alpha & \beta \\ \varsigma & \zeta \end{pmatrix} \begin{pmatrix} |p\rangle \\ |n\rangle \end{pmatrix} = \mathbf{U} \begin{pmatrix} |p\rangle \\ |n\rangle \end{pmatrix} \tag{15.3}$$

where the matrix $\mathbf{U}$ must be unitary, 2×2 and of determinant one (i.e. $\alpha\zeta - \varsigma\beta = 1$) to conserve probability. In other words, Heisenberg's idea is expressed by saying that all nuclear interactions should be invariant under $\mathbf{SU}(2)$-transformations of the neutron/proton doublet!

We call such transformations *isospin transformations.* Just as we did for rotations, we can write a general $\mathbf{SU}(2)$-transformation matrix $\mathbf{U}$ as

$$\mathbf{U} = \exp\left[-i\vec{\Upsilon} \cdot \vec{\mathfrak{J}}\right] = \begin{pmatrix} \alpha & \beta \\ \varsigma & \zeta \end{pmatrix} \quad \text{where} \quad \alpha\zeta - \varsigma\beta = 1$$
$$\Rightarrow \{\mathbf{U}(\alpha, \beta, \varsigma, \zeta)\} = \mathbf{SU}(2) \tag{15.4}$$

where there are $2^2 - 1 = 3$ generators $\vec{\mathfrak{J}} = (\mathfrak{J}_1, \mathfrak{J}_2, \mathfrak{J}_3)$. As noted in our discussion in section 3.5 of Lie groups and Lie algebras, by expanding in powers of small $\left|\vec{\Upsilon}\right|$ you can show that the generators obey

$$[\mathfrak{J}_a, \mathfrak{J}_b] = i\epsilon_{abc} \mathfrak{J}_c \tag{15.5}$$

which you might recognize from eq. (3.10) in Chapter 3 as the Lie algebra of $\mathbf{SO}(3)$. These two different groups – $\mathbf{SU}(2)$ and $\mathbf{SO}(3)$ – have the same algebra[†], so we can use the same group-theory arithmetic to describe nucleon wavefunctions.

Hence isospin is a quantum number analogous to angular momentum, except that it "rotates" protons into neutrons (and vice versa) instead of spin-↑ into spin-↓. A proton is the positive projection of a nucleon state onto the

---

[†]I emphasize that this is a very special property of these two particular Lie Groups – in general two different Lie groups have distinct Lie Algebras.

z-axis in isospin-space, whereas a neutron is the negative projection of a nucleon state onto this axis. A projection onto any other axis will be some linear combination of the proton and neutron states.

The isospin operator is denoted by $\vec{\mathfrak{I}}$. A particle undergoing nuclear interactions will thus be in a state $\Psi = |i, i_3\rangle$, where

$$\mathfrak{I}^2 \Psi = \left(\vec{\mathfrak{I}} \cdot \vec{\mathfrak{I}}\right) \Psi = i(i+1)\Psi \tag{15.6}$$

$$\mathfrak{I}_3 \Psi = i_3 \Psi \tag{15.7}$$

So the proton and neutron are the isospin states

$$|p\rangle = \begin{pmatrix} 1 \\ 0 \end{pmatrix} = \left| \frac{1}{2}, \frac{1}{2} \right\rangle \tag{15.8}$$

$$|n\rangle = \begin{pmatrix} 0 \\ 1 \end{pmatrix} = \left| \frac{1}{2}, -\frac{1}{2} \right\rangle \tag{15.9}$$

i.e. the nucleon has total isospin $i = 1/2$. The $\mathfrak{I}^2$ operator identifies the total amount of isospin, and the $\mathfrak{I}_3$ operator identifies which component total isospin the state has – in other words, whether it is a proton or a neutron.

Of course we need to eventually take electric charge into account. This is done by via the formula

$$Q = \mathfrak{I}_3 + \frac{B}{2} = \begin{cases} +1 & \text{for the proton} \\ 0 & \text{for the neutron} \end{cases} \tag{15.10}$$

where the electric charge Q of a nucleon state in units of e and B is a quantum number called baryon number. We set $B = 1$ for the nucleon in order for this formula to work. Antinucleons (antiprotons and antineutrons) have $B = -1$. Note that this formula *defines* baryon number – at this stage it does not lead to any new predictions.

In strong interactions, isospin is a conserved quantum number. The earliest evidence for this came from nuclear physics experiments that indicated that np, pp, and pn forces were all identical (except for much feebler electromagnetic effects). For example, energy level spectra in mirror nuclei nuclei which change into each other under pn interchange are almost identical: e.g. ^{13}C and ^{13}N or ^7Li and ^7Be or ^{11}B and ^{11}C.

Isospin invariance therefore means that the isospin operator $\vec{\mathfrak{I}}$ commutes with the Hamiltonian H_N governing the nuclear forces

$$\left[H_N, \vec{\mathfrak{I}}\right] = 0 \tag{15.11}$$

The presence of electromagnetism destroys this invariance, which we express as

$$\left[H_N + H_{em}, \vec{\mathfrak{I}}\right] \neq 0 \tag{15.12}$$

However, since electric charge is conserved, then we must have the charge operator commuting with the total Hamiltonian, which gives

$$[H_N + H_{em}, Q] = 0 \Rightarrow [H_N + H_{em}, \mathfrak{I}_3] = 0 \qquad (15.13)$$

where the last equality holds provided the baryon number B is a conserved quantity[‡]. This means that the 3rd component of isospin is conserved even in the presence of an electromagnetic interaction. This situation is analogous to having the z-component of spin preserved in the interactions between a spinning particle and a magnetic field.

As with spin, we can combine $i = 1/2$ states to form states of higher isospin. For example, two $i = 1/2$ states can form either an $i = 1$ triplet or an $i = 0$ singlet:

$$\left. \begin{aligned} |i = 1, i_3 = 1\rangle &= \left|\tfrac{1}{2}, \tfrac{1}{2}\right\rangle \left|\tfrac{1}{2}, \tfrac{1}{2}\right\rangle = |p\rangle\,|p\rangle \\[6pt] |i = 1, i_3 = 0\rangle &= \tfrac{1}{\sqrt{2}} \left(\left|\tfrac{1}{2}, \tfrac{1}{2}\right\rangle \left|\tfrac{1}{2}, -\tfrac{1}{2}\right\rangle + \left|\tfrac{1}{2}, -\tfrac{1}{2}\right\rangle \left|\tfrac{1}{2}, \tfrac{1}{2}\right\rangle \right) \\[6pt] &= \tfrac{1}{\sqrt{2}} \left(|p\rangle\,|n\rangle + |n\rangle\,|p\rangle \right) \\[6pt] |i = 1, i_3 = -1\rangle &= \left|\tfrac{1}{2}, -\tfrac{1}{2}\right\rangle \left|\tfrac{1}{2}, -\tfrac{1}{2}\right\rangle = |n\rangle\,|n\rangle \end{aligned} \right\} \text{triplet} \quad (15.14)$$

$$\left. \begin{aligned} |i = 0, i_3 = 0\rangle &= \tfrac{1}{\sqrt{2}} \left(\left|\tfrac{1}{2}, \tfrac{1}{2}\right\rangle \left|\tfrac{1}{2}, -\tfrac{1}{2}\right\rangle - \left|\tfrac{1}{2}, -\tfrac{1}{2}\right\rangle \left|\tfrac{1}{2}, \tfrac{1}{2}\right\rangle \right) \\[6pt] &= \tfrac{1}{\sqrt{2}} \left(|p\rangle\,|n\rangle - |n\rangle\,|p\rangle \right) \end{aligned} \right\} \text{singlet} \quad (15.15)$$

The singlet state is the deuteron, D: a stable bound state of a proton and neutron. It is the nuclear analog of para-positronium. We could call it para-deuterium, and the triplet state would then be called ortho-deuterium. However, since the triplet state is unstable we drop the "para" and "ortho" labels and call the singlet state the deuteron.

In 1947 another strongly interacting particle called the *pion* was observed [140]. In fact there were 3 pion states: π^+, π^- and π^0. The first two had identical mass, and the last of these had almost the same mass as the first two. This suggested that these 3 particles formed an isospin triplet. The charge formula

$$Q_{\text{pion}} = \mathfrak{I}_3 + \frac{B}{2} = \begin{cases} +1 & \text{for } \pi^+ \\ 0 & \text{for } \pi^0 \\ -1 & \text{for } \pi^- \end{cases} \qquad (15.16)$$

[‡]As far as we know, baryon number is conserved – there is no experimental evidence suggesting otherwise. Yet many physicists think that perhaps baryon number is only approximately conserved, and that the proton (being the lightest baryon) can actually slowly decay into other particles. Such processes are permitted in grand unified theories, and may be important in the early universe [8].

implied that $B = 0$ for the pion – indeed, it was the discovery of the pion that led to the introduction of baryon number B so that Heisenberg's formula could be satisfied in a more general context.

These assignments contain enough information to make predictions in scattering experiments using isospin conservation. For example, consider the two nuclear reactions

$$\text{(A)} \quad p + p \longrightarrow d + \pi^+ \quad \text{and} \quad \text{(B)} \quad p + n \longrightarrow d + \pi^0 \tag{15.17}$$

Reaction (A) has $i = 1$ in both the initial and final states, whereas reaction (B) has $i = 1$ in the final state, but is a 50/50 mixture of $i = 1$ and $i = 0$ in the initial state. So we expect the cross-section for (B) to be half that of (A):

$$\sigma \left(p + n \longrightarrow d + \pi^0 \right) = \frac{1}{2} \sigma \left(p + p \longrightarrow d + \pi^+ \right) \tag{15.18}$$

which is indeed observed [141]!

As another example, consider pion-nucleon scattering, $\pi + N \longrightarrow \pi + N$. There are six possible elastic scattering processes:

$$\left. \begin{array}{l} \pi^+ + p \longrightarrow \pi^+ + p \\ \pi^- + n \longrightarrow \pi^- + n \end{array} \right\} \text{ both have } i = \frac{3}{2} \tag{15.19}$$

$$\left. \begin{array}{l} \pi^- + p \longrightarrow \pi^- + p \\ \pi^- + p \longrightarrow \pi^0 + n \\ \pi^+ + n \longrightarrow \pi^+ + n \\ \pi^+ + n \longrightarrow \pi^0 + p \end{array} \right\} \text{ all have } i = \frac{1}{2} \tag{15.20}$$

along with their time-reversed counterparts[§]. To analyze these, we need to write the pion-nucleon states into isospin irreps, a job easily done using the $1 \otimes \frac{1}{2}$ Clebsch-Gordon table from figure **??**:

$$\left| \pi^+ \right\rangle \left| p \right\rangle = \left| \frac{3}{2}, \frac{3}{2} \right\rangle$$

$$\left| \pi^0 \right\rangle \left| p \right\rangle = \sqrt{\frac{2}{3}} \left| \frac{3}{2}, \frac{1}{2} \right\rangle - \sqrt{\frac{1}{3}} \left| \frac{1}{2}, \frac{1}{2} \right\rangle$$

$$\left| \pi^+ \right\rangle \left| n \right\rangle = \sqrt{\frac{1}{3}} \left| \frac{3}{2}, \frac{1}{2} \right\rangle + \sqrt{\frac{2}{3}} \left| \frac{1}{2}, \frac{1}{2} \right\rangle \tag{15.21}$$

$$\left| \pi^- \right\rangle \left| p \right\rangle = \sqrt{\frac{1}{3}} \left| \frac{3}{2}, -\frac{1}{2} \right\rangle - \sqrt{\frac{2}{3}} \left| \frac{1}{2}, -\frac{1}{2} \right\rangle$$

$$\left| \pi^0 \right\rangle \left| n \right\rangle = \sqrt{\frac{2}{3}} \left| \frac{3}{2}, -\frac{1}{2} \right\rangle + \sqrt{\frac{1}{3}} \left| \frac{1}{2}, -\frac{1}{2} \right\rangle$$

$$\left| \pi^- \right\rangle \left| n \right\rangle = \left| \frac{3}{2}, -\frac{3}{2} \right\rangle$$

[§]You can easily check these assignment by noting that for the top two reactions $|i_3| = \frac{3}{2}$ on each side, whereas $|i_3| = \frac{1}{2}$ on each side for the rest of the reactions.

The cross-section is proportional to the square of the matrix element for this process, $|\mathcal{M}_{\text{fi}}|^2 = |\langle \text{f} | H_{\pi N} | \text{i}\rangle|^2$ where $H_{\pi N}$ is the relevant part of the Hamiltonian describing the nuclear force (neglecting electromagnetic interactions). If we impose isospin conservation, then $\left[H_N, \vec{\mathfrak{I}} \right] = 0$, which implies that

$$\left\langle \text{i} = \frac{3}{2} \middle| H_{\pi N} \middle| \text{i} = \frac{1}{2} \right\rangle = 0 \tag{15.22}$$

meaning that the total isospin cannot change in any scattering process. Defining $\langle \text{i} = \frac{3}{2} | H_{\pi N} | \text{i} = \frac{3}{2}\rangle \equiv \mathcal{M}_3$ and $\langle \text{i} = \frac{1}{2} | H_{\pi N} | \text{i} = \frac{1}{2}\rangle \equiv \mathcal{M}_1$, we see that

$$\langle \pi^+ p | H_{\pi N} | \pi^+ p \rangle = \langle \pi^- n | H_{\pi N} | \pi^- n \rangle = \left\langle \frac{3}{2} \middle| H_{\pi N} \middle| \frac{3}{2} \right\rangle = \mathcal{M}_3 \tag{15.23}$$

whereas

$$
\begin{aligned}
&\langle \pi^- p | H_{\pi N} | \pi^- p \rangle \\
&= \left(\sqrt{\frac{1}{3}} \left\langle \frac{3}{2}, -\frac{1}{2} \middle| - \sqrt{\frac{2}{3}} \left\langle \frac{1}{2}, -\frac{1}{2} \middle| \right) | H_{\pi N} | \left(\sqrt{\frac{1}{3}} \middle| \frac{3}{2}, -\frac{1}{2} \right\rangle - \sqrt{\frac{2}{3}} \middle| \frac{1}{2}, -\frac{1}{2} \right\rangle \right) \\
&= \frac{1}{3} \left\langle \frac{3}{2} \middle| H_{\pi N} \middle| \frac{3}{2} \right\rangle + \frac{2}{3} \left\langle \frac{1}{2} \middle| H_{\pi N} \middle| \frac{1}{2} \right\rangle \\
&= \frac{1}{3} \mathcal{M}_3 + \frac{2}{3} \mathcal{M}_1
\end{aligned} \tag{15.24}
$$

$$
\begin{aligned}
&\langle \pi^- p | H_{\pi N} | \pi^0 n \rangle \\
&= \left(\sqrt{\frac{1}{3}} \left\langle \frac{3}{2}, -\frac{1}{2} \middle| - \sqrt{\frac{2}{3}} \left\langle \frac{1}{2}, -\frac{1}{2} \middle| \right) | H_{\pi N} | \left(\sqrt{\frac{2}{3}} \middle| \frac{3}{2}, -\frac{1}{2} \right\rangle + \sqrt{\frac{1}{3}} \middle| \frac{1}{2}, -\frac{1}{2} \right\rangle \right) \\
&= \frac{\sqrt{2}}{3} \left\langle \frac{3}{2} \middle| H_{\pi N} \middle| \frac{3}{2} \right\rangle - \frac{\sqrt{2}}{3} \left\langle \frac{1}{2} \middle| H_{\pi N} \middle| \frac{1}{2} \right\rangle \\
&= \frac{\sqrt{2}}{3} \mathcal{M}_3 - \frac{\sqrt{2}}{3} \mathcal{M}_1
\end{aligned} \tag{15.25}
$$

Summarizing, we have

$$\langle \pi^+ p | H_{\pi N} | \pi^+ p \rangle = \langle \pi^- n | H_{\pi N} | \pi^- n \rangle = \mathcal{M}_3 \tag{15.26}$$

$$\langle \pi^- p | H_{\pi N} | \pi^- p \rangle = \frac{1}{3} \mathcal{M}_3 + \frac{2}{3} \mathcal{M}_1 \tag{15.27}$$

$$\langle \pi^- p | H_{\pi N} | \pi^0 n \rangle = \frac{\sqrt{2}}{3} \mathcal{M}_3 - \frac{\sqrt{2}}{3} \mathcal{M}_1 \tag{15.28}$$

and so

$$\sigma \left(\pi^+ + p \longrightarrow \pi^+ + p \right) = \sigma \left(\pi^- + n \longrightarrow \pi^- + n \right) = \mathfrak{K} |\mathcal{M}_3|^2 \tag{15.29}$$

$$\sigma\left(\pi^- + p \longrightarrow \pi^- + p\right) = \mathfrak{K}\left|\frac{1}{3}\mathcal{M}_3 + \frac{2}{3}\mathcal{M}_1\right|^2$$

$$= \frac{1}{9}\mathfrak{K}\left|\mathcal{M}_3 + 2\mathcal{M}_1\right|^2 \tag{15.30}$$

$$\sigma\left(\pi^- + p \longrightarrow \pi^0 + n\right) = \mathfrak{K}\left|\frac{\sqrt{2}}{3}\mathcal{M}_3 - \frac{\sqrt{2}}{3}\mathcal{M}_1\right|^2$$

$$= \frac{2}{9}\mathfrak{K}\left|\mathcal{M}_3 - \mathcal{M}_1\right|^2 \tag{15.31}$$

where $\mathfrak{K}$ is a phase-space factor common to all processes. Hence we obtain the following predictions for the scattering cross-section ratios:

$$\sigma\left(\pi^+ + p \longrightarrow \pi^+ + p\right) \ : \ \sigma\left(\pi^- + p \longrightarrow \pi^- + p\right) : \sigma\left(\pi^- + p \longrightarrow \pi^0 + n\right)$$

$$= 9 : \left|1 + 2\frac{\mathcal{M}_1}{\mathcal{M}_3}\right|^2 : 2\left|1 - \frac{\mathcal{M}_1}{\mathcal{M}_3}\right|^2$$

$$\simeq 9 : 1 : 2 \qquad\qquad \text{if} \ \left|\frac{\mathcal{M}_1}{\mathcal{M}_3}\right| \ll 1$$

$$\simeq 0 : 2 : 1 \qquad\qquad \text{if} \ \left|\frac{\mathcal{M}_3}{\mathcal{M}_1}\right| \ll 1 \tag{15.32}$$

These ratios do agree with nuclear scattering data [142].

Many experimental measurements have been made of both the total and the differential cross-sections for pion-nucleon scattering. The earliest experiments measured a collimated beam of charged pions traversing a target of liquid hydrogen. For both positively and negatively charged pions there is a strong peak in the total cross-section at a pion kinetic energy of 200 MeV at an invariant mass of 1236 MeV (see fig. 15.1), discovered by Anderson, Fermi, Long and Nagle in 1952 [143]. This is an example of a *resonance* as I discussed in Chapter 9. The different resonances are given different names. The one at 1236 MeV is called the $\Delta(1236)$. The total cross-sections have a ratio of $\sigma(\pi^+ + p)/\sigma(\pi^- + p) = 3$, indicating that the $I = \frac{3}{2}$ amplitude dominates at these energies, and in good agreement with the data once we realize that we must add the cross-sections for $\pi^- + p$ to go to either of its possible final states.

Note that in general there will be several different amplitudes contributing to the total cross-section in any given energy range, which means that we can't always interpret a peak as specifying a unique resonant state.

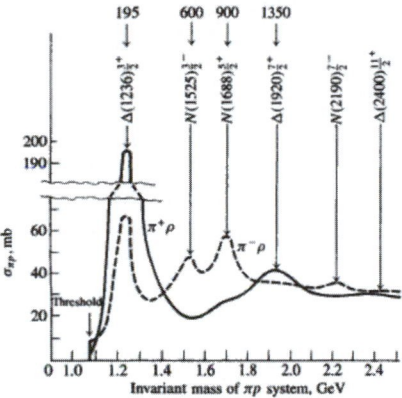

FIGURE 15.1
Variation of the total cross-section for π^+ and π^- mesons on protons as a function of the invariant mass. The positions of only a few of the known states are given. Image from Figure 4.4 of *Introduction to High Energy Physics*, 4th edition, D.H. Perkins, Cambridge University Press (2000); used with permission.

15.3   Strangeness

As nuclear physics experiments achieved higher and higher energies, new resonances called *V*-particles were observed. First observed by Rochester and Butler in 1947 [144], by 1952 many of them had been seen. They were produced in large numbers (they had large cross-sections of mb) and had long lifetimes (about 10^{-10}s), and decayed into products in a manner that did *not* conserve isospin. What were these strange things?

As an example, one *V*-particle was called the Λ-hyperon. It had no charged counterpart, suggesting $i_\Lambda = 0$. However, it decayed via

$$\Lambda \longrightarrow \pi^- + p \tag{15.33}$$

which has $i_3 = -\frac{1}{2}$ in the final state, meaning that the total final-state isospin cannot be zero. Furthermore, the lifetime for this process was $\tau_\Lambda \simeq 10^{-10}$ sec, far longer than typical strong-interaction lifetimes of 10^{-23} sec!

This suggested that these *V*-particles – such as Λ's, Σ's, K's – were produced by some strong nuclear interactions but decayed via a different weak nuclear interaction¶ (as Pais first proposed [145]). If this is so, such particles

---

¶The same could have been said for the charged pions, except that neutrinos were associated

must be produced in pairs (to conserve isospin), but can decay as singletons. This is in fact what is observed – for example we see

$$\pi^- + p \longrightarrow \begin{cases} \Sigma^- + K^+ \\ \Sigma^0 + K^0 \\ \Lambda + K^0 \end{cases} \quad \text{and} \quad \begin{matrix} \Sigma^- \longrightarrow n + \pi^- \\ \Sigma^0 \longrightarrow n + \pi^0 \end{matrix} \quad (15.34)$$

but we never see

$$\pi^- + p \longrightarrow \begin{cases} \Sigma^- + \pi^+ \\ n + K^0 \\ \Lambda + \pi^0 \end{cases} \quad (15.35)$$

Gell-Mann and Nishijima suggested that these strange phenomena be described using the formula [146]

$$Q = \Im_3 + \frac{B}{2} + \frac{S}{2} \quad (15.36)$$

where S is a new quantum number called *strangeness*.

The assignment of strangeness to a given particle follows from isospin assignments plus some conventions. By definition the strangeness of the K^+ is set equal to $+1$. All nucleons and pions have zero strangeness. As new particles continued to be discovered, isospin assignments were made on the basis of simplicity and near-degeneracy of masses of different particles when/if they occurred. Other assignments were made based on reactions that are assumed to conserve strangeness in production (but not necessarily in decay)[||]. From the reactions

$$\pi^- + p \longrightarrow \begin{cases} \Sigma^- + K^+ \\ \Sigma^0 + K^0 \\ \Lambda + K^0 \end{cases} \quad (15.37)$$

we see that the Σ^- particle has $S = -1$. Clearly the antiparticles K^- and Σ^+ have $S = +1$. A strangeness-conserving reaction is

	$K^- + p$		$\longrightarrow$	$\Lambda + \pi^0$	
S	-1	0		-1	0
$\Im$	$-\frac{1}{2}$	$+\frac{1}{2}$		0	0

implying from the above that K^0 has $S = +1$. This raises a bit of a puzzle – there are two Kaons with strangeness $+1$, but only one with $S = -1$. Gell-Mann proposed that there should be an antiparticle $\bar{K}^0$ with $S = -1$. This particle was soon seen in the reaction $\pi^+ + p \longrightarrow p + \bar{K}^0 + K^+$.

---

with their decays, and it was then thought that weak interactions were associated only with neutrinos. For the Λ and its compatriots, the decay is purely hadronic.

[||] We now understand (and will see later) that strangeness is conserved in all strong interactions, but violated in weak ones.

15.4 Flavor

The assignment of strangeness and isospin quantum numbers soon became haphazard and chaotic. It would be much nicer to use some kind of group structure to organize things, the way that isospin did for the neutron, proton and pion. In 1963 Murray Gell-Mann [147] and (independently) George Zweig [148] did just that! They enlarged the $\mathbf{SU}(2)$ isospin symmetry to include strangeness – it became an $\mathbf{SU}(3)$ symmetry. Let's look at the basic outline of how they did it.

Gell-Mann and Zweig noticed two things about the 10 lowest-mass spin-$\frac{3}{2}$ baryon states:

1. All states within the same isospin multiplet had approximately the same mass (to within a few percent)

2. States of different strangeness differed in constant-mass increments

These features are easy to see if you plot the different spin-$\frac{3}{2}$ baryon states on a graph with strangeness on the y-axis and isospin-3 component on the x-axis as in fig. 15.2. For example all four of the $\Delta(1232)$ states had $i = \frac{3}{2}$: each of them has strangeness $S = 0$, and about the same mass (namely, 1232 MeV). The lowest-mass spin-$\frac{3}{2}$ baryon states with $S = -1$ were all in an isospin triplet (i.e., $i = 1$) and each had a mass of about 1384 MeV (these are called the Σ-states). The $S = -2$ states (the Ξ^*'s) each had mass ~ 1533 MeV and were in an isospin doublet (i.e. $i = \frac{1}{2}$).

One particle was missing in the pattern: an $S = -3$ state of spin-$\frac{3}{2}$. Since the mass increments are ~ 150 MeV, as isospin decreases, this state should have a mass ~ 1680 MeV and have $i = 0$. Gell-Mann called this particle the Ω^- (since it should be the last particle to be discovered), and predicted its existence in 1961 – three years before its discovery [149]!

Similarly, the eight lowest-mass baryons with of spin-$\frac{1}{2}$ had an eight-fold pattern. Here the mass increments were $\sim (190 \pm 15)$ MeV: For example, $M_\Lambda - M_N \simeq 177$ MeV, and $M_\Xi - M_\Lambda \simeq 203$ MeV – the increments are not quite as good as the spin-$\frac{3}{2}$ collection, but are still not too bad! (Note that there are two kinds of Σ-states: a spin-$\frac{3}{2}$ isospin triplet of mass 1384 MeV and a spin-$\frac{1}{2}$ isospin triplet of mass 1193 MeV).

These regularities in strangeness and isospin were accounted for by Gell-Mann and Zweig by hypothesizing that there were three types of fermion constituents to baryons which Gell-Mann called QUARKS and Zweig called ACES. Gell-Mann's nomenclature won out over Zweig's, and persists today. Each quark had a name:

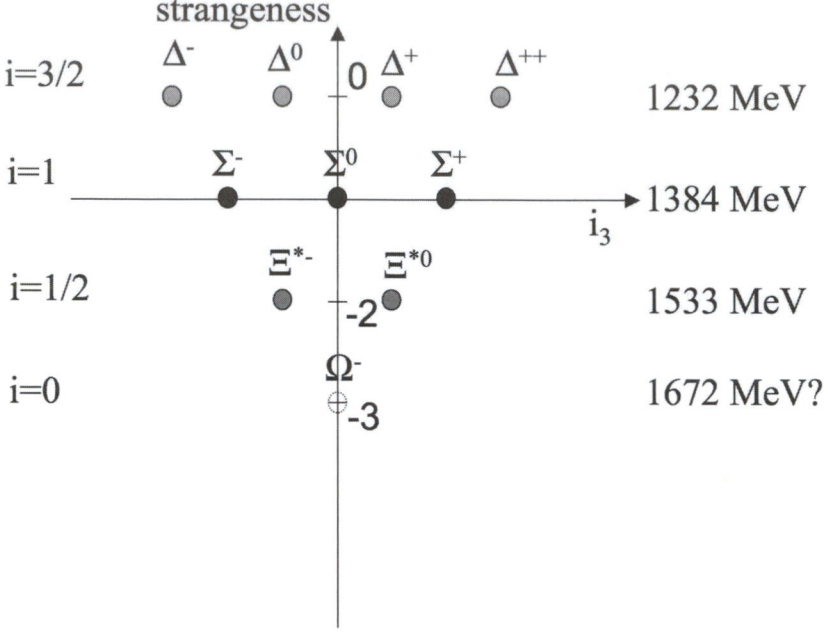

FIGURE 15.2

Isospin/strangeness relationships amongst the spin-3/2 baryon states.

u	"up" – for $i_3 = +\frac{1}{2}$	(and $S = 0$)
d	"down" – for $i_3 = -\frac{1}{2}$	(and $S = 0$)
s	"strange" – for $i = 0$	(and $S = -1$)

Baryons were assumed to each be composed of three quarks. Each quark was presumed to be "one-third" of a baryon and so were assigned $B = \frac{1}{3}$. Since

$$Q = \mathfrak{I}_3 + \frac{1}{2}(B+S) = \mathfrak{I}_3 + \frac{Y}{2} \Rightarrow Q = \begin{cases} +\frac{2}{3} & \text{for } u \\ -\frac{1}{3} & \text{for } d \\ -\frac{1}{3} & \text{for } s \end{cases} \qquad (15.38)$$

the quarks had fractional electric charge! This rather odd property of quarks prevented most physicists from accepting them as actual particles until about 1972-74. The quantity $Y = B+S$ is called *hypercharge*, and its usage persists today.

The quantum numbers isospin and strangeness were redefined into what we now call *flavor*, and the group **SU**(2) (which transforms the definitions of up and down) was enlarged to become **SU**(3) (which now transforms the

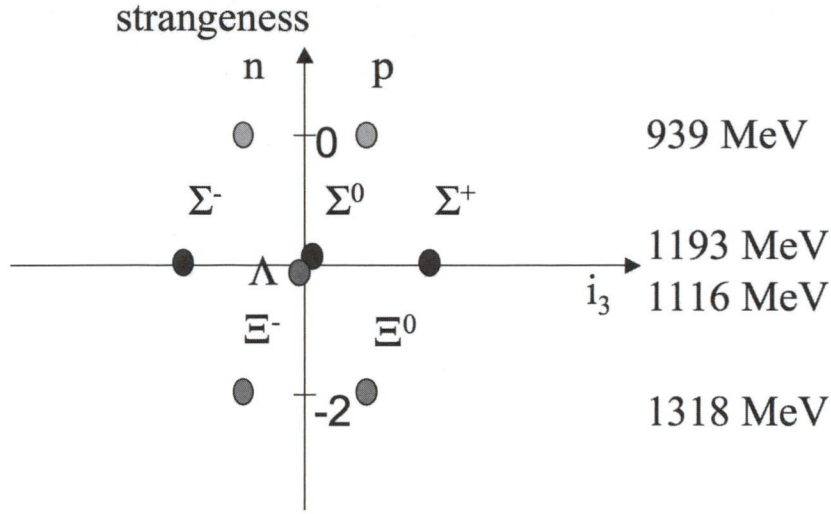

FIGURE 15.3

Isospin/strangeness relationships amongst the spin-1/2 baryon states.

definitions of up, down and strange). This simple idea explained both the patterns and the mass-splittings of the eightfold collection (the OCTET) and the tenfold collection (the DECUPLET) of baryons. As S decreases (becomes more negative), more strange quarks are present and the mass increments arise because $m_s \simeq 150$ MeV. m_u and m_d are presumed to be almost equal, explaining why proton and neutron masses are roughly equal. The isospin symmetry is an **SU**(2) subgroup of the **SU**(3) flavor symmetry, and it works so well because of the near-equality of up and down masses.

We can now understand the octet and decuplet in terms of different collections of quark flavors (i.e. u, d, and s). Working non-relativistically, we expect that a baryon wavefunction is a bound-state wavefunction of three quarks and has the form

$$\Psi_{\text{total}} = \Phi\,(\text{space})\,\psi\,(\text{spin})\,\chi\,(\text{flavor}) \tag{15.39}$$

and that the **SU**(3) symmetry allows us to interchange any two quarks[**]. Since we regard quarks of different flavors as being different states of the same particle, the Pauli principle demands that Ψ_{total} be antisymmetric under the interchange of any two quarks.

[**]Note that this **SU**(3) symmetry is an approximate symmetry: it would only be exact if the mass of the strange quark were equal to that of the up and down quarks.

Now a bound state of 3 quarks can be made from any 3 flavors, leaving $3^3 = 27$ distinct possibilities for χ (flavor). However, we saw earlier that the 8-fold and 10-fold collections (i.e., the octet and decuplet) appear to be rather special. The reason for this is that these collections are irreducible multiplets of the (approximate) $\mathbf{SU}(3)$ flavor symmetry. Let's see how this works.

One possible multiplet is completely symmetric under flavor interchange. Beginning with, say $|uuu\rangle$, we write down all possible states that we can reach from it by replacing one of the u's with either an s or a d in all possible ways that respect the full symmetry (so that the state doesn't change when we switch the positions (not the flavors) of the quarks). For example we can get $|ddd\rangle$ by replacing each u with a d. There are 10 possible states we get from this procedure, and so we get a decuplet, whose wavefunction we call χ_S, shown in fig. 15.4.

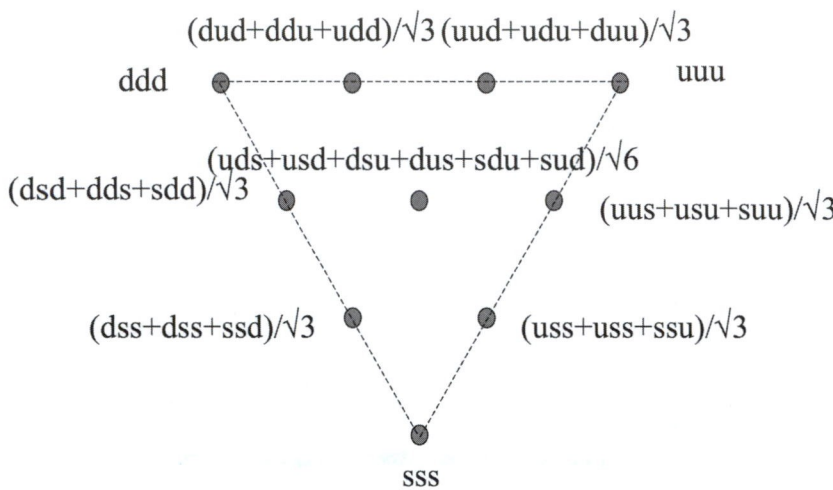

FIGURE 15.4
The decuplet χ_S.

Another multiplet must be one that is completely antisymmetric, in which the interchange of any two flavors yields a minus sign. There is only one such possibility, called χ_A, given in figure 15.5.

Finally, there are also multiplets that are combinations of mixed symmetry which are antisymmetric under the interchange of either the first two flavors (χ_{12}), the last two flavors (χ_{23}), or the first and last flavors (χ_{13}). Each multiplet has 8 states, as illustrated in fig. 15.6. The mixed-symmetry multiplet χ_{13} does not appear in fig. 15.6 because it is not independent of the other

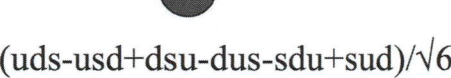

(uds-usd+dsu-dus-sdu+sud)/√6

FIGURE 15.5

The completely antisymmetric flavor state χ_A.

multiplets

$$\chi_{13} = \chi_{12} + \chi_{23} \quad \text{(since} \quad (\chi_{32} + \chi_{21}) = -(\chi_{12} + \chi_{23})\text{)} \qquad (15.40)$$

So these are all the possibilities there are: $10 + 1 + 8 + 8 = 27$ states in total. We write this as

$$3 \otimes 3 \otimes 3 = 10 \oplus 1 \oplus 8 \oplus 8$$

$$\underset{\substack{\text{all possible products of} \\ \text{3 flavors 3 times}}}{} = \underset{\substack{\text{4 distinct multiplets that} \\ \text{transform independently of each other}}}{}$$

To get the spin part of the wavefunction, we just use the Clebsch-Gordon tables ?? twice to get $\frac{1}{2} \otimes \frac{1}{2} \otimes \frac{1}{2} = (1 \oplus 0) \otimes \frac{1}{2} = \left(1 \otimes \frac{1}{2}\right) \oplus \left(0 \otimes \frac{1}{2}\right) = \frac{3}{2} \oplus \frac{1}{2} \oplus \frac{1}{2}$:

$$\left.\begin{aligned}
\left|\tfrac{3}{2}, \tfrac{3}{2}\right\rangle &= |\uparrow\uparrow\uparrow\rangle \\
\left|\tfrac{3}{2}, \tfrac{1}{2}\right\rangle &= \tfrac{1}{\sqrt{3}}(|\uparrow\uparrow\downarrow\rangle + |\uparrow\downarrow\uparrow\rangle + |\downarrow\uparrow\uparrow\rangle) \\
\left|\tfrac{3}{2}, -\tfrac{1}{2}\right\rangle &= \tfrac{1}{\sqrt{3}}(|\uparrow\downarrow\downarrow\rangle + |\downarrow\uparrow\downarrow\rangle + |\downarrow\downarrow\uparrow\rangle) \\
\left|\tfrac{3}{2}, -\tfrac{3}{2}\right\rangle &= |\downarrow\downarrow\downarrow\rangle
\end{aligned}\right\} \text{ spin } \frac{3}{2} \quad (\psi_s) \qquad (15.41)$$

$$\left.\begin{aligned}
\left|\tfrac{1}{2}, \tfrac{1}{2}\right\rangle_{12} &= \tfrac{1}{\sqrt{2}}(|\uparrow\downarrow\rangle - |\downarrow\uparrow\rangle)|\uparrow\rangle \\
\left|\tfrac{1}{2}, -\tfrac{1}{2}\right\rangle_{12} &= \tfrac{1}{\sqrt{2}}(|\uparrow\downarrow\rangle - |\downarrow\uparrow\rangle)|\downarrow\rangle
\end{aligned}\right\} \text{ spin } \frac{1}{2} \quad (\psi_{12}) \qquad (15.42)$$

$$\left.\begin{aligned}
\left|\tfrac{1}{2}, \tfrac{1}{2}\right\rangle_{23} &= \tfrac{1}{\sqrt{2}}|\uparrow\rangle(|\uparrow\downarrow\rangle - |\downarrow\uparrow\rangle) \\
\left|\tfrac{1}{2}, -\tfrac{1}{2}\right\rangle_{23} &= \tfrac{1}{\sqrt{2}}|\downarrow\rangle(|\uparrow\downarrow\rangle - |\downarrow\uparrow\rangle)
\end{aligned}\right\} \text{ spin } \frac{1}{2} \quad (\psi_{23}) \qquad (15.43)$$

where the $\uparrow$ and $\downarrow$ symbols denote spins with z-component $+\frac{1}{2}$ and $-\frac{1}{2}$ respectively. Note that we have a fully symmetric spin-$\frac{3}{2}$ state and two mixed-symmetry spin-$\frac{1}{2}$ states (as with flavor we have $\psi_{13} = \psi_{12} + \psi_{23}$, so the third mixed-symmetry state is not independent).

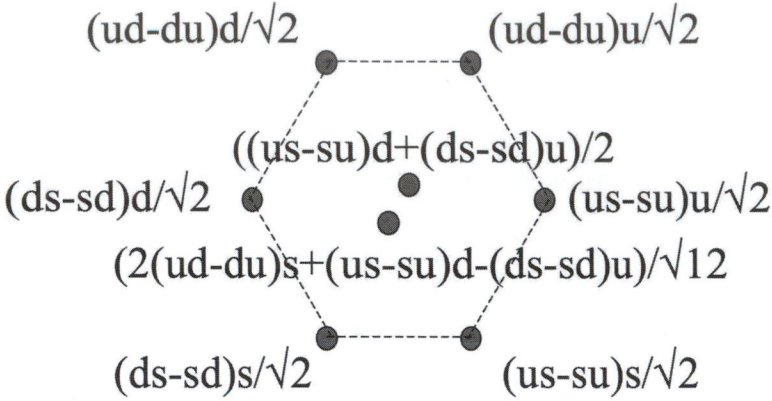

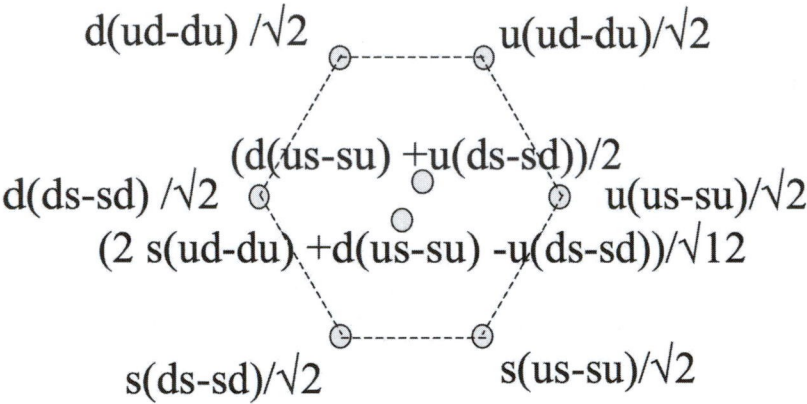

FIGURE 15.6
The mixed-symmetry states χ_{12} (top) and χ_{23} (bottom).

15.5   Color

Recall that we need $\Psi_{\text{total}} = \Phi\,(\text{space})\,\psi\,(\text{spin})\,\chi\,(\text{flavor})$ to be fully antisymmetric because of the Pauli principle. Since we regard Ψ_{total} as the wavefunction for a bound state of 3 quarks, the lowest-energy bound states will be the lowest-mass baryons. Such lowest-energy states will have no orbital angular momentum amongst the quarks (since angular momentum increases the energy). This forces Φ (space) to be symmetric under quark interchange (just as the helium atom in its ground state is symmetric under interchange of the electrons).

So it seems we need the product wavefunction ψ (spin) χ (flavor) to be fully antisymmetric so that Ψ_{total} will also be. Unfortunately this is impossible! For the decuplet, we must have ψ (spin) $= \psi_s$ and χ (flavor) $= \chi_s$ since experiment tells us there are 10 states with spin-$\frac{3}{2}$. However, these are both symmetric wavefunctions and hence their product is also symmetric. For example:

$$|\Delta^{++}\rangle = |u \uparrow u \uparrow u \uparrow\rangle \tag{15.44}$$

and we have 3 identical particles in the same state, violating the Pauli principle.

This problem was resolved by Greenberg [150], who suggested that there was a new quantum number called *color*! He postulated that each flavor of quark (u, d, s) came in three distinct types, or colors – say red, green and blue. This hypothesis implies that

$$\Psi_{\text{total}} = \Phi \,(\text{space})\, \psi \,(\text{spin})\, \chi \,(\text{flavor})\, \varphi \,(\text{color}) \tag{15.45}$$

and so now we have the possibility of making Ψ_{total} antisymmetric by making φ (color) antisymmetric under the interchange of flavors. In fact, we know how to do this: just take φ (color) $= \varphi_A$, where φ_A is given by χ_A with the replacements $u \to r$, $d \to b$, $s \to g$:

$$\varphi_A = \frac{1}{\sqrt{6}} \left(|rbg\rangle - |brg\rangle + |bgr\rangle - |gbr\rangle + |grb\rangle - |rgb\rangle \right) \tag{15.46}$$

This proved to be an enormously fruitful idea, and today we understand color symmetry – which also happens to be $\mathbf{SU}(3)$ – to be the foundational symmetry of the strong interactions. Color is to strong interactions what electric charge is to EM interactions, except that there are three types of "strong charge": r, b, and g. There are also three types of anti-strong-charge: $\bar{r}$, $\bar{b}$, and $\bar{g}$. The state φ_A is a color singlet: no matter how we switch around the definition of color amongst the quarks, this state remains the same (up to a minus sign). The physical interpretation is that φ_A is the wavefunction for a color-neutral bound state of three quarks.

The $\mathbf{SU}(3)$ color symmetry is postulated to be an exact symmetry because two given quarks of the same flavor but of different color have the same mass, charge, and all other quantum numbers. This is quite unlike the $\mathbf{SU}(3)$ flavor symmetry, in which quarks of different flavor *do not* have the same mass. This is perhaps the most confusing thing to deal with when one first encounters quarks: there are two mathematically identical symmetry groups operating in very different physical ways. To avoid confusion the color symmetry is often denoted as $\mathbf{SU}_C(3)$ and the flavor symmetry $\mathbf{SU}_F(3)$. The role played by $\mathbf{SU}_C(3)$ color symmetry is completely different than that played by the $\mathbf{SU}_F(3)$ symmetry. To say that there is an $\mathbf{SU}_C(3)$ color symmetry is to assert that the strong interaction force is the same under any unitary transformation of color charge. The $\mathbf{SU}_F(3)$ symmetry of the quark model, on the other hand, is an approximate symmetry that is used to classify the different hadrons that are made of up, down, and strange quarks. If this symmetry were exact it would mean that all hadrons of a given spin and baryon number would have the same mass.

Because color respects an $\mathbf{SU}(3)$ symmetry, we can also generate color decuplets, octets, etc. Why don't we see these? Without any compelling experimental data, we append another postulate to our list, which states:

Every naturally occuring bound state of quarks is a color singlet

which implies that we always make use of φ_A in constructing hadron wave-functions.

From these foundations we can build what is called the quark model of hadrons.

15.6 Questions

1. Do the following exist? Why or why not?

   (a) an antibaryon with charge $+2$
   (b) a positive meson with strangeness $+1$
   (c) a spin-0 baryon

2. An alpha particle has zero isospin. Can the reaction $D + D \to \alpha + \pi^0$ take place? Why or why not?

3. Consider the operator

$$\mathbb{G} = \mathbb{C} \exp\left(i\pi \mathfrak{J}_2\right)$$

   an operator known as the $\mathbb{G}$-parity operator, where $\mathbb{C}$ is the charge-conjugation operator and $\mathfrak{J}_2$ is the 2nd component of the isospin operator $\vec{\mathfrak{J}}$.

   (a) Suppose we define for the pion

$$\mathbb{C}\left|\pi^{\pm}\right\rangle = -\left|\pi^{\mp}\right\rangle$$

   Show that under $\mathbb{G}$-parity

$$\mathbb{G}\left|\pi\right\rangle = -\left|\pi\right\rangle$$

$$|\pi\rangle = \alpha |\pi^+\rangle + \beta |\pi^0\rangle + c |\pi^-\rangle$$

$$e^{i\pi J_2} = \begin{pmatrix} \cos\pi & 0 & -\sin\pi \\ 0 & 1 & 0 \\ \sin\pi & 0 & \cos\pi \end{pmatrix} = \begin{pmatrix} -1 & 0 & 0 \\ 0 & 1 & 0 \\ 0 & 0 & -1 \end{pmatrix} \qquad \mathbb{C}\begin{pmatrix} -\alpha \\ \beta \\ -c \end{pmatrix} = \begin{pmatrix} c \\ \beta \\ \alpha \end{pmatrix}$$

for the pion triplet.

(b) What is the action of $\mathbb{G}$ on a state $|\Psi(n\pi)\rangle$ of n pions?

4. Show that all non-strange non-baryonic nucleon states are eigenstates of $\mathbb{G}$.

5. A glueball is a color-singlet bound state of two gluons, something expected from QCD. The simplest glueball state $|\mathfrak{G}\rangle$ has positive parity, zero angular momentum, and is even under charge conjugation.

(a) What is the $\mathbb{G}$-parity of this glueball?

(b) What is its strangeness?

(c) Which of the following reactions are allowed

$$(a)\ \mathfrak{G} \rightarrow \pi^+ + \pi^-$$
$$(b)\ \mathfrak{G} \rightarrow \pi^0 + \pi^0$$
$$(c)\ \mathfrak{G} \rightarrow \pi^0 + \pi^0 + \pi^0$$

assuming that the glueball is sufficiently massive so that energy conservation is satisfied for each?

6. The strong interactions are invariant under charge-conjugation. For the following decays

$$(a)\ \rho^0 \rightarrow \pi^+ + \pi^-$$
$$(b)\ \omega^0 \rightarrow \pi^+ + \pi^- + \pi^0$$

how does this constrain the matrix elements $\langle p_1...p_m |M| p\rangle$ for each case?

16

The Quark Model

The basic postulate of the quark model is that all hadrons are bound states of particles called quarks. Quarks must be fermions, since only fermions can give bound states that are both bosons (if an even number of quarks bind together) and fermions (if odd number of quarks bind together).

The quark model was originally constructed using three kinds (or flavors) of quarks: up, down, and strange. Today we know that there are three more flavors of quark – charm, bottom, and top – making for six flavors in all. However, these last three flavors are much heavier than the first three, and do not play a role in our understanding of the lowest-mass baryons and mesons. So I will defer discussion of the heavier quarks to Chapter 18.

TABLE 16.1
Quark Quantum Numbers and Masses

	Q	$\Im$	$\Im_z$	S	C	B	T	B	Y	Mass (MeV)
d	$-\frac{1}{3}$	$\frac{1}{2}$	$-\frac{1}{2}$	0	0	0	0	$\frac{1}{3}$	$\frac{1}{3}$	$5.04^{+0.96}_{-1.54}$
u	$+\frac{2}{3}$	$\frac{1}{2}$	$\frac{1}{2}$	0	0	0	0	$\frac{1}{3}$	$\frac{1}{3}$	$2.55^{+0.75}_{-1.05}$
s	$-\frac{1}{3}$	0	0	-1	0	0	0	$\frac{1}{3}$	$-\frac{2}{3}$	104^{+34}_{-26}
c	$+\frac{2}{3}$	0	0	0	1	0	0	$\frac{1}{3}$	$\frac{1}{3}$	$1,270^{+170}_{-70}$
b	$-\frac{1}{3}$	0	0	0	0	-1	0	$\frac{1}{3}$	$\frac{1}{3}$	$4,200^{+70}_{-170}$
t	$+\frac{2}{3}$	0	0	0	0	0	1	$\frac{1}{3}$	$\frac{1}{3}$	$171,200 \pm 2,100$

Table 16.1 lists the rather surprising properties of quarks: their electric charge is fractional, their baryon number is 1/3, and their masses* have no discernable pattern. By convention, the sign of the quark flavor is chosen to

---

*The definition of quark mass is rather subtle, as we shall see in this and subseqent chapters. The list of masses in table 16.1 are current quark masses obtained from a broad range of experiments as discussed in the Particle Data Book [1].

be the same as that of its electric charge. Direct searches for free quarks –
in cosmic rays, in Millikan-type experiments, and in rocks from the Moon –
have been fruitless. No free quarks have ever been observed [151]. This is now
believed to be a consequence of the color force that binds them together.

The quark model reached perhaps its most sophisticated form due to the
efforts of Nathan Isgur and Gabriel Karl, who showed that all of the low energy
baryon and meson states could be understood as bound states of quarks using
the principles of atomic physics [152]. While I can't present the full details of
this model, I can give you its foundations and basic ideas [153]. Let's begin
our study of the quark model by looking at the two main kinds of hadrons:
baryons and mesons.

16.1 Baryons

A baryon is a hadron that is postulated to be a bound state of three different
quarks. The bound-state baryon wavefunction is

$$\Psi_{\text{baryon}} = \Phi\,(\text{space})\,\psi\,(\text{spin})\,\chi\,(\text{flavor})\,\varphi\,(\text{color}) \tag{16.1}$$

where

$$\varphi\,(\text{color}) = \varphi_A = \frac{1}{\sqrt{6}}\,(|rbg\rangle - |brg\rangle + |bgr\rangle - |gbr\rangle + |grb\rangle - |rgb\rangle) \tag{16.2}$$

is the fully antisymmetric color wavefunction, ensuring that all baryons are
color-neutral, since the $\mathbf{SU}_C(3)$ symmetry leaves the φ_A state invariant. The
spatial wavefunction $\Phi\,(\text{space})$ is a function of the positions of the quarks. It
must be fully symmetric under interchange of quark position because it corresponds to a state of lowest energy (i.e. we wish to describe the lowest-mass
baryons), and from atomic physics we know that such a state has principal
quantum number $n = 0$ – this is an S-wave state, which is symmetric.
The Pauli principle requires Ψ_{baryon} to be fully antisymmetric under the
interchange of any pair of quarks. Since $\Phi\,(\text{space})\,\varphi\,(\text{color})$ must be fully antisymmetric, this means that $\psi\,(\text{spin})\,\chi\,(\text{flavor})$ must be fully symmetic under
quark-pair interchange.

So if all baryons are indeed bound states of three quarks, what we must
do is combine the irreps we found in the previous chapter for $\psi\,(\text{spin})$ and
$\chi\,(\text{flavor})$ in all possible symmetric combinations. Recall that we found

$$\text{SPIN:}\ \left\{\psi_S^{3/2}, \psi_{12}^{1/2}, \psi_{23}^{1/2}\right\} \qquad\qquad \text{FLAVOR:}\ \left\{\chi_S^{10}, \chi_{12}^8, \chi_{23}^8, \chi_A^1\right\}$$

where the superscript is a reminder as to which multiplet we are considering
and the subscript denotes the symmetry/antisymmetry of the wavefunction.

All possible combinations of these wavefunction irreps that are symmetric under particle interchange will give us the lowest mass baryons.

One obvious symmetric combination is $\psi_S^{3/2}\chi_S^{10}$: we must have a decuplet of spin-$\frac{3}{2}$. We get all of the explicit states by taking a direct product of the two wave functions. For example:

$$\left|\Delta^{++}; \frac{3}{2}, \frac{3}{2}\right\rangle = |uuu\rangle \otimes |\uparrow\uparrow\uparrow\rangle = |u\uparrow u\uparrow u\uparrow\rangle \tag{16.3}$$

$$\left|\Delta^{++}; \frac{3}{2}, \frac{1}{2}\right\rangle = |uuu\rangle \otimes \left(\frac{1}{\sqrt{3}}\left(|\uparrow\uparrow\downarrow\rangle + |\uparrow\downarrow\uparrow\rangle + |\downarrow\uparrow\uparrow\rangle\right)\right)$$
$$= \frac{1}{\sqrt{3}}\left(|u\uparrow u\uparrow u\downarrow\rangle + |u\uparrow u\downarrow u\uparrow\rangle + |u\downarrow u\uparrow u\uparrow\rangle\right) \tag{16.4}$$

$$\left|\Sigma^{*+}; \frac{3}{2}, \frac{1}{2}\right\rangle = \frac{1}{\sqrt{3}}\left(|uus\rangle + |usu\rangle + |suu\rangle\right) \otimes \left(\frac{1}{\sqrt{3}}\left(|\uparrow\uparrow\downarrow\rangle + |\uparrow\downarrow\uparrow\rangle + |\downarrow\uparrow\uparrow\rangle\right)\right)$$
$$= \frac{1}{3}\left(|u\uparrow u\uparrow s\downarrow\rangle + |u\uparrow u\downarrow s\uparrow\rangle + |u\downarrow u\uparrow s\uparrow\rangle\right.$$
$$+ |u\uparrow s\uparrow u\downarrow\rangle + |u\uparrow s\downarrow u\uparrow\rangle + |u\downarrow s\uparrow u\uparrow\rangle$$
$$\left. + |s\uparrow u\uparrow u\downarrow\rangle + |s\uparrow u\downarrow u\uparrow\rangle + |s\downarrow u\uparrow u\uparrow\rangle\right) \tag{16.5}$$

and you can construct all of the others by a similar procedure of taking direct products.

The numerical coefficients allow us to compute probabilities of quark distributions inside a baryon. For example, suppose we had a Σ^{*+} in a $j_z = \frac{1}{2}$ state. We see from eq. (16.5) that there are three terms in its wavefunction that have a spin-$\downarrow$ s-quark. Each has a coefficient of $\frac{1}{3}$, and so we would have a $\left(\frac{1}{3}\right)^2 + \left(\frac{1}{3}\right)^2 + \left(\frac{1}{3}\right)^2 = \frac{1}{3}$ chance of finding a spin-$\downarrow$ s-quark. Similarly there are six terms containing a spin-$\downarrow$ u-quark and so there is a $6\times\left(\frac{1}{3}\right)^2 = \frac{2}{3}$ probability of finding a spin-$\downarrow$ u-quark in the $\Sigma^{*+}-$ if we were able to isolate the quarks!

Another symmetric combination follows using the χ_{12}^8 and χ_{23}^8 octet wavefunctions. Since $\psi_{12}^{1/2}\chi_{12}^8$ is symmetric on $(1 \leftrightarrow 2)$, we can form a fully symmetric wavefunction by taking

$$\psi\chi = \mathfrak{N}\left(\psi_{12}^{1/2}\chi_{12}^8 + \psi_{23}^{1/2}\chi_{23}^8 + \psi_{13}^{1/2}\chi_{13}^8\right)$$
$$= \mathfrak{N}\left(2\psi_{12}^{1/2}\chi_{12}^8 + 2\psi_{23}^{1/2}\chi_{23}^8 + \psi_{23}^{1/2}\chi_{12}^8 + \psi_{12}^{1/2}\chi_{23}^8\right) \tag{16.6}$$

where we recall that $\psi_{13}^{1/2} = \psi_{12}^{1/2} + \psi_{23}^{1/2}$ and $\chi_{13}^8 = \chi_{12}^8 + \chi_{23}^8$, and $\mathfrak{N}$ is some normalization factor to be determined for each state (since there will be cancellations between some terms when we look at specific spin/flavor wavefunctions). So we expect a spin-1/2 baryon octet!

Note that the observed octet is a mixture of the two flavor octets – it must be to ensure that ψ (spin) χ (flavor) is fully symmetric. Again, we find explicit states by direct multiplication. For example, a spin-up proton is

$$\left|p;\frac{1}{2},\frac{1}{2}\right\rangle = \mathfrak{N}\left\{2\left[\frac{1}{\sqrt{2}}\left(|\uparrow\downarrow\rangle - |\downarrow\uparrow\rangle\right)|\uparrow\rangle\right]\left[\frac{1}{\sqrt{2}}\left(|ud\rangle - |du\rangle\right)|u\rangle\right]\right.$$
$$+2\left[\frac{1}{\sqrt{2}}|\uparrow\rangle\left(|\uparrow\downarrow\rangle - |\downarrow\uparrow\rangle\right)\right]\left[\frac{1}{\sqrt{2}}|u\rangle\left(|ud\rangle - |du\rangle\right)\right]$$
$$+\left[\frac{1}{\sqrt{2}}|\uparrow\rangle\left(|\uparrow\downarrow\rangle - |\downarrow\uparrow\rangle\right)\right]\left[\frac{1}{\sqrt{2}}\left(|ud\rangle - |du\rangle\right)|u\rangle\right]$$
$$\left.+\left[\frac{1}{\sqrt{2}}\left(|\uparrow\downarrow\rangle - |\downarrow\uparrow\rangle\right)|\uparrow\rangle\right]\left[\frac{1}{\sqrt{2}}|u\rangle\left(|ud\rangle - |du\rangle\right)\right]\right\}(16.7)$$

which upon simplification becomes

$$\left|p;\frac{1}{2},\frac{1}{2}\right\rangle = \frac{\mathfrak{N}}{4}\left(2\left|u\uparrow d\downarrow u\uparrow\right\rangle - 2\left|u\downarrow d\uparrow u\uparrow\right\rangle - 2\left|d\uparrow u\downarrow u\uparrow\right\rangle\right.$$
$$+2\left|d\downarrow u\uparrow u\uparrow\right\rangle + 2\left|u\uparrow u\uparrow d\downarrow\right\rangle$$
$$-2\left|u\uparrow u\downarrow d\uparrow\right\rangle - 2\left|u\uparrow d\uparrow u\downarrow\right\rangle + 2\left|u\uparrow d\downarrow u\uparrow\right\rangle$$
$$+\left|u\uparrow d\uparrow u\downarrow\right\rangle - \left|u\uparrow d\downarrow u\uparrow\right\rangle$$
$$-\left|d\uparrow u\uparrow u\downarrow\right\rangle + \left|d\uparrow u\downarrow u\uparrow\right\rangle + \left|u\uparrow u\downarrow d\uparrow\right\rangle$$
$$\left.-\left|u\downarrow u\uparrow d\uparrow\right\rangle - \left|u\uparrow d\downarrow u\uparrow\right\rangle + \left|u\downarrow d\uparrow u\uparrow\right\rangle\right)$$
$$= \frac{\mathfrak{N}}{4}\left(2\left|u\uparrow d\downarrow u\uparrow\right\rangle - \left|u\downarrow d\uparrow u\uparrow\right\rangle - \left|d\uparrow u\downarrow u\uparrow\right\rangle\right.$$
$$+2\left|d\downarrow u\uparrow u\uparrow\right\rangle + 2\left|u\uparrow u\uparrow d\downarrow\right\rangle - \left|u\uparrow u\downarrow d\uparrow\right\rangle$$
$$\left.-\left|u\uparrow d\uparrow u\downarrow\right\rangle - \left|d\uparrow u\uparrow u\downarrow\right\rangle - \left|u\downarrow u\uparrow d\uparrow\right\rangle\right)\quad(16.8)$$

and the normalization coefficient can be obtained by recognizing that each state in the preceding expression is orthogonal to all the others. Hence the requirement that $\langle p;\frac{1}{2},\frac{1}{2}|p;\frac{1}{2},\frac{1}{2}\rangle = 1$ gives

$$\left|p;\frac{1}{2},\frac{1}{2}\right\rangle = \frac{1}{\sqrt{18}}\left(2\left|u\uparrow d\downarrow u\uparrow\right\rangle - \left|u\downarrow d\uparrow u\uparrow\right\rangle - \left|d\uparrow u\downarrow u\uparrow\right\rangle\right.$$
$$+2\left|d\downarrow u\uparrow u\uparrow\right\rangle + 2\left|u\uparrow u\uparrow d\downarrow\right\rangle - \left|u\uparrow u\downarrow d\uparrow\right\rangle$$
$$\left.-\left|u\uparrow d\uparrow u\downarrow\right\rangle - \left|d\uparrow u\uparrow u\downarrow\right\rangle - \left|u\downarrow u\uparrow d\uparrow\right\rangle\right)\quad(16.9)$$

and so the probability of finding, say, a spin-down d-quark in a proton (where $j_z = +\frac{1}{2}$ for the proton spin) is $\frac{1}{18} \times \left(2^2 + 2^2 + 2^2\right) = \frac{2}{3}$.

Are there any other symmetric mixtures of ψ (spin) and χ (flavor)? The answer is no: no other combination of mixed-symmetry wavefunctions can be made fully symmetric, and there is no fully antisymmetric state that can be combined with χ_A. This agrees with observation: all lowest-mass baryons are either spin-$\frac{3}{2}$ and fit into a decuplet, or are spin-$\frac{1}{2}$ and fit into an octet[†].

Constructing baryon wavefunctions is a straightforward, though somewhat tedious, job. There are three spins, three flavors, and three colors, and they all must be put together in a manner that satisfies the Pauli principle. The $\mathbf{SU}_F(3)$ flavor symmetry is approximate since the up, down and strange quarks have different masses (though the up and down masses are nearly identical). The $\mathbf{SU}_C(3)$ color symmetry is believed to be exact: a red up-quark has all the properties of a blue up-quark or a green up-quark, apart from color. Color plays an essential (but hidden) role in all of this – without it we would also be searching for antisymmetric spin-flavor wavefunctions. The only way to do this for spin-$\frac{3}{2}$ is to combine it with the antisymmetric $\mathbf{SU}_F(3)$ singlet χ_A^1 – this would mean there would be only one spin-$\frac{3}{2}$ baryon, in strong contradiction with observation.

16.2 Mesons

Another way to get color-neutral states is to combine a quark-antiquark pair into a bound state. Since there are 3 flavors, 2 particles and 2 possible spins, there are a total of $3 \times 3 \times 2 = 18$ distinct such states. These states are called *mesons.*

A meson wavefunction is of the form of eq. (16.1)

$$\Psi_{\mathrm{meson}} = \Phi\,(\mathrm{space})\,\psi\,(\mathrm{spin})\,\chi\,(\mathrm{flavor})\,\varphi\,(\mathrm{color})$$

where the Pauli principle is no longer an issue, since the particles are not identical. The spatial wavefunction will still be symmetric (since we want the lowest-energy states, which means $\Phi\,(\mathrm{space})$ is an S-wave) and the color wavefunction is guaranteed to be neutral (i.e. a singlet) since it is made from color-anticolor pairs. This color wavefunction is

$$\varphi_S = \frac{1}{\sqrt{3}}\left(|r\bar{r}\rangle + |b\bar{b}\rangle + |g\bar{g}\rangle\right) \tag{16.10}$$

where each color is paired with its anticolor.

[†]Of course excited states would be different because in that case $\varphi(\mathrm{space})$ is not necessarily symmetric since the quarks would now have some orbital angular momentum. In this case we can construct other combinations of $\psi(\mathrm{spin})$ and $\chi(\mathrm{flavor})$. The only restriction is that the bound state wavefunction be antisymmetric under interchange of any two quarks (and that it is a color singlet).

The spin wavefunctions are

$$\left.\begin{array}{l} |1,1\rangle = |\uparrow\uparrow\rangle \\ |1,0\rangle = \frac{1}{\sqrt{2}}\left(|\uparrow\downarrow\rangle + |\downarrow\uparrow\rangle\right) \\ |1,-1\rangle = |\downarrow\downarrow\rangle \end{array}\right\} \text{ triplet } \psi_T^1 \tag{16.11}$$

$$|0,0\rangle = \frac{1}{\sqrt{2}}\left(|\uparrow\downarrow\rangle - |\downarrow\uparrow\rangle\right) \Big\} \text{ singlet } \psi_S^0 \tag{16.12}$$

as we have seen before in eqs.(6.23, 6.24) for positronium. The spin-1 triplet state is the "ortho" state and the spin-0 singlet is the "para" state. These are the only ways of combining two spin-$\frac{1}{2}$ particles. Hence the quark model predicts that all (low-energy) mesons are either spin-0 (the scalar mesons) or spin-1 (the vector mesons).

The flavor wavefunction must now be made from a quark and an antiquark. The 9 possible flavor states decompose into an octet and a singlet as in fig. 16.1 where the minus sign is a (rather irritating) convention for expressing an

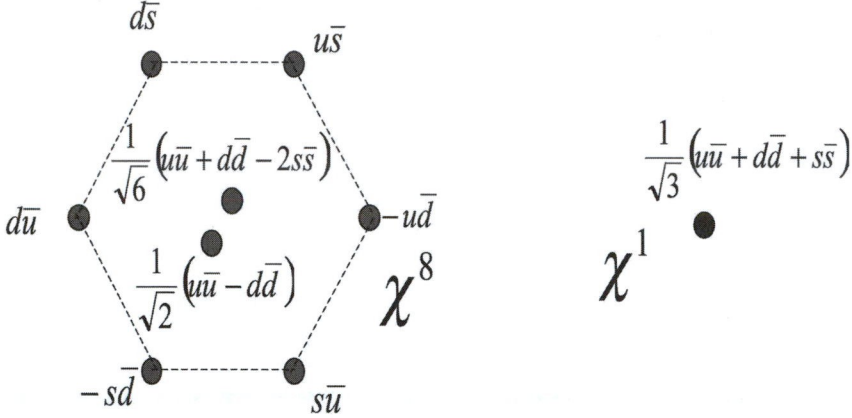

FIGURE 16.1
The octet and singlet meson flavor wavefunctions.

isospin doublet of antiquarks. We have

$$|u\rangle = \left| i = \frac{1}{2}, i_3 = \frac{1}{2} \right\rangle \qquad |d\rangle = \left| i = \frac{1}{2}, i_3 = -\frac{1}{2} \right\rangle \tag{16.13}$$

$$|\bar{d}\rangle = -\left| i = \frac{1}{2}, i_3 = \frac{1}{2} \right\rangle \qquad |\bar{u}\rangle = \left| i = \frac{1}{2}, i_3 = -\frac{1}{2} \right\rangle \tag{16.14}$$

and note that $\bar{d}$ is isospin-up (it's the antiparticle of isospin-down, after all), whereas $\bar{u}$ is isospin-down. In group-theoretic language we say that the three quarks u, d, s belong to the fundamental 3 representation of $\mathbf{SU}_F(3)$ and that the three antiquarks belong to the conjugate representation $\bar{3}$. The combination of both we write as $3 \otimes \bar{3} = 8 \oplus 1$, meaning that we have a collection of 8 objects (the octet) that transform into one another under $\mathbf{SU}_F(3)$, and one object (the singlet) that just transforms into itself under $\mathbf{SU}_F(3)$. You can easily check both statements by seeing what happens under interchange of any pair of flavors.

For the spin-0 mesons we identify these flavor wavefunctions as in fig. 16.2, or in other words

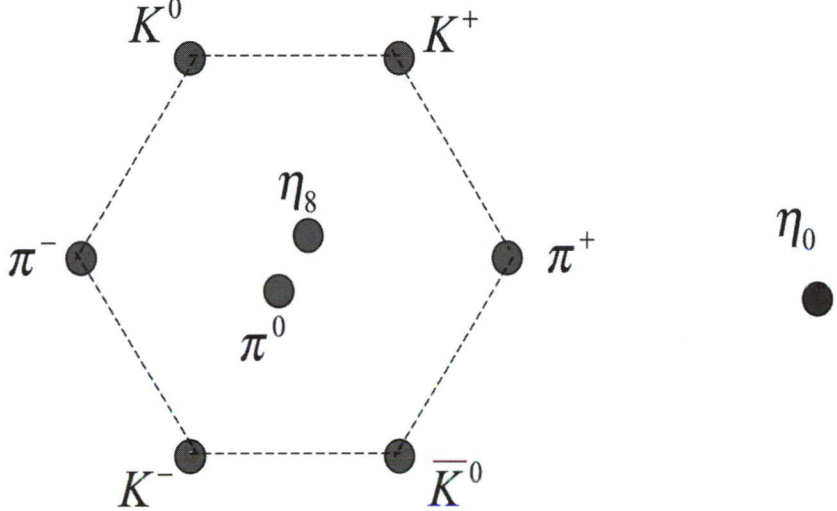

FIGURE 16.2
The spin-0 mesons.

$$|\pi^+\rangle = -|u\bar{d}\rangle\, \psi_S^0$$
$$|\pi^0\rangle = \frac{1}{\sqrt{2}}\left(|u\bar{u}\rangle - |d\bar{d}\rangle\right)\, \psi_S^0$$
$$|\pi^-\rangle = |d\bar{u}\rangle\, \psi_S^0$$
$$|K^+\rangle = |u\bar{s}\rangle\, \psi_S^0 \qquad\qquad (16.15)$$
$$|K^0\rangle = |d\bar{s}\rangle\, \psi_S^0$$
$$|\bar{K}^0\rangle = -|s\bar{d}\rangle\, \psi_S^0$$
$$|K^-\rangle = |s\bar{u}\rangle\, \psi_S^0$$

for the pion and Kaon states[‡].

The η states are a little more subtle: the flavor-spin states are

$$|\eta_0\rangle = \frac{1}{\sqrt{3}} \left(|u\bar{u}\rangle + |d\bar{d}\rangle + |s\bar{s}\rangle \right) \, \psi_S^0 \tag{16.16}$$

$$|\eta_8\rangle = \frac{1}{\sqrt{6}} \left(|u\bar{u}\rangle + |d\bar{d}\rangle - 2\,|s\bar{s}\rangle \right) \, \psi_S^0 \tag{16.17}$$

Note that the η_0 is a para-spin state and a para-isospin state (as is the η_8). This is in contrast to the pion-triplet, which is a para-spin state but an ortho-isopin state.

The experimental situation is a bit more complicated since the actual states observed in nature are orthogonal linear combinations of the η_0 and η_8:

$$|\eta\rangle = \cos\theta_\eta \, |\eta_0\rangle + \sin\theta_\eta \, |\eta_8\rangle \tag{16.18}$$

$$|\eta'\rangle = -\sin\theta_\eta \, |\eta_0\rangle + \cos\theta_\eta \, |\eta_8\rangle \tag{16.19}$$

Experiment indicates that the mixing angle $\theta_\eta \simeq 11°$. Since the η_0 and η_8 have identical quantum numbers, their wavefunctions can mix. However, the simple quark model here does not explain why they mix. The actual reason for the mixing has to do with the underlying theory of the strong $\mathbf{SU}_C(3)$ interaction – quantum chromodynamics, or QCD – which induces continuous transitions between the quark-antiquark pairs in the η and η'. It is the task of this more fundamental theory to explain the origin of this mixing angle.

For the vector mesons we have an analogous picture, given in fig. 16.3 where now

$$\begin{aligned}
|\rho^+\rangle &= -|u\bar{d}\rangle \, \psi_T^1 \\
|\rho^0\rangle &= \frac{1}{\sqrt{2}} \left(|u\bar{u}\rangle - |d\bar{d}\rangle \right) \, \psi_T^1 \\
|\rho^-\rangle &= |d\bar{u}\rangle \, \psi_T^1 \\
|K^{*+}\rangle &= |u\bar{s}\rangle \, \psi_T^1 \\
|K^{*0}\rangle &= |d\bar{s}\rangle \, \psi_T^1 \\
|\overline{K}^{*0}\rangle &= -|s\bar{d}\rangle \, \psi_T^1 \\
|K^{*-}\rangle &= |s\bar{u}\rangle \, \psi_T^1
\end{aligned} \tag{16.20}$$

[‡]The minus signs in eqs. (16.14) and (16.15) are irrelevant phase conventions that appear for technical reasons [31], but they do have the effect of modifying our linear combinations. For example, a para-isospin state is

$$\frac{1}{\sqrt{2}} \left(\left|i = \tfrac{1}{2}, i_3 = \tfrac{1}{2}\right\rangle \left|i = \tfrac{1}{2}, i_3 = -\tfrac{1}{2}\right\rangle - \left|i = \tfrac{1}{2}, i_3 = -\tfrac{1}{2}\right\rangle \left|i = \tfrac{1}{2}, i_3 = \tfrac{1}{2}\right\rangle \right)$$

$$= \frac{1}{\sqrt{2}} \left(|u\rangle \, |d\rangle + |d\rangle \, |u\rangle \right)$$

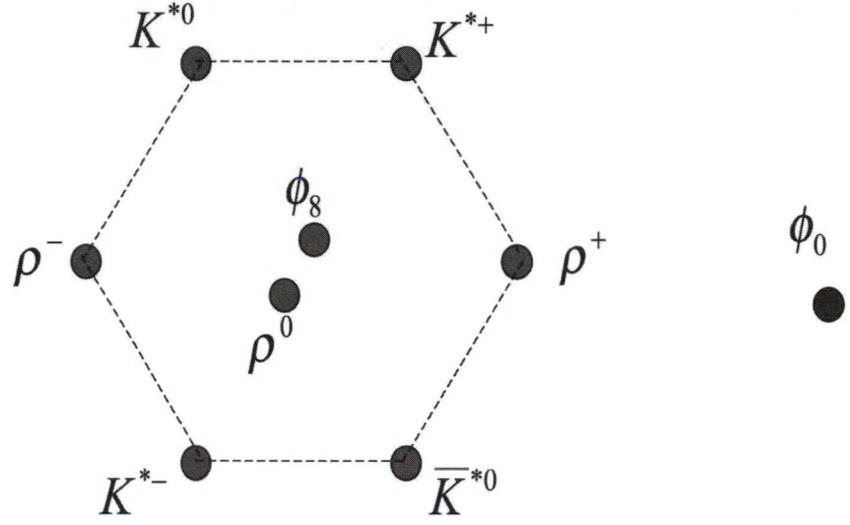

FIGURE 16.3
The spin-1 mesons.

in analogy with the spin-0 case. As with the η states, the ϕ_0 and ϕ_8 states
are

$$|\phi_0\rangle = \frac{1}{\sqrt{3}} \left(|u\bar{u}\rangle + |d\bar{d}\rangle + |s\bar{s}\rangle \right) \psi_T^1 \tag{16.21}$$

$$|\phi_8\rangle = \frac{1}{\sqrt{6}} \left(|u\bar{u}\rangle + |d\bar{d}\rangle - 2|s\bar{s}\rangle \right) \psi_T^1 \tag{16.22}$$

and so we see that the ρ-meson is an ortho-spin and ortho-isospin state,
whereas the ϕ_0 is an ortho-spin but para-isospin state. And as before with
the spin-singlet case, the actual ϕ-states observed in nature are orthogonal
linear combinations of the ϕ_0 and ϕ_8 states :

$$|\omega\rangle = \cos\theta_\phi |\phi_0\rangle + \sin\theta_\phi |\phi_8\rangle \simeq \frac{1}{\sqrt{2}} \left(|u\bar{u}\rangle + |d\bar{d}\rangle \right) \tag{16.23}$$

$$|\phi\rangle = -\sin\theta_\phi |\phi_0\rangle + \cos\theta_\phi |\phi_8\rangle \simeq |s\bar{s}\rangle \tag{16.24}$$

where now experiment indicates that the mixing angle $\theta_\phi \simeq 35^o$. This makes
the ϕ-meson almost completely $|s\bar{s}\rangle$ (explaining why it preferentially decays
into $K\overline{K}$ pairs), and the ω-meson $\frac{1}{\sqrt{2}} \left(|u\bar{u}\rangle + |d\bar{d}\rangle \right)$ (explaining why the masses
of the ρ-meson and the ω-meson are almost the same). The origin of the
mixing angle θ_ϕ is beyond the scope of the simple quark model presented
here, and requires QCD for a more complete explanation.

Mesons are hadrons with zero baryon number. It is straightforward to figure out their general properties under $\mathbb{P}$ and $\mathbb{C}$. For parity, the orbital angular momentum l contributes as for the hydrogen atom. A fermion always has a parity opposite to that of its antifermion counterpart; by convention we set all quarks to have positive parity and all antiquarks to have negative parity. Hence $\mathbb{P} = (-1)^{l+1}$ for a meson. Any system consisting of spin-$\frac{1}{2}$ particle/antiparticle pairs will emit one photon if it (a) undergoes a transition between two states where the orbital angular momentum changes by one unit or (b) if one particle undergoes a spin-flip relative to the other particle. Since photons have $\mathbb{C} = -1$, we have $\mathbb{C} = (-1)^{l+s}$ for flavorless mesons. Otherwise $\mathbb{C}$ is undefined. Summarizing:

$$\mathbb{P} = (-1)^{l+1} \quad \text{for all mesons}$$
$$\mathbb{C} = (-1)^{l+s} \quad \text{for flavorless mesons}$$

Many other mesons beyond the ground state scalar and vector mesons exist, and their quantum numbers are compatible with those of a spin-$\frac{1}{2}$ particle-antiparticle pair in an excited state with nonzero orbital angular momentum.

16.3   Mass Relations

The differing masses between the various baryon and meson states can approximately be accounted for by assuming that the strange quark has a mass $m_s \simeq 150$ MeV. For example in the decuplet we have

$$M_{\Sigma^*} - M_\Delta \simeq 152 \text{ MeV} \qquad M_{\Xi^*} - M_\Sigma \simeq 149 \text{ MeV}$$
$$M_\Omega - M_{\Xi^*} \simeq 139 \text{ MeV} \tag{16.25}$$

which can be explained reasonably well by setting $m_s \simeq 150$ MeV, but for the octet we have

$$M_\Lambda - M_N \simeq 177 \text{ MeV} \qquad M_\Xi - M_\Lambda \simeq 203 \text{ MeV} \tag{16.26}$$

which is pretty crude. We should be able to improve on this by taking the quark model more seriously to include a better description of how the quarks bind. Since we are working with constituent masses, we have already included the energy of the gluons. However, we have not taken account of interactions due to quark spin. Specifically, if mesons and baryons really are bound states of spin-$\frac{1}{2}$ quarks, presumably spin-spin interactions between the quarks should play an important role.

Why? Recall that the magnetic dipole moment for a fermion is of the form

$$\vec{\mu} = g \frac{e}{2mc} \vec{S} \tag{16.27}$$

where $g \simeq 2$ for a spin-$\frac{1}{2}$ fermion (modulo small QED corrections). Any two fermions will have a dipole-dipole interaction, whose energy is

$$H_{\text{dipole}} \approx \frac{\vec{\mu}_1 \cdot \vec{\mu}_2}{r_{12}^3} \tag{16.28}$$

where r_{12} is the separation between fermion 1 and fermion 2. In a hydrogenic atom this kind of interaction takes place between the electron and the spinning nucleus, and splits each energy level into two very close levels. Such a splitting is called a *hyperfine* splitting; for a hydrogen atom in the S-state we find that this interaction – when averaged over all directions – vanishes everywhere except at the origin. This gives

$$\Delta E_{\text{H.F.}} = \langle H_{\text{dipole}} \rangle_{\text{directions}} = \left(\frac{8\pi}{3} \right) \frac{4\pi\alpha\hbar^2}{m_e m_p c^2} \vec{S}_e \cdot \vec{S}_p \, |\Psi_{n=0}(0)|^2 \tag{16.29}$$

where $\vec{S} = \frac{\hbar}{2}\vec{\sigma}$, and the α is the fine-structure constant. Note that the energy is proportional to the square of the wavefunction at the origin due to the averaging procedure.

Quarks carry a color charge, so we expect a *color*-magnetic-dipole interaction[§]. The color potential is proportional to $1/r$ (we'll see why in Chapter 18), so we expect that for mesons

$$M_{\text{meson}} = (\text{sum of quark masses}) + (\text{color dipole interaction energy})$$

$$= m_q + m_{\bar{q}} + \frac{1}{3} \left(\frac{8\pi}{3} \right) \frac{4\pi\alpha_s}{m_q m_{\bar{q}}} \vec{S}_q \cdot \vec{S}_{\bar{q}} \, |\Psi_{\text{meson}}(0)|^2 \tag{16.30}$$

where α_s is the strong interaction constant, and the factor of $\frac{1}{3}$ is due to the three quark colors. For baryons we expect

$$M_{\text{baryon}} = m_{q_1} + m_{q_2} + m_{q_3} + \frac{1}{6} \left(\frac{8\pi}{3} \right) \sum_{i<j}^{3} \frac{4\pi\alpha_s}{m_{q_i} m_{q_j}} \vec{S}_{q_i} \cdot \vec{S}_{q_j} \, \left| \Psi_{\text{baryon}}(0) \right|^2 \tag{16.31}$$

where the factor of $\frac{1}{6}$ is also due to quark color.

We can work out the spin-spin terms as follows. Recall that

$$\vec{S}_q \cdot \vec{S}_q = \left| \vec{S}_q \right|^2 = \hbar^2 s_q (s_q + 1) = \frac{\hbar^2}{2} \left(\frac{1}{2} + 1 \right) = \frac{3}{4} \hbar^2 \tag{16.32}$$

and that $\vec{S}_q + \vec{S}_{\bar{q}} = \vec{S}_{\text{meson}}$. Hence

$$\vec{S}_q \cdot \vec{S}_{\bar{q}} \, |\text{meson}\rangle = \frac{1}{2} \left[\vec{S}_{\text{meson}} \cdot \vec{S}_{\text{meson}} - \vec{S}_q \cdot \vec{S}_q - \vec{S}_{\bar{q}} \cdot \vec{S}_{\bar{q}} \right] |\text{meson}\rangle$$

[§]Of course quarks also carry electromagnetic charge, leading to an electromagnetic dipole interaction – this is negligibly weak compared to the color dipole interaction.

$$= \frac{\hbar^2}{2} \left[s_{\text{meson}} (s_{\text{meson}} + 1) - \frac{3}{4} - \frac{3}{4} \right] |\text{meson}\rangle$$

$$= \begin{cases} +\frac{\hbar^2}{4} & \text{for } s_{\text{meson}} = 1 \text{ (vector mesons)} \\ -\frac{3\hbar^2}{4} & \text{for } s_{\text{meson}} = 0 \text{ (scalar mesons)} \end{cases} |\text{meson}\rangle \quad (16.33)$$

giving a clear energy difference between vector and scalar mesons. Using

$$m_u = m_d = 310 \text{ MeV} \quad \text{and} \quad m_s = 483 \text{ MeV} \quad (16.34)$$

we can fit all vector and scalar masses to the formula for M_{meson} above to within 1% provided

$$\frac{1}{3} \left(\frac{8\pi}{3} \right) 4\pi\alpha_s |\Psi_{\text{meson}}(0)|^2 = \left(\frac{2m_u}{\hbar} \right)^2 \times 160 \text{ MeV} \quad (16.35)$$

which is a huge improvement!

TABLE 16.2
Quark-Model Meson Masses

	Calculated	Observed
π	140	138
K	484	496
η	559	549
ρ	780	776
ω	780	783
K^*	896	892
ϕ	1032	1020

Note that the masses of the up, down, and strange quarks differ considerably from those given in table 16.1. This is because the masses in table 16.1 are the *current* quark masses: the mass of a quark "by itself." However, a bound quark is continually exchanging gluons (the force carriers of the strong interaction) with its counterparts in the bound state, which adds an appreciable amount to the total energy of the baryon or meson. It is convenient to define the *constituent* quark mass: the current quark mass plus the mass of the gluon particle field surrounding the quark. So for the strange quark its current mass is 104 MeV (a modern refinement over the original crude estimate of 150 MeV), whereas its constituent mass is 483 MeV. Similar considerations apply for the up and down quark masses.

Clearly we're on the right track. For baryons, the spin product is a bit trickier. If all quark masses are equal (as in the N, Δ, and Ω) then

$$\sum_{i<j}^{3} \frac{1}{m_{q_i} m_{q_j}} \vec{S}_{q_i} \cdot \vec{S}_{q_j} |\text{EMB}\rangle$$

$$= \frac{1}{m_q^2} \left[\vec{S}_1 \cdot \vec{S}_2 + \vec{S}_1 \cdot \vec{S}_3 + \vec{S}_2 \cdot \vec{S}_3 \right] |\text{EMB}\rangle$$

$$= \frac{1}{2m_q^2} \left[\vec{S}_\text{baryon} \cdot \vec{S}_\text{baryon} - \vec{S}_1 \cdot \vec{S}_1 - \vec{S}_2 \cdot \vec{S}_2 - \vec{S}_3 \cdot \vec{S}_3 \right] |\text{EMB}\rangle$$

$$= \frac{\hbar^2}{2m_q^2} \left[s_\text{baryon} \left(s_\text{baryon} + 1 \right) - \frac{3}{4} - \frac{3}{4} - \frac{3}{4} \right] |\text{EMB}\rangle$$

$$= \frac{\hbar^2}{m_q^2} \left\{ \begin{array}{ll} \frac{3}{4} & \text{for } s_\text{baryon} = \frac{3}{2} \quad (\text{decuplet}) \\ -\frac{3}{4} & \text{for } s_\text{baryon} = \frac{1}{2} \quad (\text{octet}) \end{array} \right. |\text{EMB}\rangle \qquad (16.36)$$

where EMB means "equal mass baryon," giving

$$M_N = 3m_u - \frac{3\hbar^2}{4m_u^2} \left(\frac{32\pi^2 \alpha_s}{18} \right) \left| \Psi_\text{baryon}(0) \right|^2 \qquad (16.37)$$

$$M_\Delta = 3m_u + \frac{3\hbar^2}{4m_u^2} \left(\frac{32\pi^2 \alpha_s}{18} \right) \left| \Psi_\text{baryon}(0) \right|^2 \qquad (16.38)$$

$$M_\Omega = 3m_s + \frac{3\hbar^2}{4m_s^2} \left(\frac{32\pi^2 \alpha_s}{18} \right) \left| \Psi_\text{baryon}(0) \right|^2 \qquad (16.39)$$

Actually, for the decuplet all spins are parallel (every pair combines to make spin-1) so

$$1(1+1)\hbar^2 |\text{decuplet}\rangle = \left(\vec{S}_{q_i} + \vec{S}_{q_j} \right)^2 |\text{decuplet}\rangle$$

$$= \left(\vec{S}_{q_i} \cdot \vec{S}_{q_i} + \vec{S}_{q_j} \cdot \vec{S}_{q_j} + 2\vec{S}_{q_i} \cdot \vec{S}_{q_j} \right) |\text{decuplet}\rangle$$

$$= \left(\frac{3}{4}\hbar^2 + \frac{3}{4}\hbar^2 + 2\vec{S}_{q_i} \cdot \vec{S}_{q_j} \right) |\text{decuplet}\rangle$$

$$\Rightarrow \vec{S}_{q_i} \cdot \vec{S}_{q_j} = \frac{1}{4}\hbar^2 \quad \text{for the decuplet} \qquad (16.40)$$

which gives

$$M_{\Sigma^*} = 2m_u + m_s + \frac{\hbar^2}{4} \left(\frac{1}{m_u^2} + \frac{2}{m_u m_s} \right) \left(\frac{32\pi^2 \alpha_s}{18} \right) \left| \Psi_\text{baryon}(0) \right|^2 \quad (16.41)$$

$$M_{\Xi^*} = 2m_s + m_u + \frac{\hbar^2}{4} \left(\frac{1}{m_s^2} + \frac{2}{m_u m_s} \right) \left(\frac{32\pi^2 \alpha_s}{18} \right) \left| \Psi_\text{baryon}(0) \right|^2 \quad (16.42)$$

for the remaining decuplet states. For the Σ and Λ octet states, the up and down quarks are respectively in isospin 1 and isospin 0 states, which means that their spins must respectively combine to give 1 and 0. Hence

$$1(1+1)\hbar^2 |\text{octet-}\Sigma\rangle = \left(\vec{S}_u + \vec{S}_d \right)^2 |\text{octet-}\Sigma\rangle$$

$$= \left(\frac{3}{4}\hbar^2 + \frac{3}{4}\hbar^2 + 2\vec{S}_u \cdot \vec{S}_d \right) |\text{octet-}\Sigma\rangle$$

$$\Rightarrow \vec{S}_u \cdot \vec{S}_d = \frac{1}{4}\hbar^2 \quad \text{for the octet } \Sigma \qquad (16.43)$$

$$0 = \left(\vec{S}_u + \vec{S}_d\right)^2 |\text{octet-}\Lambda\rangle$$

$$= \left(\frac{3}{4}\hbar^2 + \frac{3}{4}\hbar^2 + 2\vec{S}_u \cdot \vec{S}_d\right) |\text{octet-}\Lambda\rangle$$

$$\Rightarrow \vec{S}_u \cdot \vec{S}_d = -\frac{3}{4}\hbar^2 \quad \text{for the octet } \Lambda \qquad (16.44)$$

and so

$$M_\Sigma = 2m_u + m_s + \left(\frac{\vec{S}_u \cdot \vec{S}_d}{m_u m_d} + \frac{\vec{S}_1 \cdot \vec{S}_2 + \vec{S}_1 \cdot \vec{S}_3 + \vec{S}_2 \cdot \vec{S}_3 - \vec{S}_u \cdot \vec{S}_d}{m_u m_s}\right)$$

$$\times \left(\frac{32\pi^2 \alpha_s}{18}\right) \left|\Psi_{\text{baryon}}(0)\right|^2$$

$$= 2m_u + m_s + \frac{\hbar^2}{4}\left(\frac{1}{m_u^2} - \frac{4}{m_u m_s}\right)\left(\frac{32\pi^2 \alpha_s}{18}\right)\left|\Psi_{\text{baryon}}(0)\right|^2 \quad (16.45)$$

$$M_\Lambda = 2m_u + m_s - \frac{3\hbar^2}{4m_u^2}\left(\frac{32\pi^2 \alpha_s}{18}\right)\left|\Psi_{\text{baryon}}(0)\right|^2 \qquad (16.46)$$

Finally, the mass of the Ξ is computed similarly to that of the Σ, and so

$$M_\Xi = 2m_s + m_u + \frac{\hbar^2}{4}\left(\frac{1}{m_u^2} - \frac{4}{m_u m_s}\right)\left(\frac{32\pi^2 \alpha_s}{18}\right)\left|\Psi_{\text{baryon}}(0)\right|^2 \quad (16.47)$$

and so we have predictions for the masses of all states in the octet and decuplet. Using the quark masses we used for the meson case, we find that setting

$$\left(\frac{32\pi^2 \alpha_s}{18}\right)\left|\Psi_{\text{baryon}}(0)\right|^2 = \left(\frac{2m_u}{\hbar}\right)^2 \times 50 \text{ MeV} \qquad (16.48)$$

yields agreement with experimental data to within 1%, as you can see from table 16.3! The actual mass of a hadron can be can be understood as consisting of two components: (a) the current mass m_{current} from the quarks themselves and (b) a contribution Δm from the electric charge of the hadron (due to the work required to put a charge on the particle), so that $m = m_{\text{current}} + \Delta m$. If all baryons in an octet have similar charge distributions, then we expect

$$\Delta m_p = \Delta m_{\Sigma^+} \qquad \Delta m_{\Sigma^-} = \Delta m_{\Xi^-} \qquad \Delta m_{\Xi^0} = \Delta m_n \qquad (16.49)$$

Summing these equations and adding the bare masses to each side gives

$$m_p + m_{\Sigma^-} + m_{\Xi^0} = m_{\Sigma^+} + m_{\Xi^-} + m_n \qquad (16.50)$$

$$m_p - m_n = m_{\Sigma^+} - m_{\Sigma^-} + m_{\Xi^-} - m_{\Xi^0} \qquad (16.51)$$

TABLE 16.3

Quark-Model Baryon Masses

	Calculated	Observed
N	939	939
Λ	1116	1114
Σ_{octet}	1179	1193
Ξ	1327	1318
Δ	1239	1232
Σ_{decuplet}	1381	1384
Ξ^*	1529	1533
Ω	1682	1672

which is a formula due originally to Coleman and Glashow [154]. Empirically we have

$$m_p - m_n = -1.3 \text{ MeV} \tag{16.52}$$

$$m_{\Sigma^+} - m_{\Sigma^-} + m_{\Xi^-} - m_{\Xi^0} = -8.0 \text{ MeV} + 6.4 \text{ MeV}$$
$$= -1.6 \text{ MeV} \tag{16.53}$$

providing good confirmation that up and down quark masses are nearly identical.

16.4 Magnetic Moments

The magnetic moment of a baryon should be the sum of its quark magnetic moments:

$$\vec{\mu}_{\text{baryon}} = \vec{\mu}_1 + \vec{\mu}_2 + \vec{\mu}_3 \tag{16.54}$$

If the quarks are pointlike, then

$$\vec{\mu}_{\text{quark}} = g_{\text{quark}} \frac{e_{\text{quark}}}{2m_{\text{quark}}c} \vec{S}_{\text{quark}} = \frac{e_{\text{quark}}}{m_{\text{quark}}c} \vec{S}_{\text{quark}} = \frac{e_{\text{quark}}\hbar}{2m_{\text{quark}}c} \vec{\sigma}_{\text{quark}} \tag{16.55}$$

where $g_{\text{quark}} = 2$ to a high-degree of approximation (i.e. neglecting its anomalous magnetic moment). The magnitude of a baryon magnetic moment is defined to be the expectation value of the magnetic moment operator of a spin-up baryon states:

$$|\vec{\mu}_{\text{baryon}}| \equiv \mu_B = \langle \text{baryon} \uparrow | (\vec{\mu}_1 + \vec{\mu}_2 + \vec{\mu}_3)_z | \text{baryon} \uparrow \rangle \tag{16.56}$$

where

$$\vec{\mu}_j = \mu_j \vec{\sigma}_j \qquad \text{with } \mu_j = \frac{e_j \hbar}{2m_j c} \tag{16.57}$$

$$\mu_u = \frac{2}{3}\frac{e\hbar}{2m_u c} \qquad \mu_d = -\frac{1}{3}\frac{e\hbar}{2m_d c} \qquad \mu_s = -\frac{1}{3}\frac{e\hbar}{2m_s c}$$

(retaining the factors of $\hbar$ and c), and where $\vec{\sigma}_z\,|\uparrow\rangle = +\,|\uparrow\rangle$; $\vec{\sigma}_z\,|\downarrow\rangle = -\,|\downarrow\rangle$.

Let's compute this for the proton. Recall that the quark model predicts its wavefunction is given by eq. (16.9)

$$\begin{aligned}
|p\uparrow\rangle = \frac{1}{\sqrt{18}}\Big(&2\,|u\uparrow d\downarrow u\uparrow\rangle - |u\downarrow d\uparrow u\uparrow\rangle - |d\uparrow u\downarrow u\uparrow\rangle \\
&+2\,|d\downarrow u\uparrow u\uparrow\rangle + 2\,|u\uparrow u\uparrow d\downarrow\rangle - |u\uparrow u\downarrow d\uparrow\rangle \\
&- |u\uparrow d\uparrow u\downarrow\rangle - |d\uparrow u\uparrow u\downarrow\rangle - |u\downarrow u\uparrow d\uparrow\rangle \Big) \quad (16.58)
\end{aligned}$$

and we must compute $\langle p\uparrow|\,(\vec{\mu}_1 + \vec{\mu}_2 + \vec{\mu}_3)_z\,|p\uparrow\rangle$ term-by-term. For example

$$\begin{aligned}
(\vec{\mu}_1 + \vec{\mu}_2 + \vec{\mu}_3)_z\,|u\uparrow d\downarrow u\uparrow\rangle &= (\mu_u + (-1)\mu_d + \mu_u)\,|u\uparrow d\downarrow u\uparrow\rangle \\
&= (2\mu_u - \mu_d)\,|u\uparrow d\downarrow u\uparrow\rangle \quad (16.59)
\end{aligned}$$

and so the first term contributes

$$\left(\frac{2}{\sqrt{18}}\right)^2 \langle u\uparrow d\downarrow u\uparrow|\,(\vec{\mu}_1 + \vec{\mu}_2 + \vec{\mu}_3)_z\,|u\uparrow d\downarrow u\uparrow\rangle = \frac{2}{9}\times(2\mu_u - \mu_d) \quad (16.60)$$

The second term contributes

$$\left(\frac{1}{\sqrt{18}}\right)^2 \langle u\downarrow d\uparrow u\uparrow|\,(\vec{\mu}_1 + \vec{\mu}_2 + \vec{\mu}_3)_z\,|u\downarrow d\uparrow u\uparrow\rangle = \frac{1}{18}\mu_d \quad (16.61)$$

and it is easy to see that all other terms contribute either $\frac{2}{9}\times(2\mu_u - \mu_d)$ or $\frac{1}{18}\mu_d$ to the magnetic moment. Noting that all states in the proton wavefunction are orthogonal, we obtain [153] from a term-by-term calculation

$$\mu_{\text{proton}} = \left[3\times\frac{2}{9}\,(2\mu_u - \mu_d) + 6\times\frac{1}{18}\,(\mu_d)\right] = \frac{1}{3}\,(4\mu_u - \mu_d) \quad (16.62)$$

We can repeat this for all the octet states, and compare to experiment. As table 16.4 shows, agreement is good, but not outstanding. Clearly the quark model will have to be modified (by including more detailed interactions between the quarks) if we are to improve on these predictions.

The Σ^0 state has a very short lifetime of only $(7.4 \pm .07)\times 10^{-20}$ seconds (as compared to lifetimes of 10^{-10} seconds or longer for all of the other spin-1/2 baryons). Hence the magnetic moment that can be measured is a transition magnetic moment associated with its decay into a Λ particle. This quantity is not really comparable to the magnetic moments of the other octet baryons and so I have not filled in a value for it in table 16.4.

TABLE 16.4

Quark-Model Magnetic Moments

	Formula	Calculated	Observed
p	$\frac{1}{3}\left(4\mu_u - \mu_d\right)$	2.79	2.793
n	$\frac{1}{3}\left(4\mu_d - \mu_u\right)$	-1.86	-1.913
Λ	μ_s	-0.613	-0.613 ± 0.004
Σ^+	$\frac{4}{3}\mu_u - \frac{1}{3}\mu_s$	2.68	2.458 ± 0.010
Σ^0	$\frac{2}{3}\left(\mu_d + \mu_u\right) - \frac{1}{3}\mu_s$	0.82	(see note below)
Σ^-	$\frac{4}{3}\mu_d - \frac{1}{3}\mu_s$	-1.05	-1.160 ± 0.025
Ξ^0	$\frac{4}{3}\mu_s - \frac{1}{3}\mu_u$	-1.44	-1.250 ± 0.014
Ξ^-	$\frac{4}{3}\mu_s - \frac{1}{3}\mu_d$	-0.51	-0.651 ± 0.003

(all quantities in units of nuclear magnetons

$$\mu_N = \frac{e\hbar}{2m_pc} = 3.152 \times 10^{-18} \text{ MeV/Gauss})$$

16.5 Questions

1. A well-known theorist at the University of Waterloo tells you that he has been able to solve the three-body problem in quantum mechanics. As expected, a fully symmetric spatial wavefunction describes the ground state. A fully antisymmetric spatial wavefunction describes the first excited state. Assuming this is correct, how many baryon states does the quark model predict at the first excited level?

2. Verify the table

Magnetic Moments of Baryons	
	Formula
p	$\frac{1}{3}\left(4\mu_u - \mu_d\right)$
n	$\frac{1}{3}\left(4\mu_d - \mu_u\right)$
Λ	μ_s
Σ^+	$\frac{4}{3}\mu_u - \frac{1}{3}\mu_s$
Σ^0	$\frac{2}{3}\left(\mu_d + \mu_u\right) - \frac{1}{3}\mu_s$
Σ^-	$\frac{4}{3}\mu_d - \frac{1}{3}\mu_s$
Ξ^0	$\frac{4}{3}\mu_s - \frac{1}{3}\mu_u$
Ξ^-	$\frac{4}{3}\mu_s - \frac{1}{3}\mu_d$

3. Consider the spin-spin potential

$$V = \frac{K}{m_1 m_2}\vec{\sigma}_1\vec{\sigma}_2\delta\left(\vec{r}_{12}\right)$$

between a quark and an antiquark of masses m_1, m_2, separated by a distance $\vec{r}_{12}$ where K is a dimensionless constant in units where $c = \hbar = 1$.

(a) Calculate $K |\Psi(0)|^2$ from experiment for the $c\bar{c}$ system, in units where c and $\hbar$ are not set to unity.

(b) Compute the splitting of the $^3S_1$ and $^1S_0$ states for the Υ system, which is a bound state of a b-quark with its antiparticle.

(c) Does this work for the light mesons (e.g. for the ρ–π splitting)? Why or why not?

4. Calculate the masses of (a) the J/ψ (a bound state of a charm quark with its antiparticle) and the D-mesons (bound states of a charm quark with antiquarks of lighter mass) and (b) the B-mesons and the Υ. Compare your results to experiment, and comment on the implications for the quark model.

5. Suppose that there were only two colors, say red and blue, with the color symmetry being given by $\mathbf{SU}_C(2)$. Baryons would consist of colorless bound states of quarks, whereas mesons would consist of colorless bound states of a quark and an antiquark.

(a) Write down the color wavefunction for a baryon and the color wave-function for a meson.

(b) Are baryons fermions or bosons? Are mesons fermions or bosons?

(c) How do the lowest-energy baryon and meson states differ?

6. Suppose that there were N colors, with the color symmetry being given by $\mathbf{SU}_C(N)$. Baryons would consist of colorless bound states of quarks, whereas mesons would consist of colorless bound states of a quark and an antiquark. Write down the color wavefunction for a baryon and the color wavefunction for a meson.

7. Are any of these processes allowed? Why or why not?

(a) $\rho^0 \longrightarrow \pi^+ + \pi^-$ (b) $\eta^0 \longrightarrow 3\gamma$

(c) $\rho^0 \longrightarrow \pi^0 + \pi^0$ (d) $\eta^0 \longrightarrow \pi^+ + \pi^- + \pi^0$

17

Testing the Quark Model

Can we be confident that quarks actually exist? Or is the quark model nothing more than a clever classification scheme for mesons and baryons? This issue came to dominate the scientific community in the late 1960s, and many remained skeptical of the existence of quarks. This began to change in 1968 when experimental evidence was found that protons had constituents called PARTONS. For a period of time the theoretical and experimental situation was in a state of confusion. However, after a few years partons came to be identified with quarks, and a series of experiments were carried out that provided the best "direct" evidence for the existence of quarks that we have. Today quarks are an established part of the Standard Model.

Quarks have never been isolated as free particles, so it is generally thought that they are always confined to hadronic bound states. If this is the case, how could we possibly observe them? The best way is with electromagnetism. Everything in QED that applies to electrons also applies to quarks, provided the appropriate charge of $\frac{2}{3}e$ or $-\frac{1}{3}e$ is used. Since quantum electrodynamics is well understood, electromagnetic interactions furnish a useful probe of hadron substructure – if quarks exist, they should be able to interact electromagnetically with electrons, after all.

There are three basic processes that give us experimental information about quarks: vector meson decay [155], hadronic production, and deep inelastic scattering. The first two of these are (roughly speaking) time-reversals of each other, and are shown in fig. 17.1.

17.1   Vector-Meson Decay

Consider the decay of a vector meson (i.e. ϕ, ρ or ω) into a lepton-antilepton pair (i.e. μ or e). The diagram at the left in figure 17.1 gives, after doing the δ-function integrals

$$iM = \overline{v}_{\overline{q}_j}(p_2)\left(-ie_{q_j}\gamma^\mu\right)u_{q_j}(p_1)\frac{-ig_{\mu\nu}}{(p_1+p_2)^2}\overline{u}_{\ell^-}(p_2')\left(-ie\gamma^\nu\right)v_{\ell^+}(p_1')$$

$$= \frac{ie^2}{M_V^2}\left[\frac{e_{q_j}}{e}\overline{v}_{\overline{q}_j}(p_2)\left(\gamma^\mu\right)u_{q_j}(p_1)\right]\left[\overline{u}_{\ell^-}(p_2')\left(\gamma_\mu\right)v_{\ell^+}(p_1')\right] \qquad (17.1)$$

We could simply square this and put it into a cross-section formula if the initial quark and antiquark were free particles. However, they are not free particles – they are bound together into a meson. The total quark momentum is therefore $(p_1 + p_2)^2 = M_V^2$, the mass of the vector meson, and this has been put into the formula (17.1). In fact the meson is described by the bound state wavefunction $\Psi_{\text{meson}} = \Phi\,(\text{space})\,\psi\,(\text{spin})\,\chi\,(\text{flavor})\,\varphi\,(\text{color})$. We know that the color wavefunction $\varphi\,(\text{color})$ is guaranteed to be neutral (i.e. a singlet) since it is made from color-anticolor pairs and that $\Phi\,(\text{space})$ should be a 2-Body ground state wavefunction. The $\psi\,(\text{spin})$ is accounted for by the Dirac wavefunctions above, but we need the proper $\psi\,(\text{spin})\,\chi\,(\text{flavor})$ combination. So the matrix element should read

$$i\mathcal{M} = \frac{ie^2}{M_V^2}\frac{\Phi\varphi}{\sqrt{2M_V}}\left[\sum_{j=1}^{n}\frac{e_{q_j}}{e}\chi(j)\overline{v}_{\overline{q}_j}(p_2)\,(\gamma^\mu)\,u_{q_j}(p_1)\right]\left[\overline{u}_{\ell^-}(p_2')\,(\gamma_\mu)\,v_{\ell^+}(p_1')\right]$$

$$(17.2)$$

where we sum over the n flavors in the meson and a normalization factor of $\frac{1}{\sqrt{2M_V}}$ has been inserted.

To get a decay rate we need to compute the spin-averaged matrix-element and integrate over the spatial distribution of the quark-antiquark combination. Since we know that the quark and antiquark will only annihilate each other when they are at the same spatial position, this integration must give something proportional to the square of the wavefunction at the origin. Using the Casimir trick and taking traces*

$$\sum_{i_1,i_2=\uparrow,\downarrow}\left[\overline{v}_{\overline{q}_j}^{(i_2)}(p_2)\,(\gamma^\mu)\,u_{q_j}^{(i_1)}(p_1)\right]^\dagger\left[\overline{v}_{\overline{q}_k}^{(i_2)}(p_2)\,(\gamma^\mu)\,u_{q_k}^{(i_1)}(p_1)\right]$$

$$= \text{Tr}\left[\overline{\gamma}^\mu\left(\not{p}_2 + m_j\right)\gamma^\nu\left(\not{p}_1 - m_j\right)\right]\delta_{jk}$$

$$= 4\left(p_1^\mu p_2^\nu + p_1^\nu p_2^\mu - g^{\mu\nu}\left[m_j^2 + p_1\cdot p_2\right]\right)\delta_{jk}$$

$$= L^{\mu\nu}(p_1, p_2; -m_j^2)\delta_{jk}$$

$$(17.3)$$

where

$$L^{\mu\nu}\left(p_1, p_2; -m^2\right) = 4\left(p_1^\mu p_2^\nu + p_1^\nu p_2^\mu + g^{\mu\nu}\left[+m^2 - p_1\cdot p_2\right]\right)$$

$$(17.4)$$

and where the δ_{jk} appears because we get zero unless the two flavors of quark in the meson are the same. This yields for the square of the spin-averaged matrix-element

$$|\overline{\mathcal{M}}|^2 \propto \frac{e^4}{2M_V^5}|\Phi\,(0)|^2\left|\sum_{j=1}^{n}\frac{e_{q_j}}{e}\chi(j)\right|^2 L^{\mu\nu}\left(p_1, p_2; -m_j^2\right)L_{\mu\nu}\left(p_1', p_2'; -m^2\right)$$

$$(17.5)$$

---

*This procedure is exactly the same as the one we carried out in Chapter 13, so I will not repeat any details here.

where $|\varphi\,(\text{color})|^2$ yields an overall factor which is the same for all mesons. Since this expression is in an arbitrary frame, we can write as a non-relativistic approximation

$$p_1^\mu = \frac{m_{q_j}}{M_V} p_V^\mu \qquad\qquad p_2^\mu = \frac{m_{\bar{q}_j}}{M_V} p_V^\mu \qquad (17.6)$$

where we ignore the relative momentum between the quark and the antiquark. Since $m_{q_j} = m_{\bar{q}_j}$ this gives

$$
\begin{aligned}
L^{\mu\nu}\left(p_1, p_2; -m_j^2\right) &= 4\left(2\frac{m_{q_j}^2}{M_V^2} p_V^\mu p_V^\nu - g^{\mu\nu}\left[2m_{q_j}^2\right]\right) \\
&= -8m_{q_j}^2\left(g^{\mu\nu} - \frac{p_V^\mu p_V^\nu}{M_V^2}\right) \\
&= -2M_V^2\left(g^{\mu\nu} - \frac{p_V^\mu p_V^\nu}{M_V^2}\right) \qquad (17.7)
\end{aligned}
$$

since $m_{q_j} = \frac{1}{2}M_V$ – the mass of the vector meson is twice the quark mass! For the leptons emitted in the process, we have $p_1'^\mu = (E, \vec{p})$ and $p_2'^\mu = (E, -\vec{p})$ in the rest-frame of the meson, where $|\vec{p}| \simeq E$ neglecting the mass m of the leptons.

Hence we get

$$
\begin{aligned}
&L^{\mu\nu}\left(p_1, p_2; -m_j^2\right) L_{\mu\nu}\left(p_1', p_2'; -m^2\right) \\
&= -8M_V^2\left[2\left(p_1' \cdot p_2'\right) - 4\left(p_1' \cdot p_2'\right) - 2\frac{\left(p_1' \cdot p_V\right)\left(p_2' \cdot p_V\right)}{M_V^2} + \left(p_1' \cdot p_2'\right)\frac{p_V^2}{M_V^2}\right] \\
&= -8M_V^2\left[-2\left(2E^2\right) - 2\frac{\left(EM_V\right)^2}{M_V^2} + \left(2E^2\right)\frac{M_V^2}{M_V^2}\right] \\
&= 8M_V^4 \qquad (17.8)
\end{aligned}
$$

since $E = \frac{1}{2}M_V$ by energy conservation. Hence the matrix element becomes

$$|\overline{\mathcal{M}}|^2 \propto \frac{4e^4}{M_V}|\Phi(0)|^2\left|\sum_{j=1}^{n}\frac{e_{q_j}}{e}\chi(j)\right|^2 \qquad (17.9)$$

where the constant of proportionality is related to a color factor that is the same for each meson.

The two body decay rate $\Gamma_{\text{2-Body}} = \frac{S|\vec{p}|c}{8\pi\hbar(M_V c)^2}|\overline{\mathcal{M}}|^2$ becomes

$$\Gamma\left(V \longrightarrow \ell^+\ell^-\right) \propto \frac{\alpha^2}{M_V^2}\left|\sum_{j=1}^{n}\chi(j)\frac{e_{q_j}}{e}\right|^2 \qquad (17.10)$$

and will be multiplied by a function of the momenta of the initial and final states, but (apart from the factor of M_V^2) is common to all the vector mesons. Using

$$e_u = \frac{2}{3}e \text{ and } e_d = e_s = -\frac{1}{3}e \qquad (17.11)$$

along with the flavor wavefunctions for the vector mesons, we find

$$\chi\left(\rho^0\right) = \frac{1}{\sqrt{2}}\left(|u\bar{u}\rangle - |d\bar{d}\rangle\right)$$

$$\Rightarrow \left|\sum_{j=1}^{n}\chi(j)\frac{e_{q_j}}{e}\right|^2_{\rho^0} = \left|\frac{1}{\sqrt{2}}\frac{2}{3} - \frac{1}{\sqrt{2}}\left(-\frac{1}{3}\right)\right|^2 = \frac{1}{2} \qquad (17.12)$$

$$\chi\left(\phi\right) = |s\bar{s}\rangle \Rightarrow \left|\sum_{j=1}^{n}\chi(j)\frac{e_{q_j}}{e}\right|^2_{\phi} = \left|\left(-\frac{1}{3}\right)\right|^2 = \frac{1}{9} \qquad (17.13)$$

$$\chi\left(\omega\right) = \frac{1}{\sqrt{2}}\left(|u\bar{u}\rangle + |d\bar{d}\rangle\right)$$

$$\Rightarrow \left|\sum_{j=1}^{n}\chi(j)\frac{e_{q_j}}{e}\right|^2_{\omega} = \left|\frac{1}{\sqrt{2}}\frac{2}{3} + \frac{1}{\sqrt{2}}\left(-\frac{1}{3}\right)\right|^2 = \frac{1}{18} \qquad (17.14)$$

yielding

$$\Gamma\left(\rho^0 \longrightarrow \ell^+\ell^-\right) : \Gamma\left(\phi \longrightarrow \ell^+\ell^-\right) : \Gamma\left(\omega \longrightarrow \ell^+\ell^-\right)$$

$$= \begin{cases} 9 : 2\left(\frac{m_u}{m_s}\right)^2 : 1 & \text{predicted} \\ 8.8 \pm 2.6 : 1.70 \pm 0.41 : 1 & \text{observed} \end{cases} \qquad (17.15)$$

where you might recall that the ρ and ω have similar masses. This corroborates the flavor assignments for the ϕ and the ω we made in Chapter 16, and is a test both of these assignments and of the fractional charges of quarks. However, we can provide better evidence for the existence of quarks by looking at this process in reverse, as we consider in the next section.

17.2 Hadron Production

When $e^+e^- \longrightarrow \bar{q}q$ (via a photon) the quark and antiquark will fly apart as free particles until they reach a separation of about a fermi (10^{-15} m). At this separation the strong interaction is so powerful that new quarks and antiquarks are produced, which in turn will bind into hadrons. Dozens of hadrons are formed, and the resultant state is a mess. However, momentum

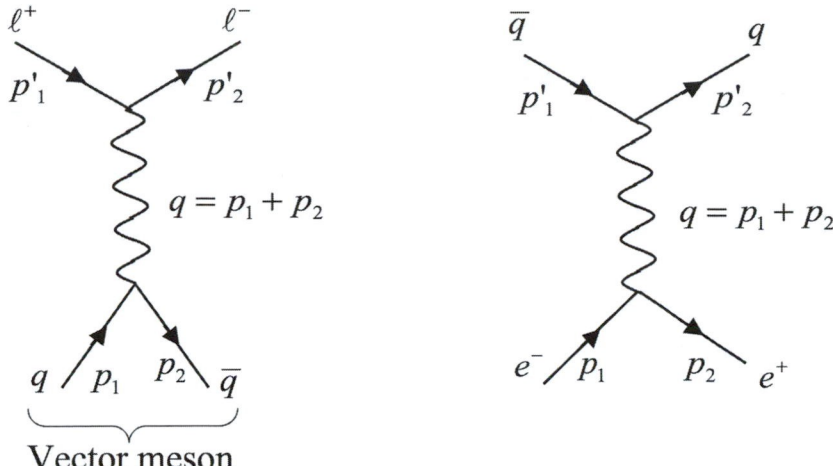

FIGURE 17.1
Diagrams for vector-meson decay (left) and Hadronic production (right).

conservation will force them (in the CMS frame) to emerge in two cone-like streams called *jets* – each jet tracks the primordial quark or antiquark from which it arose.

The cross-section for this process will be just like that for $e^+e^- \longrightarrow \mu^+\mu^-$ except that we must replace the charge of the muon (antimuon) with that of the quark (antiquark). The diagrammatic rules give from fig. 17.1

$$i\mathcal{M} = [\bar{u}_q(p'_2)(-ie_q\gamma^\mu)v_{\bar{q}}(p'_1)] \frac{-ig_{\mu\nu}}{(p_1+p_2)^2} [\bar{v}_{e^+}(p'_2)(-ie\gamma^\nu)u_{e^-}(p'_1)] \quad (17.16)$$

which upon spin-averaging and spin summing give as before

$$\overline{|\mathcal{M}|}^2 = \frac{1}{4}\left[\frac{ee_q}{(p_1+p_2)^2}\right]^2 L^{\mu\nu}(p'_1,p'_2;-m_q^2)L_{\mu\nu}(p_1,p_2;-m^2) \quad (17.17)$$

The dot products work out to give

$$\overline{|\mathcal{M}|}^2 = 8\left[\frac{ee_q}{(p_1+p_2)^2}\right]^2 ((p'_1\cdot p_1)(p'_2\cdot p_2)+(p'_1\cdot p_2)(p'_2\cdot p_1)$$

$$+m^2(p'_2\cdot p'_1)+m_q^2(p_2\cdot p_1)+2m^2m_q^2)$$

$$= (ee_q)^2\left[1+\frac{m^2}{E^2}+\frac{m_q^2}{E^2}+\left(1-\frac{m^2}{E^2}\right)\left(1-\frac{m_q^2}{E^2}\right)\cos^2\theta\right] \quad (17.18)$$

where the last line is in the CM-frame, with

$$E_1 = E_2 = E_3 = E_4 = E$$
$$\vec{p}_1 = -\vec{p}_2 = \vec{p}; \quad |\vec{p}| = \sqrt{E^2 - m^2}$$
$$\vec{p}_1' = -\vec{p}_2' = \vec{p}'; \quad |\vec{p}'| = \sqrt{E^2 - m_q^2}$$

(17.19)

Now write $e_q = Qe$, where Q is either $2/3$ or $-1/3$. The differential 2-Body cross-section is

$$
\frac{d\sigma}{d\Omega} = \left(\frac{\hbar c}{8\pi}\right)^2 \frac{\overline{|\mathcal{M}|}^2}{4E^2} \frac{|\vec{p}'|}{|\vec{p}|}
$$
$$
= \left(\frac{\hbar c}{8\pi}\right)^2 \frac{Q^2 (4\pi\alpha)^2}{8E^2} \frac{|\vec{p}'|}{|\vec{p}|}
$$
$$
\times \left[1 + \frac{m^2}{E^2} + \frac{m_q^2}{E^2} + \left(1 - \frac{m^2}{E^2}\right)\left(1 - \frac{m_q^2}{E^2}\right)\cos^2\theta\right]
$$

(17.20)

or, integrating over the solid angle $d\Omega = \sin\theta d\theta d\phi$,

$$
\sigma = (2\pi) \left(\frac{\hbar c}{8\pi}\right)^2 \frac{Q^2 (4\pi\alpha)^2}{8E^2} \frac{\sqrt{E^2 - m_q^2}}{\sqrt{E^2 - m^2}}
$$
$$
\times \left[2\left(1 + \frac{m^2}{E^2} + \frac{m_q^2}{E^2}\right) + \frac{2}{3}\left(1 - \frac{m^2}{E^2}\right)\left(1 - \frac{m_q^2}{E^2}\right)\right]
$$

(17.21)

which can be simplified to

$$
\sigma = \frac{\pi}{3}\left(\frac{Q\hbar c\alpha}{2E}\right)^2 \frac{\sqrt{1 - \frac{m_q^2}{E^2}}}{\sqrt{1 - \frac{m^2}{E^2}}} \left[\left(1 + \frac{m^2}{2E^2}\right)\left(1 + \frac{m_q^2}{2E^2}\right)\right]
$$
$$
\longrightarrow \frac{\pi}{3}\left(\frac{Q\hbar c\alpha}{2E}\right)^2 \quad \text{for } E \gg m, m_q
$$

(17.22)

Notice that for $E < m_q$ the square root is imaginary. This means that there is not enough energy to create the quark-antiquark pair and the process is kinematically forbidden.

As the beam energy E increases we will encounter a succession of thresholds: first the muon, then the light quarks (u, d, s); later (at 1500 MeV) the charm, the bottom (4700 MeV) and eventually the top quark (171,000 MeV). So we expect (for $E \gg m, m_q$)

$$
R \equiv \frac{\sigma\left(e^+ e^- \longrightarrow \text{hadrons}\right)}{\sigma\left(e^+ e^- \longrightarrow \mu^+\mu^-\right)} = \sum_{\text{colors}} \sum_{\text{flavors}} Q_{\text{flavor}}^2 = 3 \sum_{j=1}^{m_q < E} Q_j^2
$$

(17.23)

which gives (neglecting corrections due to mass)

$$R = 3 \times \left[\left(\frac{2}{3}\right)^2 + \left(-\frac{1}{3}\right)^2 + \left(-\frac{1}{3}\right)^2 \right] = 2 \qquad \text{above the } u, d, s \text{ threshold}$$

$$= 3 \times \left[\left(\frac{2}{3}\right)^2 \right] + 2 = \frac{10}{3} \qquad \text{above the } c \text{ threshold}$$

$$= 3 \times \left[\left(-\frac{1}{3}\right)^2 \right] + \frac{10}{3} = \frac{11}{3} \qquad \text{above the } b \text{ threshold}$$

$$= 3 \times \left[\left(\frac{2}{3}\right)^2 \right] + \frac{11}{3} = 5 \qquad \text{above the } t \text{ threshold} \quad (17.24)$$

There is one correction that needs to be included: the appearance of the τ lepton[†] (1784 MeV), whose discovery by Martin Perl and collaborators at SLAC in 1975 [157] came as a surprise as experimentalists checked the value of R as a function of energy. Once its presence was understood the prediction for R was found to be in pretty good agreement with the data (see fig. 17.2), especially at large E. The color factor 3 clearly needs to be there, providing experimental evidence for color! Of course we have no experimental data for the top since there is no e^+e^- collider that can reach the requisite energy threshold.

Why isn't the agreement better? First, there is a correction from QCD that corresponds to the radiation of a gluon, and shows up in the final state as a 3-jet event. The process yields

$$R = 3 \sum_{j=1}^{m_q < E} Q_j^2 \left(1 + \frac{\alpha_s \left(E^2\right)}{\pi} \right) \qquad (17.25)$$

at a center of mass-energy E, where $\alpha_s \left(E^2\right)$ is the running coupling constant of QCD, analogous to the running value for α in QED. This gives a correction of about 5%. Furthermore, in calculating the above process, we treated the quarks in the final state as free (i.e. non-interacting) particles. However, this isn't correct – they are actually virtual particles that are on their way to becoming hadrons. When the energy is at a value that is right for forming a bound state, the interactions between the quarks dominate, leading to the spikes in the above graph. Here our simple approximation fails badly, but the actual QCD calculation (the solid curve in figure 17.2) does an excellent job of matching the curve, including the resonance peaks. The flat regions of the graph show that our approach using eq. (17.24) is reasonably good for energies that are not nearby any resonance.

[†]The name comes from the Greek $\tau\rho\iota\tau o\nu$, or triton, meaning "third".

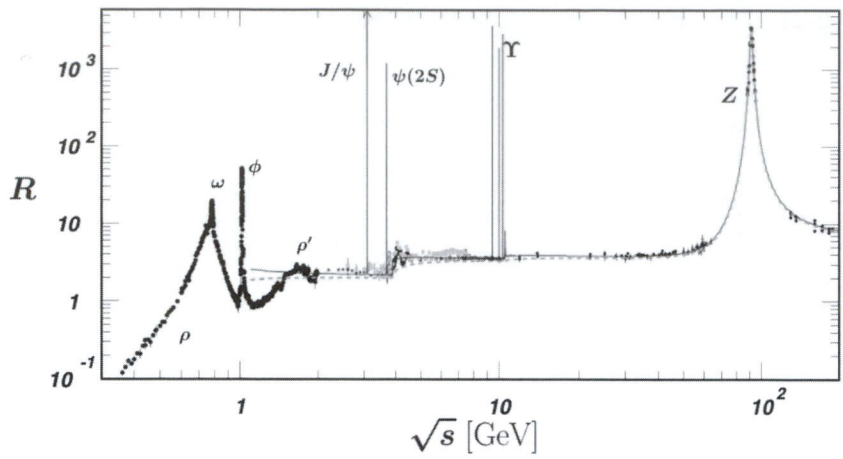

FIGURE 17.2

World data on the total cross section of $e^+e^- \longrightarrow$ hadrons and on the ratio $R \equiv \frac{\sigma(e^+e^- \longrightarrow \text{hadrons})}{\sigma(e^+e^- \longrightarrow \mu^+\mu^-)}$. The broken curve is the naive quark-parton model prediction (17.24), and the solid curve is the 3-loop QCD prediction [156]. The Breit-Wigner parameterizations of the J/ψ, $\psi(2S)$, and the $\Upsilon(nS)$, $(n = 1, 2, 3, 4)$ are also shown. Image courtesy of the Particle Data Group [1].

17.3 Elastic Scattering of Electrons and Protons

At modest energies electron-proton scattering is elastic: $e^- + p \longrightarrow e^- + p$. Based on our work in QED we might expect the spin-averaged/summed matrix element to be similar to what we got for the muon, which is

$$\overline{|\mathcal{M}|}^2 = \frac{1}{2} \left[\frac{-e^2}{q^2} \right]^2 L^{\mu\nu}(p'_1, p_1; m^2) L_{\mu\nu}(p'_2, p_2; M^2) \qquad (17.26)$$

where momentum conservation implies $q^2 = (p'_1 - p_1)^2$, and $L_{\mu\nu}$ is given by our formula above. However, this assumes that the proton (assumed to have mass M) is a structureless point particle, just like the muon.

This is not right – the proton has structure. For example, we saw in Chapter 16 that its magnetic moment is 2.79 nuclear magnetons, vastly different from a pointlike spin-$\frac{1}{2}$ particle. Consequently we shouldn't assume that photons interact with it in the same way that they would with point-like objects such as muons and electrons. The actual lowest-order diagram for this process looks something like what is in 17.3, where the blob denotes our ignorance as to how the photon and the proton interact.

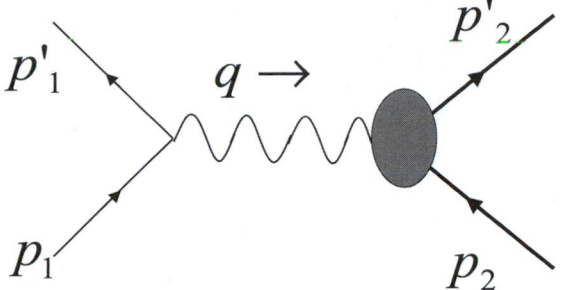

FIGURE 17.3
Scattering of an electron (thin line) off of a proton (thick line).

Except for this blob, the diagram is unchanged: in particular, the photon propagator and photon-electron vertex is unchanged, and so we expect

$$\overline{|\mathcal{M}|}^2 = \frac{1}{2}\left(\frac{e^2}{q^2}\right)^2 L^{\mu\nu}(p_1', p_1; m^2) K_{\mu\nu}(p_2', p_2; M^2) \qquad (17.27)$$

where $K_{\mu\nu}$ is an unknown quantity describing the photon-proton vertex (or, more precisely, what you get from squaring this vertex).

17.3.1 The Photon-Proton Vertex

We don't have any diagrammatic rules to tell us how to compute $K_{\mu\nu}$. However, we can make use of some basic physics that we know must be true for the proton: namely that it respects Lorentz invariance and electromagnetic gauge symmetry. This means that $K_{\mu\nu}$ must obey the following criteria.

1. $K_{\mu\nu}$ depends only only two 4-vectors, $p_2'^\mu$ and p_2^μ, or alternatively $q^\mu = p_2'^\mu - p_2^\mu = p_1^\mu - p_1'^\mu$ and p_2^μ . Since $K_{\mu\nu}$ has two Lorentz indices, it can depend only on quadratic products of these two 4-vectors or on the metric, each multiplied by a different scalar function of the momenta.

2. Since $L^{\mu\nu}$ is symmetric, $K_{\mu\nu}$ must also be symmetric. Hence, in combination with the previous criterion, its most general form is

$$K_{\mu\nu} = -K_1 g_{\mu\nu} + \frac{K_2}{M^2} p_{2\mu} p_{2\nu} + \frac{K_3}{M^2} q_\mu q_\nu + \frac{K_4}{M^2}(q_\mu p_{2\nu} + q_\nu p_{2\mu}) \quad (17.28)$$

3. The functions K_1, K_2, K_3, and K_4 must be scalar functions of q^μ and p_2^μ, and so depend only on q^2, $q \cdot p_2$ and p_2^2. From momentum conservation we have

$$p_2^2 = M^2 \quad \text{and} \quad q \cdot p_2 = -\frac{1}{2} q^2 \qquad (17.29)$$

$(p_2')^\mu = (p_2 + q)^\mu$

$p_2'^2 = p_2^2 + 2q \cdot p_2 + q^2$

$M^2 = M^2 + 2q \cdot p_2 + q^2$

$\Rightarrow q \cdot p_2 = -\frac{1}{2} q^2 .$

which means that the K_i are functions of q^2 only.

4. Gauge invariance implies conservation of the electron current $\overline{\psi}\gamma^\mu\psi$, or

$$\partial_\mu\left(\overline{\psi}\gamma^\mu\psi\right) = 0 \quad\Longrightarrow\quad q_\mu\left(\overline{\psi}\gamma^\mu\psi\right) = 0$$

$$\Longrightarrow q_\mu L^{\mu\nu}(p_1', p_1; m^2) = 0 \tag{17.30}$$

where this follows upon Fourier-transformation [97]. Eq. (17.30) can be easily checked and in turn implies that without loss of generality

$$q_\mu K^{\mu\nu}(p_2', p_2; M^2) = 0 \tag{17.31}$$

whose derivation is not obvious; I've left it as an exercise. This finally yields

$$-K_1 q_\nu + \frac{K_2}{M^2}\left(q\cdot p_2\right)p_{2\nu} + \frac{K_3}{M^2}q^2 q_\nu$$

$$+\frac{K_4}{M^2}\left(q^2 p_{2\nu} + \left(q\cdot p_2\right)q_\nu\right) = 0 \tag{17.32}$$

$$\Rightarrow K_3 = \frac{M^2}{q^2}K_1 + \frac{1}{2}K_4 \quad \text{and} \quad K_4 = \frac{1}{2}K_2 \tag{17.33}$$

so that

$$K_{\mu\nu} = K_1\left(q^2\right)\left(\frac{q_\mu q_\nu}{q^2} - g_{\mu\nu}\right) + \frac{K_2\left(q^2\right)}{M^2}\left(p_{2\mu} + \frac{1}{2}q_\mu\right)\left(p_{2\nu} + \frac{1}{2}q_\nu\right) \tag{17.34}$$

17.3.2　The Rosenbluth Formula

It is up to the experimentalist to determine the specific forms of the K's and to the theorist to find an explanation in terms of proton structure for these forms. Inserting this into the matrix element above gives

$$\overline{|\mathcal{M}|}^2 = \frac{1}{2}\left(\frac{e^2}{q^2}\right)^2 L^{\mu\nu}(p_1', p_1; m^2)$$

$$\times\left[K_1\left(q^2\right)\left(\frac{q_\mu q_\nu}{q^2} - g_{\mu\nu}\right) + \frac{K_2\left(q^2\right)}{M^2}\left(p_{2\mu} + \frac{1}{2}q_\mu\right)\left(p_{2\nu} + \frac{1}{2}q_\nu\right)\right]$$

$$= \frac{1}{2}\left(\frac{e^2}{q^2}\right)^2\left[4\left(p_1'^\mu p_1^\nu + p_1'^\nu p_1^\mu - g^{\mu\nu}\left[-m^2 + p_1'\cdot p_1\right]\right)\right]$$

$$\times\left[-K_1\left(q^2\right)g_{\mu\nu} + \frac{K_2\left(q^2\right)}{M^2}p_{2\mu}p_{2\nu}\right] \tag{17.35}$$

which simplifies to

$$\overline{|\mathcal{M}|}^2 = 4\left(\frac{e^2}{q^2}\right)^2\left[-\left(m^2 + \frac{q^2}{2}\right)K_1\left(q^2\right) + K_2\left(q^2\right)\left(\frac{\left(p_2\cdot p_1'\right)\left(p_2\cdot p_1\right)}{M^2} + \frac{q^2}{4}\right)\right] \tag{17.36}$$

In the lab frame we have

$$p_2^\mu = (M, 0) \qquad p_1^\mu = (E, \vec{p}) \qquad p_1'^\mu = (E', \vec{p}')$$ (17.37)

and so eq. (17.36) becomes

$$\overline{|\mathcal{M}|}^2 = 4 \left(\frac{e^2}{q^2}\right)^2 \left[-\left(m^2 + \frac{q^2}{2}\right) K_1\left(q^2\right) + K_2\left(q^2\right) \left(EE' + \frac{q^2}{4}\right) \right]$$ (17.38)

where $q^2 = 2\left(m^2 - EE' + |\vec{p}\,'|\,|\vec{p}|\cos\theta\right) = -4EE'\sin^2\frac{\theta}{2} + \mathcal{O}\left(\frac{m^2}{EE'}\right) < 0$.

At energies $E, E' \gg m$, the differential cross-section for $e^- + p \to e^- + p$ in the lab frame is

$$\left(\frac{d\sigma}{d\Omega}\right) = \left(\frac{\hbar}{8\pi}\right)^2 \frac{|\mathcal{M}|^2 |\vec{p}\,'|^2}{M\,|\vec{p}|\,[|\vec{p}\,'|\,(E + Mc^2) - |\vec{p}|\,E'\cos\theta]}$$ (17.39)

$$= \left(\frac{\alpha}{4EM\sin^2\frac{\theta}{2}}\right)^2 \left(\frac{E'}{E}\right) \left[2K_1\left(q^2\right)\sin^2\frac{\theta}{2} + K_2\left(q^2\right)\cos^2\frac{\theta}{2}\right]$$

which is called the *Rosenbluth formula* [158]. Note that the kinematics of the situation imply that $q \cdot p_2 = -\frac{1}{2}q^2$, or

$$E' = \frac{E}{1 + 2\frac{E}{M}\sin^2\frac{\theta}{2}}$$ (17.40)

at energies $E, E' \gg m$. This means that the Rosenbluth formula has only one independent parameter: the scattering angle θ.

By counting the number of electrons scattered in a given direction for a range of incident energies we can determine $K_1\left(q^2\right)$ and $K_2\left(q^2\right)$. However, it's customary to define

$$K_1\left(q^2\right) = \frac{q^2}{4M^2}G_M\left(q^2\right) \quad \text{and} \quad K_2\left(q^2\right) = \frac{G_E\left(q^2\right) + \frac{q^2}{4M^2}G_M\left(q^2\right)}{1 + \frac{q^2}{4M^2}}$$ (17.41)

where $G_E\left(q^2\right)$ and $G_M\left(q^2\right)$ are respectively called the electric and magnetic form factors for the proton [159]. A comparison of the measured differential cross-section $\left(\frac{d\sigma}{d\Omega}\right)_{e+p\to e+p}$ with the Rosenbluth formula above (17.39) then allows a measurement of the proton form factors. This provides us with information as to the structure of the proton.

How can we obtain direct information about the quarks inside the proton? This is the subject of the next section.

17.4   Deep Inelastic Scattering

Rutherford's experiment scattering alpha particles off of gold atoms demonstrated that the charge of an atom is concentrated in its nucleus. It provided firm evidence that the Thomson model – in which positive and negative charges were uniformly distributed throughout the atom – was not correct, paving the way for the Bohr model and, ultimately, our current understanding of the atom. The inelastic scattering of electrons off of protons has had a similar impact in particle physics.

In an *inelastic* scattering process, the particles in the final state are not the same as in the initial state. In the context of electron-proton scattering, we now have $e^- + p \longrightarrow e^- +$ junk. The diagram is now given by figure 17.4, where the "junk" represents all possible particles that can come out: pions, Kaons, Δ's, Ξ's – you name it! The main distinction between this case and

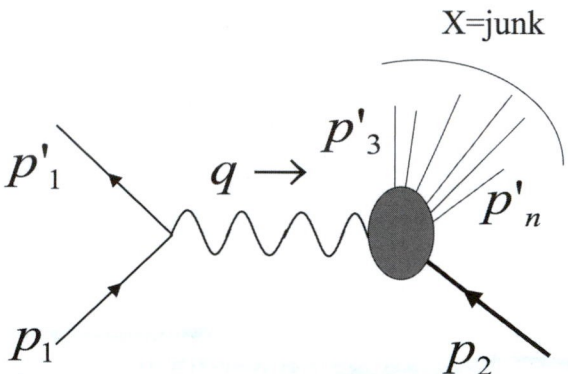

FIGURE 17.4
Deep inelastic scattering of electrons off of protons.

the previous elastic scattering case is that the final state momentum

$$p_2'^{\mu} = (p_3' + \cdots + p_n')^{\mu} \tag{17.42}$$

is no longer the square of the proton mass, i.e.

$$p_2'^2 \neq M^2 \;\Rightarrow\; q \cdot p_2 \neq -\frac{1}{2}q^2 \tag{17.43}$$

Except for this feature, the rest of the process is the same: in particular, the photon propagator and photon-electron vertex is unchanged. Hence we expect for a given final state X_n (containing n particles) in the process

$$\left|\mathcal{M}\left(X_n\right)\right|^2 = \frac{1}{2}\left(\frac{e^2}{q^2}\right)^2 L^{\mu\nu}(p_1', p_1; m^2) K_{\mu\nu}(X_n) \qquad (17.44)$$

which will give a scattering cross-section

$$d\sigma_n = \frac{\hbar^2 S}{4\sqrt{(p_1 \cdot p_2)^2 - p_1^2 p_2^2}} \left[\prod_{i=3}^{n} \frac{c\left(d^3 p_i'\right)}{2E_i'\left(2\pi\right)^3}\right] \left|\mathcal{M}\left(X_n\right)\right|^2$$

$$\times (2\pi)^4 \, \delta^{(4)}\left(p_1' + \sum_{i=3}^{n} p_i' - p_2 - p_1\right) \qquad (17.45)$$

for the state X_n.

It would be extremely difficult to compute this for a particular but arbitrary final state. Fortunately this is not necessary! Experiments typically only measure the momentum $p_1'^\mu$ of the outgoing electron. Hence all we need to do is integrate over all possible final states $X = \{X_1, X_2, \ldots X_n\}$ that could occur. This gives

$$d\sigma = \frac{\hbar^2 \pi M}{\sqrt{(p_1 \cdot p_2)^2 - p_1^2 p_2^2}} \left(\frac{e^2}{q^2}\right)^2 \frac{c\left(d^3 p_1'\right)}{2E_1'\left(2\pi\right)^3} L^{\mu\nu}(p_1', p_1; m^2) W_{\mu\nu}(p_2', p_2; M^2)$$

$$(17.46)$$

where now $W_{\mu\nu}$ is an unknown quantity

$$W_{\mu\nu}(p_2', p_2; M^2) = \frac{1}{4\pi M} \sum_{X_n} \left\{\left[\prod_{i=3}^{n} \int \frac{c\left(d^3 p_i'\right)}{2E_i'\left(2\pi\right)^3}\right] K_{\mu\nu}(X_n)\right.$$

$$\left. \times (2\pi)^4 \, \delta^{(4)}\left(p_1' + \sum_{i=3}^{n} p_i' - p_2 - p_1\right)\right\} \qquad (17.47)$$

describing the photon-proton vertex (or, more precisely, what you get from squaring this vertex). We call this the *inclusive matrix element* (and the resultant cross-section the *inclusive cross-section*) because $W_{\mu\nu}$ includes all accessible final states X.

The energies that occur in the experiment are typically much larger than the mass of the electron. If the mass M proton is at rest, then

$$\sqrt{(p_1 \cdot p_2)^2 - p_1^2 p_2^2} = M\sqrt{E^2 - m^2} \simeq ME \qquad (17.48)$$

where E is the initial electron energy. We also have $d^3 p_1' = |\vec{p}_1'|^2 \, d\,|\vec{p}_1'|\, d\Omega \simeq \left(E'\right)^2 dE' d\Omega$, where E' is the energy of the outgoing electron (if the electron has enough energy to blast apart the proton, then its mass is negligible). Hence

$$d\sigma = \frac{\pi}{E} \left(\frac{\hbar e^2}{cq^2} \right)^2 \frac{E' \, (dE' \, d\Omega)}{2 \, (2\pi)^3} L^{\mu\nu} (p'_1, p_1; m^2) W_{\mu\nu} (p'_2, p_2; M^2) \qquad (17.49)$$

or, more compactly,

$$\frac{d\sigma}{dE' \, d\Omega} = \left(\frac{\hbar\alpha}{cq^2} \right)^2 \frac{E'}{E} L^{\mu\nu} W_{\mu\nu} \qquad (17.50)$$

since $e^2 = 4\pi\alpha$ and I have inserted the relevant factors of $\hbar$ and c in the cross-section.

As with $K_{\mu\nu}$, we can make use of some basic physics, namely Lorentz invariance and electromagnetic gauge symmetry. This means that $W_{\mu\nu}$ obeys the same criteria that $K_{\mu\nu}$ did, and a similar argument gives

$$W_{\mu\nu} = W_1 \left(\frac{q_\mu q_\nu}{q^2} - g_{\mu\nu} \right) + \frac{W_2}{M^2} \left(p_{2\mu} - \frac{q \cdot p_2}{q^2} q_\mu \right) \left(p_{2\nu} - \frac{q \cdot p_2}{q^2} q_\nu \right) \qquad (17.51)$$

which is just like the expression we had before for $K_{\mu\nu}$ except that now $q \cdot p_2 \neq -\frac{1}{2} q^2$.

Inserting the form for $W_{\mu\nu}$ into the matrix element above gives[‡]

$$L^{\mu\nu} W_{\mu\nu} = 8 \left[- \left(m^2 + \frac{q^2}{2} \right) W_1 \left(q^2, q \cdot p_2 \right) \right.$$
$$\left. + W_2 \left(q^2, q \cdot p_2 \right) \left(\frac{(p_2 \cdot p'_1) \, (p_2 \cdot p_1)}{M^2} + \frac{q^2}{4} \right) \right] \qquad (17.52)$$

which in the lab frame (with $E, E' \gg m$)

$$p_2^\mu = (M, 0) \qquad p_1^\mu = (E, \vec{p}) \qquad p_1'^\mu = (E', \vec{p}') \qquad (17.53)$$

$$q^2 = 2 \left(m^2 - EE' + |\vec{p}\,'| \, |\vec{p}| \cos\theta \right) \simeq -4EE' \sin^2 \frac{\theta}{2} \qquad (17.54)$$

becomes

$$L^{\mu\nu} W_{\mu\nu} = 4EE' \left[2W_1 \left(q^2, q \cdot p_2 \right) \sin^2 \frac{\theta}{2} + W_2 \left(q^2, q \cdot p_2 \right) \cos^2 \frac{\theta}{2} \right] \qquad (17.55)$$

Hence at energies $E, E' \gg m$, the inclusive differential cross-section in the lab frame is

$$\frac{d\sigma}{dE' \, d\Omega} = \left(\frac{\hbar\alpha}{2E \sin^2 \frac{\theta}{2}} \right)^2 \left[2W_1 \left(q^2, q \cdot p_2 \right) \sin^2 \frac{\theta}{2} + W_2 \left(q^2, q \cdot p_2 \right) \cos^2 \frac{\theta}{2} \right]$$
$$(17.56)$$

[‡]Recall that $L^{\mu\nu} q_\nu = 0$ and that $p'_1 \cdot p_1 = m^2 - \frac{q^2}{2}$

This equation is the inelastic generalization of the Rosenbluth formula (17.39): there are still two structure functions but now they depend on two variables (E' and θ), since we no longer have $q \cdot p_2 = -\frac{1}{2}q^2$, or $E' \neq \frac{E}{1+2\frac{E}{M}\sin^2 \frac{\theta}{2}}$.

It's customary to write

$$x \equiv -\frac{q^2}{2q \cdot p_2} \tag{17.57}$$

so that the W_i's are functions of q^2 and x, and we have

$$\frac{d\sigma}{dE'd\Omega} = \left(\frac{\hbar\alpha}{2E\sin^2 \frac{\theta}{2}}\right)^2 \left[2W_1\left(q^2, x\right)\sin^2 \frac{\theta}{2} + W_2\left(q^2, x\right)\cos^2 \frac{\theta}{2}\right] \tag{17.58}$$

The variable x is bounded by $0 < x \leq 1$, with the elastic case the limit $x = 1$. In this limit, we should be able to recover the Rosenbluth formula by an appropriate choice of the W_i's. After all, elastic scattering is a special case of inelastic scattering in which $(p'_2)^2 = M^2$. It's not too hard to show that

$$W_{1,2}\left(q^2, x\right) = -\frac{K_{1,2}\left(q^2\right)}{2Mq^2}\delta\left(x - 1\right) \tag{17.59}$$

is the choice that recovers the elastic Rosenbluth formula once you integrate over x.

Note that so far the only physics used in obtaining the differential cross-section is our assumption of Lorentz invariance and gauge invariance. In order to proceed further we need to make some hypothesis about the W_i's to construct specific models of the proton that in turn will allow us to predict the cross-section. For example, suppose that the proton were a simple point charge of spin-$\frac{1}{2}$. We would then have

$$W_1\left(q^2, x\right) = \frac{1}{2M}\delta\left(x - 1\right) \quad W_2\left(q^2, x\right) = -\frac{2M}{q^2}\delta\left(x - 1\right) \tag{17.60}$$

At low energies – where elastic scattering occurs – this is not too bad a model. However, we know it is not fully correct – at high energies the proton scatters inelastically, and so must have some internal structure.

17.5 Quark Model Predictions

What would a quark model predict? Well, one thing that should happen is that at sufficiently high energies the virtual photon interacts with a single pointlike quark that is instantaneously free. This means that the quark structure functions should be like those for point particles as in eq. (17.59):

$$W_1^j\left(q^2, x_j\right) = \frac{Q_j^2}{2m_j}\delta\left(x_j - 1\right) \quad W_2^j\left(q^2, x\right) = -\frac{2m_jQ_j^2}{q^2}\delta\left(x_j - 1\right) \tag{17.61}$$

where m_j is the mass of the jth quark inside the proton, and

$$x_j \equiv -\frac{q^2}{2q \cdot p_j} \qquad\qquad Q_j = \frac{e_j}{e} \qquad\qquad (17.62)$$

where p_j^μ is its 4-momentum. Note that the structure functions are proportional to the quark charge because the photon couples to the individual quarks by a factor proportional to quark charge.

So we could compute the cross-section $\frac{d\sigma}{dE'\, d\Omega}$ if we knew the momentum p_j^μ of each quark in the proton. We don't know this, because the momentum of the quark inside the proton is the sum of the momentum with respect to the proton's center of mass plus the overall momentum in the lab. Let's assume that the motion with respect to the proton's center of mass is unimportant:

$$p_j^\mu = z_j p_2^\mu + \text{negligible corrections} \qquad\qquad (17.63)$$

where the total momentum p_2^μ of the proton is

$$p_2^\mu = \sum_{j=1}^{3} p_j^\mu \qquad\qquad (17.64)$$

Essentially this assumes that each component of quark momentum gets the same fraction of proton momentum – the "negligible" corrections are those that include the motion of the quarks relative to the proton's center of mass. Using this assumption we now have

$$x_j \equiv -\frac{q^2}{2q \cdot p_j} = \frac{x}{z_j} \quad \text{and} \quad m_j = z_j M$$

$$W_1^j\left(q^2, x_j\right) = \frac{Q_j^2}{2 z_j M} \delta\left(\frac{x}{z_j} - 1\right) = \frac{Q_j^2}{2M} \delta\left(x - z_j\right) \qquad\qquad (17.65)$$

$$W_2^j\left(q^2, x\right) = -\frac{2 z_j M Q_j^2}{q^2} \delta\left(\frac{x}{z_j} - 1\right) = -\frac{2 x^2 M Q_j^2}{q^2} \delta\left(x - z_j\right)$$

To get the structure functions of the proton in this model, we must multiply the (pointlike) quark structure functions $W_{1,2}^j$ by the probabilities $f_j\left(z_j\right)$ that the jth quark has momentum $z_j p_2^\mu$, and then integrate over all quarks in the proton:

$$W_1\left(q^2, x\right) = \sum_{j=1}^{3} \int_0^1 dz_j f_j\left(z_j\right) W_1^j\left(q^2, x_j\right)$$

$$= \sum_{j=1}^{3} \int_0^1 dz_j f_j\left(z_j\right) \frac{Q_j^2}{2M} \delta\left(x - z_j\right)$$

$$= \frac{1}{2M} \sum_{j=1}^{3} Q_j^2 f_j\left(x\right) \qquad\qquad (17.66)$$

$$W_2\left(q^2, x\right) = \sum_{j=1}^{3} \int_0^1 dz_j f_j\left(z_j\right) W_2^j\left(q^2, x_j\right)$$

$$= -\sum_{j=1}^{3} \int_0^1 dz_j f_j\left(z_j\right) \frac{2x^2 M Q_j^2}{q^2}\delta\left(x - z_j\right)$$

$$= -\frac{2M}{q^2}x^2\sum_{j=1}^{3} Q_j^2 f_j\left(x\right) \tag{17.67}$$

or alternatively

$$MW_1\left(q^2, x\right) = \frac{1}{2}\sum_{j=1}^{3} Q_j^2 f_j\left(x\right) \equiv F_1(x) \tag{17.68}$$

$$-\frac{q^2 W_2\left(q^2, x\right)}{2Mx} = x\sum_{j=1}^{3} Q_j^2 f_j\left(x\right) \equiv F_2(x) \tag{17.69}$$

a remarkable result known as *Bjorken Scaling* [160]: the structure functions (appropriately multiplied) are *independent* of q^2!

We also see that

$$F_2(x) = 2xF_1(x) \tag{17.70}$$

which is called the *Callen-Gross* relation [161]. Hence the prediction of the quark-parton model is that the inelastic cross-section is

$$\frac{d\sigma}{dE'd\Omega} = \left(\frac{\hbar\alpha}{2E\sin^2\frac{\theta}{2}}\right)^2\left[2W_1\left(q^2, q\cdot p_2\right)\sin^2\frac{\theta}{2} + W_2\left(q^2, q\cdot p_2\right)\cos^2\frac{\theta}{2}\right]$$

$$= \frac{F_1(x)}{2M}\left(\frac{\hbar\alpha}{E\sin\frac{\theta}{2}}\right)^2\left[1 + \frac{2EE'}{\left(E - E'\right)^2}\cos^2\frac{\theta}{2}\right] \tag{17.71}$$

The importance of these relations is that they provide a clear empirical signature for the quark model. If protons are indeed made of constituent pointlike objects, then Bjorken scaling will hold. If the constituent objects are spin-$\frac{1}{2}$, then the Callen-Gross relation holds (the function $F_1(x) = 0$ if the quarks have spin-0). You can see how these predictions fare against the data in figure 17.5. They do quite well [162], and after the deep inelastic scattering experiments [163] were performed in the early 1970s physicists accepted the existence of quarks.

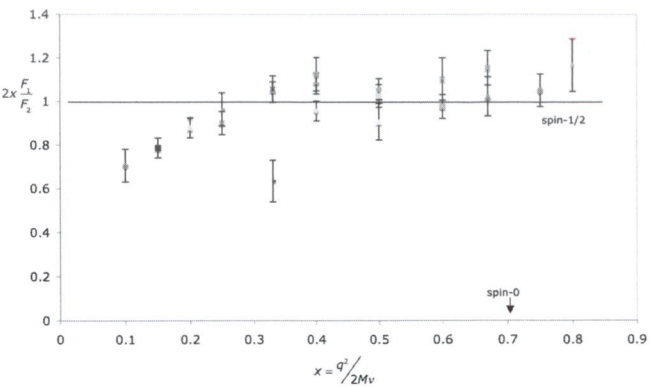

FIGURE 17.5

A plot of F_2 as a function of the energy transfer q^2, using the data from deep inelastic scattering experiments [162]. Note that it has the same value at various scattering angles for all possible q^2.

17.6   Quark Structure Functions

To finish the job, we need to compute $F_1(x)$. This is not that easy to do because we have to make assumptions about how the quark momentum is distributed inside the proton, assumptions that go beyond the naive delta-function relations given above. If we really believe that $m_j = z_j M$, then

$$f_j(x) = \delta\left(\frac{m_j}{M} - z_j\right) \tag{17.72}$$

since the momentum fraction of each quark is fixed. This would give for a proton

$$F_1(x) = \frac{1}{2}\sum_{j=1}^{3} Q_j^2 f_j(x) = \frac{1}{2}\left[2\left(\frac{2}{3}\right)^2 \delta\left(\frac{m_u}{M} - x\right) + \left(-\frac{1}{3}\right)^2 \delta\left(\frac{m_d}{M} - x\right)\right]$$

$$= \frac{1}{2}\delta\left(\frac{m_u}{M} - x\right) \tag{17.73}$$

where I have set the mass of the up and down quarks equal. This would make $F_1(x)$ a sharply peaked function, which experiment shows not to be the case.

More generally, suppose that

$$f_u(x) = \frac{1}{2}\mathfrak{u}(x) \quad \text{and} \quad f_d(x) = \mathfrak{d}(x) \tag{17.74}$$

where $\mathfrak{u}(x)$ is the probability density that momentum fraction x is carried by an up quark, and $\mathfrak{d}(x)$ is the analogous probability density for a down quark. The factor of $1/2$ ensures that $\mathfrak{u}(x)$ is the probability density for finding either up quark. This gives

$$F_1(x) = \frac{1}{2} \sum_{j=1}^{3} Q_j^2 f_j(x) = \frac{1}{2}\left[\left(\frac{2}{3}\right)^2 \mathfrak{u}(x) + \left(-\frac{1}{3}\right)^2 \mathfrak{d}(x)\right] \qquad (17.75)$$

The naive model would give $\mathfrak{u}(x) = 2\delta\left(\frac{m_u}{M} - x\right)$ and $\mathfrak{d}(x) = \delta\left(\frac{m_d}{M} - x\right)$. We want to improve on this.

One reasonable consideration in determining $\mathfrak{u}(x)$ and $\mathfrak{d}(x)$ is to require that (a) the average momentum carried by up quarks is twice that carried by down quarks (since the proton has twice as many ups as downs) and (b) that the masses of up and down are about the same. That is

$$\langle \vec{p}_{\text{up}} \rangle = 2\langle \vec{p}_{\text{down}} \rangle \Rightarrow \int_0^1 (x\vec{p}_2\mathfrak{u}(x))dx = 2\int_0^1 (x\vec{p}_2\mathfrak{d}(x))dx$$

$$\Rightarrow \int_0^1 x\mathfrak{u}(x)dx = 2\int_0^1 x\mathfrak{d}(x)dx \qquad (17.76)$$

a relationship that is supported by data from electron-neutron scattering. This implies

$$\int_0^1 F_2(x)dx = \int_0^1 x\left[\left(\frac{2}{3}\right)^2 \mathfrak{u}(x) + \left(-\frac{1}{3}\right)^2 \mathfrak{d}(x)\right]dx = \int_0^1 x\mathfrak{d}(x)dx \quad (17.77)$$

Experiment indicates that the quark structure functions are

$$\int_0^1 x\mathfrak{d}(x)dx = 0.18 \qquad \int_0^1 x\mathfrak{u}(x)dx = 0.36 \qquad (17.78)$$

which has the rather curious implication that

$$\langle p_2^\mu \rangle_{\text{all quarks}} = \int_0^1 (x p_2^\mu \mathfrak{u}(x) + x p_2^\mu \mathfrak{d}(x))\,dx$$

$$= p_2^\mu \int_0^1 (x\mathfrak{u}(x) + x\mathfrak{d}(x))\,dx = 0.54 p_2^\mu \qquad (17.79)$$

In other words, the average proton momentum carried by all of its quarks is only 54% of the total proton momentum! Where's the rest of it? Presumably in the gluons which bind the quarks together. But in fact the actual answer is quite complicated – virtual quark-antiquark pairs appear inside the proton due to gluon interactions, and these also carry some of its momentum. These virtual quarks are called "sea quarks," to distinguish them from the "valence" up, up, and down constituent quarks of the proton. The sea quarks can be any up, down or strange quark or antiquark that can be produced at these energies (the probability that heavier sea quarks are produced is very small due to their larger masses). Since the exchanged photon can couple to the sea quarks as well as the valence quarks we should modify the structure functions to include them

$$F_1(x) = \frac{1}{2}\left[\left(\frac{2}{3}\right)^2 [u_v(x) + u_s(x) + \bar{u}_s(x)]\right.$$
$$\left. + \left(-\frac{1}{3}\right)^2 [\partial_v(x) + \partial_s(x) + \bar{\partial}_s(x) + s_s(x) + \bar{s}_s(x)]\right] (17.80)$$

where the overbar denotes an antiquark. This now erodes the predictive power of the quark model – now we have six unknown functions instead of two! However, because all the sea quarks are produced by the same mechanism it is reasonable to impose

$$u_s(x) = \bar{u}_s(x) = \partial_s(x) = \bar{\partial}_s(x) = s_s(x) = \bar{s}_s(x) \qquad (17.81)$$

which implies

$$F_1(x) = \frac{1}{18}[4u_v(x) + \partial_v(x) + 12s_s(x)] \qquad (17.82)$$

The shape of these three structure functions can be inferred from experiment. However, this is a difficult job, particularly as $x \to 0$. It is one of the challenges of QCD to make a definite prediction of the quark structure functions from first principles.

17.7   Questions

1. Suppose a 4th generation of quarks and leptons is discovered, with masses on the order of 1 TeV. What do you predict for the ratio $R \equiv \frac{\sigma(e^+e^- \longrightarrow \text{hadrons})}{\sigma(e^+e^- \longrightarrow \mu^+\mu^-)}$?

2. Show that
   (a) $q_\mu L^{\mu\nu}(p_1', p_1; m^2) = 0$
   (b) $q_\mu K^{\mu\nu}(p_2', p_2; M^2) = 0$

3. Compute K_1 and K_2 for a pointlike proton that obeys the Dirac equation.

4. Show that the Rosenbluth formula becomes the Mott formula
   $$\frac{d\sigma}{d\Omega}\bigg|_{\text{Mott}} = \left(\frac{\hbar\alpha}{2|\vec{p}|^2 \sin^2\frac{\theta}{2}}\right)^2 \left[|\vec{p}|^2 \cos^2\frac{\theta}{2} + (mc)^2\right]$$

in the limit $m \ll E \ll M$, using the expressions for K_1 and K_2 from question #3.

5. Show that

$$E' = \frac{E}{1 + 2\frac{E}{M}\sin^2\frac{\theta}{2}}$$

at energies $E, E' \gg m$ for the scattering of a lepton of mass m from a proton.

6. Show that

$$W_{\mu\nu} = W_1\left(\frac{q_\mu q_\nu}{q^2} - g_{\mu\nu}\right) + \frac{W_2}{M^2}\left(p_{2\mu} - \frac{q \cdot p_2}{q^2}q_\mu\right)\left(p_{2\nu} - \frac{q \cdot p_2}{q^2}q_\nu\right)$$

7. Show that

$$L^{\mu\nu}W_{\mu\nu} = 8\left[-\left(m^2 + \frac{q^2}{2}\right)W_1\left(q^2, q \cdot p_2\right)\right.$$
$$\left. + W_2\left(q^2, q \cdot p_2\right)\left(\frac{(p_2 \cdot p_1')(p_2 \cdot p_1)}{M^2} + \frac{q^2}{4}\right)\right]$$

8. Show that setting

$$W_{1,2}\left(q^2, x\right) = -\frac{K_{1,2}\left(q^2\right)}{2Mq^2}\delta\left(x - 1\right)$$

yields the Rosenbluth formula.

9. For a proton of mass M, find the fractional momentum of the scattered quark of mass m, where neither the proton mass nor the quark mass can be neglected. Expand your result in the limit where both masses are small compared to the energy.

18

Heavy Quarks and QCD

So far we've seen that we can understand baryons as bound states of 3 quarks, both indirectly in terms of the observable properties of baryons and directly in terms of deep inelastic scattering.

Likewise, we've seen that mesons can be regarded as bound states of $q\bar{q}$ pairs. In fact we can consider a meson to be a sort of a quark-antiquark "atom," analogous to the usual nucleus-electron atom, or the e^+e^- atom (positronium). Perhaps we should call mesons "quarkonium", since one might expect that the methods used in positronium and ordinary atoms would work for mesons. They do...but with two big problems.

1. We don't know the potential.

   Unlike positronium and hydrogenic atoms, in which we know that the force is electromagnetic (and very well understood), the force binding quarks into "quarkonium" is the color force and is *not* as well understood. The (present-day) theory of color forces is called *Quantum Chromodynamics* (or QCD), which is similar to QED except for some non-linear interaction terms. It constitutes the second major part of the Standard Model, and is the foundational theory governing all that we know about quarks and their bound states.

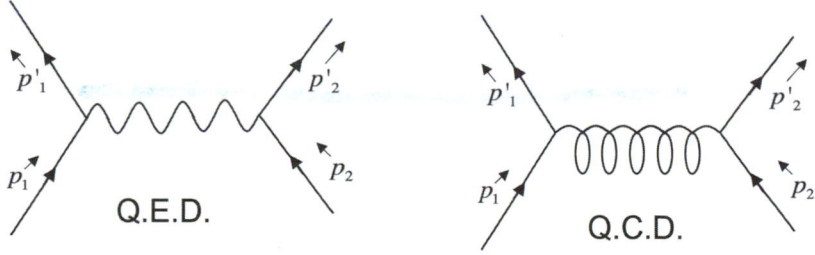

FIGURE 18.1

The one-photon and one-gluon exchange diagrams.

The curly intermediate line in figure 18.1 represents a gluon: the force carrier of the strong interactions. At short-distance separations for two quarks we expect the force to be mediated by one gluon (just like one photon in QED) and so the "one gluon exchange potential" should be

$$V_{\text{1-gluon}}(r) \sim \frac{1}{r} \qquad (18.1)$$

However, quarks are evidently confined at separations $r > 10^{-13}$ cm, so we must include another term $V_{\text{conf}}(r)$ in the potential to account for this. No one knows how to compute this from QCD, though some hints have emerged from lattice gauge theory*. At the moment we must guess it. Some physicists assume $V_{\text{conf}}(r) \sim r$ others assume $V_{\text{conf}}(r) \sim \ln(r)$, and others $V_{\text{conf}}(r) \sim V_0$ – any of these match present-day data quite well because they don't differ substantially over the narrow range of distances probed by current experiments, as shown in fig. 18.2. So we can choose

$$V_{\text{QCD}}(r) = V_{\text{1-gluon}}(r) + V_{\text{conf}}(r) = -\frac{4}{3}\frac{\alpha_s}{r} + F_0 r \qquad (18.2)$$

Experiment indicates [166] that $F_0 \simeq 16$ tons!

2. Light quarks are relativistic.

We were able to fit baryon and meson spectra reasonably well using $\mathbf{SU}_F(3)$ symmetry and a potential of the above form plus spin-orbit coupling, working in the context of a non-relativistic model. But this raises a puzzle – not that the quark model doesn't work perfectly (some arbitrary parameters had to be "fit" to the data), but rather that it works so well! It's a puzzle because the binding energies of the light (u,d,s) quarks into hadrons are typically a few hundred MeV, roughly the same size as the constituent masses of the quarks! This is a highly relativistic regime, and in strong contrast to atomic physics, where the binding energies are 10's of eV and the mass of the electron is 511 KeV, making a non-relativistic approximation reasonable. At present the reasons why the non-relativistic quark model works so well when light quarks should be relativistic is not clear.

If heavier fourth (and fifth and sixth) quarks existed, we might be tempted to use $\mathbf{SU}(4)_{\text{flavor}}$ symmetry (or $\mathbf{SU}(5)_{\text{flavor}}$ or $\mathbf{SU}(6)_{\text{flavor}}$) to explain mass splittings. Unfortunately this would not be a good idea because the different masses would break the flavor symmetry too badly. However, we might expect that for these heavier quarks, relativistic effects will be less important

---

*This is an approach proposed by Ken Wilson [164] toward understanding the strong interactions by modeling spacetime as a lattice of points and then solving the QCD equations on this lattice by brute force on a computer [165].

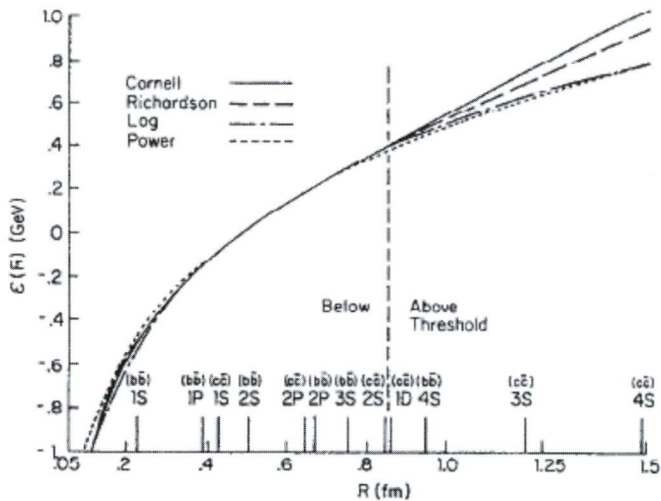

FIGURE 18.2

Form of the strong interaction potential for four distinct models, each chosen to give a best-fit to the spectrum of $c\bar{c}$ and $b\bar{b}$ bound states. Image from ref. [166]; used with permission.

(binding energies should be considerably smaller than masses) and that these heavy quarks will bind together according to the potential (18.1) to form new quarkonium states. These will be analogous to positronium even more than the light mesons were.

In fact there are heavy quarks: charm (discovered in 1974), bottom (discovered in 1977), and top (discovered in 1995). These particles (and their bound states) are all very unstable, and so must be produced in accelerators. The most straightforward way is to collide electrons and positrons, as we have seen before. If the collision energy is just right the cross-section will rapidly shoot up because a resonant bound state is produced. These bound states can make transitions to other lower-mass bound states, analogous to the way that electrons in hydrogen make transitions. Once the spectrum is empirically determined then a potential can be inferred from it.

18.1 Charm

Charm was actually first discovered in 1971 in cosmic ray experiments carried out by Niu and collaborators in Japan [167]. They used an emulsion chamber, which consists of a sandwich of emulsion sheets that can give an accurate tracking of charged particles, followed by another sandwich of emulsion sheets alternating with lead plates about 1 mm thick, which allows for detection of pions, and an identification of electrons and measurement of their energy. Events were seen corresponding to the production of two associated particles that each decayed weakly. These particles were called X particles and were found to have mass 1.8 GeV under the assumption they were mesons, and 2.9 GeV if baryons. These masses were far too large to be explained using strange quarks. They in fact were charmed hadrons, having all the characteristics we now understand for such particles, but were not recognized as such [168].

In 1974 simultaneous experiments led by Burton Richter at SLAC [169] and by Sam Ting at Brookhaven [170] observed a series of meson resonances. The Brookhaven researchers built a spectrometer to search for heavy particles with the same quantum numbers as the photon ($J = 1, \mathbb{P} = -1, \mathbb{C} = -1$). Such particles can decay into e^+e^- pairs for the same reasons a virtual photon does. The idea was to search for the reaction

$$p + N \longrightarrow J + X \longrightarrow e^+ + e^- + X \tag{18.3}$$

where N is a light nucleus – Be was used in the experiment. This seems straightforward enough, but the problem is that since the (at this stage hypothetical) J is produced by nuclear reactions, the charged particles coming out will nearly always be pions instead of e^+e^- pairs. So the spectrometer must search for particles of opposite charge but reject pions (and other hadrons). This was done using (a) threshold Cerenkov counters that could detect electrons or positrons, but not hadrons, and (b) a calorimeter that measures the longitudinal shower, which would be very different for hadrons than for electrons/positrons. A double-arm spectrometer was built, one for each charged particle, designed to accurately measure the magnitude and direction of the momentum of each charged particle. The invariant mass of the electron/positron pair is just $m^2 = (p_1 + p_2)^2$, or

$$m\left(e^+e^-\right) = \sqrt{2m_e^2 + 2E_1E_2 - 2p_1p_2\cos\left(\theta_1 + \theta_2\right)} \tag{18.4}$$

where the angles are relative to the original beam direction. If the J particle exists, it will show up as a resonance peak in $m\left(e^+e^-\right)$.

At the same time the SPEAR e^+e^- collider at SLAC was being operated at its maximum CMS energy of 8 GeV, collecting data using the Mark I detector. Some anomalies had appeared in the data at lower energies, and a decision was made to study these by varying the energy in small steps near 3 GeV. A

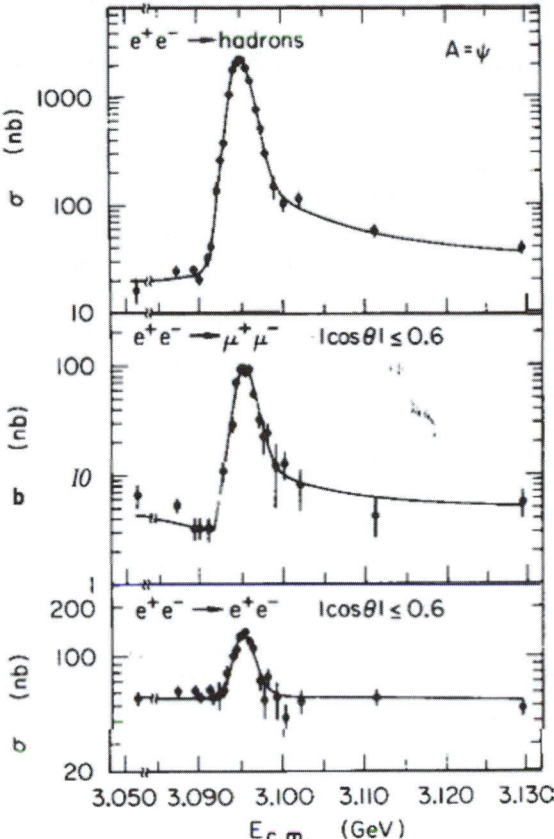

FIGURE 18.3

Observation of the J/ψ resonance at 3.1 GeV, produced by e^+e^- annihilation at SLAC. Image courtesy of Institute of Physics Publishing from ref. [171].

huge resonance (called the ψ) – more than two orders of magnitude larger than its surroundings – appeared in all cross-sections, which had the approximate form

$$\sigma(E) = \frac{3\pi}{E^2} \frac{\Gamma_i \Gamma_f}{(E - M)^2 + \frac{\Gamma^2}{4}} \qquad (18.5)$$

where M is the mass of the resonance, E the CMS energy, and Γ its width. The peak area is

$$\int \sigma(E) \, dE = \frac{6\pi^2 \Gamma_i \Gamma_f}{M^2 \Gamma} \qquad (18.6)$$

where Γ_i is the partial width of the initial e^+e^- state and Γ_f the partial width of the final state (which might be e^+e^-, $\mu^+\mu^-$ or hadrons). The smaller the width Γ, the narrower the peak and the larger its area. The SLAC experiments indicated that the width was extremely narrow: $\Gamma = 0.091$ MeV for the $M = 3097$ MeV resonance! Furthermore, these ψ states were produced from $e^+e^- \longrightarrow \gamma \longrightarrow \psi$, so the ψ had to have the same quantum numbers as the photon: $J_\psi = 1 \; \mathbb{P}_\psi = -1, \; \mathbb{C} = -1$, which we write as $J_\psi^{\mathbb{PC}} = 1^{--}$). You can see the original data in fig. 18.3.

The extreme narrowness of the ψ resonances indicated that they could not be understood as excitations of bound states of any of u, d, s with any of $\bar{u}, \bar{d}, \bar{s}$. Hence a new quark flavor called *charm* (c) was invented[†], and the ψ was identified as a $c\bar{c}$-bound state called CHARMONIUM. More ψ resononances were discovered (ψ' at 3686 MeV and ψ'' at 3770 MeV), and could be identified with various excitations of this bound state (just the way various states of the Hydrogen atom can be identified with excitations of the electron bound to the proton). Indeed, the spectrum of states is completely analogous to positronium, and the ψ was identified as the 1^3S_1 state – orthocharmonium!

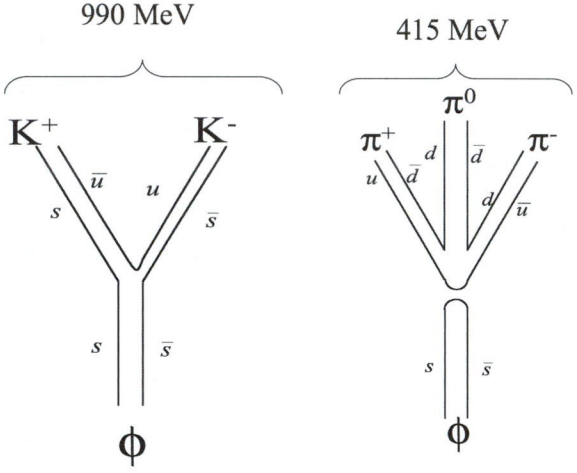

FIGURE 18.4

Quark-flow diagrams for the decay of the ϕ. The gluons mediating these processes have not been drawn.

[†]Actually charm was predicted to exist before the discovery of the ψ by Glashow, Iliopolous and Maini [172], who invented it to explain the absence of strangeness changing neutral currents in the weak interactions. We'll look at this in Chapter 21 when we discuss weak interactions. After this, Politzer and Appelquist suggested that if a heavy "charm" quark existed, it should form a nonrelativistic bound state $c\bar{c}$, with a spectrum of energy levels analogous to positronium [173]. They called this bound state charmonium.

Other combinations of quarks can also form new bound states: $c\bar{u}, c\bar{d}, c\bar{s}$ and their antiparticles. These states are called D-mesons, and they are much broader than the ψ resonance (i.e. their lifetimes are much shorter). They were observed [174] at the Mark I detector in the channels

$$e^+ + e^- \longrightarrow D^0 + \bar{D}^0 + X \qquad\qquad e^+ + e^- \longrightarrow D^+ + D^- + X \quad (18.7)$$

where the D-mesons appear as resonances in the final state (seen by decays into Kaons and pions). The D^0-meson has a mass 1865 MeV, and the $D^{\pm}$-mesons have mass 1869.5 MeV, each of which are too heavy for $\psi(3097)$ to decay into: $m_\psi < 2m_D$. Hence the ψ must decay into states consisting only of u, d, and s quarks.

It is this situation that gives the ψ its peculiar feature of a very narrow width and hence anomalously long lifetime $\tau_\psi \simeq 10^{-20}$ sec, 1000 times longer than is typical of the strong interactions, but shorter than EM decays (typically 10^{-18} sec long).

18.1.1 The OZI Rule

Why does the ψ live so long? The answer goes back to Okubo, Zweig and Iizuka and is called the OZI rule [175]. These scientists were puzzled (long before the ψ was found) by the fact that the ϕ meson ($s\bar{s}$ recall) decayed much more often into 2 Kaons (K^+K^-) instead of 3 π's. But the 3π decay is energetically favored (mass(K^+K^-) = 990 MeV whereas mass(3π) = 415 MeV).

OZI suggested that processes with unconnected quark lines were suppressed. For the decays shown in the diagram 18.4 we see that the decay into three pions has unconnected quark lines and so is suppressed. Experimentally

$$\phi \longrightarrow \left.\begin{matrix} K^+K^- \\ \bar{K}^0 K^0 \end{matrix}\right\} 90\% \qquad\qquad (18.8)$$
$$\longrightarrow \pi^+\pi^0\pi^- \} 10\%$$

In terms of gluons, the decays are given in the figure below.

Why the suppression? In the suppressed diagram on the right in fig. 18.5 there is an intermediate state of pure gluons. These must be high energy ("hard") gluons since the carry the total mass of the ϕ meson. However, on the diagram at the left, the gluon is of low energy, since most of the energy will be in the masses of the strange/antistrange pair of mesons. The OZI rule is a statement that high-energy gluon-exchange processes are suppressed in meson decay. In the context of QCD we now understand this in terms of asymptotic freedom: low-energy gluons couple much more strongly to quarks than high-energy gluons do, and so the low-energy gluon exchanges are more probable. So most often the ϕ meson will decay via the single-gluon channel, but occasionally it will decay with the OZI-suppressed 3-gluon channel.

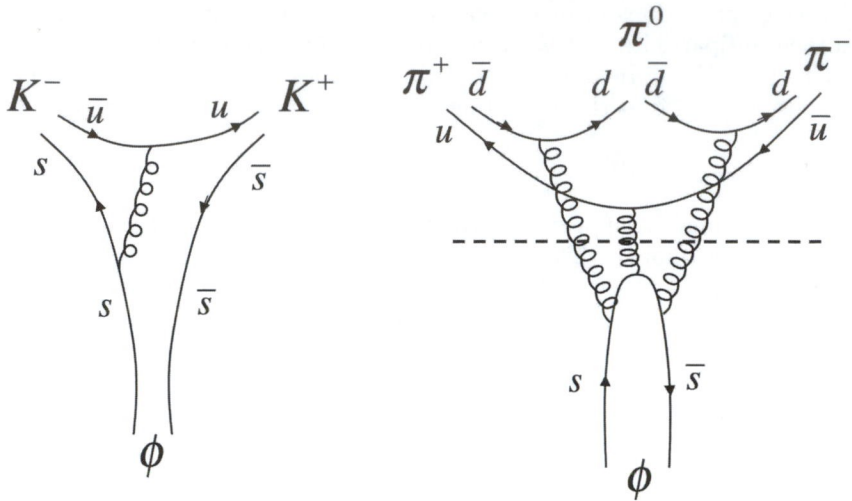

FIGURE 18.5
The OZI rule: if any diagram can be sliced in two by cutting only gluon lines, the diagram is suppressed.

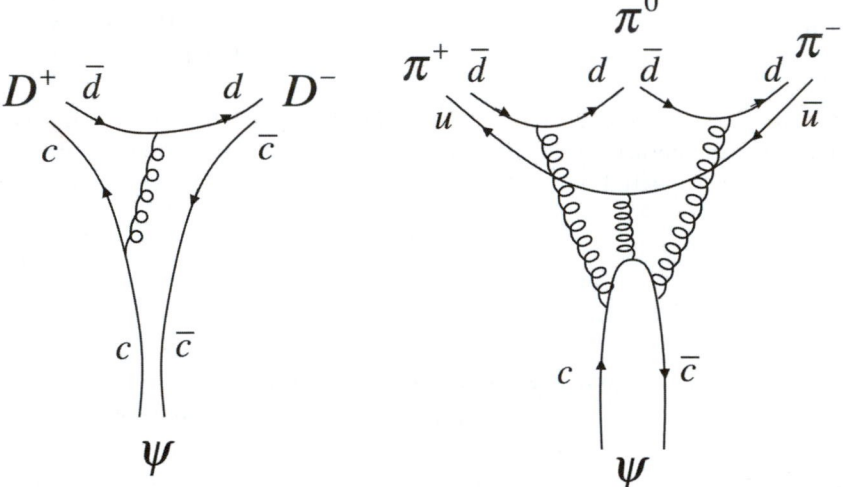

FIGURE 18.6
Why ψ-decay is suppressed: the OZI-rule favors the diagram on the left, but energy conservation forbids this. Hence the diagram on the right describes the leading-order decay of the ψ.

Now we can understand charmonium decay. Since all mesons are color singlets, the connection between the initial ψ state and its decay products must be via a color-singlet gluon combination which needs at least two gluons. But the ψ couples to the photon and has spin-1, and so must have $\mathbb{C} = -1$. Gluons also have $\mathbb{C} = -1$ (they behave just like photons in this respect) and so *three* gluons carrying all of the energy of the ψ must be exchanged. The diagrams illustrating the suppression of decay are given in fig. 18.6. Energy conservation forbids the process on the left, and so only the much slower OZI-suppressed decay process can occur, giving the ψ a relatively long lifetime.

The correlation between states of charmonium and states of positronium is remarkably good considering the aforementioned problems with quarkonium. You can see the comparison in the figure 18.7. Given the difference in energy scales – the hyperfine splitting between the $n = 1$ levels in positronium is 10^{11} times smaller than the corresponding split between the ψ and η_c particles – the relative spacings are remarkably similar [176]. A numerical fit using the Coulomb+Linear potential above can be obtained by setting $F_0 = 16$ tons $= 900$ MeV/fm.

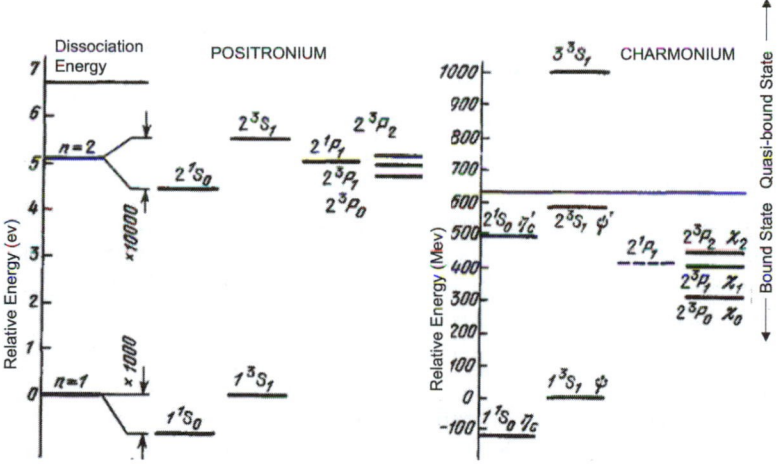

FIGURE 18.7

Comparison of the charmonium and positronium spectra[176]. Note the differing energy scales. Reprinted with permission. Copyright 1982 by Scientific American, Inc. All rights reserved.

18.2 Bottom

In 1977 a 2-arm spectrometer at Fermilab was used by Leon Lederman and collaborators [177] to study $\mu^+\mu^-$ pairs produced in hadronic collisions

$$p + (\text{Cu,Pt}) \longrightarrow \mu^+ + \mu^- + X \tag{18.9}$$

where the two arms measured the momenta of the positive and negative

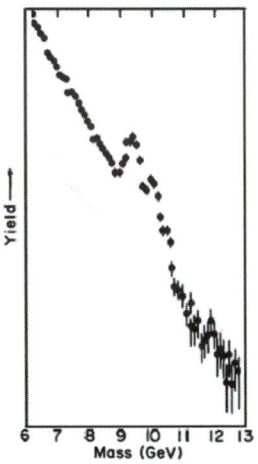
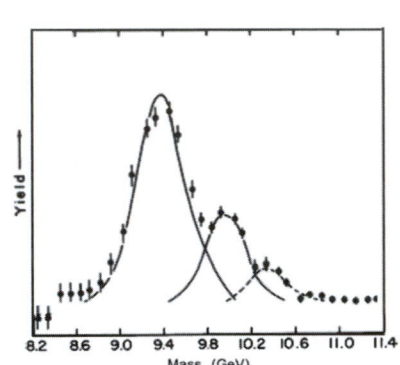

FIGURE 18.8

First evidence for the upsilon resonance from muon-antimuon annihilation. The individual states Υ, Υ' are not resolved. Reprinted figure with permission from L. Lederman, Rev. Mod. Phys. **61**, 547 (1989) figs. 12a,12b [178]. Copyright (1989) by the American Physical Society.

muon. As with the Brookhaven experiment these spectrometers must have a high hadron rejection rate, which was attained by using a block of Beryllium 18 radiation lengths thick to stop any hadrons from entering the arms of the spectrometer. A target was exposed to more than 10^{16} protons, from which about 9000 $\mu^+\mu^-$ events with energy $m(\mu^+\mu^-) > 5$ GeV were obtained.

Surprisingly, three narrow resonances (similar to the ψ) were observed at about $9.5 - 10.5$ GeV, all with $J^{\mathbb{PC}} = 1^{--}$ as shown in fig. 18.8. These could not be identified with any bound states of u, d, s or c and so another flavor of quark called b (for bottom – though some researchers preferred the term "beauty") was introduced. The lowest-energy resonance was called Υ

and was identified as a $b\bar{b}$-bound state. A precision study at DESY using an e^+e^- collider found that these three resonances had very narrow widths, and were the ground and first two radially excited states of a $b\bar{b}$-atom.

The mass of the b quark is so large (~ 4.5 GeV) that even more states of the $b\bar{b}$-system (upsilonium) should be observed than for charmonium. Of course the b quarks can also bind with the u, d, s, c quarks, giving rise to B-mesons. The masses of these mesons are now known quite accurately [1], and are given in table 18.1.

TABLE 18.1

B-Mesons

B-meson	Quark content	Measured mass (MeV)	Stat (Sys) errors	Width or Lifetime
$\Upsilon\left(1^3S_1\right)$	$b\bar{b}$	9460.30	0.26	54.02 ± 1.25 KeV
$\Upsilon\left(2^3S_1\right)$	$b\bar{b}$	10023.26	0.31	31.98 ± 2.63 KeV
$\Upsilon\left(3^3S_1\right)$	$b\bar{b}$	10355.2	0.5	20.32 ± 1.85 KeV
$\Upsilon\left(4^3S_1\right)$	$b\bar{b}$	10579.4	1.2	20.5 ± 2.5 MeV
B^+	$u\bar{b}$	5279.1	1.7(1.4)	1.638 ± 0.011 ps
B^0	$d\bar{b}$	5281.3	2.2(1.4)	1.530 ± 0.009 ps
B^0_s	$s\bar{b}$	5269.9	2.3(1.3)	1.425 ± 0.041 ps
B^0_c	$c\bar{b}$	6276	4	0.46 ± 0.07 ps

Note that the mass difference between the average value of the two lighter B-mesons and the B^0_s-meson is about 90 MeV – precise measurements yield a value of $89.7 \pm 2.7 \pm 1.2$ MeV, much lighter than the 150 MeV for the strange quark predicted in the old quark model. This value is in agreement with theoretical predictions from QCD [1]. Evidently the QCD binding energy is a significant contribution to the mass of the low-energy mesons!

B-meson physics is playing an important role in mapping out much of the remaining uncharted territory in the Standard Model. The reason for this is that the b-quark can only decay into the four lighter quarks via weak interactions. These decays (if the Standard Model is correct) violate $\mathbb{CP}$ and so B-physics provides new independent tests of $\mathbb{CP}$-violation beyond the K-meson system. I'll discuss this in more detail in Chapter 19 .

18.3 Top

After many years of searching, direct evidence for the existence of the top quark was finally obtained by the CDF group at Fermilab. The first evidence

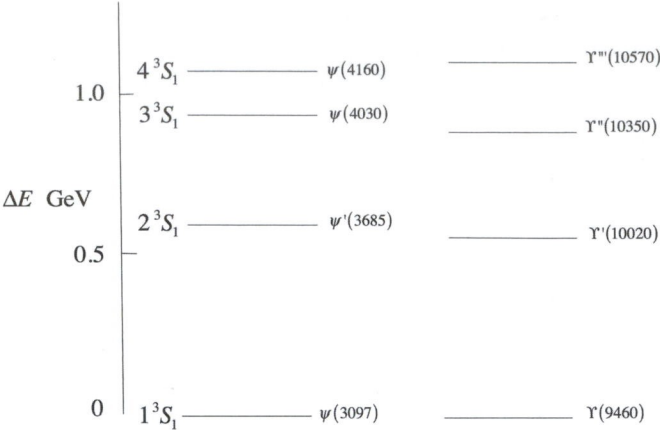

FIGURE 18.9
Comparison of the charmonium and upsilonium spectra.

came in April of 1994, when a preliminary measurement of the t-quark mass was given as 174 ± 20 GeV. This initial result was of questionable statistical significance, and so the CDF collaboration presented their work somewhat tentatively. However, in early 1995 [179], this sighting of the t-quark was confirmed by CDF and by another group at Fermilab called $D0$. The newest results from these two groups have increased the significance of the top quark signal to more than 4 standard deviations. The latest measurements are $m_t = 171.2 \pm 2.1$ GeV [1], making the top quark the heaviest known elementary particle, roughly twice the mass of the Z boson.

Observation of the top-quark is quite difficult [180]. Its lifetime is about 10^{-24} seconds, and only 1 in every 10 billion collisions produces a top. Top quarks are produced at a Tevatron CMS energy of 1.8 TeV, primarily via the process $p\bar{p} \longrightarrow t\bar{t}$. According to the Standard Model, the t-quark decays into a W-boson and a b-quark nearly 100% of the time. These events are described by diagrams of the form given in figure 18.10 and table 18.2 lists the possible decay modes of the top with the associated lowest-order branching ratios. Notice that I have listed the branching ratios for the leptons separately because they can be distinguished in the final state, unlike the quarks and antiquarks, which produce jets. A cartoon depiction of top decay is illustrated in fig. 18.11.

Since the top quark is so unstable, there are no top hadrons. When a lighter quark is produced in some process it will move rapidly in a gluon field (that it contributes to), whose energy density rapidly becomes so intense that the field materializes into quark-antiquark pairs, forming hadrons. This hadronization process does not happen for the top quark because its lifetime is shorter than

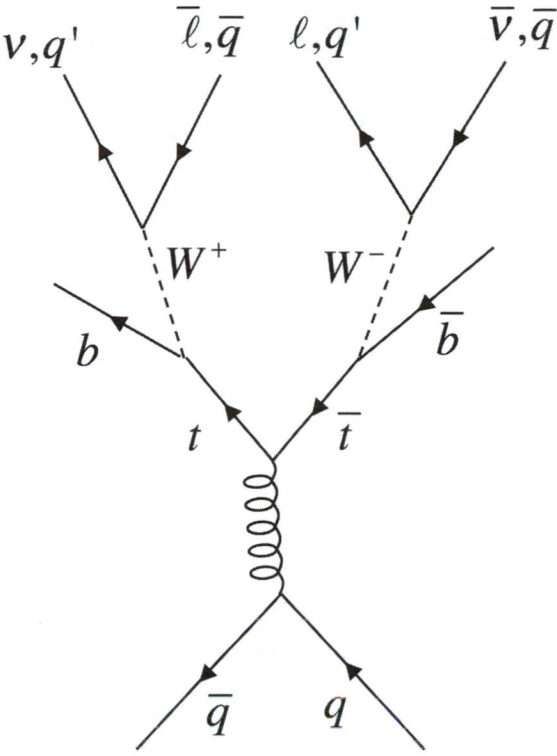

FIGURE 18.10

A generic top quark decay.

the hadronization time. So in a certain sense the top quark is a free quark, unbound to any others – but it only lives for a very short time!

The only way to really observe the top is from its decay products due to its weak interactions, which are almost always into a W-boson and a b quark:

$$p + \bar{p} \longrightarrow t + \bar{t} + X \qquad\qquad t \longrightarrow W^+ + b \qquad \bar{t} \longrightarrow W^- + \bar{b} \quad (18.10)$$

The W-bosons themselves can only be observed from their decay products, which are into leptons (33% of the time) or hadrons (66% of the time); the diagram in figure 18.11 is for a leptonic decay. In the hadronic decay of a W−boson, the W decays into two quarks, which then hadronize as jets: large numbers of particles detected in narrow but ill-defined cones along the initial directions of the quarks. Gluons produced in $p\bar{p}$ - collisions also make the same kinds of jets. These are hard to distinguish from jets produced directly by the $p\bar{p}$ collision. Consequently, leptonic W-decays are much easier to single

TABLE 18.2

Top Quark Decay Modes

Decay Mode	Branching Ratio
$t\bar{t} \rightarrow (q\bar{q}'b)(q\bar{q}'\bar{b})$	36/81
$t\bar{t} \rightarrow (q\bar{q}'b)(e\bar{\nu}_e\bar{b})$	12/81
$t\bar{t} \rightarrow (q\bar{q}'b)(\mu\bar{\nu}_\mu\bar{b})$	12/81
$t\bar{t} \rightarrow (q\bar{q}'b)(\tau\bar{\nu}_\tau\bar{b})$	12/81
$t\bar{t} \rightarrow (e^+\nu_e b)(\mu\bar{\nu}_\mu\bar{b})$	2/81
$t\bar{t} \rightarrow (e^+\nu_e b)(\tau\bar{\nu}_\tau\bar{b})$	2/81
$t\bar{t} \rightarrow (\mu^+\nu_\mu\bar{b})(\tau\bar{\nu}_\tau\bar{b})$	2/81
$t\bar{t} \rightarrow (e^+\nu_e\bar{b})(e\bar{\nu}_e\bar{b})$	1/81
$t\bar{t} \rightarrow (\mu^+\nu_\mu\bar{b})(\mu\bar{\nu}_\mu\bar{b})$	1/81
$t\bar{t} \rightarrow (\tau^+\bar{\nu}_\mu\bar{b})(\tau\bar{\nu}_\tau\bar{b})$	1/81

In the above q refers to any of u, d, s, c.

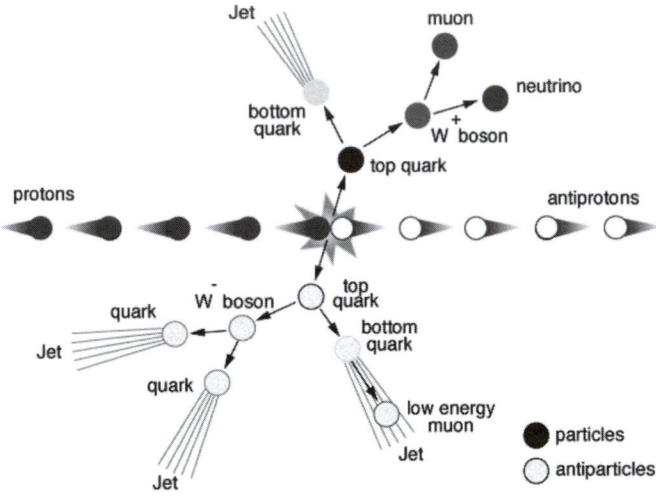

FIGURE 18.11

Pictorial depiction of Top-antitop production at the D0 Detector at Fermilab. Image courtesy of Fermilab Education Office, Topics in Modern Physics Program.

out from the background: there is a large, well isolated energy deposited by the lepton and missing energy in the transverse direction of the beam due to the undetected neutrino.

Both CDF and $D0$ require a W decay candidate from the top or antitop. Events where both W bosons decay leptonically (dilepton events) are clean, but only comprise 5% of all $t\bar{t}$- decays. Events where one W decays leptonically and the other decays into quarks (lepton plus jet events, or ℓj events) have a higher branching ratio (30%) but suffer from large backgrounds. Figure 18.3 illustrates a typical top event.

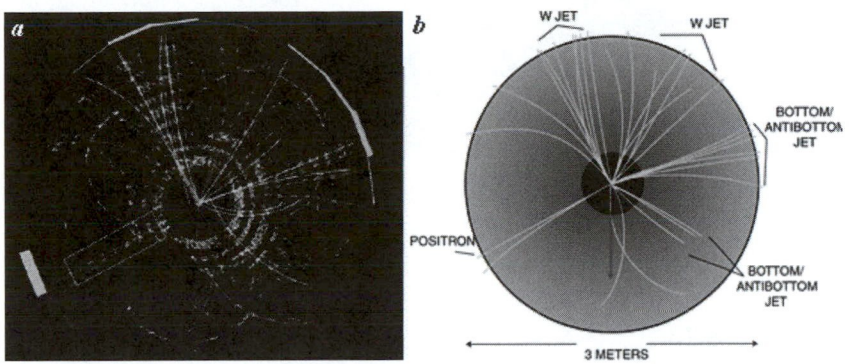

FIGURE 18.12

A proton and antiproton collide producing four distinct jets (b) and some other particles. Reprinted with permission. Copyright 1997 by Scientific American, Inc. All rights reserved.

How is the mass of the top determined? The best measurements come from the ℓj events, in which the top-antitop decay is expected to have the following products: (i) a single electron/positron or muon/antimuon (ii) missing transverse energy from a neutrino (iii) two jets from a W-decay (iv) two b-jet decays. CDF begins by selecting only events that have at least four jets. Three of them must have at least 15 GeV of transverse energy, and the 4th jet at least 8 GeV of transverse energy. Each event that survives this criterion is then fit to the hypothesis of $t\bar{t} \longrightarrow$ single lepton + jets channel. Out of all the possible jet assignments, the solution with the best fit (lowest χ^2) is kept; if this fit is too poor (i.e. the χ^2 is too large) then the event is rejected.

Less than 100 events survive this process. To improve the accuracy, the events are then subdivided as to whether or not a b-quark was "tagged" (correctly identified). There are either 2 b-tags, one, or none, and the events are then regrouped into these categories, and their likelihoods recalculated within each group. The product of these likelihoods is maximized to determine the t-quark mass.

Since these original experiments were carried out, further experiments were conducted with better data sets. Figure 18.14 illustrates an example of this, based on data from a recent run at the CDF detector at Fermilab, which obtained $m_{\text{top}} = 171.9 \pm 2.0$ GeV/c^2.

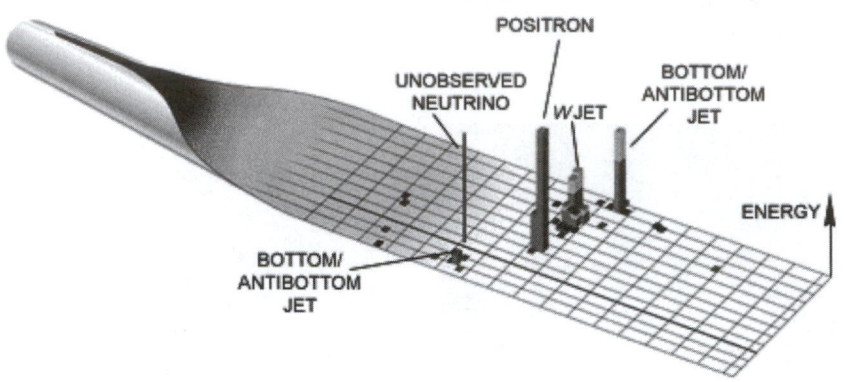

FIGURE 18.13

Analysis of a top event. In fig. 18.3 (a), a proton and antiproton collide producing four distinct jets (b) and some other particles. These are sketched in figure 18.3 (b). Multiple jets and a positron signify the possible creation of a top quark. The energies and locations of the particles measured by the calorimeter surrounding the beam line are shown in figure 18.13, with the energies released by the particles indicated by the heights of the bars [180].

18.4 QCD

The theory underlying the strong interactions that binds quarks together is called Quantum Chromodynamics, or QCD. QCD is very similar to QED, except that the one electric charge (which can be positive or negative) is replaced with three color charges – red, blue and green (and their anticolors). Table 18.3 contains a comparative chart.

QCD is a much more complicated theory than QED, mainly because it is a non-Abelian gauge theory. While the equations obeyed by the fermions

$$\left(i\gamma^\mu \left(\delta_\mathcal{A}^\mathcal{B} \partial_\mu + i g_s A_\mu^a (\lambda^a)_\mathcal{A}^\mathcal{B} \right) - m \delta_\mathcal{A}^\mathcal{B} \right) \psi_\mathcal{B} = 0 \qquad (18.11)$$

differ from the Maxwell-Dirac case only by the appearance of λ^a, which interchanges the colors, the gluons obey a generalization of Maxwell's equations

$$\partial^\mu F_{\mu\nu}^a - g_s A^{c\mu} F_{\mu\nu}^b f_{cb}^a = g_s \overline{\psi}^\mathcal{A} \gamma_\nu (\lambda^a)_\mathcal{A}^\mathcal{B} \psi_\mathcal{B} \qquad (18.12)$$

where

$$F_{\mu\nu}^a = \partial_\mu A_\nu^a - \partial_\nu A_\mu^a - g_s A_\mu^c A_\nu^b f_{cb}^a \qquad (18.13)$$

TABLE 18.3
Comparison of QED and QCD

	QED	QCD
Mediators	1 Photon (massless)	8 Gluons (all massless)
Charge	1 (+ or −) ("charge")	3 (R or $\overline{R}$, B or $\overline{B}$, G or $\overline{G}$) ("color")
Strength	$e = \sqrt{4\pi\alpha}$ $\alpha = \frac{1}{137}$	$g_s = \sqrt{4\pi\alpha_s}$ $\alpha_s \simeq 1$
Equations	$(i\gamma^\mu (\partial_\mu - ieA_\mu) - m)\psi = 0$ $\partial^\mu F_{\mu\nu} = -e\overline{\psi}\gamma_\nu\psi$ $F_{\mu\nu} = \partial_\mu A_\nu - \partial_\nu A_\mu$	$\left(i\gamma^\mu \left(\delta_{\mathcal{A}}^{\mathcal{B}}\partial_\mu + ig_s A_\mu^{\mathfrak{a}} (\lambda^{\mathfrak{a}})_{\mathcal{A}}^{\mathcal{B}}\right) - m\delta_{\mathcal{A}}^{\mathcal{B}}\right)\psi_{\mathcal{B}} = 0$ $\partial^\mu F_{\mu\nu}^{\mathfrak{a}} - g_s A^{\mathfrak{c}\mu} F_{\mu\nu}^{\mathfrak{b}} f_{\mathfrak{cb}}^{\mathfrak{a}} = g_s \overline{\psi}^{\mathcal{A}}\gamma_\nu (\lambda^{\mathfrak{a}})_{\mathcal{A}}^{\mathcal{B}}\psi_{\mathcal{B}}$ $F_{\mu\nu}^{\mathfrak{a}} = \partial_\mu A_\nu^{\mathfrak{a}} - \partial_\nu A_\mu^{\mathfrak{a}} - g_s A_\mu^{\mathfrak{c}} A_\nu^{\mathfrak{b}} f_{\mathfrak{cb}}^{\mathfrak{a}}$ $\mathcal{A}, \mathcal{B} = R, G, B$ $\mathfrak{a}, \mathfrak{b} = 1, 2, ..., 8$
Algebra/Group Manifestation	Abelian $\mathbf{U}(1)$ Binds atoms together	non-Abelian $\left[\lambda^{\mathfrak{a}}, \lambda^{\mathfrak{b}}\right] = 2if_{\mathfrak{c}}^{\mathfrak{ab}}\lambda^{\mathfrak{c}}$ $\mathbf{SU}(3)$ Binds hadrons and mesons together
Range	Long (but neutralized)	Confined – so appears to be short

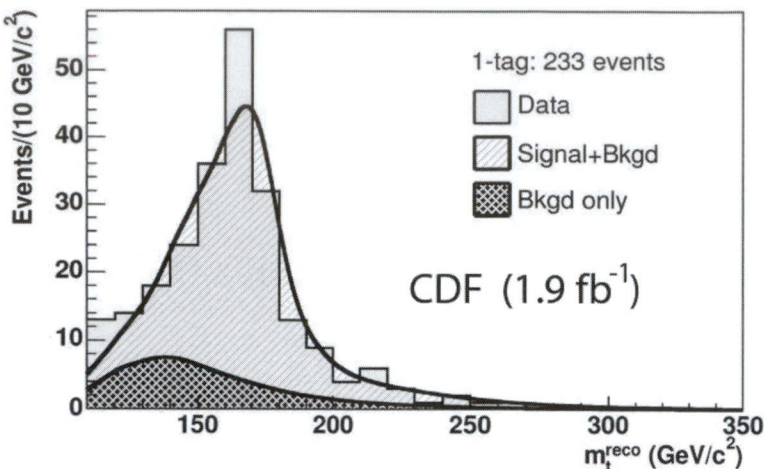

FIGURE 18.14

Reconstructed top quark mass using a data sample consisting of 233 Lepton+Jets event candidates, collected by the CDF II detector at Fermilab. The darker shaded part is the background distribution, with normalization constrained to the calculated value. Reprinted figure with permission from T. Aaltonen et al. (CDF Collaboration), Phys. Rev. **D79**, 092005 (2009) [181]. Copyright (2009) by the American Physical Society.

The general structure of such equations (with $\psi_\mathcal{B} = 0$) was first worked out by Yang and Mills [182] and so we call them Yang-Mills equations. If you're curious as to where they come from, I've provided some details in the appendix. But don't worry about solving them – my purpose here is to show you what they look like so that you can see the comparison to QED. In fact, finding general solutions to the equations of QCD is the main motivation behind lattice gauge theory, and is one of the major efforts of strong-interaction physics today [165].

The equations in the right-hand column of table 18.3 are valid for any Lie group, whose structure constants I have denoted by f^c_{ab}. They form the basis of the electroweak gauge theory, where the Lie group is $\mathbf{SU}(2)$ and of QCD, whose gauge group is the Lie group $\mathbf{SU}(3)$. For the particular case of $\mathbf{SU}(3)$ we have

$$\left[\lambda^a, \lambda^b\right] = 2if^{ab}_{\ \ c}\lambda^c \tag{18.14}$$

where the generators λ^c are generalizations of the Pauli matrices and are given by

$$\lambda^1 = \begin{pmatrix} 0 & 1 & 0 \\ 1 & 0 & 0 \\ 0 & 0 & 0 \end{pmatrix} \qquad \lambda^2 = \begin{pmatrix} 0 & -i & 0 \\ i & 0 & 0 \\ 0 & 0 & 0 \end{pmatrix} \qquad \lambda^3 = \begin{pmatrix} 1 & 0 & 0 \\ 0 & -1 & 0 \\ 0 & 0 & 0 \end{pmatrix}$$

$$\lambda^4 = \begin{pmatrix} 0 & 0 & 1 \\ 0 & 0 & 0 \\ 1 & 0 & 0 \end{pmatrix} \qquad \lambda^5 = \begin{pmatrix} 0 & 0 & -i \\ 0 & 0 & 0 \\ i & 0 & 0 \end{pmatrix} \qquad \lambda^6 = \begin{pmatrix} 0 & 0 & 0 \\ 0 & 0 & 1 \\ 0 & 1 & 0 \end{pmatrix} \quad (18.15)$$

$$\lambda^7 = \begin{pmatrix} 0 & 0 & 0 \\ 0 & 0 & -i \\ 0 & i & 0 \end{pmatrix} \qquad \lambda^8 = \frac{1}{\sqrt{3}} \begin{pmatrix} 1 & 0 & 0 \\ 0 & 1 & 0 \\ 0 & 0 & -2 \end{pmatrix}$$

often referred to as the Gell-Mann matrices. Using them, you can easily check that the structure constants $f^{ab}_{\ c}$ are fully antisymmetric, and that the only non-vanishing ones are

$$f^{12}_{\ 3} = 1 \qquad f^{12}_{\ 7} = f^{24}_{\ 6} = f^{25}_{\ 7} = f^{34}_{\ 5} = f^{51}_{\ 6} = f^{63}_{\ 7} = \frac{1}{2} \quad f^{45}_{\ 8} = f^{67}_{\ 8} = \frac{\sqrt{3}}{2} \quad (18.16)$$

along with those that are obtained from antisymmetry, e.g. $f^{21}_{\ 7} = -\frac{1}{2}$.

It is the non-vanishing of the f^{c}_{ab} that make the theory non-Abelian, giving rise to new interactions not present in the Maxwell-Dirac theory. Notice that the field strength $F^a_{\mu\nu}$ for each gluon is not linear in the gluon wavefunction, but depends on the other gluons that are present, even if there are no colored fermions. This means that the gluons act as sources for each other, which we can see by rewriting the equations for the gluons as

$$\partial^\mu \left(\partial_\mu A^c_\nu - \partial_\nu A^c_\mu \right) = g_s \overline{\psi}^A \gamma_\nu \left(\lambda^a \right)^B_A \psi_B + g_s \partial^\mu \left(A^a_\mu A^a_\nu \right) f^c_{ab} \qquad (18.17)$$
$$+ g_s A^{a\mu} \left(\partial_\mu A^b_\nu - \partial_\nu A^b_\mu - g_s A^r_\mu A^s_\nu f^b_{rs} \right) f^c_{ab}$$

The left-hand side of this equation for each gluon wavefunction A^a_ν is just like Maxwell's equations. The first term on the right-hand side is the color-current due to the quarks (analogous to the electric current $\bar{\psi}\gamma^\mu\psi$ in QED) – this gives rise to the expects quark-quark-gluon vertex. The λ^a have the effect of changing one quark color into another, both in the quark current and in the equation obeyed by the quarks. The remaining terms are the "color current" due to the gluons. Notice that there are derivative terms that are quadratic in gluon wavefunctions – these yield a triple-gluon vertex rule. There is also a term cubic in the gluon wavefunctions (the $g^2_s A^{a\mu} A^r_\mu A^s_\nu f^b_{rs} f^c_{ab}$ term), from which a four-gluon vertex rule arises. A complete list of Feynman rules for QCD is given in the appendix.

18.4.1 Basic Physical Features of QCD

So as you can see, QCD is a considerably richer physical theory than QED, and considerably more complicated. However, it has some key features that are worth noting.

1. Eight Gluons

   There are as many gauge fields as there are generators of the group **SU**(3). A unitary 3×3 matrix of determinant 1 has $3^2 - 1 = 8$ generators, and so there are eight gluons. These eight gluons form a color octet:

   $$
   \begin{array}{lll}
   |1\rangle = \tfrac{1}{\sqrt{2}}\left|R\overline{B}+B\overline{R}\right\rangle & |4\rangle = \tfrac{1}{\sqrt{2}}\left|R\overline{G}+G\overline{R}\right\rangle & |6\rangle = \tfrac{1}{\sqrt{2}}\left|G\overline{B}+B\overline{G}\right\rangle \\
   |2\rangle = -\tfrac{i}{\sqrt{2}}\left|R\overline{B}-B\overline{R}\right\rangle & & |7\rangle = \tfrac{i}{\sqrt{2}}\left|G\overline{B}-B\overline{G}\right\rangle \\
   |3\rangle = \tfrac{1}{\sqrt{2}}\left|R\overline{R}-B\overline{B}\right\rangle & |5\rangle = -\tfrac{i}{\sqrt{2}}\left|R\overline{G}-G\overline{R}\right\rangle & |8\rangle = \tfrac{1}{\sqrt{6}}\left|R\overline{R}+B\overline{B}-2G\overline{G}\right\rangle
   \end{array}
   \tag{18.18}
   $$

   and we can write the gluon wavefunction as A_μ^a where the color index $a = |1\rangle, |2\rangle, \ldots, |8\rangle$. Note that each gluon is its own antiparticle. We could also form a color singlet ($|s\rangle = \tfrac{1}{\sqrt{3}}\left|R\overline{R} + B\overline{B} + G\overline{G}\right\rangle$), which would be as common as the photon if it existed. Such a particle would couple to all baryons with the same strength, but not to all leptons, and would behave like a "fifth force" that would compete with gravity in its behavior on baryons, violating the equivalence principle. So far all experimental searches for such a force have yielded a null result [183].

2. Quark-antiquark potential

   The gluon exchange diagram on the right of fig. 18.1 gives a $\frac{1}{r}$ potential for the same reasons as QED. However, the non-abelian nature of the color charges gives an interesting answer for the coefficient of this potential due to the interaction between a quark-antiquark pair. If this pair has its color charges as part of the octet configuration above (e.g the color charges have the wavefunction $\tfrac{1}{\sqrt{2}}\left|G\overline{B} + B\overline{G}\right\rangle$), then the potential is repulsive. But if the color charges are in the singlet configuration (i.e. $\tfrac{1}{\sqrt{3}}\left|R\overline{R} + B\overline{B} + G\overline{G}\right\rangle$) then the potential is attractive! Using the rules for QCD in the appendix, one finds

   $$
   \begin{aligned}
   \mathcal{M} &= i\left[\overline{u}^{(i_1')}(p_1')c_1'^\dagger \gamma^\mu \lambda^a u^{(i_1)}(p_1)c_1\right]\frac{g_s^2 g_{\mu\nu}\delta^{ab}}{4\left(p_1' - p_1\right)^2} \\
   &\quad \times \left[\overline{u}^{(i_2')}(p_2')c_2'^\dagger \gamma^\nu \lambda^b u^{(i_2)}(p_2)c_2\right] \\
   &= i\frac{g_s^2}{4\left(p_1' - p_1\right)^2}\left[\overline{u}^{(i_1')}(p_1')\gamma^\mu u^{(i_1)}(p_1)\right]\left[\overline{u}^{(i_2')}(p_2')\gamma_\mu u^{(i_2)}(p_2)\right] \\
   &\quad \times \left[c_1'^\dagger \lambda^a c_1\right]\left[c_2'^\dagger \lambda^a c_2\right]
   \end{aligned}
   \tag{18.19}
   $$

   which yields the 1-gluon exchange potential

   $$
   V_{\text{1-gluon}}(r) = -\frac{\alpha}{r} = \frac{1}{4}\left[c_1'^\dagger \lambda^a c_1\right]\left[c_2'^\dagger \lambda^a c_2\right]\frac{\alpha_s}{r}
   \tag{18.20}
   $$

   This potential can either be attractive or repulsive, depending on how the colors of the quarks are configured. A computation of the quantity ʃ yields

$$V_{1\text{-gluon}}(r) = \begin{cases} -\frac{4}{3}\frac{1}{r} & \text{color singlet – attractive!} \\ +\frac{1}{6}\frac{1}{r} & \text{color octet – repulsive!} \end{cases} \tag{18.21}$$

Because of this, quark-antiquark pair will bind into a color singlet state, giving support to the hypothesis that only color singlets occur in nature.

3. Confinement

The preceding arguments suggest that only color singlets occur in nature, but they don't prove it. They don't explain why we can't see a free quark (or a free gluon). No rigorous proof of confinement exists yet.

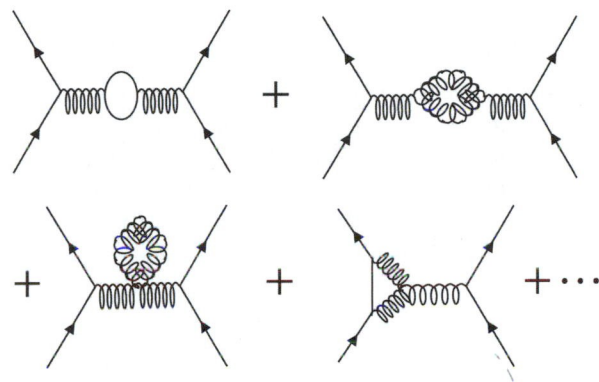

FIGURE 18.15

Some 1-loop corrections to the 1-gluon exchange diagram in figure 18.1.

4. Asymptotic Freedom

In QED we saw that quantum effects modified the QED potential so that the coupling constant varied with distance (or energy):

$$V_{\text{QED}}(r) = -\frac{\alpha(r)}{r} \quad \text{where} \quad \alpha(r) = \left[1 - \frac{2\alpha}{3\pi}Q(r)\right]^{-1} \tag{18.22}$$

which in momentum space is given by eq. (14.51)

$$\alpha(q^2) = \frac{\alpha(0)}{\left[1 - \frac{\alpha_s(0)}{3\pi}\ln\left(\frac{|q^2|}{m^2}\right)\right]} \tag{18.23}$$

and $\alpha(q^2)$ grows as q^2 increases (i.e. the momentum transfer between two particles increases, which is equivalent to the distance to between the them becoming shorter), with m the mass of the lightest charged particle in the theory (in QED this is the electron).

In QCD the analogous expression is

$$\alpha_s(q^2) = \frac{\alpha_s(\mu^2)}{\left[1 + \left(\frac{11N - 2F}{3}\right) \frac{\alpha_s(\mu^2)}{4\pi} \ln\left(\frac{q^2}{\mu^2}\right)\right]} \tag{18.24}$$

where μ^2 is some reference energy (taken to be zero in QED, but nonzero in QCD). The number of colors is N, which is 3 in QCD, and F is the number of flavors (6 that we know of for sure). Note that

$$\frac{11N - 2F}{3} = \frac{33 - 12}{3} = 7 > 0 \tag{18.25}$$

and we see that as q^2 increases, $\alpha_s(q^2)$ decreases!

This behavior is called *asymptotic freedom* – the coupling *decreases* with increasing energy or decreasing distance. This happens because the gluons antiscreen the quark. What happens is that along with a fermion loop there are now new gluon loops when one considers the exchange of a gluon between two quarks (see figure 18.15). This additional effect contributes oppositely to the fermion loops, leading to the result in eq. (18.34).

The timely discovery of asymptotic freedom in 1973 by Politzer, Gross and Wilczek [184] led to a whole new way of looking at strong interaction physics. It meant that theorists could reliably use perturbation theory in QCD at sufficiently high energies, since it allows us to treat quarks as free particles. It is a basic ingredient in constructing quarkonium, and presumably is what is responsible for the OZI rule.

18.5   Appendix: QCD and Yang-Mills Theory

18.5.1   Feynman Rules for QCD

1. NOTATION. Label the incoming (outgoing) four-momenta as $p_1, p_2, \ldots, p_n$ $(p'_1, p'_2, \ldots, p'_m)$, the incoming (outgoing) spins as $s_1, s_2, \ldots, s_n$ $(s'_1, s'_2, \ldots, s'_m)$, the incoming (outgoing) colors with $c_1, c_2, \ldots, c_n$ $(c'^\dagger_1, c'^\dagger_2, \ldots, c'^\dagger_m)$ as further discussed below, the incoming (outgoing) gluon polarizations as $\varepsilon^\mu_1, \varepsilon^\mu_2, \ldots$ $(\varepsilon^{\mu'}_1, \varepsilon^{\mu'}_2, \ldots)$, the gluon colors $\mathcal{A}^a$, and use labels $q_1, q_2, \ldots, q_j$ to denote the internal four-momenta. Assign arrows to the lines as in figure 18.16.

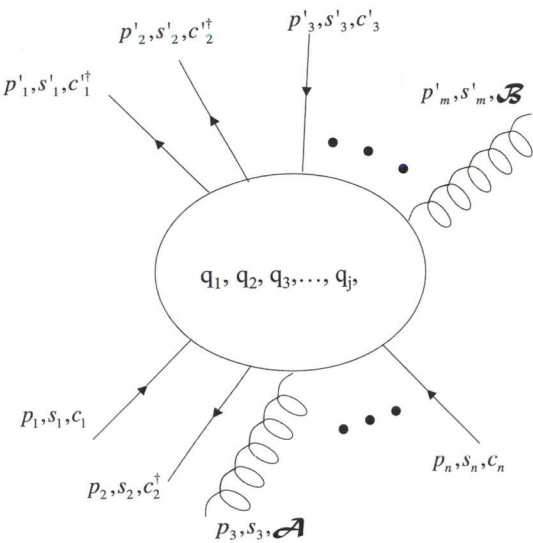

FIGURE 18.16
A typical QCD diagram.

Time flows from bottom to top, curly lines are gluons, and lines with arrows are quarks (if they point upward with time) or antiquarks (if they point downward against time). Note that no other kinds of fermions interact with gluons.

2. EXTERNAL LINES. Each external line contributes a factor as shown in figure 18.17 where the color factors are

$$c^{\mathcal{R}} = \begin{pmatrix} 1 \\ 0 \\ 0 \end{pmatrix} \qquad c^{\mathcal{B}} = \begin{pmatrix} 0 \\ 1 \\ 0 \end{pmatrix} \qquad c^{\mathcal{G}} = \begin{pmatrix} 0 \\ 0 \\ 1 \end{pmatrix}$$

for red, blue, and green respectively. As in QED, factors associated with external lines correspond to the incoming/outgoing plane-wave states, an assumption that is clearly not valid for confined quarks, but which we will take to be valid on the very short timescales in collisions that produce quarks. The gluon color factor $\mathcal{A}^a$ is one of the eight gluon states given in eq. (18.18), e.g., $\frac{1}{\sqrt{2}} \left| G\overline{B} + B\overline{G} \right\rangle$.

3. INTERNAL LINES. Each internal line contributes a factor shown in figure 18.18, where m is the mass of the quark. As before $q^2 \neq m^2$ because the particle flowing through the line is virtual (i.e. it does not obey its equations of motion). These internal lines are called *propagators*.

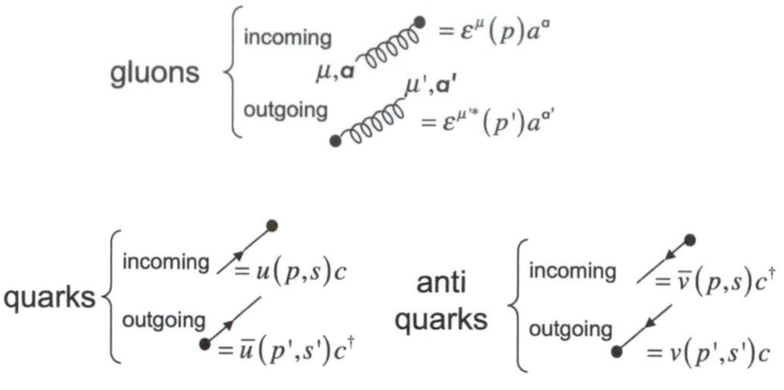

gluons $\begin{cases} \text{incoming} \\ \mu, a \\ \text{outgoing} \end{cases}$ $\begin{aligned} &= \varepsilon^{\mu}(p)a^{\sigma} \\ \mu', a' \\ &= \varepsilon^{\mu *}(p')a^{\sigma'} \end{aligned}$

quarks $\begin{cases} \text{incoming} \\ \text{outgoing} \end{cases}$ $\begin{aligned} &= u(p,s)c \\ &= \bar{u}(p',s')c^{\dagger} \end{aligned}$

anti quarks $\begin{cases} \text{incoming} \\ \text{outgoing} \end{cases}$ $\begin{aligned} &= \bar{v}(p,s)c^{\dagger} \\ &= v(p',s')c \end{aligned}$

FIGURE 18.17
External lines in QCD.

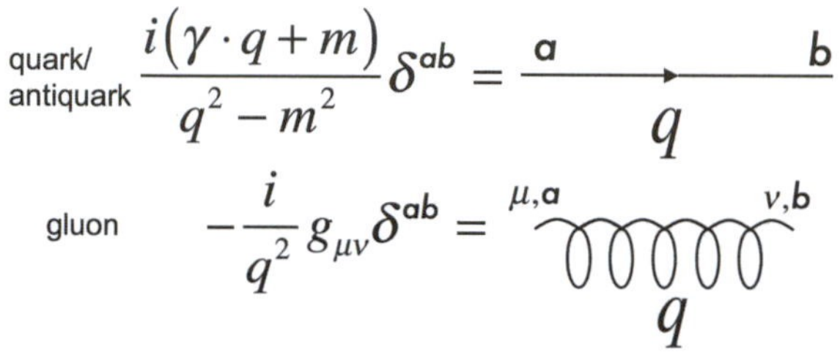

quark/ antiquark $\dfrac{i(\gamma \cdot q + m)}{q^2 - m^2}\delta^{ab} = $ $\underset{q}{\overset{a \qquad\qquad b}{\longrightarrow}}$

gluon $-\dfrac{i}{q^2}g_{\mu\nu}\delta^{ab} = $ $\underset{q}{\overset{\mu,a \qquad\qquad \nu,b}{\text{(gluon line)}}}$

FIGURE 18.18
Internal lines in QCD.

The next two rules are the same as for QED and ABB theory.

4. CONSERVATION OF ENERGY AND MOMENTUM   For each vertex, write a delta function of the form

$$(2\pi)^4 \delta^{(4)}\left(k_1 + k_2 + k_3 + \cdots + k_N\right)$$

where the k's are the four-momenta coming into the vertex (i.e. each k^{μ} will be either a q^{μ} or a p^{μ}). If the momentum leads outward, then k^{μ} is *minus* the four-momentum of that line). This factor imposes conservation of energy and momentum at each vertex (and hence for the diagram as a whole) because the delta function vanishes unless the sum of the incoming momenta equals the sum of the outgoing momenta.

5. INTEGRATE OVER INTERNAL MOMENTA For each internal momentum q, write a factor

$$\frac{d^4 q}{(2\pi)^4}$$

and integrate.

6. **VERTEX FACTOR** In QCD there are several different vertex factors, since gluons carry color. First there is the quark-gluon vertex in figure 18.19 that is similar to QED. The dimensionless coupling $g_s = \sqrt{4\pi\hbar c\alpha_s}$. There

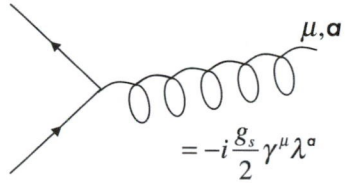

$$= -i\frac{g_s}{2}\gamma^\mu\lambda^a$$

FIGURE 18.19
The quark-gluon QCD vertex.

are also two vertices for gluons shown in figures 18.20 and 18.21

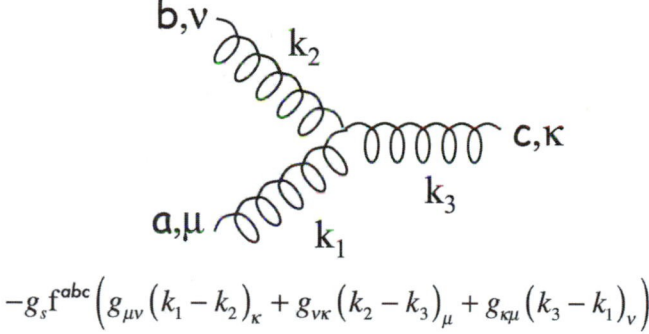

$$-g_s f^{abc}\left(g_{\mu\nu}\left(k_1 - k_2\right)_\kappa + g_{\nu\kappa}\left(k_2 - k_3\right)_\mu + g_{\kappa\mu}\left(k_3 - k_1\right)_\nu\right)$$

FIGURE 18.20
3-gluon vertex (all momenta point in).

The remaining rules are the same as for QED.

7. **TOPOLOGY** To get all contributions for a given process, draw diagrams by joining up all external points to all internal vertex points in all possible

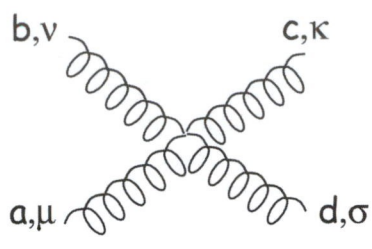

$$-ig_s^2 \left[f^{abr} f^{cdr} \left(g_{v\sigma} g_{\mu\kappa} - g_{v\kappa} g_{\mu\sigma} \right) + f^{adr} f^{bcr} \left(g_{\mu v} g_{\kappa\sigma} - g_{\mu\kappa} g_{v\sigma} \right) + f^{acr} f^{dbr} \left(g_{\mu\sigma} g_{v\kappa} - g_{\mu v} g_{\kappa\sigma} \right) \right]$$

FIGURE 18.21

4-gluon vertex (repeated indices summed over).

arrangments that are topologically inequivalent. The number of ways a given diagram can be drawn is the topological weight of the diagram. The result is equal to $-i\mathcal{M}$.

8. **ANTISYMMETRIZATION** Because fermion wavefunctions anticommute, we must include a minus sign between diagrams that differ
         (a) Only in the interchange of two incoming (or outgoing) fermions/anti-fermions of the same kind
     or
         (b) Only in the interchange of an incoming fermion with an outgoing antifermion of the same kind (or vice versa).

9. **LOOPS** Every fermion loop gets a factor of (-1).

10. **CANCEL THE DELTA FUNCTION** The result will include a factor

$$(2\pi)^4 \delta^{(4)} \left(p_1' + p_2' \cdots + p_m' - p_1 - p_2 - \cdots - p_n \right)$$

corresponding to overall energy-momentum conservation. Cancel this factor, and what remains is $-i\mathcal{M}$.

18.5.2   Yang-Mills Theory

A Yang-Mills theory is a generalization of Maxwell's theory, in which the gauge symmetry group is non-Abelian. The equations can be derived using the same reasoning that we used in applying the gauge principle to obtain Maxwell's equations.

Suppose we have a column vector Φ that transforms as

$$\Phi' = U(x)\Phi \Rightarrow \Phi'^{\mathfrak{A}} = U^{\mathfrak{A}}{}_{\mathfrak{B}}(x)\Phi^{\mathfrak{B}} \tag{18.26}$$

where U is some irreducible unitary matrix representation R of a group that in general will depend on position and time. I have put in the indices for the representation in the right-hand equation so that the transformation structure is clear. We say that Φ forms a representation of the group, by which we mean it transforms according to eq. (18.26). The entries in the column vector Φ can be thought of wavefunctions that transform into each other under this representation. For example, if the group is **SU**(3) and we let Φ be a 3-component vector, then each entry in Φ is a quark wavefunction of a given color.

We want to define a derivative $\mathbf{D}_\mu$ that has the same transformation properties as Φ does in eq. (18.26). It must be matrix since Φ is a column vector, which is why I have written it in boldface. In other words, we require

$$(\mathbf{D}_\mu \Phi)' = U(x)(\mathbf{D}_\mu \Phi) \quad \text{or} \quad (\mathbf{D}_\mu \Phi)'^{\mathfrak{A}} = U^{\mathfrak{A}}{}_{\mathfrak{B}}(x)(\mathbf{D}_\mu \Phi)^{\mathfrak{B}} \qquad (18.27)$$

where again I have put in the indices in the expression on the right. We also want this derivative to be a linear combination of the usual derivative operator plus some vector potential $\mathbf{A}_\mu$. It is clear that the vector potential must also be matrix-valued – otherwise we couldn't have U act non-trivially. So let's write

$$\mathbf{D}_\mu = \partial_\mu \mathbf{I} + ig\mathbf{A}_\mu \quad \text{or} \quad (\mathbf{D}_\mu)^{\mathfrak{A}}{}_{\mathfrak{B}} = \partial_\mu \delta^{\mathfrak{A}}{}_{\mathfrak{B}} + ig(\mathbf{A}_\mu)^{\mathfrak{A}}{}_{\mathfrak{B}} \qquad (18.28)$$

where $\delta^{\mathfrak{A}}{}_{\mathfrak{B}}$ is the Kronecker-delta function, which is simply another way of writing the identity matrix for the representation. It multiplies the partial derivative operator because the act of differentiation itself should not mix up any of the components of Φ.

Now we want to insert eq. (18.28) into eq. (18.27) and see what happens. By definition each side becomes

$$\partial_\mu \Phi' + ig\mathbf{A}'_\mu \Phi' = U(x)(\partial_\mu \Phi + ig\mathbf{A}_\mu \Phi) \qquad (18.29)$$

or, putting in the indices

$$\left(\partial_\mu \delta^{\mathfrak{A}}{}_{\mathfrak{B}} \Phi'^{\mathfrak{B}} + ig(\mathbf{A}'_\mu)^{\mathfrak{A}}{}_{\mathfrak{B}} \Phi'^{\mathfrak{B}}\right) = U^{\mathfrak{A}}{}_{\mathfrak{B}}\left(\partial_\mu \delta^{\mathfrak{B}}{}_{\mathfrak{C}} \Phi^{\mathfrak{C}} + ig(\mathbf{A}_\mu)^{\mathfrak{B}}{}_{\mathfrak{C}} \Phi^{\mathfrak{C}}\right) \qquad (18.30)$$

which simplifies to

$$\partial_\mu \Phi'^{\mathfrak{A}} + ig(\mathbf{A}'_\mu)^{\mathfrak{A}}{}_{\mathfrak{B}} \Phi'^{\mathfrak{B}} = U^{\mathfrak{A}}{}_{\mathfrak{B}} \partial_\mu \Phi^{\mathfrak{B}} + igU^{\mathfrak{A}}{}_{\mathfrak{B}} (\mathbf{A}_\mu)^{\mathfrak{B}}{}_{\mathfrak{C}} \Phi^{\mathfrak{C}} \qquad (18.31)$$

Upon inserting eq. (18.26) this becomes

$$\left(\partial_\mu U^{\mathfrak{A}}{}_{\mathfrak{B}}\right) \Phi^{\mathfrak{B}} + U^{\mathfrak{A}}{}_{\mathfrak{B}} \partial_\mu \Phi^{\mathfrak{B}} + ig(\mathbf{A}'_\mu)^{\mathfrak{A}}{}_{\mathfrak{B}} U^{\mathfrak{B}}{}_{\mathfrak{C}} \Phi^{\mathfrak{C}}$$
$$= U^{\mathfrak{A}}{}_{\mathfrak{B}} \partial_\mu \Phi^{\mathfrak{B}} + igU^{\mathfrak{A}}{}_{\mathfrak{B}} (\mathbf{A}_\mu)^{\mathfrak{B}}{}_{\mathfrak{C}} \Phi^{\mathfrak{C}} \qquad (18.32)$$

or more simply

$$\left[(\mathbf{A}'_\mu)^{\mathfrak{A}}{}_{\mathfrak{B}} U^{\mathfrak{B}}{}_{\mathfrak{C}} - U^{\mathfrak{A}}{}_{\mathfrak{B}} (\mathbf{A}_\mu)^{\mathfrak{B}}{}_{\mathfrak{C}} - \frac{i}{g}\left(\partial_\mu U^{\mathfrak{A}}{}_{\mathfrak{C}}\right)\right] \Phi^{\mathfrak{C}} = 0 \qquad (18.33)$$

Since we don't want to constrain any of the components of Φ, we must have the first term vanish, which means

$$\left(\mathbf{A}_\mu'\right)^{\mathfrak{A}}{}_{\mathfrak{B}} = U^{\mathfrak{A}}{}_{\mathfrak{C}} \left(\mathbf{A}_\mu\right)^{\mathfrak{C}}{}_{\mathfrak{D}} \left(U^{-1}\right)^{\mathfrak{D}}{}_{\mathfrak{B}} + \frac{i}{g} \left(\partial_\mu U^{\mathfrak{A}}{}_{\mathfrak{C}}\right) \left(U^{-1}\right)^{\mathfrak{C}}{}_{\mathfrak{B}} \qquad (18.34)$$

or in matrix notation

$$\mathbf{A}_\mu' = U\mathbf{A}_\mu U^{-1} + \frac{i}{g}\left(\partial_\mu U\right)\left(U^{-1}\right) \qquad (18.35)$$

Notice that if we have an Abelian group, then $U = \exp\left(i\theta\left(x\right)\right)$ and this reduces to eq. (12.14) with $g = e$.

We now know how the vector-potential transforms, but we don't know what kind of matrix it is. This is actually not too hard to find. Since Φ transforms under some irrep R, by definition this means that there exist generators $\mathbf{T}_\mathsf{R}^a$ such that

$$U\left(x\right) = \exp\left(i\theta_a\left(x\right)\mathbf{T}_\mathsf{R}^a\right) \quad \text{and} \quad \left[\mathbf{T}_\mathsf{R}^a, \mathbf{T}_\mathsf{R}^b\right] = if^{ab}{}_c\mathbf{T}_\mathsf{R}^c \qquad (18.36)$$

where the $f^{ab}{}_c$ are the structure constants of the Lie algebra. Suppose we choose $\theta_a\left(x\right) = \theta_a$ to all be constant. Then equation (18.35) becomes

$$\mathbf{A}_\mu' = \mathbf{A}_\mu + i\theta_a\left[\mathbf{T}_\mathsf{R}^a, \mathbf{A}_\mu\right] + \cdots \qquad (18.37)$$

to leading order in θ_a. Hence we need the matrix-vector potential to also be in the representation R; otherwise we won't be able to make sense of the commutator in (18.37). The most general thing we can write down is a sum over all of the generators, $\mathbf{A}_\mu = A_\mu^a \mathbf{T}_\mathsf{R}^a$, and so we have

$$\mathbf{D}_\mu = \partial_\mu\mathbf{I} + igA_\mu^a\mathbf{T}_\mathsf{R}^a \quad \text{or} \quad \left(\mathbf{D}_\mu\right)^{\mathfrak{A}}{}_{\mathfrak{B}} = \partial_\mu\delta^{\mathfrak{A}}{}_{\mathfrak{B}} + igA_\mu^a\left(\mathbf{T}_\mathsf{R}^a\right)^{\mathfrak{A}}{}_{\mathfrak{B}} \qquad (18.38)$$

for a given representation. In other words, the choice of matrix $\mathbf{A}_\mu$ depends on which representation Φ transforms under. For example, if the symmetry group is $\mathbf{SU}(3)$ and we take Φ to be the $\mathbf{3}$ of $\mathbf{SU}(3)$ then $\left(\mathbf{D}_\mu\right)^{\mathfrak{A}}{}_{\mathfrak{B}} = \partial_\mu\delta^{\mathfrak{A}}{}_{\mathfrak{B}} + igA_\mu^a\left(\lambda^a\right)^{\mathfrak{A}}{}_{\mathfrak{B}}$ where the λ^a are the Gell-Mann matrices (18.15); if the symmetry group is $\mathbf{SU}(2)$ and we take Φ to be the doublet of $\mathbf{SU}(2)$ then $\left(\mathbf{D}_\mu\right)^{\mathfrak{A}}{}_{\mathfrak{B}} = \partial_\mu\delta^{\mathfrak{A}}{}_{\mathfrak{B}} + igA_\mu^a\left(\frac{\sigma^a}{2}\right)^{\mathfrak{A}}{}_{\mathfrak{B}}$, where the σ^a are the Pauli matrices (5.31).

What is the field strength of the vector potential? It's clear that it can't be the same as in the Abelian case (i.e., it can't be simply $\partial_\mu\mathbf{A}_\nu - \partial_\nu\mathbf{A}_\mu$), since the U-derivatives in (18.35) have a non-trivial matrix structure. We could try to guess at the form of the field strength and then make sure that it had the correct transformation properties, but we aren't even sure of what those are yet. An easier way is to use the approach of question 6 in Chapter 12 – we will compute the commutator of the covariant derivatives acting on Φ and define what comes out to be the field strength. In other words, we define

$$\mathbf{F}_{\mu\nu}\Phi = ig\left[\mathbf{D}_\mu, \mathbf{D}_\nu\right]\Phi \qquad (18.39)$$

for all possible representations R. It's not too hard to show that this gives

$$\mathbf{F}_{\mu\nu} = \partial_\mu \mathbf{A}_\nu - \partial_\nu \mathbf{A}_\mu + ig\left[\mathbf{A}_\mu, \mathbf{A}_\nu\right] \tag{18.40}$$

or alternatively

$$F_{\mu\nu}^a = \partial_\mu A_\nu^a - \partial_\nu A_\mu^a - gf_{bc}^a A_\mu^b A_\nu^c \tag{18.41}$$

where I have used the property of the structure constants that $f^{bc}{}_a = f^{bca} = -f^{bac} = f^{abc} = f_{bc}^a$.

You can also use (18.39) to show that

$$\mathbf{F}'_{\mu\nu} = U\mathbf{F}_{\mu\nu}U^{-1} \tag{18.42}$$

under a gauge transformation. Unlike the Abelian case, where the field strength was invariant, here we see that the field strength is covariant – in other words the field strength itself transforms under a representation of the gauge group. In group-theory language, matrices that tranform as $M' = UMU^{-1}$ are in the adjoint representation, where the generators are the structure constants of the algebra $((\mathbf{T}^a)^b{}_c = if^{ab}{}_c)$ as discussed in question 10 of Chapter 3. So the field strength transforms under the adjoint representation.

Finally, we need differential equations for the vector potentials A_μ^a. To ensure that the Abelian case reduces to Maxwell's theory, they must be second-order differential equations. They should also be covariant under a gauge transformation, which means that we must use covariant derivatives in the equation for $\mathbf{F}_{\mu\nu}$. The only equation that satisifies both of these criteria is

$$(D^\mu F_{\mu\nu})^a = J_\nu^a \tag{18.43}$$

where J_ν^a is called the Yang-Mills current, a generalization of the electric current to Yang-Mills theory. Expanding out the left-hand side gives

$$\partial^\mu F_{\mu\nu}^a - g_s A^{c\mu} F_{\mu\nu}^b f_{cb}^a = J_\nu^a \tag{18.44}$$

and it is possible to show from this that

$$(D^\mu J_\nu)^a = 0 \tag{18.45}$$

or in other words that the Yang-Mills current is covariantly conserved.

18.6 Questions

1. Draw and label one lowest-order diagram for each of the processes

    (a) $\pi^- + p \longrightarrow \Lambda^0 + K^0$ (b) $\pi^+ + p \longrightarrow \Sigma^+ + K^+$

    (c) $\pi^+ + n \longrightarrow \pi^0 + p$ (d) $p + \bar{p} \longrightarrow K^- + K^+$

    (e) $\quad p + p \longrightarrow \Lambda^0 + K^+ + p$

Use quark-lines in your diagrams to describe all hadronic particles.

2. Find the value of $R \equiv \frac{\sigma(e^+e^- \longrightarrow \text{hadrons})}{\sigma(e^+e^- \longrightarrow \mu^+\mu^-)}$ at CM energies of 2.5 GeV and 5 GeV.

3. For the QCD λ-matrices show that $\text{Tr}(\lambda^a \lambda^b) = 2\delta^{ab}$

4. Show that 2 mesons in the color singlet state $\frac{1}{\sqrt{3}}\left|R\overline{R} + B\overline{B} + G\overline{G}\right\rangle$ experience a potential $V = -\frac{4}{3}\frac{1}{r}$.

5. What is the color singlet combination $|\mathcal{S}\rangle$ of two gluons?

6. (a) Draw and label the lowest order-diagrams for gluon-gluon scattering.

   (b) Draw and label the lowest order-diagrams for quark-gluon scattering.

7. Verify the formula

$$\lambda^a_{ij}\lambda^a_{kl} = 2\delta_{il}\delta_{jk} - \frac{2}{3}\delta_{ij}\delta_{kl}$$

   for the λ-matrices.

8. (a) Given
$$\mathbf{F}_{\mu\nu}\Phi = ig\left[\mathbf{D}_\mu, \mathbf{D}_\nu\right]\Phi$$
   show that
$$\mathbf{F}_{\mu\nu} = \partial_\mu \mathbf{A}_\nu - \partial_\nu \mathbf{A}_\mu + ig\left[\mathbf{A}_\mu, \mathbf{A}_\nu\right]$$
   or alternatively
$$F^a_{\mu\nu} = \partial_\mu A^a_\nu - \partial_\nu A^a_\mu - gf^a_{bc}A^b_\mu A^c_\nu$$

   (b) Show that for a given representation R
$$((D_\mu D_\nu - D_\nu D_\mu)\,\Phi) = igF^a_{\mu\nu}\mathbf{T}^a_R\Phi$$

9. (a) Show that
$$\mathbf{F}'_{\mu\nu} = U\mathbf{F}_{\mu\nu}U^{-1}$$
   under a gauge transformation.

   (b) Show that if
$$(D^\mu F_{\mu\nu})^a = J^a_\nu$$
   then
$$(D^\mu J_\nu)^a = 0$$

10. Show that any Hermitian 3x3 matrix can be written as a linear combination of the identity matrix and the eight λ^a matrices.

19

From Beta Decay to Weak Interactions

The first observations of weak interactions were made by accident in 1896, when Henri Bequerel was investigating phosphorescence [185], the phenomenon in which certain materials glow in the dark after exposure to light. He thought that phosphorescence might be connected with the glow produced in cathode ray tubes by X-rays, and so would place various phosphorescent minerals on a photographic plate to see what would happen. There was no effect until he tried to expose a compound of uranium salts to the sun, and photographed the emission spectrum. He stored the compound and the photographic plates inside a desk drawer. When he returned to develop the plates he found that they were overexposed. The plate was deeply blackened, and it soon became clear that the uranium compound must have emitted some new kind of penetrating radiation quite different from phosphorescence, since the plate blackened even when the mineral was in the dark. This was the first observation of natural radioactivity, a phenomenon governed by the weak interactions [186].

Further exploration showed that radioactivity was a complicated phenomenon. A magnetic field was found to split radioactive emissions into three types of beams: alpha, beta and gamma. The alpha rays were seen to carry a positive charge, beta rays a negative charge, and gamma rays were neutral. Furthermore the alpha particles were much more massive than beta particles, as determined from their deflection angle. The alpha particles were eventually found to be helium nuclei, the gamma particles photons, and the beta particles electrons.

The beta particles presented a puzzle lasting 30 years. It was clear that they were electrons that were emitted not with discrete energies, but rather as a continuum, quite unlike the emissions due to other nuclear transitions. Why was that? Furthermore, it was clear from experiments in the 1920s that there were no electrons inside nuclei. So where do the beta rays come from?

Pauli suggested that there was a very light, uncharged, and penetrating particle that he called the neutron, two years before Chadwick's discovery [137]. While it is common today to introduce and discover new particles, this was a radical step in 1930, when the electron and proton were the only known elementary particles, and Pauli wrote in a letter in December 1930 that "From now on, every solution to the issue must be discussed. Thus, dear radioactive people, look and judge." Most of them judged negatively, but Fermi was one of the few people to take Pauli's idea seriously, and renamed the particle the neutrino, or "little neutral one." Fermi proposed that a neutrino is emitted along with an electron in every beta decay. Both the electron and the neutrino share the decay energy, with the electron sometimes having a little energy and sometimes a lot, leading to the continuous spectrum.

Controlled experiments on the very slow decays of certain neutron-rich (or proton-rich) nuclei into other, more stable nuclei were able to isolate beta emission. For example in the process

$$Cs^{136} \longrightarrow Ba^{136} + e^- + \bar{\nu}_e \tag{19.1}$$

there is the emission of a beta particle (the electron), along with the anti-neutrino, which we now know is not the same as the neutrino. The initial state (Cs^{136} here) is called the parent nucleus and the final state (in this case Ba^{136}) is called the daughter nucleus. Another example is

$$C^{10} \longrightarrow B^{10} + e^+ + \nu_e \tag{19.2}$$

The atomic weight of the daughter nucleus in both cases is almost exactly the same as the parent.

Radioactive decay will change one nucleus to another whenever the binding energy of the product nucleus is larger than that of the initial decaying nucleus. The difference in binding energy determines which decays are energetically possible and which are not. The excess binding energy appears as kinetic and rest mass energy of the decay products. The Chart of the Nuclides (shown in fig. 19.1) plots the proton number, Z, of the nuclei against their neutron number, N. This chart plots all known nuclei – stable and radioactive – naturally occurring and artificially made – along with their decay properties. You can see that nuclei with an excess of protons or neutrons as compared to stable nuclei will decay toward the stable nuclei either by changing protons into neutrons (or neutrons into protons) or by emitting neutrons or protons (either singly or in combination)*.

After the discovery of the neutron, it became clear that all radioactive processes could be described by the reactions

$$n \longrightarrow p + e^- + \bar{\nu}_e \quad \text{or} \quad p \longrightarrow n + e^+ + \nu_e \tag{19.3}$$

which are referred to as β-decay processes. In terms of quark constituents, we understand these reactions to be

$$d \longrightarrow u + e^- + \bar{\nu}_e \quad \text{or} \quad u \longrightarrow d + e^+ + \nu_e \tag{19.4}$$

but we don't need to look at this level of substructure to understand the basic physics of these decays.

---

*Nuclei are also unstable if they are excited, that is, not in their lowest energy states. In this case the nucleus can decay by getting rid of its excess energy without changing Z or N by emitting a gamma ray.

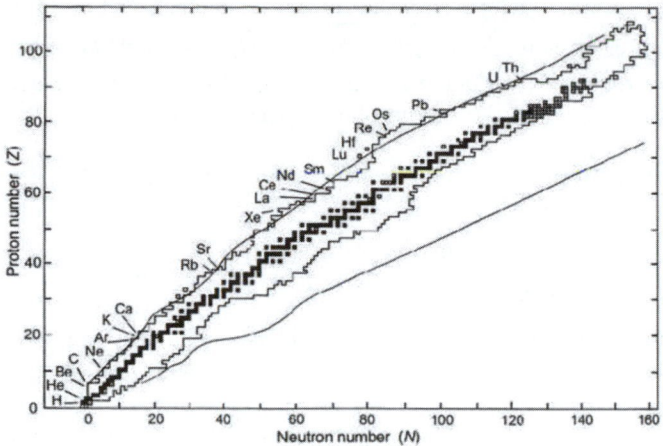

FIGURE 19.1

Chart of the Nuclides, in which horizontal rows represent the same element (constant Z) with a variable number of neutrons (N), known as isotopes. The black squares represent stable nuclei, the open squares unstable nuclei, and the smooth curves are theoretical nuclide stability limits. Image from *Radiogenic Isotope Geology* by A. P. Dickin, Cambridge University Press (1997) [187]; used with permission.

19.1   Fermi's Theory of Beta-Decay

How can we understand them? Fermi proposed the first theory of β-decay in 1934 [188]. The decay probability of a nucleus of mass M_N is given by the same decay formula that we've been using all along

$$d\Gamma = \frac{\mathcal{S}c^2}{2\hbar M_N} \left[\prod_{i=1}^{3} \frac{c\left(d^3 p_i'\right)}{2E_i'\left(2\pi\right)^3} \right] \left| \langle p_1' p_2' p_3' \left| \mathrm{M} \right| p \rangle \right|^2 \left(2\pi\right)^4 \delta^{(4)} \left(\sum_{i=1}^{3} p_i' - p \right) \quad (19.5)$$

except that we now have three particles in the final state: the stable final nucleus, the electron and the antineutrino. So we have

$$d\Gamma = \frac{c^5}{16\left(2\pi\right)^5 \hbar M_N E_N'} \left[\frac{\left(d^3 p_e'\right)\left(d^3 p_{\bar{\nu}}'\right)}{E_e' \, E_{\bar{\nu}}'} \right] \left| \mathcal{M} \right|^2 \delta\left(E_0 - E_e' - E_{\bar{\nu}}'\right) \quad (19.6)$$

where I have integrated out $d^3 p_N' \delta\left(\vec{p}_N' + \vec{p}_e' + \vec{p}_{\bar{\nu}}'\right)$, which just ensures that momentum is conserved between the electron, the anti-neutrino and the final state nucleus N'. The quantity $E_0 = E_N - E_N'$ is the energy difference between the parent nucleus and the daughter nucleus. The kinetic energy of the daughter nucleus is $\sqrt{|\vec{p}_N'|^2 + M_{N'}^2} - M_{N'} \simeq 10^{-3} M_{N'}$ and so we neglect it – we regard E_0 as being shared entirely between the electron and the antineutrino, and $E_N' \simeq M_N'$. The phase space measures are

$$d^3 p_{\bar{\nu}}' = 4\pi p_{\bar{\nu}}^2 dp_{\bar{\nu}} = 4\pi p_{\bar{\nu}}' E_{\bar{\nu}}' dE_{\bar{\nu}}' \qquad\qquad d^3 p_e' = 4\pi p_e'^2 dp_e' \qquad (19.7)$$

and so

$$d\Gamma = \frac{\pi^2 c^5 dp_e'}{(2\pi)^5 \, \hbar M_N M_N'} \frac{p_e'^2}{E_e'} \sqrt{(E_0 - E_e')^2 - m_{\bar{\nu}}^2} \, |\mathcal{M}|^2 \qquad (19.8)$$

once we integrate over the δ-function. I have included a non-zero mass for the anti-neutrino in the above expression, since we will soon see there is some interesting physics here.

The matrix element $\mathcal{M}$ is determined by what we postulate our theory of β-decay to be. Fermi's original idea was to take the square of the matrix element to be proportional to the energies of the emitted particles, i.e.

$$|\mathcal{M}|^2 = E_e' E_{\bar{\nu}}' F(Z, E_e') |\mathcal{M}_0|^2 \qquad (19.9)$$

where an additional factor $F(Z, E)$ – called the *Coulomb factor* – has been included. It takes account of the energy lost (by electrons) or gained (by positrons) as they escape from the Coulomb field of the nucleus. Non-relativistically it is [189]

$$F(Z, E_e) = \frac{2\pi\eta}{1 - \exp\left[-2\pi\eta\right]} \quad \text{where} \quad \eta = \pm \frac{Ze^2}{4\pi\epsilon_0 \hbar v} \qquad (19.10)$$

for β particles of speed v emitted from a nucleus of charge Ze. Relativistically it is complicated to calculate, but can be done exactly.

Hence we find

$$d\Gamma = \frac{c^5 |\mathcal{M}_0|^2}{(32\pi^3) \, \hbar M_N M_N'} p_e^2 \left(E_0 - E_e'\right) F(Z, E_e') \sqrt{(E_0 - E_e')^2 - m_{\bar{\nu}}^2} dp_e' \quad (19.11)$$

Now the decay rate is the probability that an electron with energy $E = E_e'$ will be emitted within the momentum range $p = p_e'$ and $p = p_e' + dp_e'$. The probability is proportional to the number of electrons emitted within this momentum range. Dropping the primes and the subscript "e" we have

$$N(p)dp \propto |\mathcal{M}_0|^2 \, p^2 \left(E_0 - E\right)^2 F(Z, E) \sqrt{1 - \frac{m_{\bar{\nu}}^2}{(E_0 - E)^2}} dp \qquad (19.12)$$

So by plotting $\sqrt{\frac{N(p)}{p^2 F(Z,E)}}$ vs. the emitted electron energy E, we can test this simple picture of β-decay. Such plots are called Kurie plots, and are found to be linear for pretty much all radioactive nuclei, indicating that the remaining matrix element factor $|\mathcal{M}_0|^2$ is approximately constant and that Fermi's idea was on the right track. In fact, experiment indicates two possible values for this quantity:

$$|\mathcal{M}_0|^2 = \left(M_p c^2\right)^2 \times \begin{cases} G_F^2 & \text{if } \Delta J = 0 \text{ (Fermi transition)} \\ 3G_F^2 & \text{if } \Delta J = 1 \text{ (Gamow-Teller transition)} \end{cases}$$
(19.13)

where M_p is the mass of the proton and G_F is a dimensional constant now known as Fermi's constant. It can be found by integrating the Kurie plot from $p = 0$ to $p = p_{\max}$, since this gives the total decay probability. The inverse of this probability is the half-life of the unstable nucleus. A standard in nuclear physics is the decay

$$O^{14} \longrightarrow N^{14} + e^+ + \nu_e \tag{19.14}$$

which yields

$$G_F = 1.166 \times 10^{-5} \, (\text{GeV})^{-2}$$

from its Kurie plot [190]. Today the value of the Fermi constant is set by accurate measurements of the lifetime of the muon [191],

$$G_F = 1.16637(1) \times 10^{-5} \, (\text{GeV})^{-2} \tag{19.15}$$

a process we will consider in Chapter 20.

Note that the details of a Kurie plot depend on whether or not the (anti)neutrino has mass. The plots in figure 19.2 show a comparison between the straight-line $m_\nu = 0$ case (dotted line), and what a Kurie plot looks like if the neutrino has a non-zero mass. Unfortunately this has not proven to be a very good way to determine if neutrinos have mass. β-decay spectrometers have finite resolution that is poor near the end point energy $E = E_0$. At present the best limits are given in table 19.1. The upper bound for the mass

TABLE 19.1
Neutrino Mass Limits

Neutrino	Upper bound on Mass (in eV)
ν_e	2
ν_μ	1.9×10^5
ν_τ	1.82×10^7

of ν_e comes from observations of tritium decay. These experiments actually estimate the quantity $m_{\nu_e}^2$ and find negative values, albeit consistent with zero within limits of error [192]. This illustrates how difficult it is to extract the mass from spectrometer experiments via Kurie plots. These experiments report $m_{\nu_e} < 2.3$ eV, a slightly higher upper bound than the Particle Data

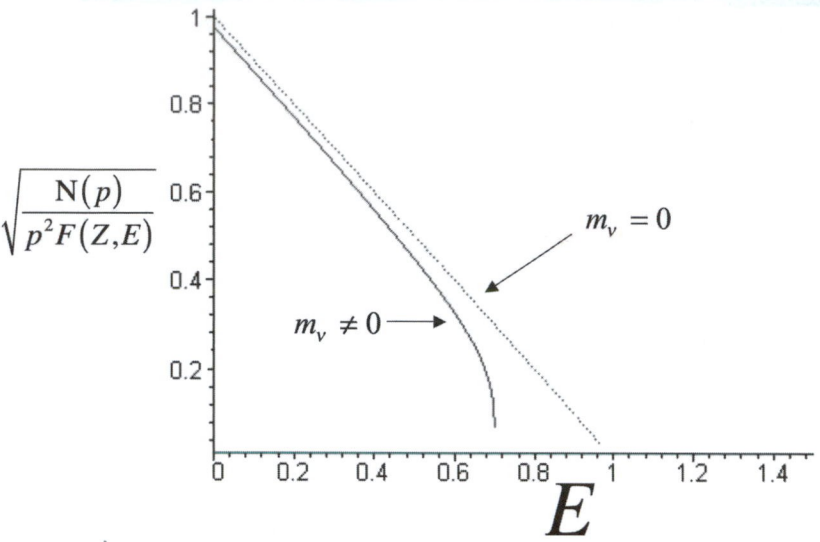

FIGURE 19.2
A typical Kurie plot, showing (in exaggerated form) the difference between zero and nonzero neutrino masses.

Group average in table 19.1. The best bounds on the m_{ν_μ} mass is also lower, $m_{\nu_\mu} < 1.7 \times 10^5$ eV [193] but the best upper bound for the m_{ν_τ} is from the ALEPH group [194] and is in table 19.1.

It is possible to get bounds on neutrino mass from cosmological observations. Neutrinos with masses of only a few tenths of an electron volt (eV) can significantly influence the formation of large-scale structures (such as galaxy clusters) in our universe. By observing galaxy distributions in the context of a the best-fit cosmological model, an upper bound of $\sum m_{\nu_i} < 0.62$ on the sum of all neutrino masses has been placed [195] using data from the 2 Degree Field Galaxy Redshift Survey (2dFGRS), which is the largest existing redshift survey [196].

We'll return to the topic of neutrino masses in more detail in chapter 25, once we've had a chance to study the structure of the weak interactions in more depth.

19.2 Neutrino Properties

Now let's use the Fermi theory to compute a cross-section for the process

$$\bar{\nu}_e + p \longrightarrow n + e^+ \tag{19.16}$$

We don't yet have any Feynman rules for this process. Fortunately we don't need them because in the formula for the differential cross section in the lab frame (proton at rest) is

$$\frac{d\sigma}{d\Omega} = \left(\frac{\hbar}{8\pi}\right)^2 \frac{S\,|\mathcal{M}|^2\,|\vec{p}_1'|^2}{M_2\,|\vec{p}_1|\,[|\vec{p}_1'|\,(E_1 + M_2c^2) - |\vec{p}_1|\,E_1'\cos\theta]} \tag{19.17}$$

and so all we need is the matrix element, which from Fermi's theory is $|\mathcal{M}|^2 = EE'\,|\mathcal{M}_0|^2 = EE'\,(M_pc^2)^2\,G_F^2$, setting the Coulomb factor to unity. Taking the antineutrino to be massless, so that $|\vec{p}_1| = E$, and neglecting the mass of the electron so that $|\vec{p}_1'| = E'$, we get

$$\frac{d\sigma}{d\Omega} = \left(\frac{\hbar}{8\pi}\right)^2 \frac{M_pc^4\,(E')^2\,G_F^2}{[(E + M_pc^2) - E\cos\theta]} \tag{19.18}$$

where $M_2 = M_p$. For incident antineutrinos in the MeV range, $M_pc^2 \gg E, E'$ and so

$$\frac{d\sigma}{d\Omega} \simeq \left(\frac{\hbar}{8\pi}\right)^2 \frac{M_pc^4\,(E')^2\,G_F^2}{M_pc^2}\left(1 + \mathcal{O}\left(\frac{E}{M_pc^2}\right)\right) \tag{19.19}$$

which implies that

$$\sigma \simeq 4\pi\left(\frac{\hbar c G_F E'}{8\pi}\right)^2 \simeq 10^{-45}\ \mathrm{cm}^2 \tag{19.20}$$

This number will change by a factor of order unity if the target is a nucleus, so we can regard $\sigma = 10^{-45}\,\mathrm{cm}^2$ as a typical cross-section for neutrino scattering off of any substance.

This cross-section is almost unimaginably tiny. If you think of a proton as a hard sphere, then you might expect it to present an area of πr_p^2 to an incoming particle, where $r_p \simeq 10^{-13}$ cm is the approximate radius of the proton. This gives $\pi r_p^2 \simeq 10^{-25}$ cm^2, about 20 orders of magnitude larger! The tiny cross-section in eq. (19.20) implies, for example, that the mean free path of the antineutrino in water is

$$L_{\bar{\nu}} = (\text{length/nucleus}) = (\#\text{ of nuclei/vol} \times \sigma)^{-1} \simeq \left((2 \times 10^8)^3 \times \sigma\right)^{-1}$$

$$= 10^{20}\mathrm{cm}\ \text{for water} \tag{19.21}$$

or 100 light years! Put another way, the probability that a single antineutrino will interact with a target of water about a meter thick is 10^{-18}.

This small cross section is the reason why neutrinos are so harmless. To put this into perspective, a human body typically contains about 20 milligrams of K^{40}, which is β-radioactive. As a consequence, we emit about 340 million neutrinos per day – but we don't even notice it because they almost never interact with any of our cells! The flux of neutrinos from the sun, the earth, and other cosmological sources likewise interacts so infrequently with our bodies and all other biomatter that we can proceed through everyday life as though they don't exist. Too bad the same can't be said for ultraviolet light!

However, the small cross section in eq. (19.20) also makes neutrinos difficult to detect, because they almost never interact with anything. Remarkably, Reines and Cowan [197] observed such interactions in 1959! Dubbing their efforts "Project Poltergeist" (because the neutrino is so elusive), they used a 1000 MW nuclear reactor as a source, which has a flux of about 10^{13} antineutrinos $/cm^2/sec$, as illustrated in fig. 19.3. The target was $CdCl_2$ and water. An antineutrino, upon interacting with a proton, will produce a neutron and a positron. The positron rapidly comes to rest via ionization loss and forms positronium, which emits γ-rays. The water thermalizes the neutrons before the cadmium captures them, a process taking several microseconds. Once captured by cadmium, the neutrons emit another γ-ray. Hence the characteristic signal for an antineutrino is two γ-rays emitted microseconds apart.

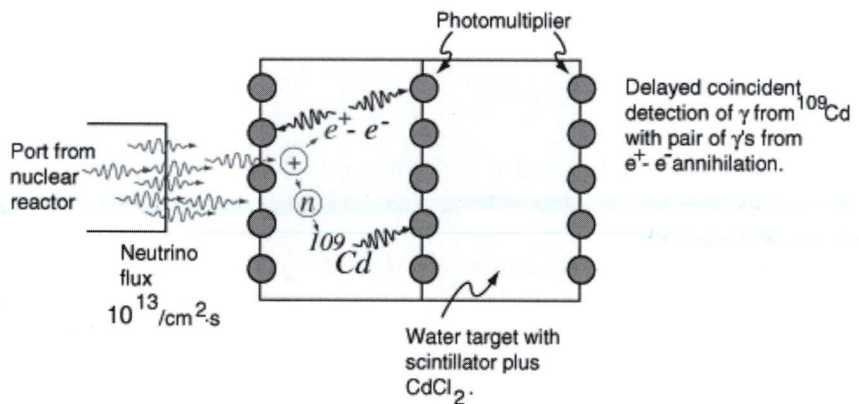

FIGURE 19.3

Schematic Diagram of the Reines-Cowan experiment. Image courtesy of the Hyperphysics Web site, copyright C.R. Nave, Georgia State University (2005).

Nowadays, a modern neutrino experiment like NOMAD [198] detects about one neutrino every 10 seconds, depositing on average 27 GeV in the detector. For the entire duration of the experiment, neutrinos will have deposited a little more than 0.03 Joules – about $1/10$ the energy in a sneeze!

Neutrinos have other unusual properties that were also discovered in the 1950s and 60s. In 1956 Lee and Yang suggested (before neutrinos had been observed in the Reines/Cowan experiment) that the weak interactions did not conserve parity [199]. They offered this as an explanation for why the decays

$$K^+ \longrightarrow \pi^+ \pi^0 \quad \text{and} \quad K^+ \longrightarrow \pi^+ \pi^0 \pi^0 \tag{19.22}$$

were both observed – the pion parity is negative and so the final states have opposite parities. If parity were conserved then one or the other of these states should not be seen.

In 1957 Wu tested this idea [39] by looking at the decay of Co^{60} nuclei at very low temperature (0.01 °K). Co^{60} decays via the reaction

$$Co^{60} \longrightarrow Ni^{60} + e^- + \bar{\nu} \tag{19.23}$$

and one can check parity violation by measuring the expectation value of the operator $\hat{J} \cdot \hat{p}$, where $\vec{J} = J\hat{J}$ is the spin of the nucleus and $\hat{p}$ is the direction of the electron momentum. The quantity is a pseudoscalar $\left(\mathbb{P} \left(\hat{J} \cdot \hat{p} \right) = -\hat{J} \cdot \hat{p} \right)$ and so if parity is conserved, the electron emission rate (which must be proportional to $\hat{J} \cdot \hat{p}$) should be the same in either direction regardless of the spin of the nucleus. This seems straightforward enough to measure, except that at room temperature the spins of the nuclei will all be randomly oriented. Consequently the experiment must be performed at ultracold temperatures in order to minimize this effect. A magnetic field $\vec{B}$ will split the energy levels of the nucleus via the Zeeman effect, so that the energy of a state with magnetic quantum number m is $E(m) = E_0 - g\mu_N m \left| \vec{B} \right|$. At a given temperature T the number of nuclei with magnetic quantum number m' as compared to magnetic quantum number m will be

$$\frac{N(m')}{N(m)} = \exp\left(-\frac{E(m') - E(m)}{k_B T} \right) = \exp\left((m' - m) \frac{g\mu_N \left| \vec{B} \right|}{k_B T} \right) \tag{19.24}$$

making the most positive magnetic quantum number (e.g., $m = +1$ for a $J = 1$ state) the most likely to be populated. At room temperature the population differences are negligible, but at low temperature $k_B T << g\mu_N \left| \vec{B} \right|$, only the most positive m state will be populated, fully polarizing the nucleus in the direction of the magnetic field, as shown in fig. 19.4. The mirror world is obtained by reversing the direction of the magnetic field.

The spin-parity of the cobalt nucleus is $J^{\mathbb{P}} = 5^+$, and the actual decay observed is

$$\left| Co^{60}; j = 5, m = 5 \right\rangle \longrightarrow \left| Ni^{**60}; j = 4, m = 4 \right\rangle + e^- + \bar{\nu} \tag{19.25}$$

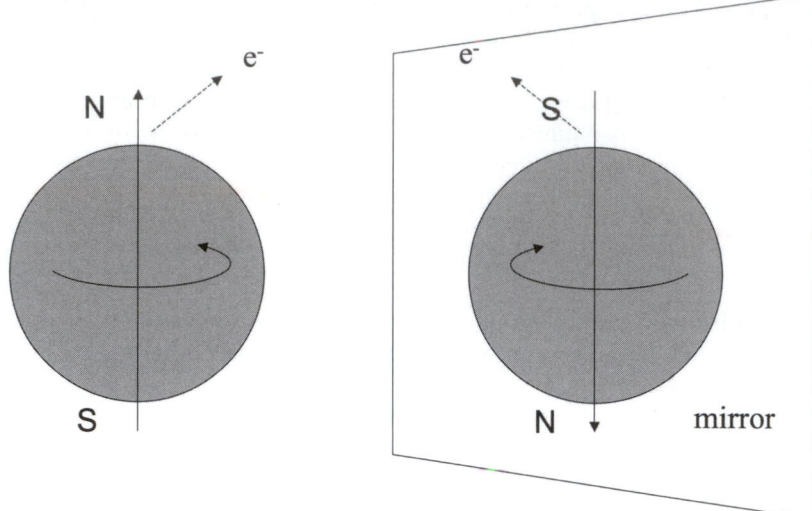

FIGURE 19.4

β-decay of Co^{60} and its mirror image.

where the excited nickel nucleus decays to its ground state via the emission of two gamma rays

$$\left|\text{Ni}^{**60}; j = 4, m = 4\right\rangle \longrightarrow \left|\text{Ni}^{*60}; j = 2, m = 2\right\rangle + \gamma \,(1.173 \text{ MeV})$$
$$\longrightarrow \left|\text{Ni}^{60}; j = 0, m = 0\right\rangle + \gamma \,(1.332 \text{ MeV}) \,(19.26)$$

where in each electromagnetic decay there is some relative orbital angular momentum between the gamma ray and the nucleus. This means that for each decay the emission probability is not isotropic but will depend on the angle of the γ emitted relative to the magnetic field. However, it won't depend on a reversal of the magnetic field since electromagnetic interactions conserve parity. Hence the degree of polarization of the cobalt nucleus can be monitored by measuring the γ anisotropy.

The experimental setup is shown in fig. 19.5. Wu used two NaI crystals placed at 90° relative to the long direction of the container and the other was placed close to the top (at nearly 0°) to count the emitted γ's . The emitted electrons are counted by a small anthracene crystal placed just 2 cm above the cobalt sample. The magnetic field is switched on to polarize the nuclei, and then switched off after a few seconds so that counting of electrons and photons can take place. As expected, the same γ anisotropy was observed regardless of the field direction, indicating the sample was polarized for a few minutes. During this time electrons were observed to be preferentially emitted when the magnetic field pointed up (i.e. from the north pole of the nucleus) – in

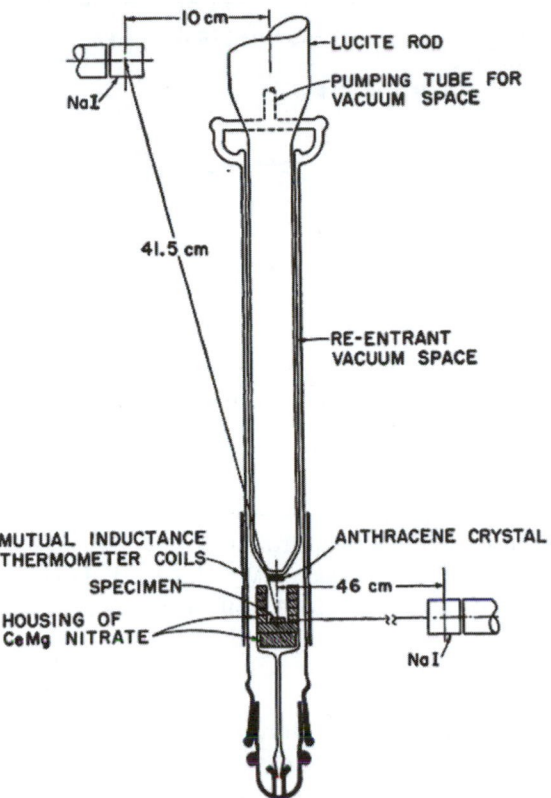

FIGURE 19.5

Setup diagram of the apparatus used measure beta decay in Co^{60}. The specimen, a cerium magnesium nitrate crystal containing a thin surface layer of radioactive cobalt-60, was supported in a cerium magnesium nitrate housing within an evacuated glass vessel (lower half of photograph). An anthracene crystal about 2 cm above the cobalt-60 source served as a scintillation counter for beta-ray detection. A plastic rod made of lucite (upper half of photograph) transmitted flashes from the counter to a photomultiplier at the top (not shown). Reprinted figure with permission from C.S. Wu, E. Ambler, R.W. Hayward, D.D. Hoppes and R.P. Hudson, *Phys. Rev.* **105**, 1413 (1957). [39]. Copyright (1957) by the American Physical Society.

a mirror world they would be observed to be preferentially emitted from the south pole (i.e. when the field pointed down). This must mean that parity is violated – the only way that parity could be conserved is if the electrons were emitted in equal numbers at both poles. The data from Wu's experiment appears in fig. 19.6.

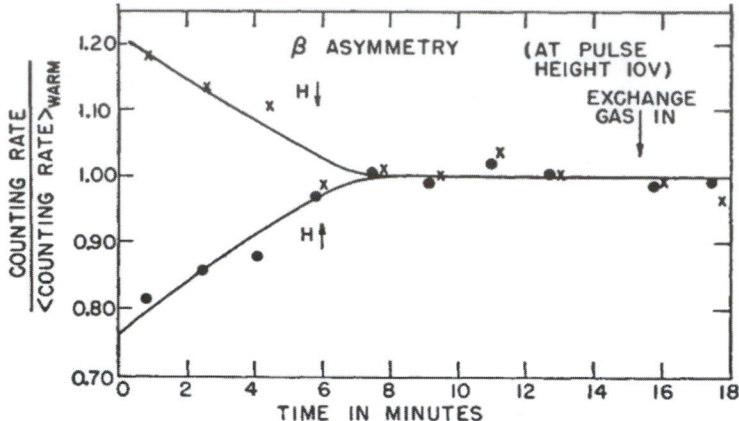

FIGURE 19.6

Wu's data showing the preferential counting rate of electrons emitted from Co^{60} nuclei as the sample heats up from 0.01 °K. Reprinted figure with permission from C.S. Wu, E. Ambler, R.W. Hayward, D.D. Hoppes and R.P. Hudson, *Phys. Rev.* **105**, 1413 (1957). [39]. Copyright (1957) by the American Physical Society.

So weak interactions violate parity – but by how much? The parity-conserving electromagnetic force suggested that the neutrino is the particle that must have parity-violating properties. A simple way to quantify this is as follows. Consider a spinning particle moving in some direction. If its spin is aligned with the direction of motion, we say that it is right-handed; if it is anti-aligned, we say that it is left-handed. You can determine this by pointing your right thumb along the direction of motion. If your fingers curl in the same sense as the particle is rotating, the particle is right-handed; if in the opposite sense then it is left-handed (see fig. 19.7). Note that if a particle is massive, this is not an invariant concept – by going to a reference frame that moves in the same direction of the particle but faster, the sense of motion will reverse direction, but not the sense of spin. Hence chirality (or helicity or handedness) is an observer-dependent concept for a massive particle. However, if a particle is massless, you can never go to a frame of reference that moves faster than the particle (since it is moving at the speed of light) – and so chirality is an invariant concept for massless particles.

If parity is violated in weak interactions, then neutrinos of a given helicity will be preferentially emitted more than the other helicity. Goldhaber, Grodznis and Sunyar [200] figured out a clever way to measure neutrino helicity by transferring it to a photon via the reaction

$$\left|Eu^{152}; j = 0, m = 0\right\rangle + \left|e^{-}; j = \frac{1}{2}, m_e\right\rangle$$

FIGURE 19.7
Pictorial representation of helicity.

$$\longrightarrow \left|\text{Sm}^{*152}; j = 1, m_{\text{Sm}}\right\rangle + \left|\nu; j = \frac{1}{2}, m_\nu\right\rangle \qquad (19.27)$$

where the Eu^{152} absorbs the electron when it is in an S-wave, i.e. no orbital angular momentum, a process called K-capture. The excited Sm^{*152} nucleus decays via

$$\left|\text{Sm}^{*152}; j = 1, m_{\text{Sm}^*}\right\rangle \longrightarrow \left|\text{Sm}^{152}; j = 0, m = 0\right\rangle + \left|\gamma; j = 1, m_\gamma\right\rangle \quad (19.28)$$

Angular momentum conservation requires the z-components of the spins to balance

$$m_e = m_{\text{Sm}^*} + m_\nu = m_\gamma + m_\nu \qquad (19.29)$$

admitting only the following relationships

$$m_e \ m_\nu \ m_\gamma$$

$$+\tfrac{1}{2} \ -\tfrac{1}{2} \ \ 1 \qquad (19.30)$$

$$-\tfrac{1}{2} \ +\tfrac{1}{2} \ -1$$

since $m_\gamma = \pm 1$ because the photon is massless. This means that the z-component of the neutrino spin – its helicity – is anticorrelated with the z-component of the photon spin.

What one needs to do is measure the z-component of the spin of the emitted photons. This was done by placing the Eu^{152} sample inside an iron slab in a magnetic field $\vec{B}$, which orients the electrons in the iron in an opposing direction. If the photon spin is aligned with $\vec{B}$ then the electrons can absorb the photons, whereas if it is antialigned then they cannot do so. In this way the iron slab acts like a filter that admits only one value of m_γ. A ring made of Sm$_2$O$_3$ surrounds the detector so that the emitted photons must encounter it. The emitted photon – if it has the right energy – will excite the Sm^{152} in the ring, which upon decay back to its ground state will emit a photon that can be detected (at least sometimes) by a NaI detector. However, the photon hitting the ring will have the right energy only if the Sm^{*152} produced from the Eu^{152} is moving in the same direction as the photon, in turn ensuring the neutrino is moving in the opposite direction. Otherwise, the recoil energy of the Sm*^{152} will change the energy of the emitted photon, forbidding excitation of the Sm in the ring. In this manner the orientation of the magnetic field controls the emission direction of the neutrino. By measuring the asymmetry in the number (or intensity) of photons at the NaI detector $I_\pm$ emitted for opposite directions of $\vec{B}$, one measures $I_\pm = I\,(m_\gamma = \pm 1)$

$$R = \frac{I_+ - I_-}{I_+ + I_-} \tag{19.31}$$

and so in turn measures the handedness of the neutrinos emitted. Since the experiment ensures that whenever a photon is detected the neutrino is moving opposite to the direction of $\vec{B}$, a measurement of R tells us how much the neutrino spin is aligned with the direction of motion (right-handed) or antialigned (left-handed).

The experiment showed that all neutrinos were left-handed – there were no right-handed neutrinos. In other words the mirror-image of a neutrino does not exist!

So the most natural interpretation of Goldhaber's experiment is that the neutrino is fully left-handed and therefore massless. However, recent observations of neutrino oscillations at Super-Kamiokande [41] and the Sudbury Neutrino Observatory [42] indicate that the different species of neutrinos can transform into each other, a process we will look at in Chapter 25. The simplest explanation of this phenomenon is that neutrinos have mass, and hence both the right-handed and left-handed kinds exist. So – more precisely – if there are right-handed neutrinos, they do not interact with any known form of matter[†], other than their left-handed counterparts into which they can oscillate.

19.3   Kaon Oscillation

The realization that $\mathbb{P}$ is not a symmetry of nature was quite disturbing to many physicists. Why should nature be lopsided, choosing between left and right? Pauli, for example, was very reluctant to believe it – "I cannot believe God is a weak left hander" he said – but after Wu's experiments, he said "God is indeed a weak left hander – the laughter is rightly with the others" [201]. But acting with $\mathbb{P}$ on a left-handed neutrino yields the (evidently) non-existent right-handed neutrino. It appears that God is completely left-handed!

However, once the properties of neutrinos were determined (by about 1960), it appeared that $\mathbb{CP}$ could be a symmetry. Acting with $\mathbb{P}$ on a left-handed

---

[†] Of course here I mean non-gravitational interactions – since all neutrinos have energy they will all respond to the gravitational influences of other bodies and vice versa.

neutrino yields the (evidently) non-existent right-handed neutrino. But acting with $\mathbb{CP}$ yields the right-handed antineutrino, which does exist! Similarly, acting with $\mathbb{CP}$ on the right-handed antineutrino yields the left-handed neutrino.

So perhaps God is more subtly ambidextrous, since all other known particles also respect this symmetry. For a few years people thought this was the case, and that perhaps $\mathbb{CP}$ is a symmetry of nature. But in 1964 James Cronin, Val Fitch and collaborators observed violation of $\mathbb{CP}$ in Kaon decay. This is a rather peculiar situation, so we'll pause to look at it carefully.

It had been noted in 1955 by Gell-Mann and Pais [202] that a K^0 (strangeness $+1$) could change into its own antiparticle, the $\bar{K}^0$ (strangeness $+1$). These mesons had been determined to be pseudoscalars by previous experiments. Hence

$$\mathbb{P}\left|K^0\right\rangle = -\left|K^0\right\rangle \ \text{ and } \ \mathbb{P}\left|\bar{K}^0\right\rangle = -\left|\bar{K}^0\right\rangle \tag{19.32}$$

$$\mathbb{C}\left|K^0\right\rangle = \left|\bar{K}^0\right\rangle \ \ \text{ and } \ \mathbb{C}\left|\bar{K}^0\right\rangle = \left|K^0\right\rangle \tag{19.33}$$

since $\mathbb{C}$ interchanges particles and antiparticles. Hence

$$\mathbb{CP}\left|K^0\right\rangle = -\left|\bar{K}^0\right\rangle \ \text{ and } \ \mathbb{CP}\left|\bar{K}^0\right\rangle = -\left|K^0\right\rangle \tag{19.34}$$

so we can construct eigenstates of $\mathbb{CP}$:

$$\left|K_1\right\rangle \equiv \frac{1}{\sqrt{2}}\left(\left|K^0\right\rangle - \left|\bar{K}^0\right\rangle\right) \ \text{ and } \ \left|K_2\right\rangle \equiv \frac{1}{\sqrt{2}}\left(\left|K^0\right\rangle + \left|\bar{K}^0\right\rangle\right) \tag{19.35}$$

$$\Rightarrow \mathbb{CP}\left|K_1\right\rangle = \left|K_1\right\rangle \ \text{ and } \ \mathbb{CP}\left|K_2\right\rangle = -\left|K_2\right\rangle \tag{19.36}$$

This provides us with a system in which we can test $\mathbb{CP}$: if $\mathbb{CP}$ is conserved then $\left|K_1\right\rangle$ can decay only into eigenstates with $\mathbb{CP} = +1$, and $\left|K_2\right\rangle$ only into eigenstates with $\mathbb{CP} = -1$. Since Kaons typically decay into pions we therefore predict

$$K_1 \longrightarrow 2\pi, 4\pi, 6\pi, \ldots \quad \text{ and } \quad K_2 \longrightarrow 3\pi, 5\pi, 7\pi, \ldots \tag{19.37}$$

since

$$\mathbb{CP}\left|\pi^0\pi^0\right\rangle = \left(\mathbb{CP}\left|\pi^0\right\rangle\right)\left(\mathbb{CP}\left|\pi^0\right\rangle\right) = (-1)^2 = +1 \tag{19.38}$$

$$\mathbb{CP}\left|\pi^+\pi^-\right\rangle = \mathbb{C}\left|\pi^+\pi^-\right\rangle\mathbb{P}\left|\pi^+\pi^-\right\rangle = (-1)^\ell(-1)^\ell = +1 \tag{19.39}$$

We can only produce beams of K^0's (or $\bar{K}^0$'s) from other scattering experiments, which means that the beam is a mixture of K_1 and K_2 states

$$\left|K^0\right\rangle \equiv \frac{1}{\sqrt{2}}\left(\left|K_1\right\rangle + \left|K_2\right\rangle\right) \tag{19.40}$$

Since it is easier (more probable) to decay into fewer particles due to the available phase space, we expect that the K_1 part of the beam will decay

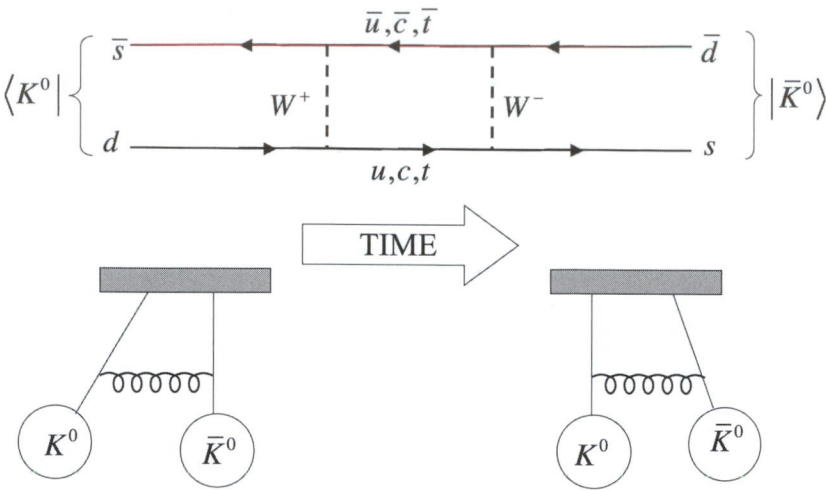

FIGURE 19.8

Feynman diagram for Kaon oscillation, analogous to coupled pendula as shown in the figure underneath.

away quickly into pairs of pions, and that after a few centimeters the beam should be pure K_2. Indeed, experiment indicates

$$\tau_{K_1} = (0.8926 \pm .0012) \times 10^{-10}\text{s} \qquad \tau_{K_2} = (5.17 \pm .04) \times 10^{-8}\text{s} \qquad (19.41)$$

Hence a beam of Kaons should decay only into 3π's after a few centimeters.

In 1964 Christensen, Fitch, Cronin and Turlay carried out this experiment [203]. By steering a proton beam extracted from the Brookhaven synchrotron onto a target, a beam of Kaons and other neutral and charged particles can be produced. The charged ones can be deflected away by magnets, leaving an undeflected neutral beam. After a few meters only the long-lived K_2 will be in the beam, which is unavoidably contaminated with neutrons and gamma rays. The K_2's enter a volume that ideally should be vacuum, but was actually a bag of helium gas that minimized interactions between beam particles that could simulate the decay. A two-arm spectrometer, adjusted to anticipate possible decays into two pions, was placed on the other side of the bag. Spark chambers placed before and after bending magnets allowed measurement of the momentum and charge of the decay products. By requiring that the angle between the sum of the momenta of the decay products and the beam axis is zero and that the mass $m(\pi\pi)$ of the two-particle system be compatible with the Kaon mass, spurious three-body decays are suppressed.

Out of 22, 700 Kaon decays, 45 were into 2 pions! The simplest interpretation of this experiment is that there is a small amount of $\mathbb{CP}$-violation in the

Kaon system. The long-lived Kaon is evidently not pure K_2, but is instead mostly K_2 with a small mixture of K_1:

$$|K_L\rangle \equiv \frac{1}{\sqrt{1+|\varepsilon|^2}} \left(\varepsilon |K_1\rangle + |K_2\rangle\right) \qquad (19.42)$$

where ε is a measure of nature's departure from perfect $\mathbb{CP}$ symmetry. Experimentally [1]

$$\varepsilon + \varepsilon' \equiv \eta_{+-} \equiv \frac{A\left(K_L \longrightarrow \pi^+\pi^-\right)}{A\left(K_S \longrightarrow \pi^+\pi^-\right)}$$

$$= (2.285 \pm .019) \times 10^{-3} \exp\left[i 44^o\right] = |\eta_{+-}| e^{i\phi^{+-}} \qquad (19.43)$$

$$\varepsilon - 2\varepsilon' \equiv \eta_{00} \equiv \frac{A\left(K_L \longrightarrow \pi^0\pi^0\right)}{A\left(K_S \longrightarrow \pi^0\pi^0\right)}$$

$$= (2.275 \pm .019) \times 10^{-3} \exp\left[i 44^o\right] = |\eta_{00}| e^{i\phi^{00}} \qquad (19.44)$$

where $|K_S\rangle \equiv \frac{1}{\sqrt{1+|\varepsilon|^2}} \left(|K_1\rangle - \varepsilon |K_2\rangle\right)$ is the short-lived orthogonal Kaon state. We see that, empirically, the magnitudes of η_{00} and η_{+-} are almost the same, and we can set $|\varepsilon| = |\eta_{+-}| = .0023$ – a small but non-zero effect! This value is about $45/22,700 = .00198$, the original Fitch-Cronin result.

Interestingly, $\mathbb{CP}$ violation cannot be explained in the old (pre-1976) Standard Model with only four quarks; you need at least six quarks in order to be able to explain this effect (you'll find out why in Chapter 21). Other models were proposed well before this to explain $\mathbb{CP}$ violation, most notably Wolfenstein's postulated "superweak" interaction: a new force in nature whose strength is 10^{-10} that of the weak interactions [204]. It predicted that $\varepsilon' = 0$, and for quite a long time was consistent with experiment. However, the most recent experiments [1] have confirmed that $\mathfrak{Re}\left(\frac{\varepsilon'}{\varepsilon}\right) = (1.65 \pm 0.26) \times 10^{-3}$, ruling out the superweak model.

$\mathbb{CP}$ violation is important and interesting to study for several reasons. Almost all extensions of the Standard Model imply that additional sources of $\mathbb{CP}$ violation exist. Futhermore, a necessary condition for baryogenesis – the dynamical generation of the matter-antimatter asymmetry observed in our universe – is that there be a sufficient amount of $\mathbb{CP}$ violation [8]. However, the Standard Model with six quarks fails to account for this asymmetry by several orders of magnitude [205], suggesting that there are additional sources of $\mathbb{CP}$ violation.

How might such additional sources be found? In general, $\mathbb{CP}$ violation may be tested by measuring

$$\mathcal{A} = \frac{\Gamma - \overline{\Gamma}}{\Gamma + \overline{\Gamma}} \qquad (19.45)$$

where $\mathcal{A}$ is the asymmetry for a given process: the difference between a given decay rate and its $\mathbb{CP}$-conjugate divided by the sum of these rates. For Kaons

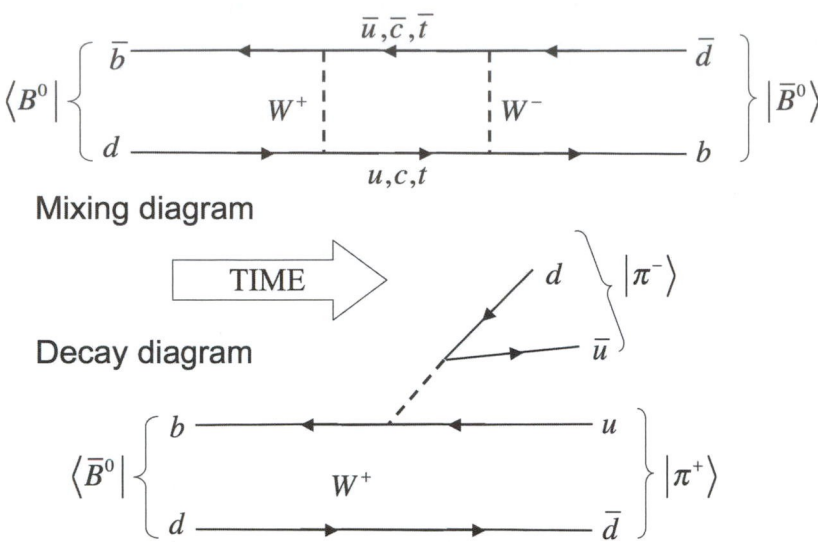

FIGURE 19.9
Diagrams for $\mathbb{CP}$ violation in the B-meson sector of the Standard Model.

$\mathcal{A}$ is naturally small – $\mathcal{A} \sim 10^{-3}$ for various Kaon decays. This makes the hunt for additional sources of $\mathbb{CP}$ violation difficult because no matter what Kaon decay you look at, $\mathcal{A}$ is always going to be this small, making it extremely difficult to experimentally check the origin of $\mathbb{CP}$ violation as explained in a given theoretical framework (Standard Model or otherwise).

Fortunately the situation is not quite so grim. The standard 6-quark model also predicts that $\mathbb{CP}$ violation will also occur in the $D^0\bar{D}^0$ and $B^0\bar{B}^0$ systems, each by the same mechanism as the Kaon system. The D^0 experimental physics is messy, but the B-meson physics is "clean": it is virtually a complete analog of the Kaon system, except for its heavier mass and shorter lifetime. But not as short as it might have been – the B^0-meson fortuitously has a relatively long lifetime of abour 10^{-12} sec and a large mixing with its antiparticle $\bar{B}^0$-meson [1]. This means that, in principle, $\mathcal{A}$ could be of order unity for B^0 decays. This would be consistent with the Standard Model in which $\mathbb{CP}$ violation arises due to the difference between various quark masses (b, s, d, etc.). In contrast to this, the superweak theory (and most other competitors) explain $\mathbb{CP}$ violation as coming from a new force independent of quark mass; hence in these models $\mathcal{A}$ should be the same for B^0-meson as for Kaons. So observation of large values of $\mathcal{A}$ would rule out (or severely constrain) such models.

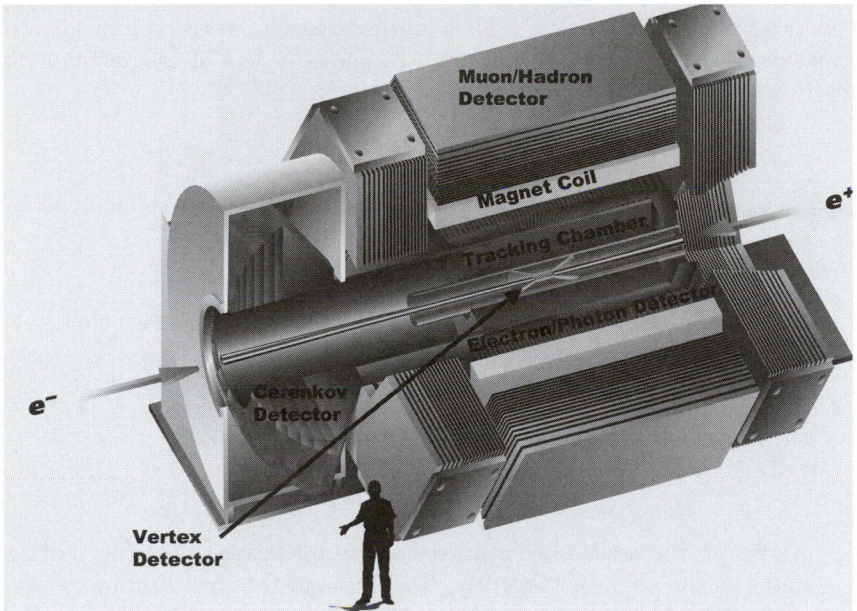

FIGURE 19.10

The BaBar detector, an important detector used to determine the decay properties of B-mesons. Image courtesy of SLAC National Accelerator Laboratory.

The BaBar experiment was the first to observe $\mathbb{CP}$ violation in the B-meson system [206]. Studying the $\Upsilon(4s)$ resonance, from which tens of millions of $B\bar{B}$ pairs are produced, a search is carried out for a $\mathbb{CP}$-violating decay of a B-meson (reconstructed from the observed decay products that typically contain charm) that is correlated with the recoil of a neutral B-meson (the tag). Early hopes were that physics beyond the Standard Model would soon be found.

Unfortunately things weren't quite that easy. Direct observation of $\mathbb{CP}$ violation in the decay $B^0 \to K^+\pi^-$ indicates that [1]

$$\mathcal{A}_{K^+\pi^-} = \frac{\Gamma\left(\bar{B}^0 \to K^-\pi^+\right) - \Gamma\left(B^0 \to K^+\pi^-\right)}{\Gamma\left(\bar{B}^0 \to K^-\pi^+\right) + \Gamma\left(B^0 \to K^+\pi^-\right)} = -0.095 \pm 0.013 \quad (19.46)$$

which is rather small effect. However, $\mathbb{CP}$ violation has also been observed in $B \to \pi^+ + \pi^-$, $B^0 \to \eta K^{*0}$, and $B^+ \to \rho^0 K^+$ decays. Searches for additional $\mathbb{CP}$ asymmetries in decays of B, D and K mesons continues. So far every result obtained has been consistent with the Standard Model.

It sounds like a "good news" confirmation of the Standard Model. However, incorporating these measurements into the Standard Model yields a prediction that there is one proton for every 10^{18} photons due to the cosmic imbalance of particles and anti-particles. This is in disagreement with cosmological observations (one proton to 10^9 photons) by many orders of magnitude [205]. Clearly we have a lot to learn about $\mathbb{CP}$ violation!

19.4 Questions

1. How many meters of iron must a ν_μ of 1 GeV penetrate so that it will scatter, on average, once? How long does this take?

2. In the decay $\left|\text{Co}^{60}; j = 5, m = 5\right\rangle \longrightarrow \left|\text{Ni}^{**60}; j = 4, m = 4\right\rangle + e^- + \bar{\nu}$ the intensity of the emitted electrons has the form

$$I(\beta, \theta) = 1 + \alpha\beta\cos\theta$$

where β is the speed of the emitted electron relative to the speed of light and θ is the angle between its direction and the direction of the Co^{60} spin. Neglecting orbital angular momenta in the final state, deduce the value of α.

3. A K^0 and a $\bar{K}^0$ beam, each of equal intensity, pass through a slab of matter. Are the beams attenuated equally? Why or why not?

4. A K_2 beam passes through a slab of matter. What will the emerging beam consist of and why? How can you experimentally test your answer?

5. You make contact with alien physicists through a wormhole into another part of the multiverse, and soon develop a common language of communication with them. They have developed methods of traveling through the wormhole in short times and are considering visiting Earth. However, before they visit, you want to be sure that they are not made of antimatter. Because of this lack of knowledge, it's too dangerous to send objects through the wormhole, but you can ask them any questions you want about experiments they have performed, and you are able to communicate with them results of experiments performed here.

   (a) Can you determine if any of $\mathbb{C}$, $\mathbb{P}$, or $\mathbb{T}$ are conserved in all interactions in their universe?

   (b) Can you determine if $\mathbb{CP}$ is conserved by $\mathbb{C}$ and $\mathbb{P}$ are both violated in their universe?

   (c) Can you determine if $\mathbb{CP}$, $\mathbb{C}$ and $\mathbb{P}$ are each violated in their universe?

6. How many kinds of neutral B-meson oscillation are there? Draw the lowest-order diagrams for each.

7. The GALLEX experiment measures the flux of ν_e from the sun by counting electrons in the reaction

$$\nu_e + \mathrm{Ga}^{71} \rightarrow e^- + \mathrm{Ge}^{71}$$

The energy threshold for this reaction is 233 KeV. The expected flux of neutrinos from the standard solar model is $\Phi = 6 \times 10^{14}/\mathrm{m}^2/\mathrm{s}$. Suppose for simplicity that the entire flux is above threshold. If detection efficiency is 35%, how many Ga^{71} nuclei are needed to have one interaction with a neutrino per day? What mass of Ga^{71} does this entail? How much mass of actual gallium is needed if the Ga^{71} isotope is 40% abundant?

20

Charged Leptonic Weak Interactions

Today Fermi's theory of beta decay has been superseded by the modern-day theory of electroweak interactions. If we remained consistent with the nomenclature for the electromagnetic and strong interactions, we would call this theory Quantum Geusidynamics [207], or QGD ("geusi" meaning "flavor" in Greek – another possibility is quantum aesthenodynamics (QAD), "aestheno" meaning "weak"). This nomenclature did not catch on, and today people simply refer to the theory as Electroweak theory.

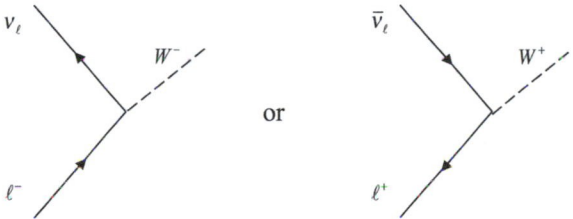

FIGURE 20.1
General form of leptonic weak interaction vertices.

All known quarks and leptons undergo weak interactions. The mediators of weak interactions are three spin-1 particles: the W^+, W^-, Z^0, the superscripts referring to the electric charges of the mediators. They are analogous to the photon in QED and to the gluons in QCD. However,, they are strikingly different in that they are extremely massive

$$M_W = 80.398 \pm 0.0259 \text{ GeV} \quad M_Z = 91.1876 \pm 0.0021 \text{ GeV} \tag{20.1}$$

and they are unstable to decay:

$$\Gamma_W = 2.141 \pm 0.041 \text{ GeV} \quad \Gamma_Z = 2.4952 \pm 0.0023 \text{ GeV} \tag{20.2}$$

The above masses are world averages of all experimental results [1].

"Charged" weak interactions (mediated by the W's) are simpler than "neutral" ones (mediated by the Z) so we'll consider them first. To start with, we'll look at weak interactions of leptons. As with QED and QCD, we can describe weak interactions using vertices and propagators. The fundamental leptonic weak interaction can be described by the vertex shown in fig. 20.1 where $\ell^- = e^-, \mu^-$, or τ^- and $\nu_\ell = \nu_e, \nu_\mu$, or ν_τ. Notice that – unlike either QED or QCD – *weak interactions change one type of particle into another!* In the case above charged leptons are changed into their neutrino partners. Note also that there are no e^-/ν_μ or τ^-/ν_e vertices – leptonic weak interactions do not "cross over" between generations!

The diagrammatic rules are similar to QED, except for two things – the vertex factor has a $\frac{1}{2}\left(1 - \gamma^5\right)$, and the propagator has terms that depend upon the mass and width of the W, as shown in fig. 20.2. The width only

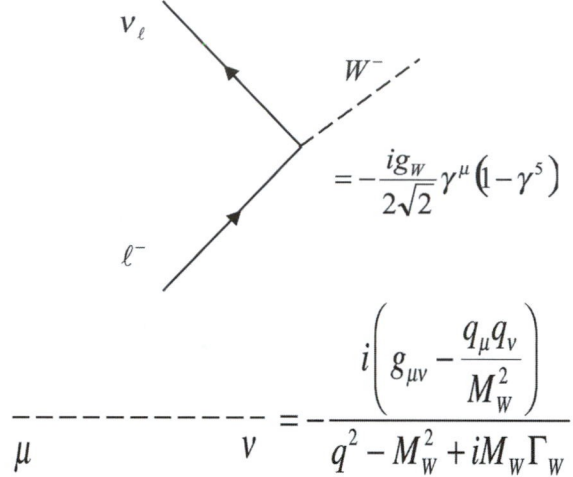

FIGURE 20.2
The leptonic vertex factor and internal-line propagator for W bosons.

becomes important if we are colliding particles at energies at or near the mass of the W and so we will neglect it. The constant g_W is the "weak charge," with $\alpha_W = \frac{g_W^2}{4\pi\hbar c}$ the weak coupling constant, analogous to α in QED and α_s in QCD.

The factor of $\frac{1}{2}\left(1 - \gamma^5\right)$ is of crucial importance, and must be inserted (at least in the lepton sector) because of Goldhaber's observation that all neutrinos are left-handed. Recall (from the appendix in Chapter 11) that

$$\left(\overline{\psi}\gamma^\mu\psi\right)' = \Lambda^\mu_{\ \nu}\left(\overline{\psi}\gamma^\nu\psi\right) \tag{20.3}$$

$$\left(\overline{\psi}\gamma^\mu\gamma^5\psi\right)' = \det\left(\Lambda\right)\Lambda^\mu_{\ \nu}\left(\overline{\psi}\gamma^\nu\gamma^5\psi\right) \tag{20.4}$$

i.e. the $(\overline{\psi}\gamma^\nu\psi)$ transforms as a vector, but $(\overline{\psi}\gamma^\nu\gamma^5\psi)$ transforms as a pseudovector (or axial vector, as it is often called) as indicated by the $\det(\Lambda)$ factor. Since the vertex has both a vector part and an axial vector part – we say that the W-boson couples to both a vector current and an axial-vector current – it must violate parity. That the parts are equal in magnitude but opposite in sign is a consequence of the purely left-handed character of the neutrino. For a fermion $\psi = \begin{pmatrix} \varphi \\ \chi \end{pmatrix}$ we have

$$\frac{1}{2}\left(1-\gamma^5\right) = \frac{1}{2}\begin{pmatrix} \mathbf{I} & -\mathbf{I} \\ -\mathbf{I} & \mathbf{I} \end{pmatrix} \Rightarrow \psi_L \equiv \frac{1}{2}\left(1-\gamma^5\right)\psi = \frac{1}{2}\begin{pmatrix} \varphi-\chi \\ -(\varphi-\chi) \end{pmatrix} \quad (20.5)$$

and we see that $\frac{1}{2}\left(1-\gamma^5\right)$ projects out one of the two spinor components (in this basis, the $\varphi+\chi$ part is eliminated). So a left-handed spinor has the form $\psi = \begin{pmatrix} \zeta \\ -\zeta \end{pmatrix}$. A right-handed spinor is obtained by projection with $\frac{1}{2}\left(1+\gamma^5\right)$

$$\frac{1}{2}\left(1+\gamma^5\right) = \frac{1}{2}\begin{pmatrix} \mathbf{I} & \mathbf{I} \\ \mathbf{I} & \mathbf{I} \end{pmatrix} \Rightarrow \psi_R \equiv \frac{1}{2}\left(1+\gamma^5\right)\psi = \frac{1}{2}\begin{pmatrix} \varphi+\chi \\ (\varphi+\chi) \end{pmatrix}$$

and so has the form $\psi = \begin{pmatrix} \zeta \\ \zeta \end{pmatrix}$.

The W-propagator differs from the usual $-i\frac{g_{\mu\nu}}{q^2}$ that appears in QED and QCD because of the mass of the W's. In almost all reactions (except those at LEP and higher-energy machines) $q^2 \ll M_W^2$, so we find

$$\lim_{q^2 \ll M_W^2}\left[-i\frac{g_{\mu\nu}-q_\mu q_\nu/M_W^2}{q^2-M_W^2+i\hbar M_W\Gamma_W}\right] \simeq +i\frac{g_{\mu\nu}}{M_W^2} \quad (20.6)$$

as a good approximation to the W-propagator*.

Finally, we still have the condition that the polarization vector ε^μ of a W is orthogonal to its 4-momentum:

$$\varepsilon \cdot p = 0 \quad (20.7)$$

which reduces the number of independent degrees of freedom of a W from 4 to 3. However, unlike the photon (and gluon), this condition fully exhausts the gauge freedom in the model of weak interactions – we do not invoke the Coulomb gauge.

---

*I should note here that the Breit-Wigner form of the propagator (20.6) is an approximation for energies near the mass of the weak bosons, and that the Γ_W term is neglected everywhere else ($\Gamma_W \ll M_W$ in general), yielding the low-energy propagator that follows from the equations in electroweak theory.

20.1   Neutrino-Electron Scattering

Consider the simple scattering process $e^- + \nu_\mu \to \mu^- + \nu_e$ as shown in fig. 20.3

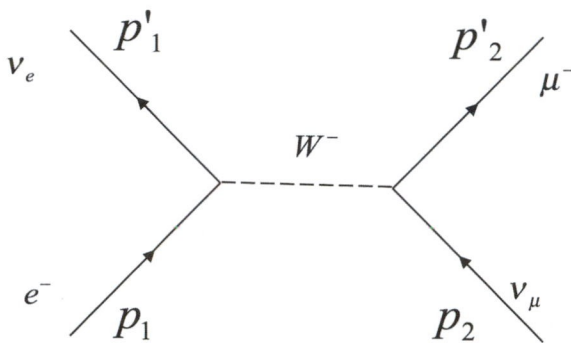

FIGURE 20.3
Electron-neutrino scattering diagram to lowest order.

The matrix element is

$$\mathcal{M} = (-i)^2 \left(\frac{g_{\mathrm{w}}}{2\sqrt{2}}\right)^2 \left[\overline{u}(p_1')\gamma^\mu \left(1 - \gamma^5\right) u(p_1)\right]$$

$$\times \left[\frac{g_{\mu\nu} - q_\mu q_\nu / M_W^2}{q^2 - M_W^2}\right] \left[\overline{u}(p_2')\gamma^\nu \left(1 - \gamma^5\right) u(p_2)\right]$$

$$\simeq \left(\frac{\sqrt{2}g_{\mathrm{w}}}{4M_W}\right)^2 \left[\overline{u}(p_1')\gamma^\mu \left(1 - \gamma^5\right) u(p_1)\right] \left[\overline{u}(p_2')\gamma_\mu \left(1 - \gamma^5\right) u(p_2)\right] \tag{20.8}$$

The Casimir trick gives, using the trace theorems in Chapter 13

$$\sum_{\text{spins}} \left[\overline{u}(p_1')\gamma^\mu \left(1 - \gamma^5\right) u(p_1)\right]^\dagger \left[\overline{u}(p_1')\gamma^\nu \left(1 - \gamma^5\right) u(p_1)\right]$$

$$= \text{Tr}\left[\overline{\gamma^\mu \left(1 - \gamma^5\right)} \left(\not{p}_1 + m\right) \gamma^\nu \left(1 - \gamma^5\right) \left(\not{p}_1' + m_{\nu_e}\right)\right]$$

$$= \text{Tr}\left[\gamma^\mu \left(1 - \gamma^5\right) \left(\not{p}_1 + m\right) \gamma^\nu \left(1 - \gamma^5\right) \left(\not{p}_1'\right)\right] \qquad (\text{neglect } m_{\nu_e})$$

$$= \text{Tr}\left[\gamma^\mu \left(1 - \gamma^5\right) \left(\not{p}_1 + m\right) \left(1 + \gamma^5\right) \gamma^\nu \left(\not{p}_1'\right)\right]$$

$$= \text{Tr}\left[\gamma^\mu \left(1 - \gamma^5\right)^2 \left(\not{p}_1\right) \gamma^\nu \left(\not{p}_1'\right)\right]$$

$$= 2\mathrm{Tr}\left[\gamma^\mu\left(1-\gamma^5\right)\left(\not{p}_1\right)\gamma^\nu\left(\not{p}'_1\right)\right]$$

$$= 2\mathrm{Tr}\left[\gamma^\mu\left(\not{p}_1\right)\gamma^\nu\left(\not{p}'_1\right)\right] - 2\mathrm{Tr}\left[\gamma^\mu\gamma^5\left(\not{p}_1\right)\gamma^\nu\left(\not{p}'_1\right)\right]$$

$$= 8\left[p_1'^\mu p_1^\nu + p_1^\mu p_1'^\nu - g^{\mu\nu}\left(p_1\cdot p'_1\right) + i\varepsilon^{\mu\alpha\nu\beta}p_{1\alpha}p'_{1\beta}\right] \tag{20.9}$$

where $\overline{\gamma^\mu\left(1-\gamma^5\right)} = \gamma^\mu\left(1-\gamma^5\right)$ (as is easy to show) and the neutrino has been taken to be massless. Hence we get

$$\overline{|\mathcal{M}|}^2 = \frac{64}{2}\left(\frac{g_\mathrm{w}}{2\sqrt{2}M_W}\right)^4\left[p_1'^\mu p_1^\nu + p_1^\mu p_1'^\nu - g^{\mu\nu}\left(p_1\cdot p'_1\right) + i\varepsilon^{\mu\alpha\nu\beta}p_{1\alpha}p'_{1\beta}\right]$$

$$\times\left[p'_{2\mu}p_{2\nu} + p_{2\mu}p'_{2\nu} - g_{\mu\nu}\left(p_2\cdot p'_2\right) + i\varepsilon_{\mu\rho\nu\sigma}p_2^\rho p_2'^\sigma\right]$$

$$= \frac{1}{2}\left(\frac{g_\mathrm{w}}{M_W}\right)^4\left[2\left(p'_1\cdot p'_2\right)\left(p_1\cdot p_2\right) + 2\left(p_1\cdot p'_2\right)\left(p'_1\cdot p_2\right)\right.$$

$$\left. +i^2\varepsilon^{\mu\alpha\nu\beta}p_{1\alpha}p'_{1\beta}\varepsilon_{\mu\rho\nu\sigma}p_2^\rho p_2'^\sigma\right] \tag{20.10}$$

where terms of the form $\varepsilon^{\mu\alpha\nu\beta}p_{1\alpha}p'_{1\beta}p_{2\mu}p'_{2\nu}$ vanish, because by momentum conservation $p'_{2\nu} = -p'_{1\nu} + p_{2\nu} + p_{1\nu}$, in which case the ε-tensor contracts over two identical objects (similarly $\varepsilon^{\mu\alpha\nu\beta}p_{1\alpha}p'_{1\beta}g_{\mu\nu} = 0$). We also have the result

$$\varepsilon^{\mu\alpha\nu\beta}\varepsilon_{\mu\rho\nu\sigma} = -2\left(\delta^\alpha_\rho\delta^\beta_\sigma - \delta^\alpha_\sigma\delta^\beta_\rho\right) \tag{20.11}$$

which (as shown in eq. (20.33)) follows from the definition of the ε-tensor. Hence

$$\overline{|\mathcal{M}|}^2 = \frac{1}{2}\left(\frac{g_\mathrm{w}}{M_W}\right)^4\left[2\left(p'_1\cdot p'_2\right)\left(p_1\cdot p_2\right) + 2\left(p_1\cdot p'_2\right)\left(p'_1\cdot p_2\right)\right.$$

$$\left. -2i^2 p_{1\alpha}p'_{1\beta}\left(\delta^\alpha_\rho\delta^\beta_\sigma - \delta^\alpha_\sigma\delta^\beta_\rho\right)p_2^\rho p_2'^\sigma\right]$$

$$= \frac{1}{2}\left(\frac{g_\mathrm{w}}{M_W}\right)^4\left[2\left(p'_1\cdot p'_2\right)\left(p_1\cdot p_2\right) + 2\left(p_1\cdot p'_2\right)\left(p'_1\cdot p_2\right) + 2\left(p'_1\cdot p'_2\right)\left(p_1\cdot p_2\right)\right.$$

$$\left. -2\left(p_1\cdot p'_2\right)\left(p'_1\cdot p_2\right)\right]$$

$$= 2\left(\frac{g_\mathrm{w}}{M_W}\right)^4\left(p'_1\cdot p'_2\right)\left(p_1\cdot p_2\right) \tag{20.12}$$

a remarkably simple expression!

Now let's look at this in the CM frame, for which

$$p_1^\mu = (E,\vec{p}) \quad p_2^\mu = (|\vec{p}|,-\vec{p}) \quad p_1'^\mu = (|\vec{p}\,'|,-\vec{p}\,') \quad p_2'^\mu = (E',\vec{p}\,')$$

$$E' + |\vec{p}\,'| = E + |\vec{p}| \Rightarrow |\vec{p}\,'| = \frac{1}{2}\left(E + |\vec{p}|\right) - \frac{M^2}{2\left(E + |\vec{p}|\right)} \tag{20.13}$$

$$(p_1\cdot p_2) = E|\vec{p}| + |\vec{p}|^2 \qquad (p'_1\cdot p'_2) = E'|\vec{p}\,'| + |\vec{p}\,'|^2$$

where M is the mass of the muon. Hence

$$\overline{|\mathcal{M}|}^2 = 2\left(\frac{g_\mathrm{w}}{M_W}\right)^4\left(E'|\vec{p}\,'| + |\vec{p}\,'|^2\right)\left(E|\vec{p}| + |\vec{p}|^2\right)$$

$$= 2 \left(\frac{g_\text{W}}{M_W} \right)^4 |\vec{p}\,'| \, |\vec{p}| \, (E + |\vec{p}|)^2$$

$$= \left(\frac{g_\text{W}}{M_W} \right)^4 \left[(E + |\vec{p}|)^2 - M^2 \right] (E + |\vec{p}|) \, |\vec{p}|$$

$$= \left(\frac{g_\text{W}}{M_W} \right)^4 \left[\left(E + \sqrt{E^2 - m^2} \right)^2 - M^2 \right] \left(E\sqrt{E^2 - m^2} + E^2 - m^2 \right)$$

$$\simeq 8 \left(\frac{E g_\text{W}}{M_W} \right)^4 \left[1 - \frac{M^2}{4E^2} \right] + \cdots \tag{20.14}$$

where the electron mass m has been neglected in the last line. The formula for the differential scattering cross-section is

$$\frac{d\sigma}{d\Omega} = \left(\frac{\hbar c}{8\pi} \right)^2 \frac{\mathcal{S} \, |\mathcal{M}|^2}{(E_1 + E_2)^2} \frac{|\vec{p}\,'|}{|\vec{p}|}$$

$$= 2 \left(\frac{\hbar c}{8\pi} \right)^2 \left(\frac{g_\text{W}}{M_W} \right)^4 E^2 \left[1 - \frac{M^2}{4E^2} \right] \left(\frac{(E + |\vec{p}|)^2 - M^2}{2 (E + |\vec{p}|) \, |\vec{p}|} \right)$$

$$\simeq \frac{1}{2} \left(\frac{\hbar c g_\text{W}^2}{4\pi M_W^2 c^4} \right)^2 E^2 \left[1 - \frac{M^2 c^4}{4E^2} \right]^2 + \cdots \tag{20.15}$$

in the limit of large momentum, where in the last line the correct speed-of-light factors have been introduced.

We see here that the cross-section increases with energy, something quite different from what we observed in QED and QCD. This is a consequence of our approximation of the W-propagator as we shall see.

20.2 Muon Decay

Neutrino-electron scattering is really hard to experimentally implement. However, muon-decay is an easy experiment to do, and the diagram is given in fig. 20.4.

The matrix element is

$$\mathcal{M} = (-i)^2 \left(\frac{g_\text{W}}{2\sqrt{2}} \right)^2 \left[\overline{u}(p_1')\gamma^\mu \left(1 - \gamma^5 \right) u(p_1) \right] \left[\frac{g_{\mu\nu} - q_\mu q_\nu / M_W^2}{q^2 - M_W^2} \right]$$
$$\times \left[\overline{u}(p_2)\gamma^\nu \left(1 - \gamma^5 \right) v(p_2) \right] \tag{20.16}$$

This is just like what we had for neutrino electron scattering, except that the electron-antineutrino has a v-spinor final state. This is nice because the Casimir-tricks don't change at all provided the neutrinos are assumed to be

⟨?⟩. only for spin-

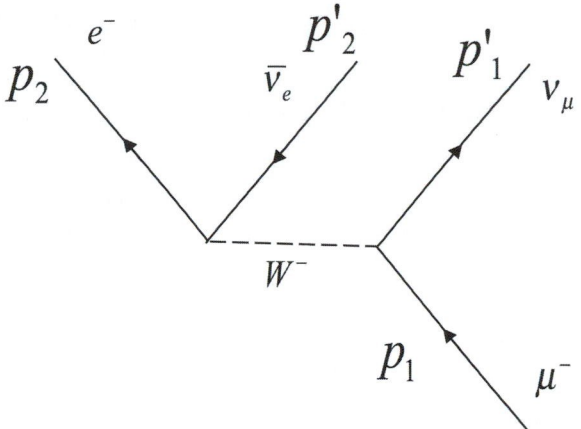

FIGURE 20.4

Muon decay diagram

massless. Hence none of the manipulations change, and we get

$$\overline{|\mathcal{M}|}^2 = 2 \left(\frac{g_{\rm w}}{M_W} \right)^4 (p_1' \cdot p_2)(p_1 \cdot p_2') \qquad (20.17)$$

in other words, the matrix element is proportional to the dot products of the lepton momenta with their counterpart neutrino partners. Note the relabeling of momenta in fig. 20.4 relative to the diagram 20.3 for electron-neutrino scattering.

Since we are concerned with muon decay, this time we want to look at this expression in the rest-frame of the muon. We have

$$p_1^\mu = (p_1' + p_2 + p_2')^\mu \quad p_1^\mu = (M, \vec{0}) \quad p_2'^\mu = (E_2', \vec{p}_2')$$

$$\Rightarrow (p_2 + p_1')^2 = m^2 + 2(p_2 \cdot p_1') = (p_1 - p_2')^2 = M^2 - 2(p_1 \cdot p_2')$$

$$\Rightarrow (p_2 \cdot p_1') = \frac{1}{2}(M^2 - m^2) - (p_1 \cdot p_2') = \frac{1}{2}(M^2 - m^2) - ME_2' \quad (20.18)$$

and so

$$\overline{|\mathcal{M}|}^2 = 2 \left(\frac{g_{\rm w}}{M_W} \right)^4 ME_2' \left(\frac{1}{2}(M^2 - m^2) - ME_2' \right)$$

$$\simeq \left(\frac{g_{\rm w}}{M_W} \right)^4 M^2 E_2' (M - 2E_2') \qquad (20.19)$$

where the electron mass m has again been neglected. The decay rate is

$$d\Gamma = \frac{c^2}{2\hbar M} \left[\frac{d^3 p_1'}{2E_1'(2\pi)^3} \frac{d^3 p_2'}{2E_2'(2\pi)^3} \frac{d^3 p_2}{2E_2(2\pi)^3} \right] |\mathcal{M}|^2$$

$$\times (2\pi)^4\, \delta^{(4)}\, (p_1' + p_2 + p_2' - p_1) \tag{20.20}$$

where now $E_2 = |\vec{p}_2| = E$ is the energy of the outgoing electron, neglecting the electron mass. The 3-Body phase-space integral is done in the appendix, and the result from eq. (20.56) is

$$d\Gamma = \left(\frac{g_\text{w}}{M_W c}\right)^4 \frac{M^2 E^2 dE}{2\hbar\,(4\pi)^3} \left(1 - \frac{4E}{3M^2}\right) \tag{20.21}$$

Momentum conservation forces $E < \frac{1}{2}Mc^2$ (see why in the appendix), so the total decay rate of the muon is

$$\begin{aligned}
\Gamma_\mu &= \int_0^{\frac{1}{2}Mc^2} dE \left(\frac{g_\text{w}}{M_W c}\right)^4 \frac{M^2 E^2}{2\hbar\,(4\pi)^3} \left(1 - \frac{4E}{3Mc^2}\right) \\
&= \left(\frac{g_\text{w}}{M_W}\right)^4 \frac{M^5 c^2}{12\hbar\,(8\pi)^3}
\end{aligned} \tag{20.22}$$

for a muon lifetime of

$$\tau_\mu = \frac{1}{\Gamma_\mu} = \left(\frac{M_W}{M g_\text{w}}\right)^4 \frac{12\hbar\,(8\pi)^3}{Mc^2} \tag{20.23}$$

which goes like the inverse fifth power of the muon mass M.

Note that in both of these problems, the coupling constant g_w always is divided by the W-mass. If we define

$$G_F = \frac{\sqrt{2}}{8} \left(\frac{g_\text{w}}{M_W c^2}\right)^2 \tag{20.24}$$

then

$$\tau_\mu = \frac{1}{\Gamma_\mu} = \frac{192\pi^3 \hbar}{G_F^2 \,(Mc^2)^5} \tag{20.25}$$

We can experimentally measure the muon lifetime – it is $\tau_\mu^{\text{expt}} = 2.197 \times 10^{-6}$ sec. The best measurements [1] then imply

$$G_F = 1.16637(1) \times 10^{-5}\,(\text{GeV})^{-2} \tag{20.26}$$

which (as we saw in eq. (19.15)) is Fermi's constant! This was an early result from the Fermi theory of β-decay. Today the muon lifetime provides the standard by which the value of Fermi's constant is empirically determined[†].

Of course in Fermi's original theory [188], there was no W-boson. Weak interaction vertices were given by direct 4-fermion couplings of the type shown in figure 20.5 but from the modern perspective we understand this diagram

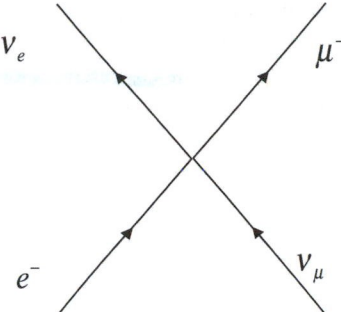

FIGURE 20.5

A vertex in Fermi's original theory.

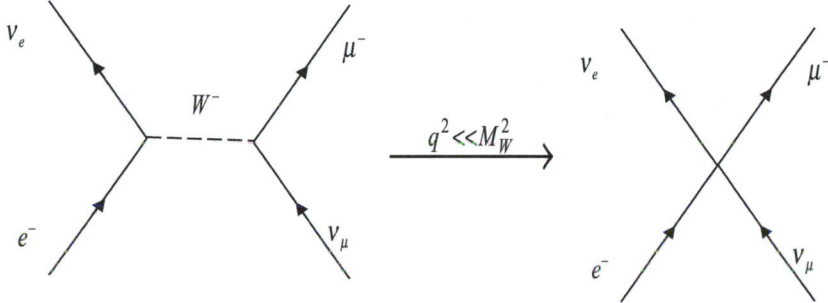

FIGURE 20.6

Weak interaction theory reduces to Fermi's theory in the limit of large M_W.

to be an approximation: That is, Fermi's theory is reproduced whenever the W-mass is much larger than any other energy in the problem! Indeed, the idea of a W-mediator was first suggested by Oscar Klein in 1938 [208].

Today it is possible to independently measure the mass of the W, and so we can deduce the strength of the weak coupling constant g_{w}. We find

$$g_{\mathrm{w}} = M_W c^2 \sqrt{4\sqrt{2}G_F} = 80 \times \sqrt{4\sqrt{2} \times 1.166 \times 10^{-5}} = 0.653$$

$$\Rightarrow \alpha_W = \frac{(g_{\mathrm{w}})^2}{4\pi} = \frac{(0.653)^2}{12.57} = .034 \simeq \frac{1}{29} \tag{20.27}$$

---

†Fermi's constant is actually defined to be $\mathbf{G}_F = \frac{\sqrt{2}}{8}\left(\frac{g_{\mathrm{w}}}{M_W c^2}\right)^2 (\hbar c)^3 = G_F (\hbar c)^3$. Throughout this book I will work with the constant G_F, defined so that it has units of inverse energy, and (hopefully without confusion) refer to this as Fermi's constant.

which is a surprise – the weak coupling strength is five times *larger* than the electromagnetic strength $\alpha = \frac{1}{137}$!!

Weak interactions are feeble – but not because the coupling is α_W small. Instead, they are feeble because the gauge bosons that mediate the interactions are very massive compared to all subatomic particles (except for the top quark). In LEP experiments – which operated at energies comparable or a bit bigger than the mass of the Z-boson – weak interactions were observed to be stronger than electromagnetic ones!

20.3   Appendix: Mathematical Tools for Weak Interactions

20.3.1   A Note on the $\varepsilon-$Tensor

The epsilon tensor is a fully antisymmetric tensor (it flips sign under interchange of any pair of indices) that has as many indices as there are dimensions. In two spacetime dimensions it is

$$\varepsilon_{\mu\nu} = 0 \quad \text{if} \quad \mu = \nu; \qquad \varepsilon_{01} = 1 = -\varepsilon_{10} \qquad (20.28)$$

Note that $\varepsilon^{01} = -1$ because of the minus sign that appears in the metric (recall $\varepsilon^{\mu\nu} = g^{\mu\rho}g^{\nu\sigma}\varepsilon_{\rho\sigma}$). The product of two ε-tensors in two dimensions is

$$\varepsilon^{\alpha\beta}\varepsilon_{\mu\nu} = -\left(\delta^{\alpha}_{\mu}\delta^{\beta}_{\nu} - \delta^{\alpha}_{\nu}\delta^{\beta}_{\mu}\right) \qquad (20.29)$$

which can be shown by brute force: we must get zero if either $\mu = \nu$ or $\alpha = \beta$. To get a nonzero answer clearly α must equal either one of μ or ν, and β must equal the other of these. This means that every term on the right-hand side must be a product of delta functions between the upper indices and the lower indices, since the magnitude of any index set is either 0 or 1. Setting $\mu = 0 = \alpha$ and $\nu = 1 = \beta$ then gives the correct signs.

The same reasoning can be applied to any dimension. In our $(3+1)$-dimensional world the ε-tensor must have 4 indices and be fully antisymmetric on all of them, and so is given by

$$\varepsilon_{\mu\nu\alpha\beta} = 0 \quad \text{if any two indices are equal;} \qquad \varepsilon_{0123} = 1$$
$$\varepsilon_{\mu\nu\alpha\beta} = \varepsilon_{[\mu\nu\alpha\beta]} = -\varepsilon_{\nu\mu\alpha\beta} = \varepsilon_{\nu\alpha\mu\beta} = -\varepsilon_{\nu\alpha\beta\mu} = \text{ etc} \qquad (20.30)$$

where the square brackets mean "fully antisymmetrize on all indices", and $\varepsilon^{0123} = -1$. The product of two ε-tensors is always either 0 or ± 1 and so as before we will get a product of Kronecker-delta-functions:

$$\varepsilon^{\rho\tau\sigma\gamma}\varepsilon_{\mu\nu\alpha\beta} = -\left(\delta^{\rho}_{\mu}\delta^{\tau}_{\nu}\delta^{\sigma}_{\alpha}\delta^{\gamma}_{\beta} + \text{ all possible signed permutations}\right)$$

$$= -\delta^\rho_{[\mu} \delta^\tau_\nu \delta^\sigma_\alpha \delta^\gamma_{\beta]}$$
$$= -\delta^\rho_\mu \delta^\tau_\nu \delta^\sigma_\alpha \delta^\gamma_\beta + \delta^\rho_\nu \delta^\tau_\mu \delta^\sigma_\alpha \delta^\gamma_\beta - \delta^\rho_\nu \delta^\tau_\alpha \delta^\sigma_\mu \delta^\gamma_\beta + \text{ etc} \tag{20.31}$$

where there are 24 terms in all. Contracting over the (ρ, μ) index pair yields

$$\varepsilon^{\rho\tau\sigma\gamma} \varepsilon_{\rho\nu\alpha\beta} = - \left(4\delta^\tau_\nu \delta^\sigma_\alpha \delta^\gamma_\beta - \delta^\tau_\nu \delta^\sigma_\alpha \delta^\gamma_\beta + \delta^\tau_\alpha \delta^\sigma_\mu \delta^\gamma_\beta - \text{ etc} \right)$$
$$= -\delta^\tau_{[\nu} \delta^\sigma_\alpha \delta^\gamma_{\beta]} \tag{20.32}$$

Contracting on another pair of indices (e.g., (τ, ν)) gives

$$\varepsilon^{\rho\tau\sigma\gamma} \varepsilon_{\rho\tau\alpha\beta} = -\delta^\tau_{[\tau} \delta^\sigma_\alpha \delta^\gamma_{\beta]}$$
$$= - \left(4\delta^\sigma_\alpha \delta^\gamma_\beta - \delta^\sigma_\alpha \delta^\gamma_\beta + \delta^\sigma_\beta \delta^\gamma_\alpha - \delta^\sigma_\alpha \delta^\gamma_\beta + \delta^\sigma_\beta \delta^\gamma_\alpha - 4\delta^\sigma_\beta \delta^\gamma_\alpha \right)$$
$$= -2 \left(\delta^\sigma_\alpha \delta^\gamma_\beta - \delta^\sigma_\beta \delta^\gamma_\alpha \right) \tag{20.33}$$

and again

$$\varepsilon^{\rho\tau\sigma\gamma} \varepsilon_{\rho\tau\sigma\beta} = -2 \left(\delta^\sigma_\sigma \delta^\gamma_\beta - \delta^\sigma_\beta \delta^\gamma_\sigma \right) = -2 \left(4\delta^\gamma_\beta - \delta^\gamma_\beta \right) = -6\delta^\gamma_\beta \tag{20.34}$$

and finally

$$\varepsilon^{\rho\tau\sigma\gamma} \varepsilon_{\rho\tau\sigma\gamma} = -24 = -4! \tag{20.35}$$

Note that

$$\gamma^5 = i\gamma^0 \gamma^1 \gamma^2 \gamma^3 = \frac{i}{4!} \varepsilon_{\mu\nu\alpha\beta} \gamma^\mu \gamma^\nu \gamma^\alpha \gamma^\beta \tag{20.36}$$

because the γ-matrices all anticommute.

20.4 Appendix: 3-Body Phase Space Decay

Consider the decay rate of a body of mass M into three other bodies, as in fig. 20.4:

$$\Gamma = \frac{1}{2M} \left[\int \frac{d^3 p'_1}{2E'_1 (2\pi)^3} \frac{d^3 p'_2}{2E'_2 (2\pi)^3} \frac{d^3 p_2}{2E_2 (2\pi)^3} \right] |\mathcal{M}|^2$$
$$\times (2\pi)^4 \delta^{(4)} (p'_1 + p_2 + p'_2 - p_1) \tag{20.37}$$

where we'll assume that $m'_2 = 0$ (i.e. one of the final state particles is massless) so that

$$E'_1 = \sqrt{|\vec{p}_1'|^2 + m_1^2} \quad E'_2 = |\vec{p}_2'| \quad E_2 = \sqrt{|\vec{p}_2|^2 + m_2^2} \tag{20.38}$$

Integrating over d^3p_1' gives

$$\Gamma = \frac{2}{(4\pi)^5 M} \left[\int \frac{d^3p_2' d^3p_2}{E_1 E_2' E_2} \right] |\mathcal{M}|^2 \delta \left(E_1' + E_2 + E_2' - M \right) \qquad (20.39)$$

where now $E_1' = \sqrt{|\vec{p}_2' + \vec{p}_2|^2 + m_1^2}$.

Next let's do the p_2'-integral. We write the coordinate axes so that $\vec{p}_2' \cdot \vec{p}_2 = |\vec{p}_2'| \, |\vec{p}_2| \cos\theta = |\vec{p}_2'| \, E_2' \cos\theta$, which fixes the polar axis along the direction of $\vec{p}_2$. This gives

$$d^3p_2' = |\vec{p}_2'|^2 \, d\, |\vec{p}_2'| \sin\theta d\theta d\phi = E_2'^2 dE_2' \sin\theta d\theta d\phi \qquad (20.40)$$

The ϕ integration is easy (it gives 2π), but the θ integration requires more care because E_1' depends on θ, and so the δ-function depends non-trivially on θ. Explicitly

$$E_1' = \sqrt{E_2'^2 + |\vec{p}_2|^2 + 2\, |\vec{p}_2|\, E_2' \cos\theta + m_1^2} \equiv X\,(\theta)$$

$$\Rightarrow \frac{E_2'}{E_1'} \sin\theta d\theta = -\frac{dX}{|\vec{p}_2|} \qquad (20.41)$$

and we can rewrite the θ integration as an X integration, giving

$$\Gamma = \frac{2}{(4\pi)^5 M} \left[\int \frac{d^3p_2}{E_2} \int \frac{E_2'^2 dE_2' \sin\theta d\theta d\phi}{E_1' E_2'} \right] |\mathcal{M}|^2 \delta \left(E_1' + E_2 + E_2' - M \right)$$

$$= \frac{2\,(2\pi)}{(4\pi)^5 M} \left[\int \frac{d^3p_2 dE_2'}{|\vec{p}_2| E_2} |\mathcal{M}|^2 \int_{X_-}^{X_+} dX\, \delta \left(X + E_2 + E_2' - M \right) \right] \qquad (20.42)$$

where the range of $X\,(\theta)$ is bounded by the cosine function:

$$X_\pm = X\,(\cos\theta = \pm 1) = \sqrt{\left(E_2' \pm |\vec{p}_2| \right)^2 + m_1^2} \qquad (20.43)$$

This yields

$$\int_{X_-}^{X_+} dX\, \delta \left(X + E_2 + E_2' - M \right) = \begin{cases} 1 & \text{if } X_- < M - E_2 - E_2' < X_+ \\ 0 & \text{otherwise} \end{cases}$$

$$(20.44)$$

Noting that $X_+^2 > \left(M - E_2 - E_2' \right)^2 > X_-^2$ we obtain a range for E_2' which is

$$E_- < E_2' < E_+ \quad \text{where } E_\pm = \frac{\frac{1}{2} \left(M^2 - m_1^2 + m_2^2 \right) - M E_2}{M - E_2 \mp |\vec{p}_2|} \qquad (20.45)$$

Hence we integrate

$$\int dE_2' |\mathcal{M}|^2 \int_{X_-}^{X_+} dX\, \delta \left(X + E_2 + E_2' - M \right) = \int_{E_-}^{E_+} dE_2' |\mathcal{M}|^2 \qquad (20.46)$$

where $|\mathcal{M}|^2$ is a function of E_2' as determined from the diagrams.

In a 3-Body Decay with one massless particle, we have

$$\overline{|\mathcal{M}|}^2 = 2\left(\frac{g_{\rm w}}{M_W}\right)^4 (p_1' \cdot p_2)(p_1 \cdot p_2') \quad \text{where } p_1^\mu = (p_1' + p_2 + p_2')^\mu \quad (20.47)$$

and conservation of momentum implies

$$(p_2 + p_1')^2 = m_2^2 + m_1^2 + 2(p_2 \cdot p_1')$$

$$= (p_1 - p_2')^2 = M^2 - 2(p_1 \cdot p_2') \quad (20.48)$$

$$(p_2 \cdot p_1') = \frac{1}{2}\left(M^2 - m_2^2 - m_1^2\right) - ME_2' \quad (20.49)$$

simplifying the matrix element to

$$\overline{|\mathcal{M}|}^2 = 2\left(\frac{g_{\rm w}}{M_W}\right)^4 ME_2'\left[\frac{1}{2}\left(M^2 - m_2^2 - m_1^2\right) - ME_2'\right] \quad (20.50)$$

which gives

$$\int_{E_-}^{E_+} dE_2' \, |\mathcal{M}|^2$$

$$= M\left(\frac{g_{\rm w}}{M_W}\right)^4 \int_{E_-}^{E_+} dE_2' E_2' \left[\left(M^2 - m_2^2 - m_1^2\right) - 2ME_2'\right]$$

$$= M\left(\frac{g_{\rm w}}{M_W}\right)^4 \left[\frac{1}{2}\left(M^2 - m_2^2 - m_1^2\right)\left(E_+^2 - E_-^2\right) - \frac{2}{3}M\left(E_+^3 - E_-^3\right)\right]$$

$$\equiv M\left(\frac{g_{\rm w}}{M_W}\right)^4 J(E_2) \quad (20.51)$$

since from (20.45) $E_\pm$ are both functions of E_2.

Finally we do the p_2-integral. There is no further angular dependence, so writing

$$d^3p_2 = |\vec{p}_2|^2 \, d\,|\vec{p}_2| \, d\Omega = |\vec{p}_2| \, E_2 dE_2 d\Omega \Rightarrow \frac{d^3p_2}{|\vec{p}_2| \, E_2} = dE_2 d\Omega \quad (20.52)$$

and using $\int d\Omega = 4\pi$, we get

$$\Gamma = \frac{2(2\pi)}{(4\pi)^5 M}\left[\int \frac{d^3p_2}{|\vec{p}_2| \, E_2} \int dE_2' \, |\mathcal{M}|^2 \int_{X_-}^{X_+} dX \, \delta\left(X + E_2 + E_2' - M\right)\right]$$

$$= \frac{2(2\pi)(4\pi)}{(4\pi)^5 M}\int dE_2 M\left(\frac{g_{\rm w}}{M_W}\right)^4 J(E_2) \quad (20.53)$$

We can write

$$\frac{d\Gamma}{dE} = \frac{J(E)c^2}{\hbar \, (4\pi)^3}\left(\frac{g_{\rm w}}{M_W c^2}\right)^4 \quad (20.54)$$

as the decay rate per unit energy of particle 2, which is usually taken to be the electron. I have put in the factors of $\hbar$ and c, which you can deduce by noting that $J(E)$ has units of $(\text{Energy})^4/c^2$.

For the muon decay problem $m_1 = 0$, since the $1'$ particle is also a neutrino. Neglecting the electron mass m_2 gives

$$E_{\pm} = \frac{\frac{1}{2}M^2 - ME}{M - E \mp E} \Rightarrow E_- = \frac{1}{2}M - E \text{ and } E_+ = \frac{1}{2}M \tag{20.55}$$

yielding

$$J(E) = \left[\frac{1}{2}\left(M^2\right)\left(\frac{1}{4}M^2 - \left(\frac{1}{2}M - E\right)^2\right) - \frac{2}{3}M\left(\frac{1}{8}M^3 - \left(\frac{1}{2}M - E\right)^3\right) \right]$$

$$= \frac{1}{2}M^2E^2\left(1 - \frac{4E}{3M}\right) \tag{20.56}$$

which is what we obtained for muon decay.

20.5 Questions

1. (a) Compute the relative rate for the τ lepton to decay into a muon as compared to an electron.

   (b) Estimate the lifetime of the τ lepton, neglecting the mass of the muon and the electron. How well does this agree with experiment?

2. (a) Compute the ratio

$$\Re = \frac{\sigma\left(\nu_{\mu} + e^- \longrightarrow \nu_{\mu} + e^-\right)}{\sigma\left(\overline{\nu}_{\mu} + e^- \longrightarrow \overline{\nu}_{\mu} + e^-\right)}$$

   assuming that $m_e \ll E \ll M_W$.

   (b) Experimentally $\Re = 1.38^{+0.57}_{-0.48}$. Use this information to determine the value of $\sin^2 \theta_W$. How does it compare to the best present-day value?

3. For the scattering process $\mu^- + \nu_{\tau} \to \tau^- + \nu_{\mu}$, compute the cross-section in the CM frame to lowest-order in the couplings. Do not neglect the masses of the neutrinos.

4. Consider muonium, a bound state of μ^+ with e^-.

(a) What are the possible spins of the two lowest-energy states? Which has the lowest energy?

(b) There are two possible ways that muonium can decay. What are they?

(c) Draw a diagram for each decay mode in part (b).

(d) Only one decay mode is possible for the lowest energy states. Which one is it, and why?

(e) Compute the ratio of the decay rates for the state in which both decay modes are allowed. Which is the more likely decay?

5. Compute the following for the 3-index Levi-Civita tensor ε^{abc}:

$$(a) \quad \varepsilon^{abc}\varepsilon_{def} \qquad (b) \quad \varepsilon^{abk}\varepsilon_{dek} \qquad (c) \quad \varepsilon^{ajk}\varepsilon_{bjk} \qquad (d) \quad \varepsilon^{ijk}\varepsilon_{ijk}$$

21

Charged Weak Interactions of Quarks and Leptons

The structure of the weak interactions for quarks is somewhat different than for leptons. Before getting to the quark interactions themselves, let's first see what we can learn by applying the same methods to the decay of hadrons. After studying a few key weak hadronic decay processes, we'll then go on to explore the charged weak interactions of the quarks, comparing them to those of the leptons.

21.1 Neutron Decay

Since we experimentally observe

$$n \longrightarrow p + e^- + \overline{\nu}_e \tag{21.1}$$

we expect that charged weak interactions (mediated by the W's) amongst neutrons and protons are described by the vertex in figure 21.1.

If the neutron and proton behave like leptons, then we expect the axial current coupling c_A to equal the vector current coupling c_V and for both to equal unity. Of course we only expect this to be valid at low energies, where the substructure of the neutron and proton are irrelevant and we can treat them both as point particles. The fundamental theory will have a vertex containing only quarks as we shall see. For now let's use the vertex in figure 21.1, treating it as an effective theory, and see how far we get.

So neutron decay at low energies should (to lowest order) be represented by the diagram which is just like the diagram 20.4 for muon decay, with $\mu^- \longrightarrow n$ and $\nu_\mu \longrightarrow p$. Hence we obtain

$$\mathcal{M} = (-i)^2 \, c_V \left(\frac{g_{\mathrm{w}}}{2\sqrt{2}} \right)^2 \left[\overline{u}(p_1')\gamma^\mu \left(1 - \epsilon\gamma^5 \right) u(p_1) \right] \left[\frac{g_{\mu\nu} - q_\mu q_\nu / M_W^2}{q^2 - M_W^2} \right]$$
$$\times \left[\overline{u}(p_2)\gamma^\nu \left(1 - \gamma^5 \right) v(p_2') \right]$$
$$\simeq c_V \left(\frac{\sqrt{2}g_{\mathrm{w}}}{4M_W} \right)^2 \left[\overline{u}(p_1')\gamma^\mu \left(1 - \epsilon\gamma^5 \right) u(p_1) \right]$$

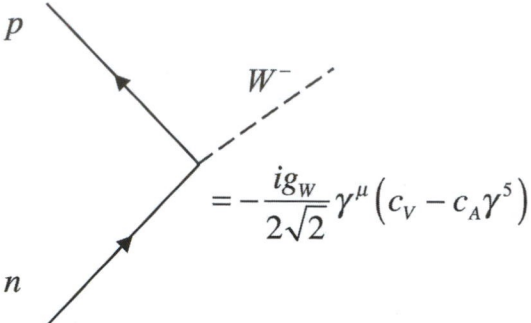

FIGURE 21.1

General form of the neutron-proton-W vertex.

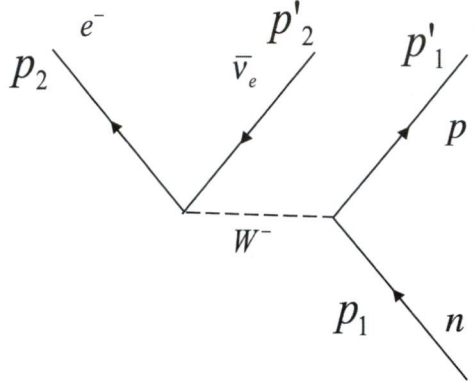

FIGURE 21.2

Lowest order neutron decay diagram.

$$\times \left[\overline{u}(p_2)\gamma_\mu \left(1 - \gamma^5\right) v(p_2') \right] \tag{21.2}$$

where $\epsilon = \frac{c_A}{c_V}$. The Casimir trick now gives

$$\sum_{\text{spins}} \left[\overline{u}(p_1')\gamma^\mu \left(1 - \epsilon\gamma^5\right) u(p_1) \right]^\dagger \left[\overline{u}(p_1')\gamma^\nu \left(1 - \epsilon\gamma^5\right) u(p_1) \right]$$

$$= 4 \left[\left(1 + \epsilon^2\right) \left(p_1'^\mu p_1^\nu + p_1^\mu p_1'^\nu - g^{\mu\nu} \left(p_1 \cdot p_1'\right) \right) \right.$$
$$\left. + 2i\epsilon\varepsilon^{\mu\alpha\nu\beta} p_{1\alpha} p_{1\beta}' + \left(1 - \epsilon^2\right) m_n m_p g^{\mu\nu} \right] \tag{21.3}$$

and so we get

$$\overline{|\mathcal{M}|}^2 = \frac{32c_V^2}{2}\left(\frac{g_W}{2\sqrt{2}M_W}\right)^4 \left[(1+\epsilon^2)\left(p_1'^\mu p_1^\nu + p_1^\mu p_1'^\nu - g^{\mu\nu}(p_1\cdot p_1')\right)\right.$$
$$+2i\epsilon\varepsilon^{\mu\alpha\nu\beta}p_{1\alpha}p_{1\beta}' + (1-\epsilon^2)m_n m_p g^{\mu\nu}\right]$$
$$\times \left[p_{2\mu}'p_{2\nu} + p_{2\mu}p_{2\nu}' - g_{\mu\nu}(p_2\cdot p_2') + i\varepsilon_{\mu\rho\nu\sigma}p_2'^\rho p_2^\sigma\right]$$
$$= \frac{c_V^2}{4}\left(\frac{g_W}{M_W}\right)^4 \left[2(1+\epsilon^2)\left((p_1'\cdot p_2')(p_1\cdot p_2) + (p_1\cdot p_2')(p_1'\cdot p_2)\right)\right.$$
$$+2i^2\epsilon\varepsilon^{\mu\alpha\nu\beta}p_{1\alpha}p_{1\beta}'\varepsilon_{\mu\rho\nu\sigma}p_2'^\rho p_2^\sigma - 2(1-\epsilon^2)m_n m_p(p_2\cdot p_2')\right]$$
$$= \frac{c_V^2}{2}\left(\frac{g_W}{M_W}\right)^4 \left[(1+\epsilon)^2(p_1\cdot p_2')(p_1'\cdot p_2) + (1-\epsilon)^2(p_1'\cdot p_2')(p_1\cdot p_2)\right.$$
$$\left.- (1-\epsilon^2)m_n m_p(p_2\cdot p_2')\right] \tag{21.4}$$

where as before terms vanish whenever the ε-tensor contracts over two identical objects and we have used eq. (20.33)

$$\varepsilon^{\mu\alpha\nu\beta}\varepsilon_{\mu\rho\nu\sigma} = -2\left(\delta^\alpha_\rho\delta^\beta_\sigma - \delta^\alpha_\sigma\delta^\beta_\rho\right)$$

to obtain the last line. We see that if $\epsilon = 1$ (i.e. if the neutron and proton have the same vector/axial-vector coupling to the W boson as the leptons do) then the matrix element is the same as for the muon decay problem.

In the rest-frame of the neutron we have $p_1^\mu = (m_n, \vec{0})$. Assuming the antineutrino is massless, from momentum conservation $p_1^\mu = (p_1' + p_2 + p_2')^\mu$ we obtain the following relations

$$(p_2\cdot p_1') = \frac{1}{2}\left(m_n^2 - m_e^2 - m_p^2\right) - m_n E_2'$$
$$(p_2'\cdot p_1') = \frac{1}{2}\left(m_n^2 + m_e^2 - m_p^2\right) - m_n E_2 \tag{21.5}$$
$$(p_2\cdot p_2') = \frac{1}{2}\left(m_p^2 - m_n^2 - m_e^2\right) + m_n E_2' + m_n E_2$$

which you can show by simplifying relations such as $(p_2 + p_1')^2 = (p_1 - p_2')^2 = m_n^2 - 2(p_1\cdot p_2')$. Hence we get

$$\overline{|\mathcal{M}|}^2 = \frac{c_V^2}{2}\left(\frac{g_W}{M_W}\right)^4 \left[(1+\epsilon)^2 m_n E_2'\left(\frac{1}{2}\left(m_n^2 - m_e^2 - m_p^2\right) - m_n E_2'\right)\right.$$
$$+ (1-\epsilon)^2 m_n E_2\left(\frac{1}{2}\left(m_n^2 - m_p^2 + m_e^2\right) - m_n E_2\right)$$
$$\left.- (1-\epsilon^2)m_n m_p\left(\frac{1}{2}\left(m_p^2 - m_n^2 - m_e^2\right) + m_n E_2' + m_n E_2\right)\right] \tag{21.6}$$

which we now insert into the 3-Body Decay rate we computed in the previous chapter:

$$\Gamma\left(n \longrightarrow p + e^- + \bar{\nu}_e\right) = \frac{1}{(4\pi)^4 m_n} \left[\int \frac{d^3 p_2}{|\vec{p}_2| E_2} \int_{E_-}^{E_+} dE_2' \overline{|\mathcal{M}|}^2\right] \qquad (21.7)$$

where now

$$E_{\pm} = \frac{\frac{1}{2}\left(m_n^2 - m_p^2 + m_e^2\right) - m_n E}{m_n - E \mp |\vec{p}_2|} \qquad (21.8)$$

with $E_2 = E$ the energy of the outgoing electron. Note that we can no longer neglect the electron mass since $m_n \simeq m_p$. Carrying out the integration (as is done in detail in the appendix of Chapter 20) gives

$$\frac{d\Gamma}{dE} = \frac{J(E)c^2}{\hbar (4\pi)^3} \left(\frac{g_w}{M_W c^2}\right)^4 \qquad (21.9)$$

where

$$J(E) = \left[\frac{1}{2}\left(m_n^2 - m_e^2 - m_p^2\right)c^2\left(E_+^2 - E_-^2\right) - \frac{2}{3}m_n\left(E_+^3 - E_-^3\right)\right] \qquad (21.10)$$

putting in the correct factors of c.

To get the total decay rate of the neutron we need to integrate $J(E)$ over E. Rather than do this integral in full detail, we can make use of the fact that $\frac{d\Gamma}{dE}$ contains a lot of small numbers. Let's define

$$\varsigma \equiv \frac{m_n - m_p}{m_n} = 0.001293 \qquad \qquad \delta \equiv \frac{m_e}{m_n} = 0.000511$$

$$\eta \equiv \frac{E}{m_n c^2} \qquad \qquad \phi \equiv \sqrt{\eta^2 - \delta^2} = \frac{|\vec{p}_2|}{m_n c} \quad (21.11)$$

and then approximate $\frac{d\Gamma}{dE}$ to leading order in these small parameters. The result is

$$\Gamma = \frac{m_n\left(c_V^2 + 3c_A^2\right)}{(4\pi)^3}\left(\frac{m_n g_w}{M_W}\right)^4 \int_\delta^\varsigma d\eta\, \eta\,(\varsigma - \eta)^2 \sqrt{\eta^2 - \delta^2} \qquad (21.12)$$

The last integral is straightforward to do, and we obtain

$$\Gamma\left(n \longrightarrow p + e^- + \bar{\nu}_e\right) = \frac{\left(c_V^2 + 3c_A^2\right)}{4} \frac{m_e c^2}{(4\pi)^3 \hbar}\left(\frac{m_e g_w}{M_W}\right)^4 \qquad (21.13)$$

$$\times \left[\frac{2\kappa^4 - 9\kappa^2 - 8}{15}\sqrt{\kappa^2 - 1} + \kappa \ln\left(\kappa + \sqrt{\kappa^2 - 1}\right)\right]$$

where

$$\kappa = \frac{\varsigma}{\delta} = \frac{m_n - m_p}{m_e} = \frac{12.93}{5.11} = 2.530 \qquad (21.14)$$

and this gives

$$\Gamma\left(n \longrightarrow p + e^- + \overline{\nu}_e\right) = 6.533 \frac{(c_V^2 + 3c_A^2)}{4} \frac{m_e c^2}{(4\pi)^3 \hbar} \left(\frac{m_e g_w}{M_W}\right)^4$$

$$= 8.0732 \times 10^{-4} \frac{(c_V^2 + 3c_A^2)}{4} \tag{21.15}$$

for a neutron lifetime of

$$\tau_n = 1239 \times \frac{4}{(c_V^2 + 3c_A^2)} \text{ sec} \tag{21.16}$$

The actual lifetime of the neutron is $\tau_n^{\text{expt}} = 885.7 \pm 0.8$ seconds [1]. It is clear that if neutrons and protons couple the same way to the W-boson as leptons do (i.e. if $c_V = c_A = 1$) then we get only very crude agreement between experiment and theory. In fact, we can independently measure c_V and c_A (from, for example, the decay of ^{14}O), and the results are

$$c_V = 1.000 \pm 0.003 \qquad\qquad c_A = 1.26 \pm 0.02$$

and the most precise measurements give [209]

$$c_A/c_V = 1.27200.0018 \tag{21.17}$$

which means we obtain

$$\tau_n = 1239 \times .694 = 859.8 \text{ sec} \tag{21.18}$$

which is better, but still out by about 3%. More accurate calculations have today yielded an agreement between theory an experiment to within about 1%.

Why can't we do much better than 1%? And why do c_V and c_A differ? The answer presumably lies with the strong interactions amongst the quarks that make up the proton. It is the task of sophisticated QCD-modified quark-model calculations – perhaps via lattice gauge theory [210] – to predict these values.

21.2 Pion Decay

Pions are bound states of up and down quarks, known in the charged case to decay into a lepton and its corresponding antineutrino. The decay of a π^- is a weak-interaction analog of positronium decay. We don't know the bound state wavefunction for a $\overline{u}$ and d to form a π^-, so let's write the diagram as in figure 21.3. The matrix element is

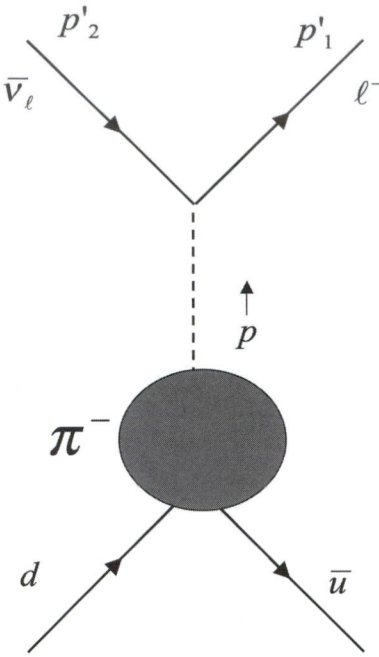

FIGURE 21.3
Lowest-order pion decay diagram.

$$\mathcal{M} = (-i)^2 \left(\frac{g_{\mathrm{w}}}{2\sqrt{2}M_W} \right)^2 \left[\overline{u}(p_1')\gamma^\mu \left(1 - \gamma^5\right) v(p_2') \right] F_\mu(p) \qquad (21.19)$$

where the quantity $F_\mu(p)$ is called the pion form factor.

 We are in a similar situation to the one we faced with deep inelastic scattering: we don't know what the vertex is between the pion and the photon. So let's proceed as we did in Chapter 17. By Lorentz covariance F_μ must be a 4-vector. Since it can only depend upon the pion 4-momentum, we must have

$$F_\mu(p) = f\left(p^2\right) p_\mu = f\left(m_\pi^2\right) p_\mu \equiv f_\pi p_\mu \qquad (21.20)$$

[handwritten: $p^2 = m_\pi^2$]

since $p^2 = m_\pi^2$. The quantity f_π is called the pion decay constant, and must be determined by experiment. Summing over the final spin states gives

$$\overline{|\mathcal{M}|}^2 = f_\pi^2 \left(\frac{g_{\mathrm{w}}}{2\sqrt{2}M_W} \right)^4 \sum_{\mathrm{spins}} \left[\overline{u}(p_1')\not{p} \left(1 - \gamma^5\right) v(p_2') \right]^\dagger \left[\overline{u}(p_1')\not{p} \left(1 - \gamma^5\right) v(p_2') \right]$$

$$= \frac{f_\pi^2}{64} \left(\frac{g_{\mathrm{w}}}{M_W} \right)^4 \mathrm{Tr} \left[\not{p}(1 - \gamma^5)\not{p}_2'\not{p} \left(1 - \gamma^5\right) \left(\not{p}_1' + m_\ell \right) \right]$$

$$= \frac{f_\pi^2}{64} \left(\frac{g_w}{M_W} \right)^4 \mathrm{Tr} \left[\not{p} \left(1 - \gamma^5 \right)^2 \not{p}_2' \not{p} \not{p}_1' \right]$$

$$= \frac{f_\pi^2}{8} \left(\frac{g_w}{M_W} \right)^4 \left[2 \left(p \cdot p_2' \right) \left(p \cdot p_1' \right) - p^2 \left(p_2' \cdot p_1' \right) \right] \qquad (21.21)$$

Assuming a massless antineutrino, momentum conservation implies

$$(p_2' \cdot p_1') = \frac{1}{2} \left((p_2' + p_1')^2 - m_\ell^2 \right) = \frac{1}{2} \left(p^2 - m_\ell^2 \right) = \frac{1}{2} \left(m_\pi^2 - m_\ell^2 \right)$$

$$(p \cdot p_2') = p_2'^2 + (p_2' \cdot p_1') = (p_2' \cdot p_1') = \frac{1}{2} \left(m_\pi^2 - m_\ell^2 \right) \qquad (21.22)$$

$$(p \cdot p_1') = p_1'^2 + (p_2' \cdot p_1') = m_\ell^2 + 2 \left(p_2' \cdot p_1' \right) = m_\pi^2$$

giving the expression

$$\overline{|\mathcal{M}|}^2 = m_\ell^2 \left(m_\pi^2 - m_\ell^2 \right) \frac{f_\pi^2}{16} \left(\frac{g_w}{M_W} \right)^4$$

This is the matrix-element for a 2-Body decay, which is really easy to compute:

$$\Gamma = \frac{S |\vec{p}|}{8\pi \hbar m_\pi^2 c} \overline{|\mathcal{M}|}^2 = \frac{f_\pi^2 m_\pi^3}{256 \pi \hbar} \left(\frac{g_w}{M_W} \right)^4 \frac{m_\ell^2}{m_\pi^2} \left(1 - \frac{m_\ell^2}{m_\pi^2} \right) \qquad (21.23)$$

[handwritten:] $\vec{p} = ?$

[handwritten:] $|\vec{p}| c = m_a c^2$ for $\vec{p}$ in Fig 21.3

From this we learn two things

[handwritten:] $\therefore \; |\vec{p}| = \frac{1}{2} m_\pi c$ for $\vec{p}$ in Eq. (21.23).

1. If $m_\ell > m_\pi$, then $\Gamma = 0$ because energy and momentum cannot be conserved. This is as we expect – the pion cannot decay into any particle more massive than itself, such as a τ-lepton.

2. Much more surprisingly, if m_ℓ were to vanish, then the pion could not decay either!

This last situation is a bit counter intuitive, and merits a bit more investigation. Computing the ratio between the decays into the electron channel and the muon channel, we find

$$\frac{\Gamma \left(\pi^- \longrightarrow e^- + \bar{\nu}_e \right)}{\Gamma \left(\pi^- \longrightarrow \mu^- + \bar{\nu}_\mu \right)} = \frac{m_e^2 \left(1 - \frac{m_e^2}{m_\pi^2} \right)^2}{m_\mu^2 \left(1 - \frac{m_\mu^2}{m_\pi^2} \right)^2} = 1.28 \times 10^{-4} \qquad (21.24)$$

which means that the π^- is about 10,000 times more likely to decay into a muon than an electron! The experimental value for this ratio is $(1.230 \pm 0.004) \times 10^{-4}$ [1]. Why is this, when phase-space considerations make the lightest particle the most probable one to decay into?

The reason is that the pion is spinless, so whenever it decays it must emit particles that spin in opposite directions. Since the antineutrino is always

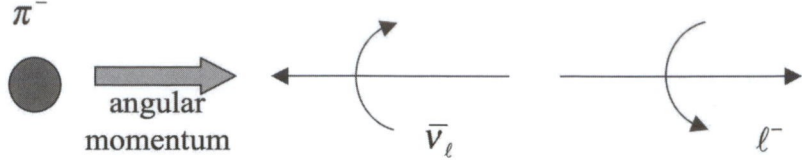

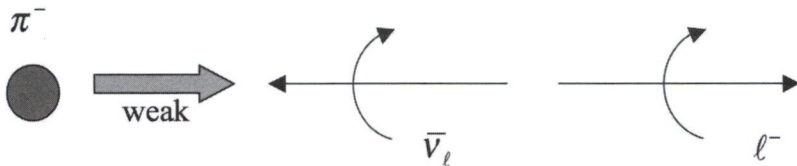

FIGURE 21.4

Final state spins in pion decay as respectively dictated by angular momentum conservation and the weak interaction.

right-handed, this means that the lepton ℓ^- must also come out in a right-handed state since it is moving in the opposite direction, as shown in the top part of figure 21.4. However, if the mass of the lepton were zero, then weak interactions would dictate that the lepton must come out left-handed, in violation of angular momentum conservation. The only way to reverse the spin is through the lepton's mass, which couples its left-handed and right-handed parts. The smaller the lepton mass, the more suppressed this spin-reversal is, and so the decay is heavily suppressed for the lighter electron as compared to the heavier muon.

21.3   Quark and Lepton Vertices

We are now ready to collect together our knowledge of the charged weak interactions, putting together in a coherent whole what we know about leptons and quarks.

Let's begin with the leptons. While we know that electrons can couple to neutrinos via a W-boson, we have never observed

$$\mu^- \longrightarrow W^- + \nu_e \qquad \nu_\mu \longrightarrow W^+ + \tau^- \qquad e^- \longrightarrow W^- + \nu_\tau \quad \text{etc.} \quad (21.25)$$

This indicates that the lepton/W-boson vertex of the weak interactions, shown in fig. 21.5, always couples leptons within a generation:

$$e^- \longrightarrow W^- + \nu_e \qquad \mu^- \longrightarrow W^- + \nu_\mu \qquad \tau^- \longrightarrow W^- + \nu_\tau \qquad (21.26)$$

but never across generations. We refer to this state of affairs as the *conser-*

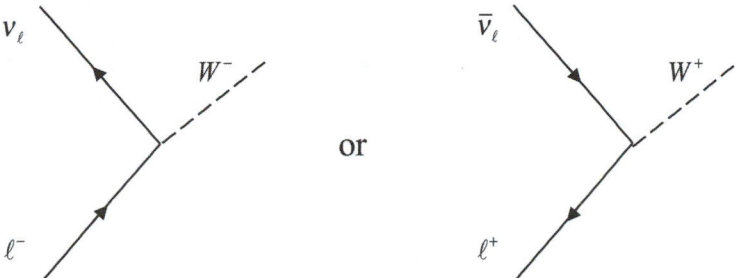

or

FIGURE 21.5

The charged leptonic weak vertex

vation of lepton number: "electron-ness" is always conserved in any physical interaction (as is "muon-ness" and "tau-ness").

The weak interaction symmetry group (as we will see) is $\mathbf{SU}(2)$. This symmetry group performs a role in the weak interactions that is analogous to the one that $\mathbf{SU}_C(3)$ performs for the strong interactions, though with some non-trivial subtle features as we shall see. All known lepton wavefunctions transform as irreducible representations of this $\mathbf{SU}(2)$ group.

Experiment also indicates that all neutrinos are left-handed. A simple interpretation of this from a group-theory viewpoint is that the right-handed leptons (i.e. the right-handed parts of the electron, muon and tau wavefunctions) are singlets in this group. Physically this means that they don't couple to charged weak bosons. The left-handed leptons (all neutrinos, plus the left-handed parts of the electron, muon and tau wavefunctions) group into irreducible doublet representations of $\mathbf{SU}(2)$:

$$\mathfrak{E}_L = \begin{pmatrix} \nu_e \\ e^- \end{pmatrix}_L, \quad \mathfrak{M}_L = \begin{pmatrix} \nu_\mu \\ \mu^- \end{pmatrix}_L, \quad \mathfrak{T}_L = \begin{pmatrix} \nu_\tau \\ \tau^- \end{pmatrix}_L \qquad (21.27)$$

and where, for example

$$\begin{pmatrix} \nu_\tau \\ \tau^- \end{pmatrix}_L = \begin{pmatrix} \psi_L\left(\nu_\tau\right) \\ \psi_L\left(\tau^-\right) \end{pmatrix}$$

and from eq. (20.5) the Dirac wavefunctions are $\psi_L \equiv \frac{1}{2}\left(1-\gamma^5\right)\psi$. It's common in writing the expressions for weak doublets to just use the names of the leptons and quarks, and I will continue this practice. But keep in mind each entry is really a Dirac wavefunction.

So a "spin-up" $\mathfrak{E}_L$-particle is an electron-neutrino; a "spin-down" $\mathfrak{T}_L$-particle is a tau lepton. In other words, what we generally regard as two distinct particles, the weak interactions regard as two states of a single particle!

This feature of the weak interactions takes some getting used to. However, we have seen something like it before in both electromagnetism and the strong interactions. In electromagnetism we have a vertex corresponding to the process $e^- \to e^- + \gamma$. Since the photon is spin-1, angular momentum conservation forces the electron to undergo a spin-flip in this process, for example $e^{-(\uparrow)} \to e^{-(\downarrow)} + \gamma$. We normally don't regard the spin-up electron as a distinct particle from the spin-down electron – instead we regard both as different states of one particle. In QCD there is a vertex in which a quark emits a gluon – for example $u \to u + g$. Again angular momentum conservation – and also color symmetry – imply that the quark must undergo a spin-flip and that its color must change. So the vertex is really something like $u^{R(\uparrow)} \to u^{B(\downarrow)} + g^{R\bar{B}}$ – again, we regard $u^{R(\uparrow)}$ and $u^{B(\downarrow)}$ as different states of the up quark, and not different particles.

From the perspective of the weak interactions, a lepton and its neutrino partner are likewise different states of the same particle. Historically we have regarded them as different particles because their masses and charges differ. However, as far as the charged weak interactions are concerned, they are different states of one particle, with vertices given in figure 21.6. For example,

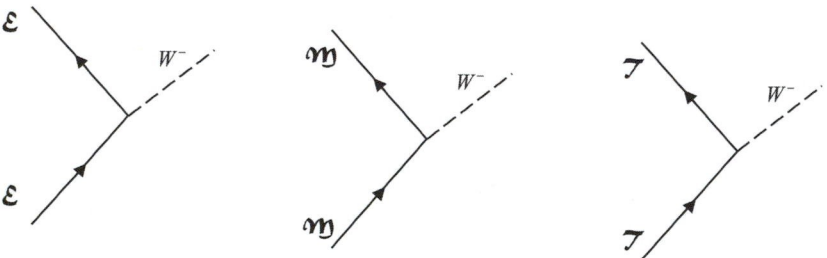

FIGURE 21.6

Generalized weak vertices for the electron, muon and tau leptons

a left-handed $\mathfrak{T}_L$ wavefunction, say, emits a W^- boson and continues on its way. In the process it flips from a τ^- state to a ν_τ state, analogous to the way a spin-$\frac{1}{2}$ particle flips its spin when it emits a photon, or a quark changes

Margin note: $S_z = 0$ is impossible for photon because it's massless.

color when it emits a gluon.

What about the quarks? Since we know that the process underlying neutron decay

$$n \longrightarrow p + e^- + \bar{\nu}_e \tag{21.28}$$

is

$$d \longrightarrow u + e^- + \bar{\nu}_e \tag{21.29}$$

as shown in figure 21.7 we might also guess that left-handed quarks group

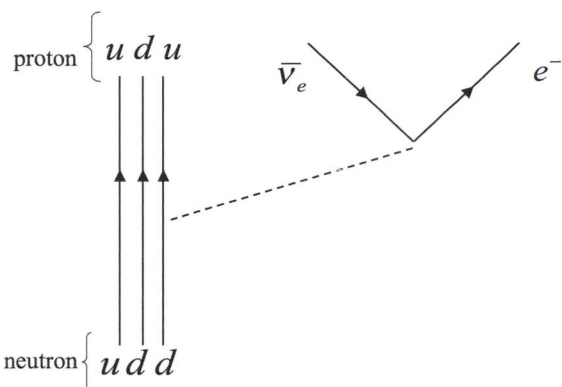

FIGURE 21.7

Neutron decay in terms of quarks

themselves into irreducible doublet representations of **SU**(2):

$$\begin{pmatrix} u \\ d \end{pmatrix}_L, \quad \begin{pmatrix} c \\ s \end{pmatrix}_L, \quad \begin{pmatrix} t \\ b \end{pmatrix}_L \tag{21.30}$$

However, experiment tells us this is wrong because we also observe

$$\Lambda \longrightarrow p + e^- + \bar{\nu}_e \tag{21.31}$$

which corresponds to

$$s \longrightarrow u + e^- + \bar{\nu}_e \tag{21.32}$$

as is easily seen by looking at the quark content on both sides of this reaction. We also know that

$$B^- \longrightarrow \pi^0 + e^- + \bar{\nu}_e \tag{21.33}$$

which in quark terms is

$$b \longrightarrow u + e^- + \bar{\nu}_e \tag{21.34}$$

and that there are many other flavor-changing decays. If this were not the case, then we would have three more conservation laws: conservation of "up/down-ness," "charm/strange-ness" and "top/bottom-ness." This would mean that the lightest strange meson (the K^-), the lightest strange baryon (the Λ) and the lightest B-meson would be stable and never decay. We don't observe this – so somehow we must take the crossover between generations into account.

We will proceed in stages, following the historical path that led to our current understanding. Some terminology will be helpful here. In general there are three kinds of weak interaction processes for hadrons:

1. **Leptonic:** all decay products are leptons (e.g. $\pi^- \longrightarrow \ell^- + \overline{\nu}_\ell$)

2. **Semi-leptonic:** some decay products are leptons (e.g. $\Sigma^0 \to \Sigma^+ + \ell^- + \overline{\nu}_\ell$)

3. **Non-leptonic:** no decay products are leptons (e.g. $\Lambda \longrightarrow p + \pi^-$)

These are progressively harder to analyze because of the progressively greater importance of the strong interactions in each case. The leptonic decays are the easiest to understand because strong interaction effects are minimized. Hence to understand cross-over between generations, the simplest thing to look at is the leptonic decay of the Kaon. This is the process Nicola Cabbibo concentrated on when he wanted to understand the structure of weak hadronic decays.

In 1963 Cabbibo suggested [211] (shortly after the 3-quark model had been proposed) that quark charged weak-vertices were of the form given in figure 21.8 This makes the $u/d/W$ vertex just like the $\nu_e/e/W$ vertex except for the constant factor of $\cos\theta_c$; likewise the $u/s/W$ vertex is just like the $\nu_e/e/W$ vertex except for the $\sin\theta_c$ factor. If θ_c were zero, then the weak interactions would respect conservation of quark generations.

Experimentally, the decays of Kaons and neutrons together imply [1]

$$\theta_c = 13.04^o \tag{21.35}$$

and we now call this angle the *Cabbibo angle*. This small value of θ_c indicates that, empirically, we almost have conservation of quark generations – but not quite! To see how this works, consider the decay

$$K^- \longrightarrow \overline{\nu}_\ell + \ell^- \tag{21.36}$$

where $\ell^- = e^-$ or μ^-. The diagram 21.9 is just like pion decay (fig. 21.3), so we obtain

$$\mathcal{M} = (-i)^2 \left(\frac{g_{\rm w}}{2\sqrt{2}M_W}\right)^2 \left[\overline{u}(p_1')\gamma^\mu \left(1 - \gamma^5\right) v(p_2')\right] f_K p_\mu \tag{21.37}$$

as the matrix element. All that is different is that f_K replaces f_π. Hence the decay rate is

$$\Gamma = \frac{f_K^2 m_K^3}{256\pi\hbar} \left(\frac{g_{\rm w}}{M_W}\right)^4 \frac{m_\ell^2}{m_K^2} \left(1 - \frac{m_\ell^2}{m_K^2}\right)^2 \tag{21.38}$$

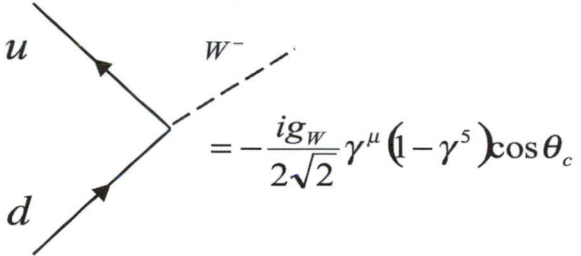

FIGURE 21.8
Cabbibo's proposed W-boson/quark vertices.

The strong coupling strength binding s and $\bar{u}$ into a Kaon is presumably the same as that which binds d and $\bar{u}$ into a pion. The only difference should be in the weak couplings: $\cos\theta_c$ for the pion and $\sin\theta_c$ for the Kaon. Hence we expect

$$\frac{f_K}{f_\pi} = \frac{\sin\theta_c}{\cos\theta_c} = \tan\theta_c \tag{21.39}$$

yielding

$$\frac{\Gamma\left(K^- \longrightarrow \ell^- + \bar{\nu}_\ell\right)}{\Gamma\left(\pi^- \longrightarrow \ell^- + \bar{\nu}_\ell\right)} = \tan^2\theta_c \frac{m_K}{m_\pi} \frac{\left(1 - \frac{m_\ell^2}{m_K^2}\right)^2}{\left(1 - \frac{m_\ell^2}{m_\pi^2}\right)^2} \tag{21.40}$$

which for $\theta_c = 13.04^\circ$ gives

$$\frac{\Gamma\left(K^- \longrightarrow \ell^- + \bar{\nu}_\ell\right)}{\Gamma\left(\pi^- \longrightarrow \ell^- + \bar{\nu}_\ell\right)} = \begin{cases} .19 & \text{for } \ell^- = e^- \\ .96 & \text{for } \ell^- = \mu^- \end{cases} \tag{21.41}$$

and compared to experiment

$$\frac{\Gamma^{\text{expt}}\left(K^- \longrightarrow \ell^- + \bar{\nu}_\ell\right)}{\Gamma^{\text{expt}}\left(\pi^- \longrightarrow \ell^- + \bar{\nu}_\ell\right)} = \begin{cases} .26 & \text{for } \ell^- = e^- \\ 1.34 & \text{for } \ell^- = \mu^- \end{cases} \tag{21.42}$$

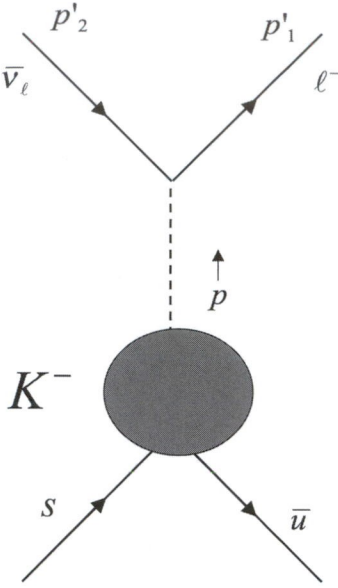

FIGURE 21.9
Decay of a Kaon into a lepton and the associated antineutrino.

we see that there is a crude agreement. The mismatch between the theoretical and experimental values is due to strong interaction effects, which are reasonably well understood [212].

21.4   The GIM Mechanism

Cabbibo's model worked reasonably well for many strangeness-changing decays. But it also predicted the decay $K^0 \longrightarrow \mu^- + \mu^+$ via the diagram* given in fig. 21.10. This seems harmless enough – experiment indicated that the branching ratio was 7.3×10^{-9} (today this is known more precisely to be $7.18 \pm 0.17 \times 10^{-9}$ [213]) – but the problem was that Cabbibo's theory predicted from the "box" diagram in fig. 21.10 a known quantity $\mathcal{I}\left(p_i^\mu; m_u^2\right)$ from a well-determined (and finite) loop integral mulitplied by $\sin\theta_c \cos\theta_c$. This theoretical result yielded a quantity much larger than what was observed.

---

*There is another diagram with the internal W's crossed over that I have left out.

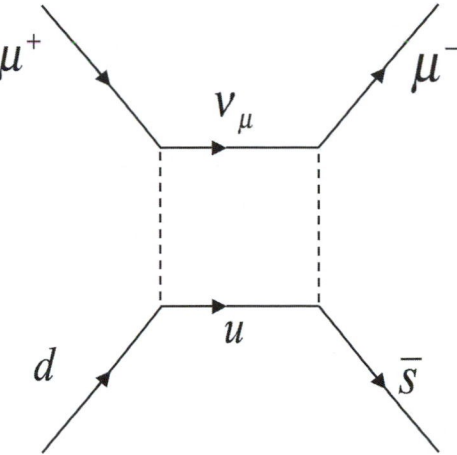

FIGURE 21.10
Expected lowest order diagram $K^0 \longrightarrow \mu^+\mu^-$ decay.

More generally there were some other strange features in Kaon decays that were not satisfactorily explained by the Cabbibo theory. For example the decays

$$K^+ \longrightarrow \pi^+ + \nu_e + \bar{\nu}_e \qquad\qquad K^0 \longrightarrow \pi^0 + \nu_e + e^+ \qquad (21.43)$$

might be expected to proceed with equal rates – after all, there is a similar change in strangeness for each. The only difference between them is that the lepton charge in the final state is neutral for the K^+ decay (a so-called neutral current process). Yet experiment indicated that

$$\text{BR}\left(K^+ \longrightarrow \pi^+ + \nu_e + \bar{\nu}_e\right) = 1.5 \pm 1.3 \times 10^{-10}$$
$$\text{BR}\left(K^0 \longrightarrow \pi^0 + \nu_e + e^+\right) = 4.98 \pm 0.07 \times 10^{-2} \qquad (21.44)$$

a difference of many orders of magnitude[†]. Evidently strangeness-changing processes prefer charged leptons in the final state and strangeness-changing neutral currents are suppressed – but why?

In 1970 Glashow, Iliopoulos and Maiani (GIM) proposed that the solution to these problems was the existence of a new quark that was a counterpart of the up quark [172]. It had the same charge, but orthogonal couplings to the down and strange quarks, as shown in fig. 21.11. If this new quark exists, and

---

[†]I've used present-day experimental values [1], but the problem was clear when these measurements were much less accurate.

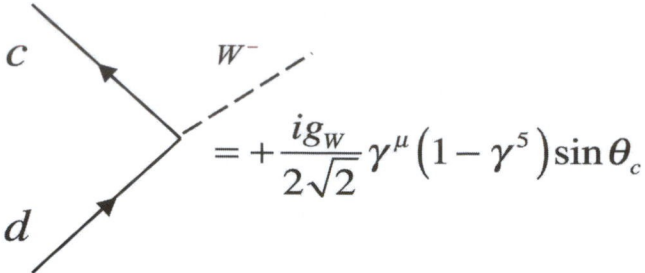

$$= +\frac{ig_W}{2\sqrt{2}}\gamma^\mu\left(1-\gamma^5\right)\sin\theta_c$$

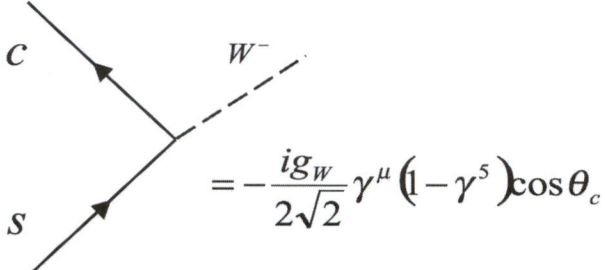

$$= -\frac{ig_W}{2\sqrt{2}}\gamma^\mu\left(1-\gamma^5\right)\cos\theta_c$$

FIGURE 21.11
The GIM vertices.

if its couplings are as in fig. 21.11, then the actual lowest-order diagrams for
$K^0 \longrightarrow \mu^- + \mu^+$ are given in fig. 21.12. These diagrams thus add as follows

$$\mathcal{M} = \sin\theta_c \cos\theta_c \mathcal{I}\left(p_i^\mu; m_u^2\right) - \cos\theta_c \sin\theta_c \mathcal{I}\left(p_i^\mu; m_c^2\right) \simeq \frac{m_u^2}{m_c^2}\mathcal{F}\left(p_i^\mu\right) \quad (21.45)$$

where the p_i^μ are the external momenta in the problem. Both diagrams give
the same integral with m_u^2 replaced with m_c^2, and so (because of the orthogonal
couplings in fig. 21.11) the integrals cancel up to a factor of $\frac{m_u^2}{m_c^2}$. Setting the
observed rate equal to the theoretical value yields $m_c \simeq 1.5$ GeV.

GIM thought that this would be a charming solution to the puzzle of $K^0 \longrightarrow$
$\mu^- + \mu^+$, and so they named the new quark the charm quark. Nearly four
years later it was discovered [169, 170], and had just the mass predicted by
the GIM mechanism!

So in the GIM-Cabbibo scheme, the "correct" quarks to use in the charged
weak interactions are not d and s, but rather

$$\begin{pmatrix} u \\ d' \end{pmatrix}_L = \begin{pmatrix} u \\ d\cos\theta_c + s\sin\theta_c \end{pmatrix}_L \quad \text{and} \quad \begin{pmatrix} c \\ s' \end{pmatrix}_L = \begin{pmatrix} c \\ -d\sin\theta_c + s\cos\theta_c \end{pmatrix}_L$$
$$(21.46)$$

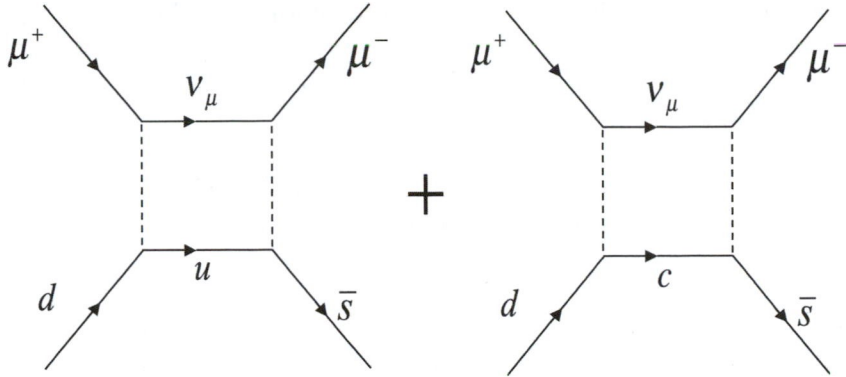

FIGURE 21.12
Lowest order diagrams for $K^0 \longrightarrow \mu^+ + \mu^-$.

In other words, the "elementary particles" of the electromagnetic and strong interactions are d and s, but the "elementary particles" of the weak interactions are d' and s', where

$$\begin{pmatrix} d' \\ s' \end{pmatrix}_L = \begin{pmatrix} \cos\theta_c & \sin\theta_c \\ -\sin\theta_c & \cos\theta_c \end{pmatrix} \begin{pmatrix} d \\ s \end{pmatrix}_L \qquad (21.47)$$

and the matrix transforming between the two is called the Cabbibo matrix. This mechanism suppresses the puzzling strangeness-changing neutral current processes noted above. Since the right-handed parts of the quark wavefunctions don't experience the weak interaction, we can make the above redefinition for these parts too.

21.5 The CKM Matrix

Before charm was discovered in the "November Revolution" of 1974, Makoto Kobayashi and Toshihide Maskawa turned their attention to the problem of $\mathbb{CP}$ violation in Kaon decay in 1973 [214]. They wanted to see if a GIM-type mechanism could be used to explain it.

To do this, they needed a complex number in the Cabbibo matrix. However, if you put one in, you can always remove it by redefining the phases of the down and strange wavefunctions. This is because the most general relationship between the weak (primed) eigenstates and the mass (unprimed) eigenstates is given by a 2×2 unitary matrix $\mathcal{V}$, whose most general form is

$$\begin{pmatrix} d' \\ s' \end{pmatrix} = \mathcal{V} \begin{pmatrix} d \\ s \end{pmatrix} = \begin{pmatrix} e^{i\phi_1} \cos\theta & e^{i(\phi_1+\phi_2)} \sin\theta \\ -e^{i\phi_3} \sin\theta & e^{i(\phi_2+\phi_3)} \cos\theta \end{pmatrix} \begin{pmatrix} d \\ s \end{pmatrix} \qquad (21.48)$$

which has 4 parameters, as expected for a unitary matrix. You can easily check (or deduce!) that this matrix is unitary. However, the 4 parameters are not all physically relevant, which can be seen by writing the above as

$$\begin{pmatrix} d' \\ s' \end{pmatrix} = \begin{pmatrix} e^{i\phi_1} & 0 \\ 0 & e^{i\phi_3} \end{pmatrix} \begin{pmatrix} \cos\theta & \sin\theta \\ -\sin\theta & \cos\theta \end{pmatrix} \begin{pmatrix} 1 & 0 \\ 0 & e^{i\phi_2} \end{pmatrix} \begin{pmatrix} d \\ s \end{pmatrix} \qquad (21.49)$$

Redefining the quark wavefunctions so that

$$\begin{pmatrix} \tilde{d} \\ \tilde{s} \end{pmatrix} = \begin{pmatrix} 1 & 0 \\ 0 & e^{i\phi_2} \end{pmatrix} \begin{pmatrix} d \\ s \end{pmatrix} \qquad (21.50)$$

$$\begin{pmatrix} \tilde{d}' \\ \tilde{s}' \end{pmatrix} = \begin{pmatrix} e^{-i\phi_1} & 0 \\ 0 & e^{-i\phi_3} \end{pmatrix} \begin{pmatrix} d' \\ s' \end{pmatrix} \qquad (21.51)$$

we see that the only relevant physical parameter is θ – the Cabbibo angle!

In a 6-quark model this strategy does not work because there are not enough quark wavefunctions to absorb all of the phases in a 3×3 unitary matrix. So their idea was to consider a 6-quark model in which the charge $\left(-\frac{1}{3}\right)$ quarks that couple to the W-bosons are related to their strong/electromagnetic counterparts via

$$\begin{pmatrix} d' \\ s' \\ b' \end{pmatrix}_L = \begin{pmatrix} V_{ud} & V_{us} & V_{ub} \\ V_{cd} & V_{cs} & V_{cb} \\ V_{td} & V_{ts} & V_{tb} \end{pmatrix} \begin{pmatrix} d \\ s \\ b \end{pmatrix}_L \qquad (21.52)$$

so that the quark weak-doublet structure is

$$\mathfrak{D}_L = \begin{pmatrix} u \\ d' \end{pmatrix}_L, \quad \mathfrak{S}_L = \begin{pmatrix} c \\ s' \end{pmatrix}_L, \quad \mathfrak{B}_L = \begin{pmatrix} t \\ b' \end{pmatrix}_L \qquad (21.53)$$

with diagrams as shown in fig. 21.13.

From this perspective the weak interactions conserve up/down'-ness, charm/strange'-ness, and top/bottom'-ness! Somewhat analogous to the leptonic case, a "spin-up" $\mathfrak{S}_L$-particle is a charm quark; a "spin-down" $\mathfrak{B}_L$-particle is a b' quark, which is a mixture of the d, s, and b quarks (and not just a b quark).

The matrix relating primed quark wavefunctions to the unprimed ones is a unitary matrix known as the CKM matrix (for Cabbibo-Kobayashi-Maskawa). It is a 3×3 generalization of the Cabbibo matrix. By redefining quark wavefunction phases, it can be written in the standard form

$$\mathcal{V} \equiv \begin{pmatrix} V_{ud} & V_{us} & V_{ub} \\ V_{cd} & V_{cs} & V_{cb} \\ V_{td} & V_{ts} & V_{tb} \end{pmatrix} = \begin{pmatrix} c_1 & s_1 c_3 & s_1 s_3 \\ -s_1 c_2 & c_1 c_2 c_3 - s_2 s_3 e^{i\delta} & c_1 c_2 s_3 + s_2 c_3 e^{i\delta} \\ -s_1 s_2 & c_1 s_2 c_3 + c_2 s_3 e^{i\delta} & c_1 s_2 s_3 - c_2 c_3 e^{i\delta} \end{pmatrix} \qquad (21.54)$$

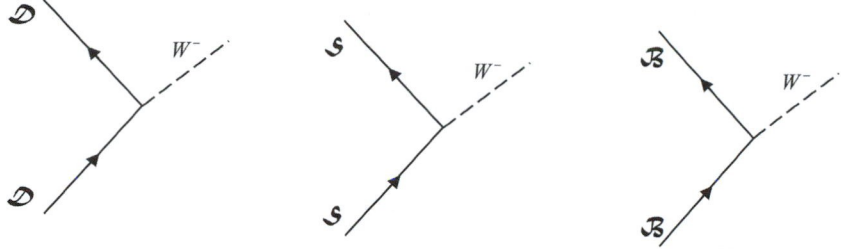

FIGURE 21.13

Generalized weak vertices for the down-type, strange-type and bottom-type quarks.

where $c_i = \cos\theta_i$, $s_i = \sin\theta_i$.

Suppose we set $\theta_2 = \theta_3 = 0$. It's easy to see from eq. (21.54) that the 3rd generation doesn't mix with the other two, and we recover the original Cabbibo matrix, with $\theta_1 = \theta_c$. Empirically we do observe mixing of the third generation with the other two (in decays of B^0-mesons; recall $B^- \longrightarrow \pi^0 + e^- + \bar{\nu}_e$), but it is small, as we infer empirically from the relatively long lifetime of the B meson and the small branching ratio of decays into uncharmed mesons.

The parametrization of the CKM matrix given in eq. (21.54) is the original one used by Kobayashi and Maskawa. It is more common now to use Euler angles $(\theta_{12}, \theta_{23}, \theta_{13})$ and one CP-violating phase (δ) [1], in which case

$$\mathcal{V} = \begin{pmatrix} c_{12}c_{13} & s_{12}c_{13} & s_{13}e^{-i\delta} \\ -s_{12}c_{23} - c_{12}s_{23}s_{13}e^{i\delta} & c_{12}c_{23} - s_{12}s_{23}s_{13}e^{i\delta} & s_{23}c_{13} \\ s_{12}s_{23} - c_{12}c_{23}s_{13}e^{i\delta} & -c_{12}s_{23} - s_{12}c_{23}s_{13}e^{i\delta} & c_{23}c_{13} \end{pmatrix} \quad (21.55)$$

where again $c_{ij} = \cos\theta_{ij}$ and $s_{ij} = \sin\theta_{ij}$. This form of the CKM matrix makes it easy to see how the different generations couple to each other, since couplings between quark generations i and j vanish if $\theta_{ij} = 0$.

The values of the four angles in the CKM matrix are constants of nature, just the way that the coupling constant of electromagnetism, Planck's constant, and the speed of light are constants of nature. These angles tell us how strongly the different quark generations couple to each other. They must be empirically measured. Experimentally, we know only the magnitudes of the matrix values, which are [1]

$$|V| \equiv \begin{pmatrix} |V_{ud}| & |V_{us}| & |V_{ub}| \\ |V_{cd}| & |V_{cs}| & |V_{cb}| \\ |V_{td}| & |V_{ts}| & |V_{tb}| \end{pmatrix}$$

$$= \begin{pmatrix} .97383 \pm 0.00024 & .2272 \pm 0.0010 & .00396 \pm 0.00009 \\ .2271 \pm 0.0010 & .97296 \pm 0.00024 & .04221 \pm 0.00045 \\ .00814 \pm 0.00048 & .04161 \pm 0.00012 & .999100 \pm 0.000034 \end{pmatrix} \quad (21.56)$$

We see that, empirically, the third generation hardly mixes at all with the other two – $s_{12} \gg s_{23} \gg s_{13}$. This is why Cabbibo's original idea worked so well.

The Standard model does not explain the origin of the values of the entries in the CKM matrix. So far there have been no successful extensions of the Standard model that have offered a convincing rationale for why they have the values they do. At this point in time the best we can do is to experimentally check these values for consistency. This is done via something called the *unitarity triangle*, which exploits the fact that the CKM matrix is unitary, i.e. $V^\dagger V = 1$. For example, the third row and first column entry gives $(V^\dagger V)_{31} = 0$, or

$$V_{ud}V_{ub}^* + V_{cd}V_{cb}^* + V_{td}V_{tb}^* = 0 \tag{21.57}$$

which is a sum of three complex numbers that add up to zero. These numbers can be drawn as vectors in the complex plane, and since they sum up to zero they form a closed triangle, as shown in fig. 21.14. The relative angles between two sides are the arguments of the ratios of these complex numbers

$$\alpha = \arg\left(-\frac{V_{td}V_{tb}^*}{V_{ud}V_{ub}^*}\right) \quad \beta = \arg\left(-\frac{V_{cd}V_{cb}^*}{V_{td}V_{tb}^*}\right) \quad \gamma = \arg\left(-\frac{V_{ud}V_{ub}^*}{V_{cd}V_{cb}^*}\right) \tag{21.58}$$

and since it's a triangle we must have $\alpha + \beta + \gamma = 180°$. Other unitarity

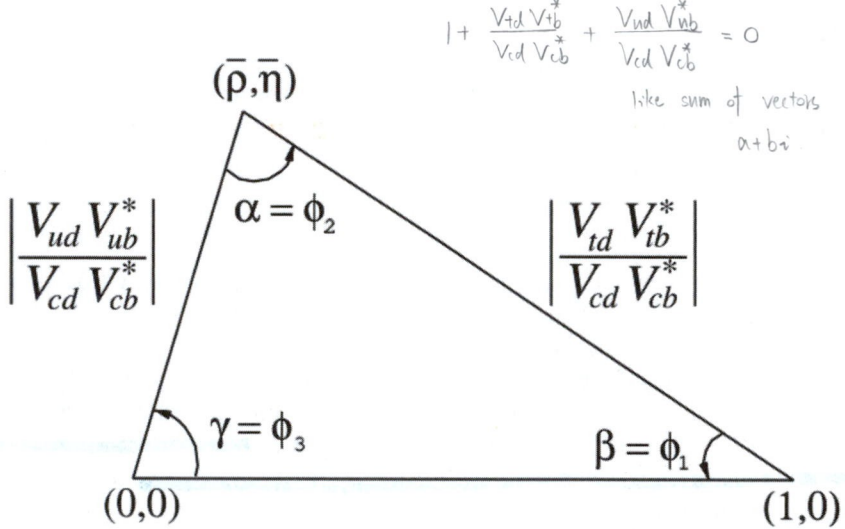

FIGURE 21.14

The unitarity triangle. Image courtesy of the Particle Data Group [1].

triangles can be defined from the CKM matrix, but the one in fig. 21.14 is the most commonly used since its matrix elements are the best known.

The areas of all unitarity triangles are the same, and have half the value of the Jarlskog invariant $\mathfrak{J}$, defined as [215]

$$Im\left(V_{ij}V_{kl}V_{il}^{*}V_{kj}^{*}\right) = \mathfrak{J}\sum_{m,n}\varepsilon_{ikm}\varepsilon_{jln} \qquad (21.59)$$

One computes the Jarlskog invariant by simply choosing particular components of the CKM matrix; for example, $\mathfrak{J} = Im\left(V_{ud}V_{cs}V_{us}^{*}V_{cd}^{*}\right)$. The unitarity of the CKM matrix ensures that the value is numerically the same no matter what the choice. It is a phase-independent measure of $\mathbb{CP}$-violation, and has the experimental value [1]

$$\mathfrak{J} = (3.08 \pm 0.17) \times 10^{-5} \qquad (21.60)$$

One of the important goals of modern particle physics is to overconstrain the CKM matrix. Finding that its parameters do not yield unitarity triangles would be a clear signal of physics beyond the Standard Model.

21.6 Questions

1. Calculate the decay rate for $\tau^{-} \longrightarrow \pi^{-} + \nu_{\tau}$. Using the measured value for f_{π} and the lifetime of the τ^{-}, what is the branching ratio for this decay? How does it compare to experiment?

2. Find the branching ratio for the decays $\tau^{-} \longrightarrow \pi^{-} + \nu_{\tau}$ and $\tau^{-} \longrightarrow K^{-} + \nu_{\tau}$.

3. In the decay $\pi^{-} \longrightarrow \mu^{-} + \bar{\nu}_{\mu}$, suppose that the weak coupling vertex factor is $\frac{g_{W}}{2\sqrt{2}}\gamma^{\mu}\left(1 - \varepsilon\gamma_{5}\right)$.

   (a) Calculate the spin-averaged and spin-summed matrix element for this process.

   (b) For what value of ε is the ratio

   $$\mathfrak{R} = \frac{\Gamma\left(\pi^{-} \longrightarrow \mu^{-} + \bar{\nu}_{\mu}\right)}{\Gamma\left(\pi^{-} \longrightarrow e^{-} + \bar{\nu}_{e}\right)}$$

   maximal?

   (c) How might you determine ε from experiment?

4. Consider the decays $D_{s}^{-} \rightarrow \tau^{-} + \bar{\nu}_{\tau}$ and $D_{s}^{-} \rightarrow \mu^{-} + \bar{\nu}_{\mu}$. The decay rates can can computed in terms of a decay constant $f_{D_{s}}$.

   (a) What is the ratio of the decay rates?

   (b) Calculate the expected branching ratio for the decay into $\tau^{-} + \bar{\nu}_{\tau}$ if $f_{D_{s}} = 280$ MeV.

5. For the process in figure 21.15 show that

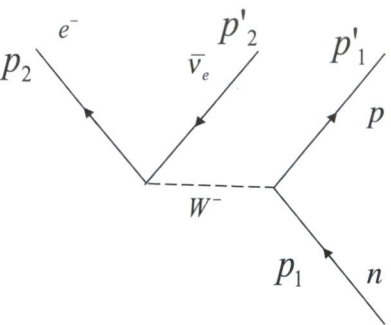

FIGURE 21.15
Neutron decay diagram.

$$\sum_{\text{spins}} \left[\overline{u}(p_1')\gamma^\mu \left(1 - \epsilon\gamma^5\right) u(p_1)\right]^\dagger \left[\overline{u}(p_1')\gamma^\nu \left(1 - \epsilon\gamma^5\right) u(p_1)\right]$$
$$= 4\left[\left(1 + \epsilon^2\right)\left(p_1'^\mu p_1^\nu + p_1^\mu p_1'^\nu - g^{\mu\nu}\left(p_1 \cdot p_1'\right)\right)\right.$$
$$\left. + 2i\epsilon\varepsilon^{\mu\alpha\nu\beta} p_{1\alpha}p_{1\beta}' + \left(1 - \epsilon^2\right) m_n m_p g^{\mu\nu}\right]$$

6. Suppose that the muon neutrino has a mass $\mathfrak{m}$.

   (a) Compute the ratio $\mathfrak{R}$, where

$$\mathfrak{R} = \frac{\Gamma\left(\pi^+ \longrightarrow e^+ + \nu_e\right)}{\Gamma\left(\pi^+ \longrightarrow \mu^+ + \nu_\mu\right)}$$

   (b) For what value of $\mathfrak{m}$ is this quantity maximized? For what value $\mathfrak{m}$ of is it minimized?

   (c) Could the experimental value of $\mathfrak{R}$ be used to determine an empirical value for $\mathfrak{m}$? Why or why not?

7. Show that for the momenta defined in figure 21.15

$$(p_2 \cdot p_1') = \frac{1}{2}\left(m_n^2 - m_e^2 - m_p^2\right) - m_n E_2'$$
$$(p_2' \cdot p_1') = \frac{1}{2}\left(m_n^2 + m_e^2 - m_p^2\right) - m_n E_2$$
$$(p_2 \cdot p_2') = \frac{1}{2}\left(m_p^2 - m_n^2 - m_e^2\right) + m_n E_2' + m_n E_2$$

8. Beginning with the expression for the matrix element $\overline{|\mathcal{M}|}^2$ for the decay of the neutron

$$
\begin{aligned}
\overline{|\mathcal{M}|}^2 = \frac{c_V^2}{2} \left(\frac{g_W}{M_W} \right)^4 &\left[(1+\epsilon)^2 \, m_n E_2' \left(\frac{1}{2} \left(m_n^2 - m_e^2 - m_p^2 \right) - m_n E_2' \right) \right. \\
&+ (1-\epsilon)^2 \, m_n E_2 \left(\frac{1}{2} \left(m_n^2 - m_p^2 + m_e^2 \right) - m_n E_2 \right) \\
&\left. - \left(1 - \epsilon^2 \right) m_n m_p \left(\frac{1}{2} \left(m_p^2 - m_n^2 - m_e^2 \right) + m_n E_2' + m_n E_2 \right) \right]
\end{aligned}
$$

given in eq. (21.6), find the decay rate of the neutron in terms of the parameters

$$
\varsigma \equiv \frac{m_n - m_p}{m_n} \qquad\qquad \delta \equiv \frac{m_e}{m_n}
$$

$$
\eta \equiv \frac{E_2}{m_n} \qquad\qquad \phi \equiv \sqrt{\eta^2 - \delta^2} = \frac{|\vec{p}_2|}{m_n}
$$

to lowest order in these parameters. Insert the appropriate factors of c.

22

Electroweak Unification

Electroweak theory – sometimes referred to as the Glashow-Salam-Weinberg model – is a model that unifies weak and electromagnetic interactions. In other words, from the perspective of the Standard Model these two forces are part of a more unified whole. The first paper on unification of electromagnetic and weak interactions by Glashow in 1961 [216] required such neutral interactions, and in 1968 Glashow's model was modified by Weinberg and Salam [217] to include a physical mechanism that made the force carriers (the weak vector bosons W and Z) massive. In 1971 Gerard 't Hooft and Martinus Veltman showed that this model is renormalizable [218] (with Ben Lee carrying out similar work independently [219]), and in 1983 the W and Z particles were discovered [73]. All subsequent particle physics experiments have confirmed this theory.

To properly lay out the basic structure of this theory we must first turn our attention of an aspect of the weak interactions that I have so far ignored, namely neutral currents. The Z boson is the force-carrier of the weak neutral force. Let's begin by looking at its properties.

22.1 Neutral Currents

Neutral currents were first postulated by Bludman in 1958 [220], shortly after the W-boson was proposed as the force carrier underlying the weak interactions. He suggested that the W-boson had an electrically neutral partner called the Z. Its basic interaction, shown in fig. 22.1, indicates that the Z couples to any fermion f. Note that the identity of f is preserved in the interaction. There is no vertex for which $e^- \longrightarrow Z + \mu^-$ or $s \longrightarrow Z + t$, for example. Such processes would not be consistent with low-energy meson decays, such as $K^0 \longrightarrow \mu^- + \mu^+$. This is what made neutral currents so difficult to observe – they would be swamped by electromagnetic effects, which are also neutral currents: the photon is electrically neutral!. For this reason many people doubted that Bludman's postulate was correct.

The first evidence experimentally for weak neutral currents came in 1973 at CERN in the Gargamelle bubble chamber, shown in fig. 22.2. The searches for neutral currents in previous neutrino experiments had resulted in discour-

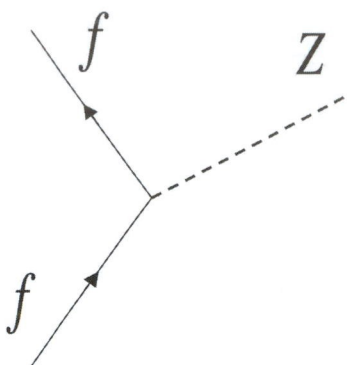

FIGURE 22.1

Bludman's general proposal for the coupling of the Z to fermions.

agingly low limits, and by the late 1960s it was commonly assumed that no weak neutral currents existed. People were more excited about the possibility of proton substructure. This provoked the question as to what structure would be revealed by the W in neutrino experiments, analogous to what the photon revealed in deep-inelastic electron-proton scattering.

André Lagarrigue, André Rousset and Paul Musset worked out a proposal for a neutrino experiment that aimed to increase the event rate by an order of magnitude, which meant building a large heavy-liquid bubble chamber. The key challenge for the experiment was to deal with the unavoidable background of events in which a charged hadron leaves the visible volume of the chamber without visible interaction. This "fakes" a muon event. Events with a muon candidate were collected in one category, A, while events consisting of secondaries that were all identified as hadrons were collected in a second category, B. Category-B events (referred to as neutron stars (n^*) – don't confuse them with the astrophysical objects) were thought to arise when undetected upstream neutrino interactions emitted a neutron that interacted in the chamber. It was then easy to deduce from these events the fraction that did not interact, thus simulating a muon, and to subtract them from the observed number of events in category A.

If weak neutral currents indeed existed, they would have induced events consisting of hadrons only, just as the n^*s did, and they would be waiting to be discovered as part of category B. The main task was then to find ways of distinguishing neutrino-induced events from neutron-induced events. A neutrino beam was formed from decaying pions ($\pi^- \longrightarrow \mu^- + \bar{\nu}_\mu$, recall). The muons were filtered out by a thick iron shield and the neutrino reactions were observed in a bubble chamber. The purpose of Gargamelle was to "see neutrinos" by making visible any charged particles set in motion by the interaction

FIGURE 22.2

The Gargamelle heavy-liquid bubble chamber, installed into the magnet coils, at CERN in 1970 (copyright CERN; used with permission).

of neutrinos in the liquid. Neutrinos interact very rarely (recall eq. (19.20)), so Gargamelle was designed not only to be as big as possible, but also to work with a dense liquid - Freon (CF_3Br) - in which neutrinos would be more likely to interact. The final chamber was a cylinder, 4.8 m long and 1.85 m wide, with a volume of 12 cubic meters.

Events of the form

$$\nu_\mu + N \longrightarrow \nu_\mu + X \tag{22.1}$$

$$\overline{\nu}_\mu + N \longrightarrow \overline{\nu}_\mu + X \tag{22.2}$$

where N is a nucleon and X is a hadron were observed [221]. Since there is no charged lepton in the final state, such interactions must be due to a weak neutral current. By 1973 the number of neutral-current candidates was encouragingly large, as table 22.1 indicates. Spatial distributions of the events

TABLE 22.1

Neutral Current Candidates from the Gargamelle
Experiment

	ν-exposure	$\overline{\nu}$-exposure
# of Neutral current candidates	102	64
# of Charged current candidates	428	148

suggested that the vertex distribution of the neutral-current candidates is neutrino-like, since they were flat like the charged-current events. Furthermore, there was no indication of an exponentially falling distribution at the beginning of the chamber, which is what one would expect if the neutral-current candidates were dominantly induced by neutrons. These arguments were both corroborated by Monte Carlo simulations. A careful check of the neutron-induced background (that took several months to complete) indicated that only a small fraction of the neutral-current candidates could be explained by neutron-induced events. An independent check exploited the spatial distributions of neutral current and charged-current candidates, providing further evidence that the neutral-current sample was not dominated by $n*$s.

The data indicated that the branching ratios were

$$\frac{\sigma\left(\nu_\mu + N \longrightarrow \nu_\mu + X\right)}{\sigma\left(\nu_\mu + N \longrightarrow \mu^- + \tilde{X}\right)} \simeq .25 \quad \text{and} \quad \frac{\sigma\left(\bar{\nu}_\mu + N \longrightarrow \bar{\nu}_\mu + X\right)}{\sigma\left(\bar{\nu}_\mu + N \longrightarrow \mu^- + \tilde{X}\right)} \simeq .45$$

(22.3)

which meant that weak neutral-current reactions (due to Z exchange) were comparable to weak charged-current interactions (due to $W^\pm$ exchange). This meant that they could not be explained as a higher order effect. There was also evidence for $\bar{\nu}_\mu + e^- \longrightarrow \bar{\nu}_\mu + e^-$ at a similar rate: a purely leptonic neutral current event!

This was good news to theorists, who had years earlier postulated such weak neutral currents and used them to unify weak interactions with electromagnetism, most notably in what is sometimes called the Glashow-Salam-Weinberg model. Together with QCD, this model forms what has come to be called the Standard Model.

22.2 Electroweak Neutral Scattering Processes

The coupling of fermions to Z's is somewhat more complicated than to W's. The basic vertex is given in fig. 22.3 and contains both vector and axial-vector couplings. These depend on the species of fermion f; the values of c_V^f and c_A^f appear in table 22.2. The rather haphazard-looking structure is a consequence of electroweak unification as we see in Chapter 23.

Note that all the couplings depend on another angle, θ_W, called the weak mixing angle. This angle also relates the weak and electromagnetic coupling constants:

$$g_\text{w} \sin \theta_W = g_e = e \quad \text{and} \quad g_Z \cos \theta_W = g_\text{w} = \frac{g_e}{\sin \theta_W} \quad (22.4)$$

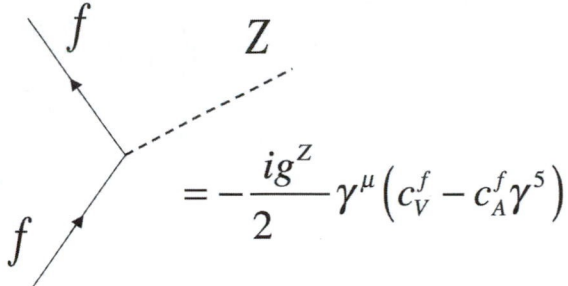

$$= -\frac{ig^Z}{2} \gamma^\mu \left(c_V^f - c_A^f \gamma^5 \right)$$

FIGURE 22.3
The fermion Z-vertex.

TABLE 22.2
Neutral Current Couplings to
Fermions

f	c_V^f	c_A^f
ν_e, ν_μ, ν_τ	$\frac{1}{2}$	$\frac{1}{2}$
e^-, μ^-, τ^-	$-\frac{1}{2} + 2\sin^2\theta_W$	$-\frac{1}{2}$
u, c, t	$\frac{1}{2} - \frac{4}{3}\sin^2\theta_W$	$\frac{1}{2}$
d, s, b	$-\frac{1}{2} + \frac{2}{3}\sin^2\theta_W$	$-\frac{1}{2}$

As with the CKM matrix, the value of the weak angle is unexplained in the Standard Model. It must be determined from experiment, which indicates [1]

$$\sin^2\theta_W = 0.23119 \pm .00014 \quad \Rightarrow \quad \theta_W^{\text{expt}} = 28.74^o \qquad (22.5)$$

The Z propagator is similar to that of the W, as shown in fig. 22.4 where the masses of the W and Z are related to one another via

$$M_Z \cos\theta_W = M_W \qquad (22.6)$$

The predictions for the values of g_w, g_Z, M_W, and M_Z in terms of the weak angle θ_W and the electric charge coupling e follow from the electroweak theory of Glashow, Salam and Weinberg.

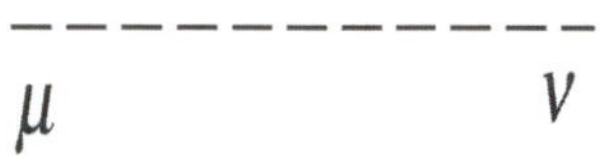

$$\frac{}{\mu \qquad\qquad\qquad\qquad \nu} = -\frac{i\left(g_{\mu\nu} - \dfrac{q_\mu q_\nu}{M_Z^2}\right)}{q^2 - M_Z^2 + iM_Z\Gamma_Z}$$

FIGURE 22.4
The Z propagator.

22.2.1 Neutrino-Electron Neutral Current Scattering

Before we get to that, we can see how the weak neutral current mediates scattering processes since we now have enough information to compute them, say $\nu_\mu + e^- \longrightarrow \nu_\mu + e^-$. The diagram is given in fig. 22.5 and the matrix

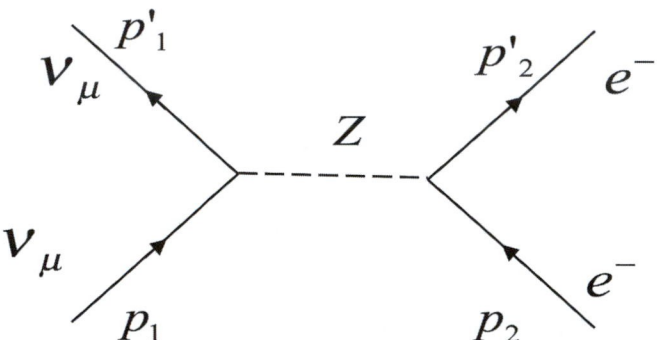

FIGURE 22.5
Lowest-order electron neutrino scattering.

element is

$$\mathcal{M} = (-i)^2 \, c_V \left(\frac{g_Z}{2}\right)^2 \left[\overline{u}(p_1')\gamma^\mu \left(\frac{1}{2} - \frac{1}{2}\gamma^5\right) u(p_1)\right] \left[\frac{g_{\mu\nu} - q_\mu q_\nu/M_Z^2}{q^2 - M_Z^2 + iM_Z\Gamma_Z}\right]$$
$$\times \left[\overline{u}(p_2')\gamma^\nu \left(1 - \epsilon\gamma^5\right) u(p_2)\right] \qquad (22.7)$$
$$\simeq -\frac{c_V^e}{2}\left(\frac{g_Z}{2M_Z}\right)^2 \left[\overline{u}(p_2')\gamma^\mu \left(1 - \epsilon\gamma^5\right) u(p_2)\right] \left[\overline{u}(p_1')\gamma^\mu \left(1 - \gamma^5\right) u(p_1)\right]$$

where $\epsilon = \frac{c_A^e}{c_V^e}$. The procedure is completely analogous to that for neutron decay

$$\sum_{\text{spins}} \left[\bar{u}(p_2')\gamma^\mu \left(1 - \epsilon\gamma^5\right) u(p_2)\right]^\dagger \left[\bar{u}(p_2')\gamma^\nu \left(1 - \epsilon\gamma^5\right) u(p_2)\right]$$

$$= \text{Tr}\left[\gamma^\mu \left(1 - \epsilon\gamma^5\right) \left(\not{p}_2\right) \gamma^\nu \left(1 - \epsilon\gamma^5\right) \left(\not{p}_2'\right)\right]$$

$$+ m_e^2 \text{Tr}\left[\gamma^\mu \left(1 - \epsilon\gamma^5\right) \gamma^\nu \left(1 - \epsilon\gamma^5\right)\right]$$

$$= 4\left[\left(1 + \epsilon^2\right)\left(p_2'^\mu p_2^\nu + p_2^\mu p_2'^\nu - g^{\mu\nu}\left(p_2 \cdot p_2'\right)\right) + 2i\epsilon\varepsilon^{\mu\alpha\nu\beta} p_{2\alpha}p_{2\beta}'\right]$$

$$+ 4m_e^2\left(1 - \epsilon^2\right) g^{\mu\nu} \tag{22.8}$$

and so

$$\overline{|\mathcal{M}|}^2 = \frac{(c_V^e)^2}{2}\left(\frac{g_Z}{M_Z}\right)^4 \left[\left(1 + \epsilon\right)^2 \left(p_1' \cdot p_2'\right)\left(p_1 \cdot p_2\right) + \left(1 - \epsilon\right)^2 \left(p_1 \cdot p_2'\right)\left(p_1' \cdot p_2\right)\right.$$

$$\left. - \left(1 - \epsilon^2\right) m_e^2 \left(p_1 \cdot p_1'\right)\right] \tag{22.9}$$

just as before. Neglecting the electron mass, in the CM frame we have

$$p_1^\mu = (E, \vec{p}) \quad p_2^\mu = (E, -\vec{p}) \quad \text{and} \quad p_1'^\mu = (E, \vec{p}\,') \quad p_2'^\mu = (E, -\vec{p}\,')$$

$$\text{where} \quad |\vec{p}|^2 = |\vec{p}\,'|^2 = E^2 \quad \text{and} \quad \vec{p} \cdot \vec{p}\,' = E^2 \cos\theta \tag{22.10}$$

and so obtain

$$\left(p_1' \cdot p_2'\right) = \left(p_1 \cdot p_2\right) = 2E^2$$

$$\left(p_1 \cdot p_2'\right) = \left(p_1' \cdot p_2\right) = E^2 \left(1 + \cos\theta\right) = 2E^2 \cos^2\frac{\theta}{2} \tag{22.11}$$

yielding

$$\overline{|\mathcal{M}|}^2 \simeq 2E^4 \left(c_V^e\right)^2 \left(\frac{g_Z}{M_Z}\right)^4 \left[\left(1 + \epsilon\right)^2 + \left(1 - \epsilon\right)^2 \cos^4\frac{\theta}{2}\right] \tag{22.12}$$

The differential cross-section is

$$\frac{d\sigma}{d\Omega} = \left(\frac{\hbar c}{8\pi}\right)^2 \frac{|\mathcal{M}|^2}{(2E)^2}$$

$$= \frac{E^2 \left(c_V^e\right)^2}{2} \left(\frac{\hbar c}{8\pi}\right)^2 \left(\frac{g_Z}{M_Z c^2}\right)^4 \left[\left(1 + \epsilon\right)^2 + \left(1 - \epsilon\right)^2 \cos^4\frac{\theta}{2}\right] \tag{22.13}$$

and integrating over angles yields

$$\sigma = (4\pi)\frac{E^2 \left(c_V^e\right)^2}{2} \left(\frac{\hbar c}{8\pi}\right)^2 \left(\frac{g_Z}{M_Z c^2}\right)^4 \left[\left(1 + \epsilon\right)^2 + \frac{1}{3}\left(1 - \epsilon\right)^2\right]$$

$$= \frac{(\hbar c E)^2 \left(c_V^e\right)^2}{24\pi} \left(\frac{g_Z}{M_Z c^2}\right)^4 \left[1 + \epsilon + \epsilon^2\right] \tag{22.14}$$

Let's compare this to the result for $\nu_\mu + e^- \longrightarrow \nu_e + \mu^-$, which we found from eq. (20.15) to be

$$\frac{d\sigma}{d\Omega} \simeq \frac{1}{2}\left(\frac{\hbar c g_w^2}{4\pi M_W^2 c^4}\right)^2 E^2 \left[1 - \frac{m_\mu^2 c^4}{4E^2}\right]^2 + \cdots$$

where m_μ is the mass of the muon. Neglecting this and integrating over the angles gives

$$\sigma \simeq \frac{1}{8\pi}\left(\frac{\hbar c g_w^2}{M_W^2 c^4}\right)^2 E^2 \tag{22.15}$$

Putting it all together we find that

$$\frac{\sigma\left(\nu_\mu + e^- \longrightarrow \nu_\mu + e^-\right)}{\sigma\left(\nu_\mu + e^- \longrightarrow \nu_e + \mu^-\right)}$$

$$= \frac{(\hbar c E)^2 (c_V^e)^2 \left(\frac{g_Z}{M_Z c^2}\right)^4 \left[1 + \epsilon + \epsilon^2\right]}{24\pi \quad \frac{1}{8\pi}\left(\frac{\hbar c g_w^2}{M_W^2 c^4}\right)^2 E^2}$$

$$= \frac{1}{3}\left(\frac{g_Z M_W}{g_w M_Z}\right)^4 \left[(c_V^e)^2 + (c_V^e)(c_A^e) + (c_A^e)^2\right]$$

$$= \frac{1}{3}\left[\left(-\frac{1}{2} + 2\sin^2\theta_W\right)^2 - \frac{1}{2}\left(-\frac{1}{2} + 2\sin^2\theta_W\right) + \left(-\frac{1}{2}\right)^2\right]$$

$$= \frac{1}{4} - \sin^2\theta_W + \frac{4}{3}\sin^4\theta_W$$

$$= \begin{cases} 0.09 & \text{Theory} \\ 0.08 & \text{Expt} \end{cases} \tag{22.16}$$

The experimental value is 0.08 [222], good to within 10%!

The reason why it took about 15 years before experiment could confirm Bludman's idea is that at low energies neutral current processes compete with electromagnetic ones. An example is given by electron-positron annihilaton into some other fermion-antifermion pair, as shown in fig. 22.6. Since the mass of the Z is so heavy, the neutral current interaction is very feeble compared to the electromagnetic one. That's why neutrino beams were needed to see weak neutral currents — neutrinos don't couple to photons!

22.2.2 Electron-Positron Neutral Current Scattering

Let's consider electron-positron scattering into a fermion-antifermion pair. The lowest-order diagrams in electroweak theory are given in fig. 22.6. The fermion can be anything except for another electron, since in that case we'd have to add two other diagrams. The matrix element for the Z diagram in

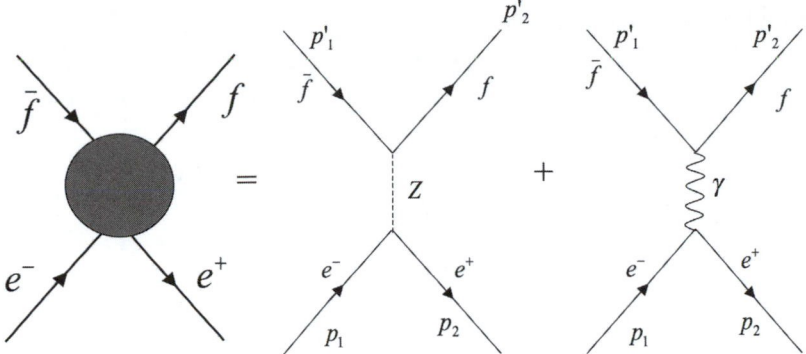

FIGURE 22.6

Electron-positron annihilation to some other fermion-antifermion pair to lowest order in electroweak theory.

22.6 is

$$\mathcal{M} = (-i)^2 \, c_V^e c_V^f \left(\frac{g_Z}{2}\right)^2 \left[\bar{v}(p_2)\gamma^\mu \left(1 - \epsilon^e \gamma^5\right) u(p_1)\right] \left[\frac{g_{\mu\nu} - q_\mu q_\nu / M_Z^2}{q^2 - M_Z^2 + iM_Z\Gamma_Z}\right]$$
$$\times \left[\bar{u}(p_2')\gamma^\nu \left(1 - \epsilon^f \gamma^5\right) v(p_1')\right] \qquad (22.17)$$

where $q^2 = (p_2 + p_1)^2$. We want to work at energies comparable to the mass of the Z, and so the modification of the propagator to the Breit-Wigner form will be important here, and we will no longer be able to neglect the q_μ terms. However, note that

$$\bar{v}(p_2)\gamma^\mu \left(1 - \epsilon^e \gamma^5\right) u(p_1)q_\mu = 2m_e \left[\bar{v}(p_2) \left(\epsilon^e \gamma^5\right) u(p_1)\right] \qquad (22.18)$$

which you can show using $\left(\not{p}_1 - m_e\right) u(p_1) = 0$ and $\bar{v}(p_2) \left(\not{p}_2 + m_e\right) = 0$. Similiarly

$$\left[\bar{u}(p_2')\gamma^\nu \left(1 - \epsilon^f \gamma^5\right) v(p_1')\right] q_\nu = 2m_f \left[\bar{u}(p_2')\gamma^\nu \left(\epsilon^f \gamma^5\right) v(p_1')\right] \qquad (22.19)$$

So the q_μ terms contribute factors proportional to the electron and fermion masses. These are negligible relative to the mass of the Z and so we neglect them.

Hence

$$\mathcal{M} = (-i)^2 \, c_V^e c_V^f \left(\frac{g_Z}{2}\right)^2 \left[\bar{v}(p_2)\gamma^\mu \left(1 - \epsilon^e \gamma^5\right) u(p_1)\right] \left[\frac{1}{q^2 - M_Z^2 + iM_Z\Gamma_Z}\right]$$
$$\times \left[\bar{u}(p_2')\gamma_\mu \left(1 - \epsilon^f \gamma^5\right) v(p_1')\right] \qquad (22.20)$$

and the Casimir trick implies

$$\overline{|\mathcal{M}|}^2 = \frac{1}{4}\left[\frac{c_V^e c_V^f \left(\frac{g_Z}{2}\right)^2}{|q^2 - M_Z^2 + iM_Z\Gamma_Z|}\right]^2 \text{Tr}\left[\gamma^\mu \left(1 - \epsilon^e\gamma^5\right)\not{p}_1\gamma^\nu \left(1 - \epsilon^e\gamma^5\right)\not{p}_2\right]$$

$$\times \text{Tr}\left[\gamma_\mu \left(1 - \epsilon^f\gamma^5\right)\not{p}_1'\gamma_\nu \left(1 - \epsilon^f\gamma^5\right)\not{p}_2'\right]$$

$$= \frac{1}{2}\left[\frac{\left(c_V^e c_V^f g_Z^2\right)^2}{(q^2 - M_Z^2)^2 + (M_Z\Gamma_Z)^2}\right]$$

$$\times \left\{\left(1 + (\epsilon^e)^2\right)\left(1 + (\epsilon^f)^2\right)[(p_1' \cdot p_1')(p_2 \cdot p_2') + (p_1 \cdot p_2')(p_1' \cdot p_2)]\right.$$

$$\left. + 4\epsilon^e\epsilon^f [(p_1' \cdot p_1')(p_2 \cdot p_2') - (p_1 \cdot p_2')(p_1' \cdot p_2)]\right\} \qquad (22.21)$$

where in the CM frame (22.11)

$$\overline{|\mathcal{M}|}^2 = \frac{\left(c_V^e c_V^f g_Z^2 E\right)^2 \left(\left(1 + (\epsilon^e)^2\right)\left(1 + (\epsilon^f)^2\right)(1 + \cos^2\theta) - 8\epsilon^e\epsilon^f\cos\theta\right)}{\left((2E)^2 - M_Z^2\right)^2 + (M_Z\Gamma_Z)^2} \qquad (22.22)$$

with θ the scattering angle. The differential cross section is therefore

$$\frac{d\sigma}{d\Omega} = \left[\frac{(\hbar c)}{16\pi}\right]^2 \frac{\left(c_V^e c_V^f g_Z^2 E\right)^2 \left(\left(1 + (\epsilon^e)^2\right)\left(1 + (\epsilon^f)^2\right)(1 + \cos^2\theta) - 8\epsilon^e\epsilon^f\cos\theta\right)}{\left((2E)^2 - M_Z^2\right)^2 + (M_Z\Gamma_Z)^2} \qquad (22.23)$$

which integrates to a total cross-section of

$$\sigma = \frac{\left(\hbar c g_Z^2 E\right)^2}{48\pi}\left[\frac{\left[(c_V^e)^2 + (c_A^e)^2\right]\left[\left(c_V^f\right)^2 + \left(c_A^f\right)^2\right]}{\left((2E)^2 - M_Z^2\right)^2 + (M_Z\Gamma_Z)^2}\right] \qquad (22.24)$$

When mediated by a photon, the cross-section for the same process is, from eq. (17.22)

$$\sigma = \frac{\left(\hbar c g_e^2\right)^2}{48\pi}\frac{\left(Q^f\right)^2}{E^2} \qquad (22.25)$$

where Q^f is the charge of the fermion in units of e. So the ratio is

$$\frac{\sigma_Z}{\sigma_\gamma} \equiv \frac{\sigma\left(e^+e^- \longrightarrow Z \longrightarrow f\bar{f}\right)}{\sigma\left(e^+e^- \longrightarrow \gamma \longrightarrow f\bar{f}\right)}$$

$$= \frac{\left[\frac{1}{2} - 2\sin^2\theta_W + 4\sin^4\theta_W\right]\left[\left(c_V^f\right)^2 + \left(c_A^f\right)^2\right]}{\left(\sin^2\theta_W \cos^2\theta_W Q^f\right)^2}$$

$$\times \left[\frac{E^4}{\left((2E)^2 - (M_Z c^2)^2 \right)^2 + (\hbar \Gamma_Z M_Z c^2)^2} \right] \qquad (22.26)$$

inserting the appropriate factors of $\hbar$ and c.

At energies much less than the mass of the Z (but still much greater than the fermion masses) we have

$$\frac{\sigma_Z}{\sigma_\gamma} = \frac{\left[\frac{1}{2} - 2\sin^2 \theta_W + 4\sin^4 \theta_W \right] \left[\left(c_V^f \right)^2 + \left(c_A^f \right)^2 \right]}{\left(\sin^2 \theta_W \cos^2 \theta_W Q^f \right)^2} \left(\frac{E}{M_Z c^2} \right)^4 \qquad (22.27)$$

and the electromagnetic interaction dominates, since even for $E = \frac{1}{4} M_Z c^2$, we have $\left(\frac{E}{M_Z c^2} \right)^4 \simeq 0.3\%$. However, when the electron and positron collide at exactly the rest-mass of the Z (i.e. at $E = \frac{1}{2} M_Z c^2$), we have

$$\frac{\sigma_Z}{\sigma_\gamma} = \frac{\left[\frac{1}{2} - 2\sin^2 \theta_W + 4\sin^4 \theta_W \right] \left[\left(c_V^f \right)^2 + \left(c_A^f \right)^2 \right]}{16 \left(\sin^2 \theta_W \cos^2 \theta_W Q^f \right)^2} \left(\frac{M_Z c^2}{\hbar \Gamma_Z} \right)^2 \qquad (22.28)$$

which shows that the neutral weak interaction dominates because $\frac{1}{16} \left(\frac{M_Z c^2}{\hbar \Gamma_Z} \right)^2 \simeq 84$.

22.3 The SU(2) × U(1) Model

The general form of the vertex for fermions coupling to W's is one we looked at in the previous chapter and is shown in fig. 22.7 where the wavefunction χ can be any one of

$$\mathfrak{E}_L = \begin{pmatrix} \nu_e \\ e^- \end{pmatrix}_L, \quad \mathfrak{M}_L = \begin{pmatrix} \nu_\mu \\ \mu^- \end{pmatrix}_L, \quad \mathfrak{T}_L = \begin{pmatrix} \nu_\tau \\ \tau^- \end{pmatrix}_L \qquad (22.29)$$

$$\mathfrak{D}_L = \begin{pmatrix} u \\ d' \end{pmatrix}_L, \quad \mathfrak{S}_L = \begin{pmatrix} c \\ s' \end{pmatrix}_L, \quad \mathfrak{B}_L = \begin{pmatrix} t \\ b' \end{pmatrix}_L \qquad (22.30)$$

where in the above $\nu_e = \overline{\psi}^{\nu_e}$, $\mu^- = \overline{\psi}^{\mu^-}$, etc and we recall that $\psi_L = \frac{1}{2} \left(1 - \gamma^5 \right) \psi$ (and $\overline{\psi}_L = \left(\frac{1}{2} \left(1 - \gamma^5 \right) \psi \right)^\dagger \gamma^0 = \overline{\psi} \left(\frac{1}{2} \left(1 + \gamma^5 \right) \right)$). The primed quark wavefunctions are related to the unprimed ones familiar from QED and QCD via the CKM matrix as in eq. (21.52). What is needed to describe this is an underlying theory that accounts for this vertex as well as for the vertex for the Z boson in fig. 22.3.

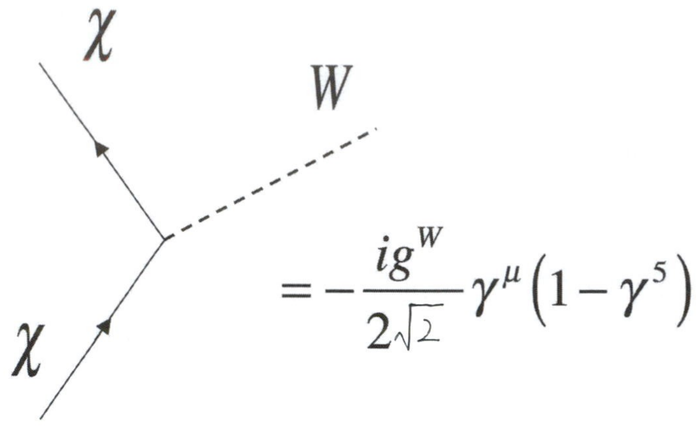

$$= -\frac{ig^W}{2\sqrt{2}}\gamma^\mu\left(1-\gamma^5\right)$$

FIGURE 22.7

The general form of the W-boson/fermion vertex.

However, we will need more – since the $W^\pm$ bosons are charged, they will need to couple to the photon – in other words, there should be a $WW\gamma$ vertex. This suggests that we need a Yang-Mills theory as our underlying theory because (as we have seen with the gluons) it provides for vertices with 3 gauge particles.

A number of attempts were made to describe the weak interactions using a Yang-Mills theory once it became clear from experiments on parity violation in the 1950s that the weak nuclear force was quite different from the strong one. Many models were proposed over the following two decades until experiment finally winnowed out all but the one theory we have today, namely the electroweak theory of Glashow, Salam and Weinberg. This theory has a non-abelian symmetry $\mathbf{SU}(2) \otimes \mathbf{U}(1)$, with 3 gauge bosons

$$\{W_\mu^1, W_\mu^2, W_\mu^3\} = \vec{W}_\mu$$

associated with the $\mathbf{SU}(2)$ and 1 gauge boson called B_μ that is associated with the $\mathbf{U}(1)$. Physically the non-abelian symmetry means that, analogous to gluon interactions, the W and B bosons will interact with each other as well as with the quarks and leptons, resulting in additional vertex rules for electroweak theory.

But the symmetry is "broken" – the lowest-energy state of the theory does not reflect the full symmetry of the theory. We will see that the two neutral bosons – the W_μ^3 and the B_μ – mix together to form a massless boson (the photon) and a very heavy boson (the Z). The remaining W bosons will combine into the charged $W^\pm$ bosons. But how? And why do the W_μ^3 and B_μ boson wavefunctions get mixed up into the Z-boson and the photon? And why are the W-bosons and Z-boson so heavy, but the photon massless? You'll have to go to the next chapter to find out.

22.4 Questions

1. Compute the decay widths Γ_W and Γ_Z, neglecting fermion masses.

2. (a) Compute the decay rate $Z \to f + \bar{f}$, where f is any quark or lepton lighter than the Z. Neglect the masses of the fermions.

   (b) Find the branching ratio for each species of quark and lepton, assuming that these are the dominant decay modes.

3. Compute the ratio $\mathfrak{R}$ of quark pair production to muon pair production due to e^+e^- scattering when the process is mediated by a Z_0 particle. How does it compare to the photon-mediated case at low energies? How does it compare at energies equal to the mass of the Z_0? Take into account the finite lifetime of the Z_0.

4. For electron-positron scattering, as depicted in the diagram in fig. 22.8, show that

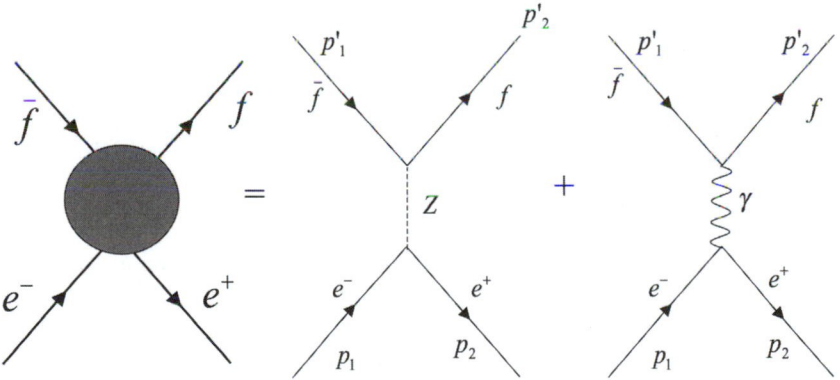

FIGURE 22.8
Diagrams for electron positron scattering in electroweak theory.

$$\left[\bar{u}(p_2')\gamma^\nu \left(1 - \epsilon^f \gamma^5\right) v(p_1')\right] q_\nu = 2m_f \left[\bar{u}(p_2')\gamma^\nu \left(\epsilon^f \gamma^5\right) v(p_1')\right]$$

and

$$\bar{v}(p_2)\gamma^\mu \left(1 - \epsilon^e\gamma^5\right) u(p_1)q_\mu = 2m_e \left[\bar{v}(p_2)\left(\epsilon^e\gamma^5\right) u(p_1)\right]$$

5. Draw all relevant Feynman diagrams for the following processes to lowest order.

   (a) $\gamma + \gamma \longrightarrow \nu_\tau + \bar{\nu}_\tau$ (b) $\gamma + \gamma \longrightarrow Z + Z$ (c) $W^+ + W^- \longrightarrow Z + \gamma$

6. Consider the process $e^+ + e^- \to Z \to f + \bar{f}$ for both right-handed and left-handed electron wavefunctions. Compute

$$A = \frac{\sigma_L - \sigma_R}{\sigma_L + \sigma_R}$$

for this process.

23

Electroweak Symmetry Breaking

The Higgs mechanism provides a means for the weak boson gauge fields of electroweak theory to acquire their mass. The gauge group is $\mathbf{SU}(2) \times \mathbf{U}(1)$, which means the equations of the theory have two kinds of gauge fields: the three W_μ^a wavefunctions, which obey a set of Yang-Mills equations, and the B_μ wavefunction, which obeys a Maxwell equation. This is kind of like a mixture of QCD and QED, except that (a) in addition to fermions we will also have the scalar Higgs wavefunction and (b) the symmetry will be spontaneously broken.

Before considering how this mechanism works in the full electroweak theory, it will be helpful to consider how it works in the $\mathbf{U}(1)$ subsector of the theory. Let's begin there.

23.1 The Higgs Mechanism

The idea that the weak bosons had to be very heavy was considered as long ago as 1961 when Glashow first suggested that the weak and electromagnetic interactions were unified [216]. The disparity in strength between the two interactions could be accounted for if the weak bosons had large masses. But nobody knew how to properly do this.

The problem is not that we don't know how to give a particle a mass – for example in the Klein-Gordon equation

$$\left(\partial_\mu \partial^\mu + m^2\right) \phi = 0 \tag{23.1}$$

the $(\text{mass})^2$ term appears as a constant multiplying the wavefunction – indeed, we can regard this as a definition of mass. However, if we do this for a gauge boson, say the B_μ, then we would modify eq. (12.27) to read

$$\partial^\mu \left(\partial_\mu B_\nu - \partial_\nu B_\mu\right) + m_B^2 B_\nu = j_\nu^Y \tag{23.2}$$

which would give the B_μ a mass but destroy the $\mathbf{U}(1)$ gauge invariance of our theory (recall the discussion in section 12.4.1). This in turn renders the theory unrenormalizable, and therefore of no predictive power. So it appears that if we want to give a mass to a gauge boson, we destroy renormalizability of the theory!

The resolution to this dilemma was found by Higgs in 1963 [223], and then applied in 1967-8 to the weak interactions by Salam and Weinberg [217]. It involves introducing a scalar particle that couples in a particular way to the electroweak bosons and that has a non-trivial ground state.

To see how this works, consider first a modification of the Klein-Gordon equation (23.1):

$$\partial_\mu \partial^\mu \phi + \mathcal{V}'(\phi) = 0 \qquad (23.3)$$

where we can regard $\mathcal{V}(\phi)$ as the potential energy of the ϕ particle; $\mathcal{V}'(\phi) = \frac{\partial \mathcal{V}(\phi)}{\partial \phi}$ is like a self-force that acts on ϕ. For example, if $\mathcal{V}(\phi) = \frac{1}{2}m^2\phi^2$ then we obtain $\left(\partial_\mu \partial^\mu + m^2\right)\phi = 0$, which is the Klein-Gordon equation for a massive particle. In this sense we can understand mass as being like a spring constant in a harmonic oscillator potential!

More generally, we can make the potential anything we want it to be, but renormalizability demands* that it grow no faster than ϕ^4. Hence we can write the most general renormalizable potential as

$$\mathcal{V}(\phi) = V_0 + V_1\phi + V_2\phi^2 + V_3\phi^3 + V_4\phi^4 \qquad (23.4)$$

The constant term V_0 sets the zero of energy; we can remove it if we like, though we won't do so here. The cubic term can be removed by redefining ϕ by a constant shift: $\phi = \tilde{\phi} - 4\frac{V_3}{V_4}$. The linear term simply adds a constant to the modified Klein-Gordon equation, so for simplicity we'll set $V_1 = 0$.

Hence we take

$$\mathcal{V}(\phi) = V_0 - \frac{1}{2}\mu^2\phi^2 + \frac{1}{4}\lambda^2\phi^4 \qquad (23.5)$$

which yields

$$\partial_\mu \partial^\mu \phi - \mu^2\phi + \lambda^2\phi^3 = 0 \qquad (23.6)$$

as the general form of our modified Klein-Gordon equation.

Although an exact solution to this equation (for arbitrary boundary conditions) would tell us everything we want to know about ϕ, we don't know how obtain such a solution. Fortunately this is not necessary. We can consider perturbative solutions by writing $\phi(x) = \phi_0 + \hat{\phi}(x)$, where ϕ_0 is the lowest-energy state (i.e. the ground state) of ϕ, and $\hat{\phi}$ is a wavefunction describing small excitations of ϕ. Inserting this into eq. (23.6), to leading order in $\hat{\phi}$ we have

$$\partial_\mu \partial^\mu \hat{\phi} + \left(-\mu^2 + 3\lambda^2\phi_0^2\right)\hat{\phi} + \lambda^2\phi_0^3 - \mu^2\phi_0 \simeq 0 \qquad (23.7)$$

But what should we take for ϕ_0? If we want to expand about the ground state, we need to expand about the minimal value of $\mathcal{V}(\phi)$, i.e. the value

---

*A proof of this is beyond the scope of this textbook; you'll have to take my word for it or else read a more advanced book on the subject [97].

$\phi(x) = \phi_0$ such that $\mathcal{V}'(\phi_0) = 0$. One solution is clearly $\phi_0 = 0$; however this gives

$$\partial_\mu \partial^\mu \hat{\phi} - \mu^2 \hat{\phi} \simeq 0 \qquad (23.8)$$

which is a Klein-Gordon equation with the wrong sign for the mass term. This is strange – the excitations have imaginary mass[†], so the momenta of a $\hat{\phi}$ particle would be larger than its energy, implying that relativistically it would move faster than light! However, upon closer inspection we see that $\phi_0 = 0$ is really a local maximum of $\mathcal{V}(\phi)$, so we perhaps should not be surprised that these excitations are unphysical.

Another solution to $\mathcal{V}'(\phi_0) = 0$ is $\phi_0 = \pm\frac{\mu}{\lambda}$; this gives

$$\partial_\mu \partial^\mu \hat{\phi} + 2\mu^2 \hat{\phi} \simeq 0 \qquad (23.9)$$

which is a Klein-Gordon equation for a particle of mass $\sqrt{2}\mu$! So excitations about $\phi(x) = \pm\frac{\mu}{\lambda}$ are (to leading order) those of a free (and massive) Klein-Gordon particle! The full equation for $\hat{\phi}$ is

$$\partial_\mu \partial^\mu \hat{\phi} + 2\mu^2 \hat{\phi} + 3\mu\lambda\hat{\phi}^2 + \lambda^2\hat{\phi}^3 = 0 \qquad (23.10)$$

and solving this equation (say perturbatively in λ) will give us the wavefunctions corresponding to states excited above the ground state $\phi_0 = \frac{\mu}{\lambda}$. Of course we could have chosen $\phi_0 = -\frac{\mu}{\lambda}$ as the ground state of the system, in which case

$$\partial_\mu \partial^\mu \hat{\phi} + 2\mu^2 \hat{\phi} - 3\mu\lambda\hat{\phi}^2 + \lambda^2\hat{\phi}^3 = 0 \qquad (23.11)$$

describes the excitations.

Note that there is no new physics here – the equation for $\hat{\phi}$ represents exactly the same physical system as the equation for ϕ. However, the $\hat{\phi}$ version is the one better suited for describing excitations about the ground state of the system.

The phenomenon described above is referred to as *spontaneous symmetry breaking* because it breaks the original symmetry of the theory. You can see that

$$\mathcal{V}(\phi) = V_0 - \frac{1}{2}\mu^2\phi^2 + \frac{1}{4}\lambda^2\phi^4 \qquad (23.12)$$

is invariant under $\phi \longleftrightarrow -\phi$, but that

$$\mathcal{V}(\hat{\phi}) = V_0 + \frac{1}{4}\left(\frac{\mu^2}{\lambda}\right)^2 + \mu^2\hat{\phi}^2 \pm \mu\lambda^2\hat{\phi}^3 + \frac{1}{4}\lambda^2\hat{\phi}^4 \qquad (23.13)$$

is not invariant under $\hat{\phi} \longleftrightarrow -\hat{\phi}$. This happens because the ground state of the system does not respect the original symmetry of the theory. It is caused

---

[†]They are known as *tachyons* in the literature.

by the system tending to its lowest-energy state with nothing external to the system being responsible (hence the term "spontaneous").

We can now make use of this phenomenon to give a mass to the gauge bosons – this is called the Higgs mechanism. In the Standard Model it works by coupling a scalar field to the vector bosons in such a way that the photon remains massless but the Z does not.

Let's consider how it works for the B_μ boson. If we want to couple this to a scalar wavefunction φ then we must modify the φ equation so that the derivatives in it are gauge-covariant derivatives

just like in Chapter 12,

$$D_\mu \varphi = \partial_\mu \varphi + i g_Y B_\mu \varphi \tag{23.14}$$

(12.13) says $D_\mu \Psi = \partial_m \Psi + i e A_m \Psi$

thereby modifying the Klein-Gordon equation to

to make it gauge invariant.

$$\frac{1}{2} D_\mu D^\mu \varphi + \frac{\partial \mathcal{V}}{\partial \varphi^*} = 0 \tag{23.15}$$

where we take[‡]

$$\mathcal{V} = V_0 - \frac{1}{2}\mu^2 \varphi^* \varphi + \frac{1}{4}\lambda^2 (\varphi^* \varphi)^2 \tag{23.16}$$

We also need the equation for the gauge field, which is

$$\partial^\mu (\partial_\mu B_\nu - \partial_\nu B_\mu) = \frac{1}{2} i g_Y (\varphi^* D_\nu \varphi - (D_\nu \varphi)^* \varphi) \tag{23.17}$$

The quantity on the right-hand side is the current induced by the scalar wavefunction φ.

Note that we have taken the scalar wavefunction to be a complex linear combination of two wavefunctions, i.e.

$$\varphi = \phi_1 + i\phi_2 \tag{23.18}$$

This is necessary because we want our system to be locally **U**(1) gauge invariant, i.e. invariant under

$$\varphi(x) \to e^{i\alpha(x)} \varphi(x) \tag{23.19}$$

and the functional degree of freedom $\alpha(x)$ is sufficient to eliminate one of the wavefunctions ϕ_1 or ϕ_2 (hence the need for two wavefunctions, or a single complex φ). It is not hard to show that the current induced by the scalar wavefunction is conserved:

$$\partial^\nu \left[\frac{1}{2} i g_Y (\varphi^* D_\nu \varphi - (D_\nu \varphi)^* \varphi) \right] = 0 \tag{23.20}$$

as it must be to ensure the **U**(1) gauge invariance.

[‡]Note that $\frac{\partial}{\partial \varphi^*}(\varphi^* \varphi) = \varphi$. The conjugate equation is $\frac{1}{2}(D_\mu D^\mu \varphi)^* + \frac{\partial \mathcal{V}}{\partial \varphi} = 0$.

The φ equation is therefore a pair of equations:

$$\partial_\mu \partial^\mu \phi_1 - \mu^2 \phi_1 + \lambda^2 \left(\phi_1^2 + \phi_2^2\right) \phi_1 - g_Y^2 B_\mu B^\mu \phi_1$$
$$-g_Y \partial^\mu B_\mu \phi_2 - 2g_Y B_\mu \partial^\mu \phi_2 = 0 \quad (23.21)$$

$$\partial_\mu \partial^\mu \phi_2 - \mu^2 \phi_2 + \lambda^2 \left(\phi_1^2 + \phi_2^2\right) \phi_2 - g_Y^2 B_\mu B^\mu \phi_2$$
$$+g_Y \partial^\mu B_\mu \phi_1 + 2g_Y B_\mu \partial^\mu \phi_1 = 0 \quad (23.22)$$

The minimum of $\mathcal{V}(\varphi) = 0$ is now located at $(\phi_{01})^2 + (\phi_{02})^2 = \varphi_0^* \varphi_0 = \left(\frac{\mu}{\lambda}\right)^2$. There are many ways of solving this equation, since the minimum lies on a circle – in φ space – of radius $\frac{\mu}{\lambda}$. We can expand about any ground state satisfying this condition, so we might as well write

$$\phi_1(x) = -\frac{\mu}{\lambda} + \hat{\phi}_1(x) \qquad \phi_2(x) = \hat{\phi}_2(x) \quad (23.23)$$

in which case we get

$$\partial_\mu \partial^\mu \hat{\phi}_1 + 2\mu^2 \hat{\phi}_1 - 3\mu\lambda \hat{\phi}_1^2 - \mu\lambda \hat{\phi}_2^2 + \lambda^2 \left(\hat{\phi}_1^2 + \hat{\phi}_2^2\right) \hat{\phi}_1$$
$$-g_Y^2 B_\mu B^\mu \hat{\phi}_1 + g_Y^2 \frac{\mu}{\lambda} B_\mu B^\mu - g_Y \partial^\mu B_\mu \hat{\phi}_2 - 2g_Y B_\mu \partial^\mu \hat{\phi}_2 = 0 \quad (23.24)$$

$$\partial_\mu \partial^\mu \hat{\phi}_2 - 2\mu\lambda \hat{\phi}_1 \hat{\phi}_2 + \lambda^2 \left(\hat{\phi}_1^2 + \hat{\phi}_2^2\right) \hat{\phi}_2 - g_Y^2 B_\mu B^\mu \hat{\phi}_2$$
$$+g_Y \partial^\mu B_\mu \hat{\phi}_1 + 2g_Y B_\mu \partial^\mu \hat{\phi}_1 = 0 \quad (23.25)$$

Notice that the mass of $\hat{\phi}_1$ is $\sqrt{2}\mu$, but the mass of $\hat{\phi}_2$ is zero. This isn't a coincidence – in fact it is a general phenomenon that always accompanies spontaneous symmetry breaking, as shown by Jeffrey Goldstone [224]. Spontaneous symmetry breaking of a *continuous* global symmetry is always accompanied by the presence of one or more massless scalar particles. This result is known as Goldstone's theorem and the the massless scalars are known as Goldstone bosons. In our example here, the continuous global symmetry (made local via the gauge principle) is the **U**(1) phase symmetry, and the Goldstone boson is the $\hat{\phi}_2$.

What happens to the gauge field equation? In components it is

$$\partial^\mu \left(\partial_\mu B_\nu - \partial_\nu B_\mu\right) = g_Y \left(\phi_2 \partial_\nu \phi_1 - \phi_1 \partial_\nu \phi_2 - g_Y \left(\phi_1^2 + \phi_2^2\right) B_\nu\right) \quad (23.26)$$

which, when expanded about the ground state becomes

$$\partial^\mu \left(\partial_\mu B_\nu - \partial_\nu B_\mu\right) = g_Y \left(\hat{\phi}_2 \partial_\nu \hat{\phi}_1 - \hat{\phi}_1 \partial_\nu \hat{\phi}_2 - g_Y \left(\hat{\phi}_1^2 + \hat{\phi}_2^2\right) B_\nu\right)$$
$$-\frac{(g_Y \mu)^2}{\lambda^2} B_\nu + \frac{\mu g_Y}{\lambda} \partial_\nu \hat{\phi}_2 + \frac{2\mu (g_Y)^2}{\lambda} B_\nu \hat{\phi}_1 \quad (23.27)$$

or alternatively

$$\partial^\mu \left(\partial_\mu B_\nu - \partial_\nu B_\mu\right) + \frac{(g_Y \mu)^2}{\lambda^2} B_\nu = g_Y \left(\hat{\phi}_2 \partial_\nu \hat{\phi}_1 - \hat{\phi}_1 \partial_\nu \hat{\phi}_2 - g_Y \left(\hat{\phi}_1^2 + \hat{\phi}_2^2\right) B_\nu\right)$$
$$+\frac{\mu g_Y}{\lambda} \partial_\nu \hat{\phi}_2 + 2\frac{\mu (g_Y)^2}{\lambda} B_\nu \hat{\phi}_1 \quad (23.28)$$

We now see that a miracle has happened – the gauge field has acquired a mass $m_B = \frac{\mu g_Y}{\lambda}$!

We can actually eliminate the $\hat{\phi}_2$ wavefunction by using the local gauge invariance mentioned above; just choose $\alpha(x) = -\tan^{-1}(\phi_2/\phi_1)$ in eq. (23.19). The mechanism still works, and we are left with a massive gauge field with three degrees of freedom, and a single scalar particle $\hat{\phi}_1$. Note that this is the same number of degrees of freedom as in the original system: the massless gauge field had two degrees of freedom and we had two scalar particles. By promoting the global phase invariance to a local gauge invariance, the massless Goldstone boson is eliminated, replaced by the extra degree of freedom in the now-massive gauge boson. Note that the physical systems are the same both before and after expanding about the ground state – what has changed is the description of the physics relative to the ground state.

23.2   Breaking the SU(2) Symmetry

To implement the Higgs mechanism in Electroweak theory, the Higgs wavefunction is taken to be a complex doublet under **SU**(2), which I will call Φ. We take the covariant derivative to be

$$D_\mu \Phi = \partial_\mu \Phi - i\frac{g_W}{2}\sigma^a \cdot W_\mu^a \Phi + i\frac{g_Y}{2}B_\mu \Phi \tag{23.29}$$

and repeat the above procedure, setting

$$\Phi = \frac{1}{\sqrt{2}}\begin{pmatrix} \phi_1 + i\phi_2 \\ \phi_3 + i\phi_4 \end{pmatrix} = \begin{pmatrix} \Phi^u \\ \Phi^0 \end{pmatrix}$$

and introducing a potential $\mathcal{V}(\Phi) = V_0 - \frac{1}{2}\mu^2 \Phi^\dagger \Phi + \frac{1}{4}\lambda^2 (\Phi^\dagger \Phi)^2$. Note that the Higgs wavefunction has both hypercharge ($Y = +1$) and weak isospin of $\frac{1}{2}$, as it must since we want it to mix the W_μ^3 and B_μ wavefunctions after spontaneous symmetry breaking. The quantities σ^a are none other than the Pauli matrices, which generate the **SU**(2) group (recall eq. (5.17))

$$\left[\frac{\sigma^a}{2}, \frac{\sigma^b}{2}\right] = i\epsilon^{abc}\frac{\sigma^c}{2} \tag{23.30}$$

where ϵ^{abc} is the familiar fully antisymmetric Levi-Civita symbol. This means that the field strength for the $W^{a\mu}$ is

$$F_{\mu\nu}^a = \partial_\mu W_\nu^a - \partial_\nu W_\mu^a - g_W W_\mu^c W_\nu^b e_{cb}^a \tag{23.31}$$

which sometimes is written as

$$\vec{F}_{\mu\nu} = \partial_\mu \vec{W}_\nu - \partial_\nu \vec{W}_\mu - g_W \vec{W}_\mu \times \vec{W}_\nu \tag{23.32}$$

taking advantage of the cross-product structure from the ϵ^c_{ab} tensor. It will also be useful to define

$$\sigma^\pm = \frac{1}{2}\left(\sigma^1 \pm i\sigma^2\right) \tag{23.33}$$

$$\Rightarrow \sigma^+ = \begin{pmatrix} 0 & 1 \\ 0 & 0 \end{pmatrix} \qquad \sigma^- = \begin{pmatrix} 0 & 0 \\ 1 & 0 \end{pmatrix}$$

23.2.1 The Gauge Equations

The equations of motion for the gauge particles of the electroweak theory are

$$\partial^\mu F^a_{\mu\nu} - g_{\rm w} W^{c\mu} F^b_{\mu\nu}\epsilon^a_{cb} = \frac{g_{\rm w}}{2}\overline{\chi^{\mathfrak{L}}_L}\gamma_\nu\left(\sigma^a\right)\chi^{\mathfrak{L}}_L + \frac{g_{\rm w}}{2}\overline{\chi^{\mathfrak{Q}}_L}\gamma_\nu\left(\sigma^a\right)\chi^{\mathfrak{Q}}_L$$

$$-i\frac{g_{\rm w}}{4}\left(\Phi^\dagger\sigma^a D_\nu\Phi - \left(D_\nu\Phi\right)^\dagger\sigma^a\Phi\right) \tag{23.34}$$

$$\partial^\mu\left(\partial_\mu B_\nu - \partial_\nu B_\mu\right) = i\frac{g_{\rm Y}}{4}\left(\Phi^\dagger D_\nu\Phi - \left(D_\nu\Phi\right)^\dagger\Phi\right)$$

$$+\frac{g_{\rm Y}}{2}\left[Y^{\mathfrak{L}}_L\overline{\chi^{\mathfrak{L}}_L}\gamma_\nu\chi^{\mathfrak{L}}_L + Y^{\mathfrak{L}}_R\overline{\chi^{\mathfrak{L}}_R}\gamma_\nu\chi^{\mathfrak{L}}_R + Y^{\mathfrak{Q}}_L\overline{\chi^{\mathfrak{Q}}_L}\gamma_\nu\chi^{\mathfrak{Q}}_L + Y^{\mathfrak{Q}}_R\overline{\chi^{\mathfrak{Q}}_R}\gamma_\nu\chi^{\mathfrak{Q}}_R\right] \tag{23.35}$$

where we sum over the three lepton families $\mathfrak{L} = \mathfrak{E}, \mathfrak{M}$, and $\mathfrak{T}$, and the three quark families $\mathfrak{Q} = \mathfrak{D}, \mathfrak{S}$, and $\mathfrak{B}$. Note that for the $\mathbf{U}(1)$ wavefunction B_μ each fermion current in eq. (23.35) is multiplied by $Y^{\mathfrak{F}}$, which signifies its hypercharge. As promised this differs from the QCD/QED case by the presence of the Higgs wavefunction Φ.

Before writing down the Klein-Gordon-type equations for Φ and the Dirac-type equations for the fermions, let's look at this set of equations more closely. We see from (23.35) that the $\mathbf{U}(1)$ wavefunction B_μ has a current coming from both the left-handed and right-handed fermions, as well as a current from the Higgs. The $\mathbf{SU}(2)$ wavefunctions W^c_μ also have a Higgs current, but only a left-handed fermion current as discussed above. There is also a current carried by the W^c_μ themselves, which comes from the left-hand side of the equation and is due to the quadratic and cubic terms in the W's. As with QCD, this last current will give rise to triple-W and quadruple-W vertices. Note also that the fermion current for the B_μ does not change the flavor of the fermions as they emit a B_μ particle, but that the fermion current for the $W^{c\mu}$ *does* change the flavor (e.g., turning a muon into a muon-neutrino) due to the presence of the σ^a operator in $\overline{\chi^{\mathfrak{F}}_L}\gamma_\nu\left(\sigma^a\right)\chi^{\mathfrak{F}}_L$.

The symmetry of electroweak theory is broken because the minimum of the potential

$$\mathcal{V}\left(\Phi\right) = V_0 - \frac{1}{2}\mu^2\Phi^\dagger\Phi + \frac{1}{4}\lambda^2\left(\Phi^\dagger\Phi\right)^2$$

is not at $\Phi = 0$, but rather at

$$\Phi^\dagger\Phi = \phi_1^2 + \phi_2^2 + \phi_3^2 + \phi_4^2 = v^2 \tag{23.36}$$

where $v = \frac{\mu}{\lambda}$. Under a local $\mathbf{SU}(2)$ transformation $\Phi' = \exp\left[i\theta^a\left(x\right)\frac{\sigma^a}{2}\right]\Phi$, the potential remains invariant and the equations of motion are covariant (provided the gauge and fermion wavefunctions are also correspondingly transformed). Since there are three arbitrary functions $\theta^a\left(x\right)$ we can use these to eliminate three of the ϕ's, allowing us to write

$$\Phi = \begin{pmatrix} 0 \\ v + \mathsf{h}\left(x\right) \end{pmatrix} \tag{23.37}$$

where $\mathsf{h}\left(x\right) = 0$ now signifies the ground state, i.e. the minimum of the potential. Making this choice restricts us to a specific choice of gauge, but it is one where the physical degrees of freedom are explicit.

To see how the gauge bosons become massive all we need to do is to insert this form for Φ into the equations of electroweak theory. It is straightforward to show that when (23.37) holds then

$$D_\mu\Phi = -ig_{\mathrm{w}}\frac{v}{2}\begin{pmatrix} \left(W_\mu^1 - iW_\mu^2\right) \\ -\frac{g_{\mathrm{Y}}}{g_{\mathrm{w}}}B_\mu - W_\mu^3 \end{pmatrix} + \mathcal{O}\left(\mathsf{h}\right) \tag{23.38}$$

and so the Higgs currents to this order are

$$\frac{ig_{\mathrm{w}}}{4}\left(\Phi^\dagger\sigma^a D_\mu\Phi - \left(D_\mu\Phi\right)^\dagger\sigma^a\Phi\right) = \left(\frac{g_{\mathrm{w}}v}{2}\right)^2\left\{W_\mu^1, W_\mu^2, \frac{g_{\mathrm{Y}}}{g_{\mathrm{w}}}B_\mu + W_\mu^3\right\} \tag{23.39}$$

$$\frac{ig_{\mathrm{Y}}}{4}\left(\Phi^\dagger D_\nu\Phi - \left(D_\nu\Phi\right)^\dagger\Phi\right) = -\left(\frac{v}{2}\right)^2 g_{\mathrm{w}}g_{\mathrm{Y}}\left(\frac{g_{\mathrm{Y}}}{g_{\mathrm{w}}}B_\mu + W_\mu^3\right) \tag{23.40}$$

and we see that only the linear combination $\frac{g_{\mathrm{Y}}}{g_{\mathrm{w}}}B_\mu + W_\mu^3$ appears (and not B_μ and W_μ^3 independently). Hence the gauge equations (23.34, 23.35) become

$$\partial^\mu F_{\mu\nu}^1 - g_{\mathrm{w}}W^{c\mu}F_{\mu\nu}^b\epsilon_{cb}^1 = \frac{g_{\mathrm{w}}}{2}\overline{\chi}_L^{\ell}\gamma_\nu\left(\sigma^1\right)\chi_L^{\ell} + \frac{g_{\mathrm{w}}}{2}\overline{\chi}_L^{\Omega}\gamma_\nu\left(\sigma^1\right)\chi_L^{\Omega}$$
$$- \left(\frac{g_{\mathrm{w}}v}{2}\right)^2 W_\mu^1 + \mathcal{O}\left(\mathsf{h}\right) \tag{23.41}$$

$$\partial^\mu F_{\mu\nu}^2 - g_{\mathrm{w}}W^{c\mu}F_{\mu\nu}^b\epsilon_{cb}^2 = \frac{g_{\mathrm{w}}}{2}\overline{\chi}_L^{\ell}\gamma_\nu\left(\sigma^2\right)\chi_L^{\ell} + \frac{g_{\mathrm{w}}}{2}\overline{\chi}_L^{\Omega}\gamma_\nu\left(\sigma^2\right)\chi_L^{\Omega}$$
$$- \left(\frac{g_{\mathrm{w}}v}{2}\right)^2 W_\mu^2 + \mathcal{O}\left(\mathsf{h}\right) \tag{23.42}$$

$$\partial^\mu F_{\mu\nu}^3 - g_{\mathrm{w}}W^{c\mu}F_{\mu\nu}^b\epsilon_{cb}^3 = \frac{g_{\mathrm{w}}}{2}\overline{\chi}_L^{\ell}\gamma_\nu\left(\sigma^3\right)\chi_L^{\ell} + \frac{g_{\mathrm{w}}}{2}\overline{\chi}_L^{\Omega}\gamma_\nu\left(\sigma^3\right)\chi_L^{\Omega}$$
$$- \left(\frac{g_{\mathrm{w}}v}{2}\right)^2\left(\frac{g_{\mathrm{Y}}}{g_{\mathrm{w}}}B_\mu + W_\mu^3\right) + \mathcal{O}\left(\mathsf{h}\right) \tag{23.43}$$

$$\partial^\mu\left(\partial_\mu B_\nu - \partial_\nu B_\mu\right) = \frac{g_{\mathrm{Y}}}{2}\left[Y_L^{\ell}\overline{\chi}_L^{\ell}\gamma_\nu\chi_L^{\ell} + Y_R^{\ell}\overline{\chi}_R^{\ell}\gamma_\nu\chi_R^{\ell}\right.$$
$$\left. + Y_L^{\Omega}\overline{\chi}_L^{\Omega}\gamma_\nu\chi_L^{\Omega} + Y_R^{\Omega}\overline{\chi}_R^{\Omega}\gamma_\nu\chi_R^{\Omega}\right]$$
$$- \left(\frac{v}{2}\right)^2 g_{\mathrm{w}}g_{\mathrm{Y}}\left(\frac{g_{\mathrm{Y}}}{g_{\mathrm{w}}}B_\mu + W_\mu^3\right) + \mathcal{O}\left(\mathsf{h}\right) \tag{23.44}$$

where we see mass-type terms for the gauge fields emerging. It's still not quite clear where the photon wavefunction is. Since the fermion current on the right-hand side of equation (23.44) is not the current $(j^{\mathrm{em}})^\mu = \sum_f g_e Q^f \overline{\psi}^f \gamma^\mu \psi^f$ of electromagnetism, we know that the photon can't be B_μ.

23.2.2 Gauge-Field Mixing

What then is the relationship between the bosons W_μ^3 and B_μ and the Z-boson and the photon A_μ? And what is the relationship between $W_\mu^{1,2}$ and the weak bosons $W_\mu^\pm$? Let's write

$$W_\mu^\pm \equiv \frac{1}{\sqrt{2}} \left(W_\mu^1 \mp i W_\mu^2 \right) \qquad W_\mu^3 = \cos\theta_W Z_\mu - \sin\theta_W A_\mu$$

$$B_\mu = \sin\theta_W Z_\mu + \cos\theta_W A_\mu \qquad (23.45)$$

and insert these expressions into the gauge equations (23.41-23.44). After some algebra these become

$$\partial^\mu \left(\partial_\mu W_\nu^\pm - \partial_\nu W_\mu^\pm \right) + \left(\frac{g_w v}{2} \right)^2 W_\nu^\pm$$

$$= \pm g_w \partial^\mu \left(W_\mu^\pm W_\nu^3 - W_\nu^\pm W_\mu^3 \right) - g_w W^{\pm\mu} F_{\mu\nu}^3$$

$$+ \frac{g_w}{\sqrt{2}} \overline{\chi}_L^\mathfrak{L} \gamma_\nu \left(\sigma^\pm \right) \chi_L^\mathfrak{L} + \frac{g_w}{\sqrt{2}} \overline{\chi}_L^\mathfrak{Q} \gamma_\nu \left(\sigma^\pm \right) \chi_L^\mathfrak{Q} + \mathcal{O}\left(\mathsf{h} \right) \qquad (23.46)$$

$$\partial^\mu \left(\partial_\mu Z_\nu - \partial_\nu Z_\mu \right) + \left(\frac{g_w v}{2} \right)^2 \left(\frac{g_Y}{g_w} \sin\theta_w + \cos\theta_w \right)^2 Z_\nu$$

$$+ \left(\frac{g_w v}{2} \right)^2 \left(\frac{g_Y}{g_w} \cos\theta_w - \sin\theta_w \right) \left(\frac{g_Y}{g_w} \sin\theta_w + \cos\theta_w \right) A_\nu$$

$$= g_w \cos\theta_w \left(W^{+\mu} F_{\mu\nu}^- - W^{-\mu} F_{\mu\nu}^+ \right)$$

$$+ g_w \cos\theta_w \left(\partial^\mu \left(W_\mu^- W_\nu^+ - W_\nu^- W_\mu^+ \right) \right)$$

$$- \frac{g_Y \sin\theta_w}{2} \left(Y_L^\mathfrak{L} \overline{\chi}_L^\mathfrak{L} \gamma_\nu \chi_L^\mathfrak{L} + Y_R^\mathfrak{L} \overline{\chi}_R^\mathfrak{L} \gamma_\nu \chi_R^\mathfrak{L} \right)$$

$$- \frac{g_Y \sin\theta_w}{2} \left(Y_L^\mathfrak{Q} \overline{\chi}_L^\mathfrak{Q} \gamma_\nu \chi_L^\mathfrak{Q} + Y_R^\mathfrak{Q} \overline{\chi}_R^\mathfrak{Q} \gamma_\nu \chi_R^\mathfrak{Q} \right)$$

$$+ \frac{g_w \cos\theta_w}{2} \left(\overline{\chi}_L^\mathfrak{L} \gamma_\nu \left(\sigma^3 \right) \chi_L^\mathfrak{L} + \overline{\chi}_L^\mathfrak{Q} \gamma_\nu \left(\sigma^3 \right) \chi_L^\mathfrak{Q} \right) + \mathcal{O}\left(\mathsf{h} \right) \qquad (23.47)$$

$$\partial^\mu \left(\partial_\mu A_\nu - \partial_\nu A_\mu \right) + \left(\frac{g_w v}{2} \right)^2 \left(\frac{g_Y}{g_w} \cos\theta_w - \sin\theta_w \right)^2 A_\nu$$

$$+ \left(\frac{g_w v}{2} \right)^2 \left(\frac{g_Y}{g_w} \cos\theta_w - \sin\theta_w \right) \left(\frac{g_Y}{g_w} \sin\theta_w + \cos\theta_w \right) Z_\nu$$

$$= \frac{g_Y \cos\theta_w}{2} \left(Y_L^\mathfrak{L} \overline{\chi}_L^\mathfrak{L} \gamma_\nu \chi_L^\mathfrak{L} + Y_R^\mathfrak{L} \overline{\chi}_R^\mathfrak{L} \gamma_\nu \chi_R^\mathfrak{L} \right)$$

$$+ \frac{g_Y \cos\theta_w}{2} \left(Y_L^\mathfrak{Q} \overline{\chi}_L^\mathfrak{Q} \gamma_\nu \chi_L^\mathfrak{Q} + Y_R^\mathfrak{Q} \overline{\chi}_R^\mathfrak{Q} \gamma_\nu \chi_R^\mathfrak{Q} \right)$$

$$+\frac{g_{\mathrm{w}}\sin\theta_{\mathrm{w}}}{2}\left(\overline{\chi}_L^{\mathfrak{L}}\gamma_\nu\left(\sigma^3\right)\chi_L^{\mathfrak{L}}+\overline{\chi}_L^{\mathfrak{Q}}\gamma_\nu\left(\sigma^3\right)\chi_L^{\mathfrak{Q}}\right)+\mathcal{O}\left(\mathrm{h}\right)\tag{23.48}$$

where $F_{\mu\nu}^{\pm}\equiv\frac{1}{\sqrt{2}}\left(F_{\mu\nu}^1\mp iF_{\mu\nu}^2\right)$.

This is a pretty complicated set of equations, so to understand them we will have to look at them in pieces. First notice that there is a mass term in the last equation for A_ν. If we want to identify this with the photon wavefunction, we must eliminate this term, along with the Z_ν term because the photon does not couple to the Z_ν. This we can do by setting $g_{\mathrm{w}}\sin\theta_{\mathrm{w}}=g_{\mathrm{Y}}\cos\theta_{\mathrm{w}}$; this also eliminates the A_ν term in the equation for Z_ν. We also must require that the photon current is of the form $g_e j_\nu^{em}=Q^{\mathfrak{F}}\overline{\chi}_L^{\mathfrak{F}}\gamma_\nu\chi_L^{\mathfrak{F}}$. For the leptons we have

$$g_e\left(j_\nu^{em}\right)^{\mathfrak{L}}=\frac{g_{\mathrm{w}}\sin\theta_{\mathrm{w}}}{2}\left(\overline{\chi}_L^{\mathfrak{L}}\gamma_\nu\left(\sigma^3\right)\chi_L^{\mathfrak{L}}\right)$$

$$+\frac{g_{\mathrm{Y}}\cos\theta_{\mathrm{w}}}{2}\left[Y_L^{\mathfrak{L}}\overline{\chi}_L^{\mathfrak{L}}\gamma_\nu\chi_L^{\mathfrak{L}}+Y_R^{\mathfrak{L}}\overline{\chi}_R^{\mathfrak{L}}\gamma_\nu\chi_R^{\mathfrak{L}}\right]$$

$$=\frac{1}{2}\left(g_{\mathrm{w}}\sin\theta_{\mathrm{w}}+g_{\mathrm{Y}}Y^{\nu_\ell}\cos\theta_{\mathrm{w}}\right)\overline{\chi}_L^{\nu_\ell}\gamma_\nu\chi_L^{\nu_\ell}$$

$$+\frac{1}{2}\left(-g_{\mathrm{w}}\sin\theta_{\mathrm{w}}+g_{\mathrm{Y}}Y^{e_L}\cos\theta_{\mathrm{w}}\right)\overline{\chi}_L^{e_\ell}\gamma_\nu\chi_L^{e_\ell}$$

$$+\frac{g^{\mathrm{Y}}\cos\theta_{\mathrm{w}}}{2}Y^{e_R}\overline{\chi}_R^{e_\ell}\gamma_\nu\chi_R^{e_\ell}\tag{23.49}$$

with an identical structure for the μ and τ leptons. Since neutrinos have $Q^\nu=0$ the first term on the right-hand side of eq. (23.49) must vanish, and we must also have $Y^{\nu_\ell}=Y^{e_L}$ since both states in the left-handed $\mathfrak{E}_L$ doublet must have the same hypercharge. Finally, the coefficients of both $\overline{\chi}_L^{e_\ell}\gamma_\nu\chi_L^{e_\ell}$ and $\overline{\chi}_R^{e_\ell}\gamma_\nu\chi_R^{e_\ell}$ must equal $-e$, where e is the electron charge. Putting this together we obtain the solution

$$g_{\mathrm{w}}\sin\theta_{\mathrm{w}}=g_{\mathrm{Y}}\cos\theta_{\mathrm{w}}=g_e=e\qquad Y^{\nu_\ell}=Y^{e_L}=\frac{1}{2}Y^{e_R}=-1\tag{23.50}$$

The hypercharge assignments for the μ and τ leptons are respectively the same, and so we obtain $-e\overline{\chi}^{\mathfrak{L}}\gamma_\nu\chi^{\mathfrak{L}}$ for the electric current for the leptons.

For the quarks

$$g_e\left(j_\nu^{em}\right)^{\mathfrak{Q}}=\frac{g_{\mathrm{w}}\sin\theta_{\mathrm{w}}}{2}\left(\overline{\chi}_L^{\mathfrak{Q}}\gamma_\nu\left(\sigma^3\right)\chi_L^{\mathfrak{Q}}\right)$$

$$+\frac{g_{\mathrm{Y}}\cos\theta_{\mathrm{w}}}{2}\left[Y_L^{\mathfrak{Q}}\overline{\chi}_L^{\mathfrak{Q}}\gamma_\nu\chi_L^{\mathfrak{Q}}+Y_R^{\mathfrak{Q}}\overline{\chi}_R^{\mathfrak{Q}}\gamma_\nu\chi_R^{\mathfrak{Q}}\right]$$

$$=\frac{g_e}{2}\left[\left(1+Y^{u_L}\right)\overline{\chi}_L^{u}\gamma_\nu\chi_L^{u}+Y^{u_R}\overline{\chi}_R^{u}\gamma_\nu\chi_R^{u}-\left(1-Y^{d_L}\right)\overline{\chi}_L^{d}\gamma_\nu\chi_L^{d}\right]$$

$$+\frac{g_e}{2}Y^{d_R}\overline{\chi}_R^{d}\gamma_\nu\chi_R^{d}\tag{23.51}$$

with analogous expressions for the c, s and t, b quarks. Since $Q^u = \frac{2}{3}$ and $Q^d = -\frac{1}{3}$, we obtain

$$Y^{u_L} = Y^{d_L} = Y^{c_L} = Y^{s_L} = Y^{t_L} = Y^{b_L} = \frac{1}{3} \tag{23.52}$$

$$Y^{u_R} = Y^{c_R} = Y^{t_R} = \frac{4}{3} \qquad Y^{d_R} = Y^{s_R} = Y^{b_R} = -\frac{2}{3} \tag{23.53}$$

for the quark hypercharge assignments.

23.2.3 Gauge Boson Masses

We see from eqs. (23.46 – 23.48) that the $W^\pm$ and the Z equations now have mass terms

$$M_W = \frac{g_W v}{2} \qquad M_Z = \left(\frac{g_W v}{2}\right)\left(\frac{g_Y}{g_W}\sin\theta_W + \cos\theta_W\right) = \frac{g_W v}{2\cos\theta_W} \tag{23.54}$$

yielding the relationship

$$M_W = M_Z \cos\theta_W \tag{23.55}$$

between the $W^\pm$ and the Z masses. Since we can measure the mass of the $W^\pm$ bosons, and since we can infer the value of g_W from the measurement of Fermi's constant (recall eq. (20.27)), it's not hard to show that $v = 246$ GeV.

We can read off the fermion current for the $W^\pm$ boson as

$$j_\nu^{W\pm} = \frac{g_W}{\sqrt{2}}\overline{\chi_L^{\mathfrak{F}}}\gamma_\nu\left(\sigma^\pm\right)\chi_L^{\mathfrak{F}}$$

the same for each quark and lepton doublet. Note that this current mixes together the up and down isospin components of a doublet. For example

$$\sigma^+\mathfrak{E}_L^- = \begin{pmatrix} 0 & 1 \\ 0 & 0 \end{pmatrix}\begin{pmatrix} \nu_e \\ e^- \end{pmatrix}_L = \begin{pmatrix} e^- \\ 0 \end{pmatrix}_L = \frac{1}{2}\begin{pmatrix} \left(1-\gamma^5\right)\psi^{e^-} \\ 0 \end{pmatrix} \tag{23.56}$$

$$\sigma^-\mathfrak{E}_L^- = \begin{pmatrix} 0 & 0 \\ 1 & 0 \end{pmatrix}\begin{pmatrix} \nu_e \\ e^- \end{pmatrix}_L = \begin{pmatrix} 0 \\ \nu_e \end{pmatrix}_L = \frac{1}{2}\begin{pmatrix} 0 \\ \left(1-\gamma^5\right)\psi^{\nu_e} \end{pmatrix} \tag{23.57}$$

and so

$$\frac{g_W}{\sqrt{2}}\overline{\chi_L^{\mathfrak{E}}}\gamma_\nu\left(\sigma^+\right)\chi_L^{\mathfrak{E}} = \frac{g_W}{\sqrt{2}}\overline{\nu}_L\gamma_\nu e_L \qquad \frac{g_W}{\sqrt{2}}\overline{\chi_L^{\mathfrak{E}}}\gamma_\nu\left(\sigma^-\right)\chi_L^{\mathfrak{E}} = \frac{g_W}{\sqrt{2}}\overline{e}_L\gamma_\nu\nu_L \tag{23.58}$$

This is very much like the isospin symmetry we looked at for the strong interactions, in which the proton and neutron were placed in an isospin doublet. The general form of a left-handed flavor wavefunction is

$$\chi_L^{\mathfrak{F}} = \begin{pmatrix} u_L \\ \mathfrak{d}_L \end{pmatrix} \tag{23.59}$$

where $\mathfrak{u}$ refers to the isospin-up wavefunction of $\mathfrak{F}$ (i.e. $\mathfrak{u} = \nu_e, \nu_\mu, \nu_\tau, u, c, t$) and $\mathfrak{d}$ refers to the isospin-down wavefunction of $\mathfrak{F}$ (i.e. respectively $\mathfrak{d} = e, \mu, \tau, d', s', b'$). The full interaction term is then given by summing over all $\mathfrak{F}$, analogous to the sum over all charged particles in QED.

Note that the right-handed parts $\chi_R^{\mathfrak{F}}$ have no isospin – they are singlets. I will use the notation $\mathfrak{u}_R$ and $\mathfrak{d}_R$ where $\mathfrak{u}$ and $\mathfrak{d}$ refer to up-type and down-type wave functions.

The current for the Z is a bit more complicated. If we write

$$j_\nu^3 = \frac{1}{2}\overline{\chi_L^{\mathfrak{F}}}\gamma_\nu \left(\sigma^3\right)\chi_L^{\mathfrak{F}}$$

then using eqs. (23.50) we get

$$g_Z j_\nu^{z} = -\frac{g_Y \sin\theta_{\rm w}}{2}\left[Y_L^{\mathfrak{e}}\overline{\chi_L^{\mathfrak{e}}}\gamma_\nu \chi_L^{\mathfrak{e}} + Y_L^{\mathfrak{e}}\overline{\chi_R^{\mathfrak{e}}}\gamma_\nu \chi_R^{\mathfrak{e}} + Y_L^{\mathfrak{d}}\overline{\chi_L^{\mathfrak{d}}}\gamma_\nu \chi_L^{\mathfrak{d}} + Y_R^{\mathfrak{d}}\overline{\chi_R^{\mathfrak{d}}}\gamma_\nu \chi_R^{\mathfrak{d}}\right]$$

$$+\frac{g_{\rm w}\cos\theta_{\rm w}}{2}j_\nu^3$$

$$= g_{\rm w}\cos\theta_{\rm w}j_\nu^3 - \frac{\sin\theta_{\rm w}}{\cos\theta_{\rm w}}\left[g_e j_\nu^{\rm em} - g^{\rm w}\sin\theta_{\rm w}j_\nu^3\right]$$

$$= \frac{g_e}{\sin\theta_W \cos\theta_W}\left((j^3)^\mu - \sin^2\theta_W (j^{\rm em})^\mu\right) \tag{23.60}$$

and so we identify

$$(j^z)^\mu = \overline{\chi}\left(c_V - c_A \gamma^5\right)\chi$$

$$g_Z = \frac{g_e}{\sin\theta_W \cos\theta_W} \tag{23.61}$$

$$(j^z)^\mu = (j^3)^\mu - \sin^2\theta_W (j^{\rm em})^\mu \tag{23.62}$$

From this expression for $(j^z)^\mu$ we can pick out all of the weak neutral couplings to the Z-boson, which we wrote down before in table 22.2! We see that in general

$$c_V^{\mathfrak{F}} = \left(I_{\mathfrak{F}}^3 - 2Q^{\mathfrak{f}}\sin^2\theta_W\right) \qquad c_A^{\mathfrak{F}} = I_{\mathfrak{F}}^3 \tag{23.63}$$

For a fermion doublet $\chi^{\mathfrak{F}}$ we have

$$\left(j^{z\mathfrak{F}}\right)^\mu = \frac{1}{2}\overline{\chi_L^{\mathfrak{F}}}\gamma^\mu \sigma^3 \chi_L^{\mathfrak{F}} - Q^{\mathfrak{F}}\sin^2\theta_W \left(\overline{\chi^{\mathfrak{F}}}\gamma^\mu \chi^{\mathfrak{F}}\right) \tag{23.64}$$

and for the total neutral current we sum over $\mathfrak{F}$.

23.3   Fermion Masses

The equations for the fermions are Dirac-type equations with gauge fields. We could add in mass terms directly, but why not make use of the Higgs wavefunction to generate masses for the fermions as well? This would provide a unified picture of the origin of mass, in which all particles attain their inertial masses through their interactions with the Higgs particle.

How might we do this? First notice that the Dirac equation for a wavefunction ψ

$$(i\gamma^\mu (\partial_\mu - ieA_\mu) - m) \psi = 0 \qquad (23.65)$$

can be separated into its left-handed and right-handed parts

$$i\gamma^\mu (\partial_\mu - ieA_\mu) \psi_R = m\psi_L \qquad i\gamma^\mu (\partial_\mu - ieA_\mu) \psi_L = m\psi_R \qquad (23.66)$$

by multiplying eq. (23.65) by $\frac{1}{2}\left(1 \pm \gamma^5\right)$ respectively. So the way to introduce a mass in electroweak theory – where left-handed and right-handed particles are distinguished at the outset – is to insert a constant times ψ_L in the equation for ψ_R and the same constant times ψ_R in the equation for ψ_L. A mass-generating term must have the same effect when the symmetry is broken.

For electroweak theory we have left-handed fermion doublets $\chi_L^{\mathfrak{F}}$ but right-handed fermion singlets $\chi_R^{\mathfrak{d}}$. This means that in the equation for $\chi_R^{\mathfrak{d}}$ we will need to multiply the doublet $\chi_L^{\mathfrak{F}}$ by something that will change it into a singlet. The only thing that will do this is another doublet, which we fortunately have – the Higgs wavefunction! Similarly, the equation for $\chi_L^{\mathfrak{F}}$ is a 2-component doublet equation, and so we need a doublet muliplying $\chi_R^{\mathfrak{d}}$ to give us a doublet structure on the right-hand side.

In terms of physical processes, we want our mass terms to have the effect that

$$\chi_R^{\mathfrak{d}} + \begin{pmatrix} \Phi^{\mathfrak{u}} \\ \Phi^{\mathfrak{d}} \end{pmatrix} \rightarrow \begin{pmatrix} \chi_L^{\mathfrak{u}} \\ \chi_L^{\mathfrak{d}} \end{pmatrix} \quad \text{and} \quad \begin{pmatrix} \chi_L^{\mathfrak{u}} \\ \chi_L^{\mathfrak{d}} \end{pmatrix} \rightarrow \chi_R^{\mathfrak{d}} + \begin{pmatrix} \Phi^{\mathfrak{u}} \\ \Phi^{\mathfrak{d}} \end{pmatrix} \qquad (23.67)$$

which balances isospin and hypercharge on both sides for both quarks and leptons, as you can easily check using equations (23.52) and (23.53). So in other words we put a term of the form $\Phi^\dagger \chi_L^{\mathfrak{F}}$ on the right-hand side of the equation for $\chi_R^{\mathfrak{d}}$. This term isn't constant, but if we expand about the ground state we have $\Phi^\dagger \chi_L^{\mathfrak{F}} = \frac{1}{\sqrt{2}}\left(v\chi_L^{\mathfrak{d}} + h^\dagger \chi_L^{\mathfrak{F}}\right)$, and the $v\chi_L^{\mathfrak{d}}$ term will generate a mass.

What about the equation for $\chi_R^{\mathfrak{u}}$? Here we have what appears to be a problem: for the right-handed up-type quarks $Y = \frac{4}{3}$, so if we just replace $\chi_R^{\mathfrak{d}}$ with $\chi_R^{\mathfrak{u}}$ in the above equations we find that $Y = \frac{7}{3}$ on one side but $Y = \frac{1}{3}$ on the other. The reason for this is that the Higgs doublet has $Y = +1$.

What we need is a doublet with $Y = -1$. Fortunately this can be done without introducing any new particles! Using the familiar ϵ-tensor (but now in two dimensions) we construct the charge conjugate $\tilde{\Phi}$ of the Higgs

$$\tilde{\Phi}^i = \epsilon^{ij}\Phi^{j*} \Rightarrow \begin{pmatrix} \tilde{\Phi}^{\mathfrak{u}} \\ \tilde{\Phi}^{\mathfrak{d}} \end{pmatrix} = \begin{pmatrix} \Phi^{\mathfrak{d}*} \\ -\Phi^{\mathfrak{u}*} \end{pmatrix} \qquad (23.68)$$

which has $Y = -1$ and isospin $\frac{1}{2}$, so that the processes

$$\chi_R^u + \begin{pmatrix} \tilde{\Phi}^u \\ \tilde{\Phi}^0 \end{pmatrix} \rightarrow \begin{pmatrix} \chi_L^u \\ \chi_L^0 \end{pmatrix} \quad \text{and} \quad \begin{pmatrix} \chi_L^u \\ \chi_L^0 \end{pmatrix} \rightarrow \chi_R^u + \begin{pmatrix} \tilde{\Phi}^u \\ \tilde{\Phi}^0 \end{pmatrix} \tag{23.69}$$

can be consistently mediated.

As a last step we also need terms multiplying χ_R^u and χ_R^0 on the right-hand side of the equation for $\chi_L^\mathfrak{F}$; clearly these must be of the form $\Phi\chi_R^0$ and $\tilde{\Phi}\chi_R^u$ (each of which have hypercharge $Y = \frac{1}{3}$) to balance isospin and hypercharge on both sides.

The equations for the fermions in electroweak theory generate the Higgs-fermion-fermion vertices needed to ensure the above processes take place, and can be written as

$$i\gamma^\mu \left(\partial_\mu - i\frac{g_W}{2}\sigma^a \cdot \mathbf{W}_\mu^a + i\frac{g_Y}{2}Y^{\mathfrak{F}L}B_\mu\right)\chi_L^\mathfrak{F} = G^{\mathfrak{F}u}\tilde{\Phi}\hat{\chi}_R^u + H^{\mathfrak{F}0}\Phi\hat{\chi}_R^0 \tag{23.70}$$

$$i\gamma^\mu \left(\partial_\mu + i\frac{g_Y}{2}Y^{uR}B_\mu\right)\chi_R^u = G^{u\mathfrak{K}\dagger}\left(\tilde{\Phi}^\dagger\chi_L^\mathfrak{K}\right)$$
$$i\gamma^\mu \left(\partial_\mu + i\frac{g_Y}{2}Y^{0R}B_\mu\right)\chi_R^0 = H^{0\mathfrak{K}\dagger}\left(\Phi^\dagger\chi_L^\mathfrak{K}\right) \tag{23.71}$$

where the quantities $G^{\mathfrak{F}\mathfrak{K}}$ and $H^{\mathfrak{F}\mathfrak{K}}$ are constants, known as Yukawa couplings, that parametrize how strongly the Higgs couples to the up-type fermions (via G) and down-type fermions (via H). Except for the hypercharge operator Y, repeated indices are summed over – so when a $\mathfrak{d}$ is repeated in a given term it means sum over the down-type objects, and when a $\mathfrak{u}$ is repeated it means sum over the up-type objects.

The most general thing would be to sum over all six values of $\mathfrak{d}$ and $\mathfrak{u}$. But this would lead to processes like $\mu_R + \Phi^u \rightarrow c_L$, which violate lepton and baryon number conservation. So let's make G and H reducible matrices in the quark (q) and lepton (l) sectors

$$G^{\mathfrak{F}\mathfrak{K}} = \begin{pmatrix} G_q^{FK} & 0 \\ 0 & G_l^{FK} \end{pmatrix} \qquad H^{f\ell} = \begin{pmatrix} H_q^{FK} & 0 \\ 0 & H_l^{FK} \end{pmatrix} \tag{23.72}$$

where each of G_q^{FK}, H_q^{FK}, G_l^{FK}, and H_l^{FK}, are 3×3 matrices.

The above structure is still pretty complicated – if we expand about the ground state we will get all kinds of mixings between the different flavors. However, things are not quite so messy as they might seem because we have a lot of freedom to redefine the flavors. For example in the quark sector we can define

$$u_L^I = \mathcal{L}_{qL}^{\Uparrow IJ} \mathfrak{v}\hat{u}_R^J \qquad u_R^I = \mathcal{R}_{qL}^{\Uparrow IJ} \mathfrak{v}\hat{\mathfrak{d}}_R^J \tag{23.73}$$

$$\mathfrak{d}_L^I = \mathcal{L}_{qL}^{\Downarrow IJ} \mathfrak{v}\hat{\mathfrak{d}}^J \qquad \mathfrak{d}_R^I = \mathcal{R}_{qL}^{\Downarrow IJ} \mathfrak{v}\hat{\mathfrak{d}}_R^J \tag{23.74}$$

for both the right-handed and left-handed wavefunctions, where for convenience I have defined

$$\{\psi^{\text{up}}, \psi^{\text{charm}}, \psi^{\text{top}}\} = \{u^1, u^2, u^3\}$$

and

$$\{\psi^{down}, \psi^{strange}, \psi^{bottom}\} = \left\{\mathfrak{d}^1, \mathfrak{d}^2, \mathfrak{d}^3\right\}$$

so that the generations are labeled 1,2,3 in order of increasing mass. Here $\mathcal{L}_q^\Uparrow$ is a 3×3 unitary matrix that interchanges the left-handed-up-type flavors; $\mathcal{R}_q^\Downarrow$ is a 3×3 unitary matrix that interchanges the right-handed-down-type flavors, etc.

To see what we can do with this freedom, let's unpack the fermion equations (23.71) into up-type and down-type parts

$$i\gamma^\mu \left(\partial_\mu - i\frac{g_w}{2}\mathbf{W}_\mu^3 + i\frac{g_Y}{6}B_\mu\right)u_L^I + \frac{g_w}{\sqrt{2}}\gamma^\mu\mathbf{W}_\mu^+\mathfrak{d}_L^I$$
$$= \Phi^{\mathfrak{d}*}G_q^{IK}u_R^K + \tilde{\Phi}^u H_q^{IK}\mathfrak{d}_R^K \qquad (23.75)$$

$$\frac{g_w}{\sqrt{2}}\gamma^\mu\mathbf{W}_\mu^- u_L^I + i\gamma^\mu\left(\partial_\mu + i\frac{g_w}{2}\mathbf{W}_\mu^3 + i\frac{g_Y}{6}B_\mu\right)\mathfrak{d}_L^I$$
$$= -\tilde{\Phi}^{u*}G_q^{IK}u_R^K + \Phi^{\mathfrak{d}}H_q^{IK}\mathfrak{d}_R^K \qquad (23.76)$$

$$i\gamma^\mu\left(\partial_\mu + i\frac{2g_Y}{3}B_\mu\right)u_R^I = G_q^{IK\dagger}\left(\Phi^{\mathfrak{d}}u_L^K - \Phi^u\mathfrak{d}_L^K\right) \qquad (23.77)$$

$$i\gamma^\mu\left(\partial_\mu - i\frac{g_Y}{3}B_\mu\right)\hat{\mathfrak{d}}_R^I = H_q^{IK\dagger}\left(\Phi^{u*}u_L^K + \Phi^{\mathfrak{d}*}\mathfrak{d}_L^K\right) \qquad (23.78)$$

and then multiply by the $\mathcal{L}_q$ and $\mathcal{R}_q$ matrices so that we obtain equations in terms of the hatted wavefunctions

$$i\gamma^\mu\left(\partial_\mu - i\frac{g_w}{2}\mathbf{W}_\mu^3 + i\frac{g_Y}{6}B_\mu\right)\hat{u}_L^I + \frac{g_w}{\sqrt{2}}\gamma^\mu\mathbf{W}_\mu^+\left(\mathcal{L}_q^{\Uparrow\dagger}\mathcal{L}_q^{\Downarrow}\right)^{IK}\hat{\mathfrak{d}}_L^K$$
$$= \Phi^{\mathfrak{d}*}\left(\mathcal{L}_q^{\Uparrow\dagger}G_q\mathcal{R}_q^{\Uparrow}\right)^{IK}\hat{u}_R^K + \tilde{\Phi}^u\left(\mathcal{L}_q^{\Uparrow\dagger}H_q\mathcal{R}_q^{\Downarrow}\right)^{IK}\hat{\mathfrak{d}}_R^K \quad (23.79)$$

$$\frac{g_w}{\sqrt{2}}\gamma^\mu\mathbf{W}_\mu^-\left(\mathcal{L}_q^{\Downarrow\dagger}\mathcal{L}_q^{\Uparrow}\right)^{IK}\hat{u}_L^K + i\gamma^\mu\left(\partial_\mu + i\frac{g_w}{2}\mathbf{W}_\mu^3 + i\frac{g_Y}{6}B_\mu\right)\hat{\mathfrak{d}}_L^I$$
$$= -\tilde{\Phi}^{u*}\left(\mathcal{L}_q^{\Downarrow\dagger}G_q\mathcal{R}_q^{\Uparrow}\right)^{IK}\hat{u}_R^K + \Phi^{\mathfrak{d}}\left(\mathcal{L}_q^{\Downarrow\dagger}H_q\mathcal{R}_q^{\Downarrow}\right)^{IK}\hat{\mathfrak{d}}_R^K \quad (23.80)$$

$$i\gamma^\mu\left(\partial_\mu + i\frac{2g_Y}{3}B_\mu\right)\hat{u}_R^I$$
$$= \Phi^{\mathfrak{d}}\left(\mathcal{R}_q^{\Uparrow\dagger}G_q^\dagger\mathcal{L}_q^{\Uparrow}\right)^{IK}\hat{u}_L^K - \Phi^u\left(\mathcal{R}_q^{\Uparrow\dagger}G_q^\dagger\mathcal{L}_q^{\Downarrow}\right)^{IK}\hat{\mathfrak{d}}_L^K \quad (23.81)$$

$$i\gamma^\mu\left(\partial_\mu - i\frac{g_Y}{3}B_\mu\right)\hat{\mathfrak{d}}_R^I$$
$$= \Phi^{u*}\left(\mathcal{R}_q^{\Downarrow\dagger}H_q^\dagger\mathcal{L}_q^{\Uparrow}\right)^{IK}\hat{u}_L^K + \Phi^{\mathfrak{d}*}\left(\mathcal{R}_q^{\Downarrow\dagger}H_q^\dagger\mathcal{L}_q^{\Downarrow}\right)^{IK}\hat{\mathfrak{d}}_L^K \quad (23.82)$$

This doesn't look like much of an improvement until we realize that any complex matrix M when multiplied on the right and left by different unitary matrices U_1, U_2 becomes $U_1^\dagger M U_2 = M_D$, where M_D can be chosen to be real and diagonal§. So let's choose

$$\left(\mathcal{L}_q^{\Uparrow\dagger} G_q \mathcal{R}_q^{\Uparrow}\right)^{IK} = \frac{1}{v} m_u^I \delta^{IK} \qquad \left(\mathcal{L}_q^{\Downarrow\dagger} H_q \mathcal{R}_q^{\Downarrow}\right)^{IK} = \frac{1}{v} m_{\mathfrak{d}}^I \delta^{IK} \qquad (23.83)$$

to obtain

$$i\gamma^\mu \left(\partial_\mu - i\frac{g_w}{2}\mathbf{W}_\mu^3 + i\frac{g_Y}{6}B_\mu\right)\widehat{u}_L^I + \frac{g_w}{\sqrt{2}}\gamma^\mu \mathbf{W}_\mu^+ \left(\mathcal{L}_q^{\Uparrow\dagger}\mathcal{L}_q^{\Downarrow}\right)^{IK}\widehat{\mathfrak{d}}_L^K$$

$$= \frac{1}{v}\Phi^{\mathfrak{d}*}m_u^I\widehat{u}_R^I + \tilde{\Phi}^u\left(\mathcal{L}_q^{\Uparrow\dagger}H_q\mathcal{R}_q^{\Downarrow}\right)^{IK}\widehat{\mathfrak{d}}_R^K \quad (23.84)$$

$$\frac{g_w}{\sqrt{2}}\gamma^\mu \mathbf{W}_\mu^- \left(\mathcal{L}_q^{\Downarrow\dagger}\mathcal{L}_q^{\Uparrow}\right)^{IK}\widehat{u}_L^K + i\gamma^\mu \left(\partial_\mu + i\frac{g_w}{2}\mathbf{W}_\mu^3 + i\frac{g_Y}{6}B_\mu\right)\widehat{\mathfrak{d}}_L^I$$

$$= -\tilde{\Phi}^{u*}\left(\mathcal{L}_q^{\Downarrow\dagger}G_q\mathcal{R}_q^{\Uparrow}\right)^{IK}\widehat{u}_R^K + \frac{1}{v}\Phi^{\mathfrak{d}}m_{\mathfrak{d}}^I\widehat{\mathfrak{d}}_R^I \quad (23.85)$$

$$i\gamma^\mu \left(\partial_\mu + i\frac{2g_Y}{3}B_\mu\right)\widehat{u}_R^I = \frac{1}{v}\Phi^{\mathfrak{d}}m_u^I\widehat{u}_L^I - \Phi^u\left(\mathcal{R}_q^{\Uparrow\dagger}G_q^\dagger\mathcal{L}_q^{\Downarrow}\right)^{IK}\widehat{\mathfrak{d}}_L^K \quad (23.86)$$

$$i\gamma^\mu \left(\partial_\mu - i\frac{g_Y}{3}B_\mu\right)\widehat{\mathfrak{d}}_R^I = \Phi^{u*}\left(\mathcal{R}_q^{\Downarrow\dagger}H_q^\dagger\mathcal{L}_q^{\Uparrow}\right)^{IK}\widehat{u}_L^K + \frac{1}{v}\Phi^{\mathfrak{d}*}m_{\mathfrak{d}}^I\widehat{\mathfrak{d}}_L^I \quad (23.87)$$

where there is no sum over the index I in the above equations. Expanding about the ground state $\Phi^u = 0$, $\Phi^{\mathfrak{d}} = (v + h)$ gives

$$i\gamma^\mu \left(\partial_\mu - i\frac{g_w}{2}\mathbf{W}_\mu^3 + i\frac{g_Y}{6}B_\mu\right)\widehat{u}_L^I + \frac{g_w}{\sqrt{2}}\gamma^\mu \mathbf{W}_\mu^+ \left(\mathcal{L}_q^{\Uparrow\dagger}\mathcal{L}_q^{\Downarrow}\right)^{IK}\widehat{\mathfrak{d}}_L^K$$

$$= m_u^I\widehat{u}_R^I + \mathcal{O}(h) \quad (23.88)$$

$$i\gamma^\mu \left(\partial_\mu + i\frac{g_w}{2}\mathbf{W}_\mu^3 + i\frac{g_Y}{6}B_\mu\right)\widehat{\mathfrak{d}}_L^I + \frac{g_w}{\sqrt{2}}\gamma^\mu \mathbf{W}_\mu^- \left(\mathcal{L}_q^{\Downarrow\dagger}\mathcal{L}_q^{\Uparrow}\right)^{IK}\widehat{u}_L^K$$

$$= m_{\mathfrak{d}}^I\widehat{\mathfrak{d}}_R^I + \mathcal{O}(h) \quad (23.89)$$

$$i\gamma^\mu \left(\partial_\mu + i\frac{2g_Y}{3}B_\mu\right)\widehat{u}_R^I = m_u^I\widehat{u}_L^I + \mathcal{O}(h) \quad (23.90)$$

$$i\gamma^\mu \left(\partial_\mu - i\frac{g_Y}{3}B_\mu\right)\widehat{\mathfrak{d}}_R^I = m_{\mathfrak{d}}^I\widehat{\mathfrak{d}}_L^I + \mathcal{O}(h) \quad (23.91)$$

and so we see that we have also used the Higgs mechanism to generate the masses of the quarks!

I could write these equations in terms of the physical $W^\pm$, Z and photon wavefunctions, but I won't do that here. It's a straightforward exercise since equations (23.88) – (23.91) are linear in each of these quantities. Instead, let's turn our attention to the leptons. We can repeat the same exercise in this case, setting $u^I \to \nu^I$ and $\mathfrak{d}^I \to \ell^I$. However, there is a crucial difference from

---

§To show this, the polar decomposition theorem indicates that for any complex matrix M we can write $M = HU$, where H is Hermitian and U unitary. Choosing $U_1 = S^\dagger$ and $U_2 = U^\dagger S$ where S diagonalizes H gives the result.

the quark case because in the Standard Model the right-handed neutrino does not appear. So we might as well set $\nu_R^{\mathrm{I}} = G_1^{\mathrm{IK}} = 0$ in the above equations, in which case repeating these derivations gives

$$i\gamma^\mu \left(\partial_\mu - i\frac{g_{\mathrm{w}}}{2}\mathbf{W}_\mu^3 + i\frac{g_{\mathrm{Y}}}{6}B_\mu\right)\widehat{\nu}_L^{\mathrm{I}} + \frac{g_{\mathrm{w}}}{\sqrt{2}}\gamma^\mu\mathbf{W}_\mu^+ \left(\mathcal{L}_1^{\Uparrow\dagger}\mathcal{L}_1^{\Downarrow}\right)^{\mathrm{IK}}\widehat{\ell}_L^{\mathrm{K}} = \mathcal{O}\,(\mathsf{h})$$

$$\frac{g_{\mathrm{w}}}{\sqrt{2}}\gamma^\mu\mathbf{W}_\mu^- \left(\mathcal{L}_1^{\Downarrow\dagger}\mathcal{L}_1^{\Uparrow}\right)^{\mathrm{IK}}\widehat{\nu}_L^{\mathrm{K}} + i\gamma^\mu \left(\partial_\mu + i\frac{g_{\mathrm{w}}}{2}\mathbf{W}_\mu^3 + i\frac{g_{\mathrm{Y}}}{6}B_\mu\right)\widehat{\ell}_L^{\mathrm{I}} = m_\ell^{\mathrm{I}}\widehat{\ell}_R^{\mathrm{I}}$$

$$i\gamma^\mu \left(\partial_\mu - i\frac{g_{\mathrm{Y}}}{3}B_\mu\right)\widehat{\ell}_R^{\mathrm{I}} = m_\ell^{\mathrm{I}}\widehat{\ell}_L^{\mathrm{I}} + \mathcal{O}\,(\mathsf{h})$$

which we can rewrite as

$$i\gamma^\mu \left(\partial_\mu - i\frac{g_{\mathrm{w}}}{2}\mathbf{W}_\mu^3 + i\frac{g_{\mathrm{Y}}}{6}B_\mu\right)\widetilde{\nu}_L^{\mathrm{I}} + \frac{g_{\mathrm{w}}}{\sqrt{2}}\gamma^\mu\mathbf{W}_\mu^+\widehat{\ell}_L^{\mathrm{I}} = \mathcal{O}\,(\mathsf{h}) \tag{23.92}$$

$$\frac{g_{\mathrm{w}}}{\sqrt{2}}\gamma^\mu\mathbf{W}_\mu^-\widetilde{\nu}_L^{\mathrm{I}} + i\gamma^\mu \left(\partial_\mu + i\frac{g_{\mathrm{w}}}{2}\mathbf{W}_\mu^3 + i\frac{g^{\mathrm{Y}}}{6}B_\mu\right)\widehat{\ell}_L^{\mathrm{I}} = m_\ell^{\mathrm{I}}\widehat{\ell}_R^{\mathrm{I}} \tag{23.93}$$

$$i\gamma^\mu \left(\partial_\mu - i\frac{g_{\mathrm{Y}}}{3}B_\mu\right)\widehat{\ell}_R^{\mathrm{I}} = m_\ell^{\mathrm{I}}\widehat{\ell}_L^{\mathrm{I}} + \mathcal{O}\,(\mathsf{h}) \tag{23.94}$$

where I have defined

$$\widetilde{\nu}_L^{\mathrm{I}} = \left(\mathcal{L}_1^{\Downarrow\dagger}\mathcal{L}_1^{\Uparrow}\right)^{\mathrm{IK}}\widehat{\nu}_L^{\mathrm{I}} \tag{23.95}$$

This means that the e, μ, τ leptons get mass, but not the neutrinos – and it is the $\widetilde{\nu}_L^{\mathrm{I}}$ that should be identified with the distinct neutrino flavors, i.e. $\widetilde{\nu}_L^1 = \nu_e, \widetilde{\nu}_L^2 = \nu_\mu$ and $\widetilde{\nu}_L^3 = \nu_\tau$.

There is a price to be paid for this – if we switch to hatted fermion wavefunctions here, we must also do it in the equations for the gauge particles. This will only modify the currents

$$\begin{aligned}\left(j^{\mathrm{z},\mathfrak{F}}\right)^\mu &= \frac{1}{2}\overline{\chi}_L^{\mathfrak{F}}\gamma^\mu\sigma^3\chi_L^{\mathfrak{F}} - Q^{\mathfrak{F}}\sin^2\theta_W\left(\overline{\chi}^{\mathfrak{F}}\gamma^\mu\chi^{\mathfrak{F}}\right)\\ &= \frac{1}{2}\overline{\widehat{\mathfrak{u}}}_L^{\mathrm{K}}\gamma^\mu\widehat{\mathfrak{u}}_L^{\mathrm{K}} - \frac{1}{2}\overline{\widehat{\mathfrak{d}}}_L^{\mathrm{K}}\gamma^\mu\widehat{\mathfrak{d}}_L^{\mathrm{K}} - \sin^2\theta_W\left(\frac{2}{3}\overline{\widehat{\mathfrak{u}}}^{\mathrm{K}}\gamma^\mu\widehat{\mathfrak{u}}^{\mathrm{K}} - \frac{1}{3}\overline{\widehat{\mathfrak{d}}}^{\mathrm{K}}\gamma^\mu\widehat{\mathfrak{d}}^{\mathrm{K}}\right)\\ &\quad + \frac{1}{2}\overline{\widetilde{\nu}}_L^{\mathrm{K}}\gamma^\mu\widetilde{\nu}_L^{\mathrm{K}} - \frac{1}{2}\overline{\widehat{\ell}}_L^{\mathrm{K}}\gamma^\mu\widehat{\ell}_L^{\mathrm{K}} + \sin^2\theta_W\left(\overline{\widehat{\ell}}^{\mathrm{K}}\gamma^\mu\widehat{\ell}^{\mathrm{K}}\right) \end{aligned} \tag{23.96}$$

$$\begin{aligned}\left(j^{\mathrm{em},\mathfrak{F}}\right)^\mu &= Q^{\mathfrak{F}}\left(\overline{\chi}^{\mathfrak{F}}\gamma^\mu\chi^{\mathfrak{F}}\right)\\ &= \frac{2}{3}\overline{\widehat{\mathfrak{u}}}^{\mathrm{K}}\gamma^\mu\widehat{\mathfrak{u}}^{\mathrm{K}} - \frac{1}{3}\overline{\widehat{\mathfrak{d}}}^{\mathrm{K}}\gamma^\mu\widehat{\mathfrak{d}}^{\mathrm{K}} - \overline{\widehat{\ell}}^{\mathrm{K}}\gamma^\mu\widehat{\ell}^{\mathrm{K}} \end{aligned} \tag{23.97}$$

$$\begin{aligned}j_\nu^{\mathrm{w}+} &= \frac{g_{\mathrm{w}}}{\sqrt{2}}\overline{\chi}_L^{\mathfrak{F}}\gamma_\nu\left(\sigma^+\right)\chi_L^{\mathfrak{F}}\\ &= \frac{g_{\mathrm{w}}}{\sqrt{2}}\overline{\widehat{\mathfrak{u}}}_L^{\mathrm{K}}\gamma_\nu\left(\mathcal{L}_{\mathfrak{q}}^{\Uparrow\dagger}\mathcal{L}_{\mathfrak{q}}^{\Downarrow}\right)^{\mathrm{KJ}}\widehat{\mathfrak{d}}_L^{\mathrm{J}} + \frac{g^{\mathrm{w}}}{\sqrt{2}}\overline{\widetilde{\nu}}_L^{\mathrm{K}}\gamma_\nu\widehat{\ell}_L^{\mathrm{K}} \end{aligned} \tag{23.98}$$

$$j_\nu^{\mathrm{w}-} = \frac{g_{\mathrm{w}}}{\sqrt{2}}\overline{\chi}_L^{\mathfrak{F}}\gamma_\nu\left(\sigma^-\right)\chi_L^{\mathfrak{F}}$$

$$= \frac{g_{\rm w}}{\sqrt{2}} \widehat{\eth}_L^{\rm K} \left(\mathcal{L}_{\rm q}^{\Downarrow\dagger} \mathcal{L}_{\rm q}^{\Uparrow} \right)^{\rm KJ} \gamma_\nu \widehat{u}_L^{\rm J} + \frac{g_{\rm w}}{\sqrt{2}} \overline{\widehat{\ell}}_L^{\rm K} \gamma_\nu \widetilde{\nu}_L^{\rm K} \tag{23.99}$$

which we see is not much of a modification at all! The only parts that look different are the charged quark currents in equations (23.98) and (23.99), which have the form

$$\frac{g_{\rm w}}{\sqrt{2}} \overline{\widehat{u}}_L^{\rm K} \gamma_\nu \widehat{\eth}_L^{\prime \rm K} \quad \text{where} \quad \widehat{\eth}_L^{\prime \rm K} = \left(\mathcal{L}_{\rm q}^{\Uparrow\dagger} \mathcal{L}_{\rm q}^{\Downarrow} \right)^{\rm KJ} \widehat{\eth}_L^{\rm J} = \mathcal{V}^{\rm KJ} \widehat{\eth}_L^{\rm J} \tag{23.100}$$

The matrix $\mathcal{V}^{\rm KJ}$ is a unitary matrix, which superficially has nine independent components. However, since its only appearance is in the charged currents and in the couplings of the $W^\pm$ to the quarks, we can eliminate some of its components by redefining the phases of the $\widehat{u}_L^{\rm J}$ and the $\widehat{\eth}_L^{\rm J}$. This is six phases in all, but one of these is an overall common phase that will cancel out in the charged current, leaving us with five phases that can eliminate five of the nine components of $\mathcal{V}$.

So the hatted wavefunctions are none other than the familiar quark and lepton wavefunctions that we have encountered in QED and QCD. We see from the manipulations in this section that while the Higgs mechanism provides an origin for the masses of all of these particles, it does not explain the values of these masses – they remain as arbitrary parameters. A large coupling of the Higgs particle to the top quark becomes the large top quark mass, and a small coupling of the Higgs particle to the electron becomes the small electron mass. The mysterious large ratio of the top mass to the electron mass is replaced by an equally mysterious large ratio between the Higgs coupling to the top and the Higgs coupling to the electron. You might also have guessed by now that $\mathcal{V}$ is none other than the CKM matrix (21.54):

$$\mathcal{V} \equiv \begin{pmatrix} V_{ud} & V_{us} & V_{ub} \\ V_{cd} & V_{cs} & V_{cb} \\ V_{td} & V_{ts} & V_{tb} \end{pmatrix} = \begin{pmatrix} c_1 & s_1 c_3 & s_1 s_3 \\ -s_1 c_2 & c_1 c_2 c_3 - s_2 c_3 e^{i\delta} & c_1 c_2 s_3 + s_2 c_3 e^{i\delta} \\ -s_1 s_2 & c_1 s_2 c_3 + c_2 s_3 e^{i\delta} & c_1 s_2 s_3 - c_2 c_3 e^{i\delta} \end{pmatrix} \tag{23.101}$$

whose components are also arbitrary parameters. We have no explanation for the values of these different parameters. At this point in time we can only measure them.

However, there is one scalar particle left over whose mass is undetermined. This is the Higgs particle, the last particle in the Standard model yet to be discovered. Finding it will not only be a crucial test of Standard Model physics, but also provide essential information on the origin of mass.

23.4   Appendix: Feynman Rules for Electroweak Theory

1. NOTATION. Label the incoming (outgoing) four-momenta as $p_1, p_2, \ldots, p_n$
$(p'_1, p'_2, \ldots, p'_m)$, the incoming (outgoing) spins as $s_1, s_2, \ldots, s_n$ $(s'_1, s'_2, \ldots, s'_m)$,
the incoming (outgoing) weak boson polarizations as $\varepsilon_1^\mu, \varepsilon_2^\mu, \ldots$ $(\varepsilon_1^{\mu'}, \varepsilon_2^{\mu'}, \ldots)$,
and label the internal four-momenta $q_1, q_2, \ldots, q_j$. Assign arrows to the lines
as in figure 23.1, in which time flows from bottom to top, dashed lines are

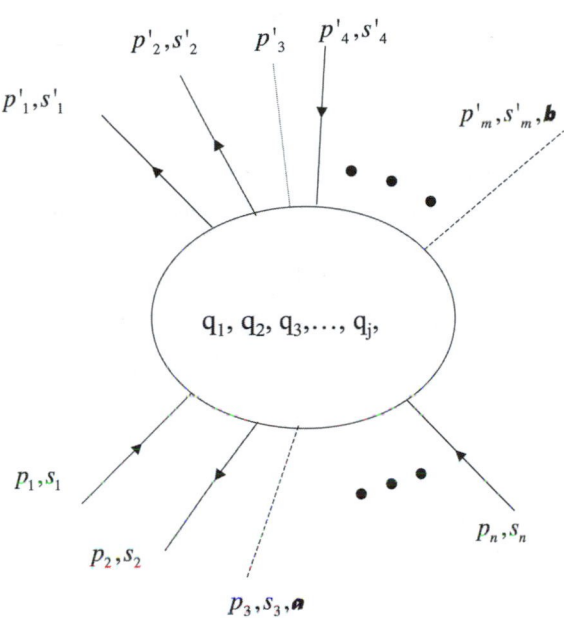

FIGURE 23.1
A typical diagram in electroweak theory. The dotted line in the outgoing state
is a Higgs particle.

weak bosons (with $\mathfrak{a}, \mathfrak{b} = +, -, 0$ labeling the W^+, W^-, Z^0 bosons respectively), Higgs particles are dotted lines, and lines with arrows are fermions
(if they point upward with time) or antifermions (or if they point downward
against time).

2. EXTERNAL LINES. Each external line contributes a factor as illustrated
in figure 23.2, and all incoming/outgoing Higgs particles have a factor of unity.

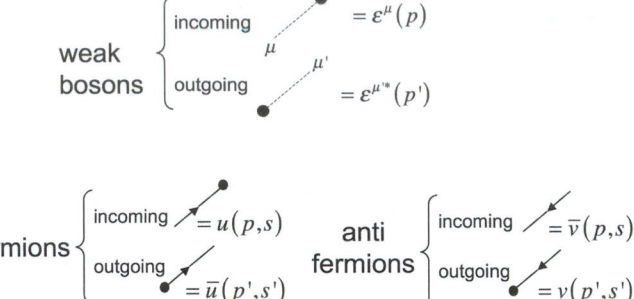

FIGURE 23.2

External lines in electroweak theory.

As in QED, factors associated with external lines correspond to the incoming/outgoing plane-wave states, an assumption that is clearly not valid for the unstable weak bosons and Higgs particle (as well as for all but the lightest fermions), but which we will take to be valid on the very short timescales in collisions.

3. **INTERNAL LINES.** Each internal line contributes a factor as shown in figure 23.3, where m is the mass of the quark or lepton, $M_\mathfrak{a}$ is the mass of

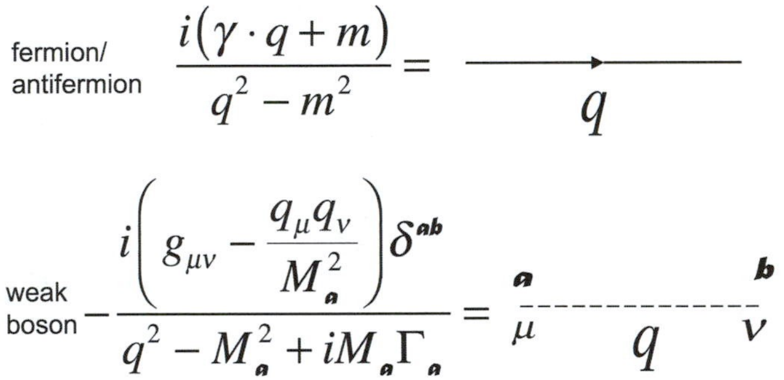

$$\text{fermion/}\atop\text{antifermion} \qquad \frac{i(\gamma \cdot q + m)}{q^2 - m^2} = \quad \xrightarrow{\qquad\qquad} \atop q$$

$$\text{weak}\atop\text{boson} \qquad -\frac{i\left(g_{\mu\nu} - \dfrac{q_\mu q_\nu}{M_\mathfrak{a}^2}\right)\delta^{\mathfrak{a}\mathfrak{b}}}{q^2 - M_\mathfrak{a}^2 + iM_\mathfrak{a}\Gamma_\mathfrak{a}} = \quad \overset{\mathfrak{a}}{\mu}\; \underset{q}{\text{- - - - - -}}\; \overset{\mathfrak{b}}{\nu}$$

FIGURE 23.3

Internal lines in electroweak theory.

the weak boson and $\Gamma_\mathfrak{a}$ is its decay width. As before, $\mathfrak{a}, \mathfrak{b} = +, -, 0$ label the W^+, W^-, Z^0 bosons respectively, and the $\delta^{\mathfrak{a}\mathfrak{b}}$ in the propagator simply ensures that the identity of the weak boson is maintained in every internal line. Once again $q^2 \neq m^2$ because the particle flowing through the line is virtual (i.e. it does not obey its equations of motion). These internal lines are called *propagators*.

The next two rules are the same as for QED, QCD and ABB theory.

4. **CONSERVATION OF ENERGY AND MOMENTUM** For each vertex, write a delta function of the form

$$(2\pi)^4 \delta^{(4)} \left(k_1 + k_2 + k_3 + \cdots + k_N\right)$$

where the k's are the four-momenta coming into the vertex (i.e. each k^μ will be either a q^μ or a p^μ). If the momentum leads outward, then k^μ is *minus* the four-momentum of that line). This factor imposes conservation of energy and momentum at each vertex (and hence for the diagram as a whole) because the delta function vanishes unless the sum of the incoming momenta equals the sum of the outgoing momenta.

5. **INTEGRATE OVER INTERNAL MOMENTA** For each internal momentum q, write a factor

$$\frac{d^4 q}{(2\pi)^4}$$

and integrate.

6. **VERTEX FACTORS** There are many different vertex factors in electroweak theory because the symmetry breaking yields a lot of distinct interactions between the various particles. They can be listed in any order, of course. I have chosen to list them proceeding from the experimentally most-studied interactions to the least-studied.

First there are the fermion-W vertices in figure 23.4 which differ between quarks and leptons. Next are the fermion-Z vertices in figure 23.5 whose couplings depend on the identity of the fermion in the process, as indicated.

f	c_V^f	c_A^f
ν_e, ν_μ, ν_τ	$\frac{1}{2}$	$\frac{1}{2}$
e^-, μ^-, τ^-	$-\frac{1}{2} + 2\sin^2\theta_W$	$-\frac{1}{2}$
u, c, t	$\frac{1}{2} - \frac{4}{3}\sin^2\theta_W$	$\frac{1}{2}$
d, s, b	$-\frac{1}{2} + \frac{2}{3}\sin^2\theta_W$	$-\frac{1}{2}$

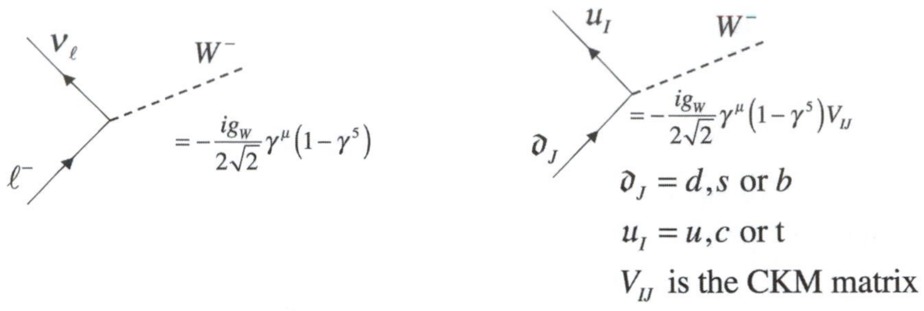

$$= -\frac{ig_w}{2\sqrt{2}} \gamma^\mu (1-\gamma^5)$$

$$= -\frac{ig_w}{2\sqrt{2}} \gamma^\mu (1-\gamma^5) V_{IJ}$$

$$\partial_J = d, s \text{ or } b$$

$$u_I = u, c \text{ or } t$$

V_{IJ} is the CKM matrix

FIGURE 23.4

Charged weak interaction vertices.

$$= -\frac{ig_Z}{2\sqrt{2}} \gamma^\mu \left(c_V^f - c_A^f \gamma^5\right)$$

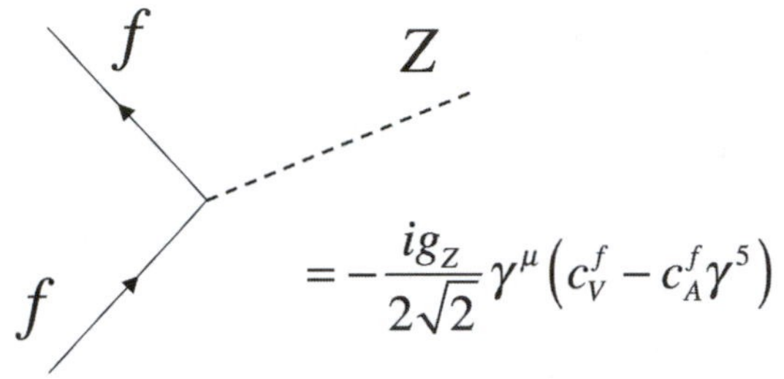

FIGURE 23.5

Neutral weak interaction vertices.

Since the weak bosons themselves have weak charge, they couple to each other, and since the W bosons carry electric charge they also couple to the photon. This gives the vertices for 3-boson couplings shown in in figure 23.6 and for 4-boson couplings in in figures 23.7 and 23.8.

Next we have the vertices describing the interaction of the Higgs particle with all of the other particles, shown in figure 23.9.

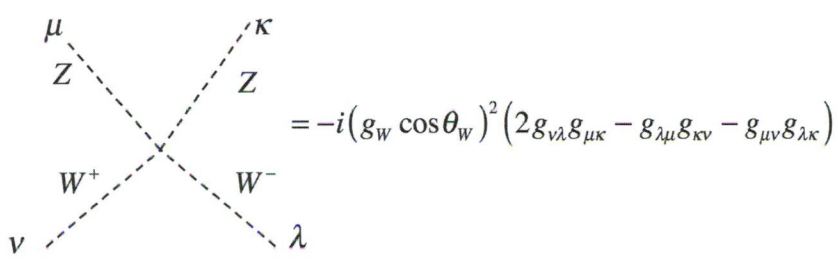

$$= ig_W \cos\theta_W \left(g_{v\lambda}\left(k_1 - k_2\right)_\mu + g_{\lambda\mu}\left(k_2 - k_3\right)_v + g_{\mu v}\left(k_3 - k_1\right)_\lambda\right)$$

$$= ie\left(g_{v\lambda}\left(k_1 - k_2\right)_\mu + g_{\lambda\mu}\left(k_2 - k_3\right)_v + g_{\mu v}\left(k_3 - k_1\right)_\lambda\right)$$

FIGURE 23.6

Triple gauge-boson vertices in electroweak theory.

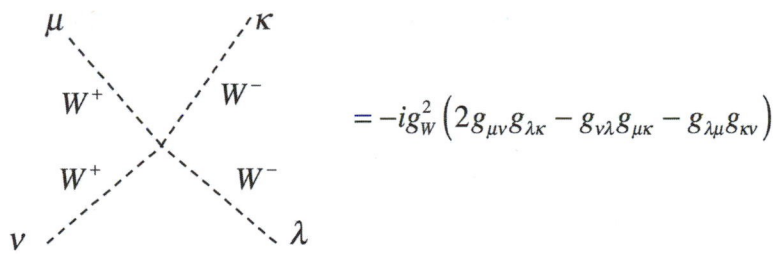

$$= -i\left(g_W \cos\theta_W\right)^2 \left(2g_{v\lambda}g_{\mu\kappa} - g_{\lambda\mu}g_{\kappa v} - g_{\mu v}g_{\lambda\kappa}\right)$$

$$= -ig_W^2 \left(2g_{\mu v}g_{\lambda\kappa} - g_{v\lambda}g_{\mu\kappa} - g_{\lambda\mu}g_{\kappa v}\right)$$

FIGURE 23.7

Quadruple gauge-boson vertices in electroweak theory.

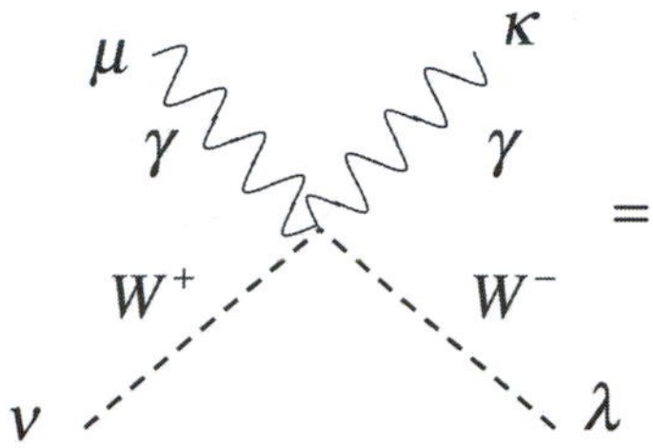

$$= -ie^2 \left(2g_{\nu\lambda}g_{\mu\kappa} - g_{\lambda\mu}g_{\kappa\nu} - g_{\mu\nu}g_{\lambda\kappa} \right)$$

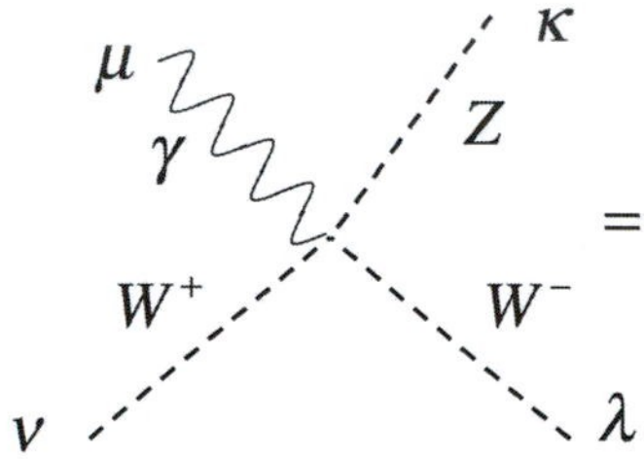

$$= -ieg_W \cos\theta_W \left(2g_{\nu\lambda}g_{\mu\kappa} - g_{\lambda\mu}g_{\kappa\nu} - g_{\mu\nu}g_{\lambda\kappa} \right)$$

FIGURE 23.8

Quadruple gauge-boson vertices in electroweak theory with at least one photon.

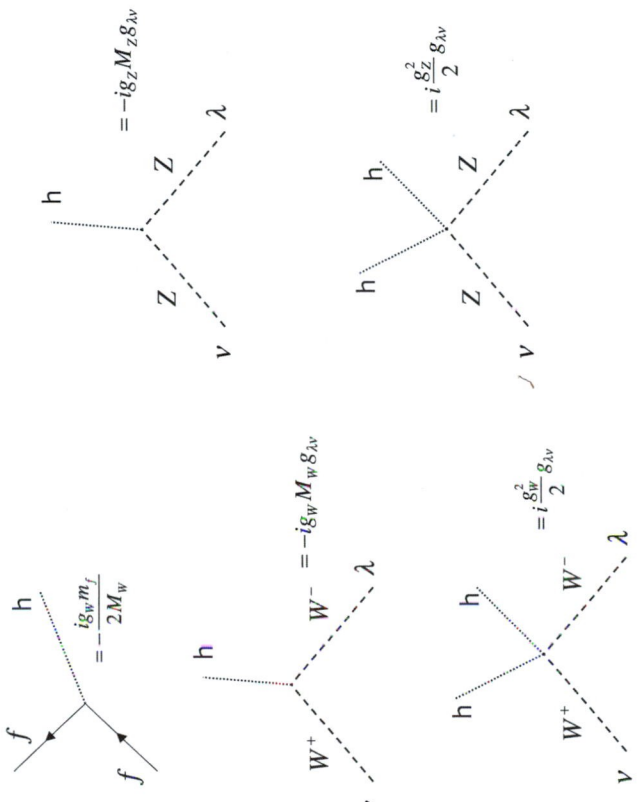

FIGURE 23.9

Higgs interaction vertices in electroweak theory.

Finally we have the vertices describing the self-interaction of the Higgs in figure 23.10.

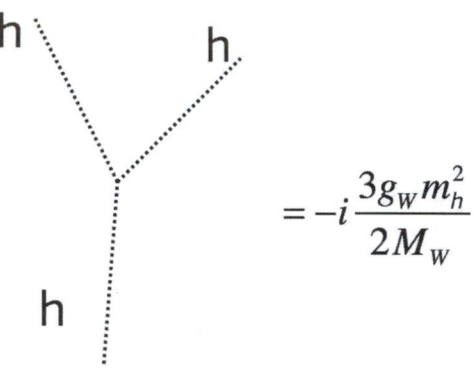

$$= -i\,\frac{3g_w m_h^2}{2M_w}$$

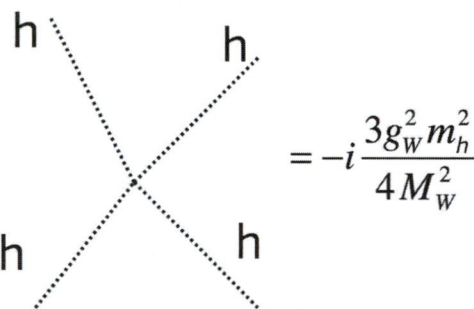

$$= -i\,\frac{3g_w^2 m_h^2}{4M_w^2}$$

FIGURE 23.10
Higgs self-interaction vertices in electroweak theory.

The remaining rules are the same as for QED and QCD.

7. **TOPOLOGY** To get all contributions for a given process, draw diagrams by joining up all external points to all internal vertex points in all possible

arrangments that are topologically inequivalent. The number of ways a given diagram can be drawn is the topological weight of the diagram. The result is equal to $-i\mathcal{M}$.

8. **ANTISYMMETRIZATION** Because fermion wavefunctions anticommute, we must include a minus sign between diagrams that differ

(a) Only in the interchange of two incoming (or outgoing) fermions/anti-fermions of the same kind

or

(b) Only in the interchange of an incoming fermion with an outgoing antifermion of the same kind (or vice versa).

9. **LOOPS** Every fermion loop gets a factor of (-1).

10. **CANCEL THE DELTA FUNCTION** The result will include a factor

$$(2\pi)^4 \delta^{(4)} \left(p_1' + p_2' \cdots + p_m' - p_1 - p_2 - \cdots - p_n \right)$$

corresponding to overall energy-momentum conservation. Cancel this factor, and what remains is $-i\mathcal{M}$.

23.5   Questions

1. Show that $v = 246$ GeV.

2. (a) Show that the potential

$$V\left(\Phi\right) = V_0 - \frac{1}{2}\mu^2\Phi^\dagger\Phi + \frac{1}{4}\lambda^2\left(\Phi^\dagger\Phi\right)^2$$

   is minimized at $\Phi^\dagger\Phi = \frac{\mu^2}{\lambda^2} = \frac{v^2}{2}$.

   (b) Show that a general $\mathbf{SU}(2)$ transformation $\Phi' = \exp\left[i\theta^a\left(x\right)\frac{\sigma^a}{2}\right]\Phi$ can be used to write

$$\Phi' = \begin{pmatrix} 0 \\ v + \mathsf{h}\left(x\right) \end{pmatrix}$$

   where $\mathsf{h}\left(x\right)$ is a real scalar function. (Hint: use the Euler form of the $\mathbf{SU}(2)$ transformation).

3. Show that the current

$$\frac{1}{2}ig_Y\left(\varphi^* D_\nu\varphi - \left(D_\nu\varphi\right)^*\varphi\right)$$

   is conserved for the scalar Higgs field in the Abelian case.

4. Show that if

$$\Phi = \begin{pmatrix} 0 \\ v + \mathsf{h}\left(x\right) \end{pmatrix}$$

   then

$$D_\mu\Phi = -ig_w\frac{v}{2}\begin{pmatrix} \left(W_\mu^1 - iW_\mu^2\right) \\ -\frac{Z_\mu}{\cos\theta_w} \end{pmatrix} + \mathcal{O}\left(\mathsf{h}\right)$$

   and

$$\frac{ig_w}{2}\left(\Phi^\dagger\sigma^a D_\mu\Phi - \left(D_\mu\Phi\right)^\dagger\sigma^a\Phi\right) = \left(\frac{g_w v}{2}\right)^2\left\{W_\mu^1, W_\mu^2, \frac{1}{\cos\theta_w}Z_\mu\right\} + \mathcal{O}\left(\mathsf{h}\right)$$

$$\frac{ig_Y}{4}\left(\Phi^\dagger D_\nu\Phi - \left(D_\nu\Phi\right)^\dagger\Phi\right) = \left(\frac{v}{2}\right)^2\frac{g_w g_Y}{\cos\theta_w}Z_\mu + \mathcal{O}\left(\mathsf{h}\right)$$

5. (a) How many independent *real* parameters are there in a unitary $N \times N$ matrix? How does this answer change if the matrix is orthogonal?

   (b) Consider an extension of the Standard Model with N generations of quarks. Since phases of quark wavefunctions are arbitrary, how many independent parameters are there in the CKM matrix of this model? For how many generations can the CKM matrix be made real?

(c) Show for $N = 3$ that

$$J_{iajb} \equiv \text{Im} \left(V_{ia} V_{jb} V_{ib}^* V_{ja}^* \right) \quad \text{(no sum over repeated indices)}$$

provides a measure of $\mathbb{CP}$-violation regardless of the choice of indices.

6. Show that the Higgs conjugate wavefunction $\tilde{\Phi}$

$$\tilde{\Phi}^i = \epsilon^{ij} \Phi^{j*}$$

transforms as an **SU**(2) doublet, but has opposite hypercharge $Y = -1$.

7. Write equations (23.88) – (23.91) and (23.92) – (23.94) in terms of the physical $W^\pm$, Z and photon wavefunctions.

8. Consider a different model for electroweak interactions, one in which the Higgs wavefunction is an **SU**(2) triplet Ψ of real scalars (analogous to the way that the W's are a triplet) instead of an **SU**(2) doublet Φ of complex scalars.

(a) Write down the potential $\mathcal{V}(\Psi)$ that is **SU**(2) invariant and renormalizable.

(b) Find the minimum of this potential.

(c) Write down the gauge equations for the W bosons coupled to the Higgs wavefunction. Set all fermion wavefunctions to zero. How many W bosons can gain a mass this way?

(d) Why isn't this Higgs triplet used to break the symmetry of the electroweak model?

24

Testing Electroweak Theory

The electroweak theory has been scrutinized in great detail for more than 25 years, with experimentalists seeking to check as many of its predictions as possible. Today we have great confidence in this theory, which, along with QCD, forms the Standard Model. In this chapter we will consider some of the main predictions of the theory and the experimental tests that confirmed it.

24.1 Discovery of the W and Z Bosons

The observation of the weak bosons at CERN in 1983 was a triumph for the electroweak theory. The search for these particles began in 1976 when Rubbia, Cline, and McIntyre proposed that the CERN Super-Proton-Synchrotron (SPS) be converted into a storage ring in which protons and antiprotons would counter-accelerated up to high speeds and then collide [225]. Energies of 270 GeV per beam appeared feasible, yielding a CM energy large enough to produce the W and Z particles, if they indeed existed. Two experiments – UA1 and UA2 – were set up at the redesigned SPS collider to search for the weak bosons.

The beam of antiprotons was obtained via the stochastic cooling technique [72] developed by Simon van der Meer as discussed in chapter 7. The proton and antiproton can be thought of as two groups of quarks and antiquarks, each carrying roughly 1/6 of the total momentum. Since the expected masses of the W and Z particles were somewhere between 70 and 100 GeV, a CM energy of six times this was needed. The initial design was for 540 GeV, to be upgraded to 640 GeV later on.

The calculation of the cross-sections begins at the quark level, then takes into account the gluons and the quark distribution functions. At these energies the sea quark contribution to the cross-sections is negligbly small, and so the annihilating quark is in the proton and the annihilating antiquark is in the antiproton. The calculations indicated that

$$\sigma\left(p\bar{p} \longrightarrow W^- + \cdots \longrightarrow e^- \nu_e + \cdots\right) \simeq 530 \text{ pb}$$
$$\sigma\left(p\bar{p} \longrightarrow Z + \cdots \longrightarrow e^- e^+ + \cdots\right) \simeq 35 \text{ pb} \tag{24.1}$$

at the design phase. These are very tiny cross-sections – about nine orders of magnitude smaller than the total $p\bar{p}$ cross-section of 60 mb. This means that the detector needs a discriminating power of about 10^{10}. Although hadronic decays of the W and Z are more frequent, they are very difficult to distinguish from the background, and so only leptonic channels were considered.

The actual decay channels used in the search are listed in table 24.1 The

TABLE 24.1

Decay Channels Used in the Search for Weak Bosons

$W^{\pm} \longrightarrow e^{\pm}\nu_e$	isolated electron at high p_T and high missing p_T
$W^{\pm} \longrightarrow \mu^{\pm}\nu_{\mu}$	isolated muon at high p_T and high missing p_T
$Z \longrightarrow e^- e^+$	two isolated electrons at high p_T of opposite sign
$Z \longrightarrow \mu^- \mu^+$	two isolated muons at high p_T of opposite sign

leptons emerging from the decays must be isolated, i.e. not appearing inside a jet. Otherwise, it is not possible to be sure that they came from a W or Z instead of the decay of some other quark state. The crucial quantity of experimental interest is the transverse momentum p_T, which is the component of momentum perpendicular to the colliding beams. If this quantity is small then we cannot be sure that the lepton in question was produced by W or Z and not some other particle in the beam. Consequently the only events retained in the large collection of data are those with isolated leptons with large transverse momentum. Of course if a neutrino is produced then it cannot be detected directly. However, if there is a large amount of missing p_T in conjunction with an isolated lepton of high p_T, then it is virtually certain that a neutrino was emitted, signalling the presence of a W boson.

Any given event will have large numbers of tracks that come from the various particles produced that are different from the W or Z. After eliminating (or "cutting") all data with low p_T, the remaining events having some track(s) with large p_T are analyzed by neglecting all tracks with p_T smaller than 1 GeV. To measure the mass of the W the "Jacobian peak" method is used. When the W is produced its momentum is almost entirely in the direction of the beam; it will have only a very small component of transverse momentum. The electron and neutrino will each have large transverse momentum, meaning that the angle between the direction of the W and that of the electron is large. If we Lorentz transform to the rest frame of the W, these transverse components remain the same (see fig. 24.1), and will be related to the angle θ^* relative to the original direction of motion via

$$p_T = |\vec{p}_e| \sin \theta^* \simeq E_e \sin \theta^* = \frac{M_W}{2} \sin \theta^* \qquad (24.2)$$

where we can neglect the mass of the electron at these high energies. The

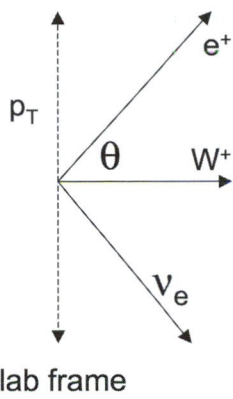
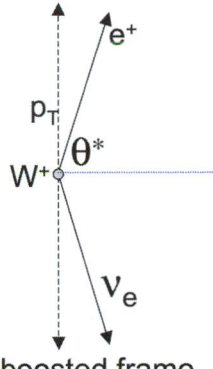

FIGURE 24.1
Decay of the W^+ in the lab frame and in its rest frame. Note that the transverse momentum p_T of the positron is the same in both frames.

transverse momentum distribution, $\frac{dn}{dp_T}$ is given by

$$\frac{dn}{dp_T} = \frac{dn}{d\theta^*}\frac{d\theta^*}{dp_T} = \frac{dn}{d\theta^*}\frac{2}{M_W\cos\theta^*} = \frac{1}{\sqrt{\left(\frac{M_W}{2}\right)^2 - p_T^2}}\frac{dn}{d\theta^*} \tag{24.3}$$

where $\frac{dn}{d\theta^*}$ is the angular distribution of the electrons in the rest frame of the W. The crucial point is that is that the Jacobian $\frac{d\theta^*}{dp_T}$ diverges when $p_T = \frac{M_W}{2}$. Since the transverse momentum of the W is not exactly zero this divergence becomes a sharp peak in the p_T distribution at $\frac{M_W}{2}$. By counting the number of events as a function of p_T and locating this maximum, the mass of the W can be determined. The UA1 experiment found $M_W = 83 \pm 3$ GeV [226], and UA2 found $M_W = 80 \pm 1.5$ GeV [227].

The mass of the Z boson can be measured by more conventional means, since the energies of the emitted lepton and antilepton can each be measured by the calorimeter and the angles $(\theta_{\ell+}, \theta_{\ell-})$ between the beam tracks can be determined. If the Z exists, it will show up as a peak in the invariant mass $m\left(\ell^+\ell^-\right)$, where

$$m\left(\ell^+\ell^-\right) = \sqrt{2m_\ell^2 + 2E_{\ell+}E_{\ell-} - 2p_{\ell+}p_{\ell-}\cos\left(\theta_{\ell+} + \theta_{\ell-}\right)} \tag{24.4}$$

and the location of the maximum determines its mass. UA1 found $M_Z = 93 \pm 3$ GeV [226], and UA2 found $M_W = 91.5 \pm 1.7$ GeV [227]. Today we know these values with considerably greater precision [1]

$$M_W = 80.398 \pm 0.0259 \text{ GeV} \qquad M_Z = 91.1876 \pm 0.0021 \text{ GeV} \tag{24.5}$$

and the electroweak model with its predictions for the W and Z have been spectacularly confirmed.

Another important prediction of electroweak theory is the lifetimes – or decay widths – of the W and Z. Let's begin first with the W. The decay diagram for $W^- \rightarrow \ell^- + \overline{\nu}_\ell$ is given in fig. 24.2 and is straightforward to

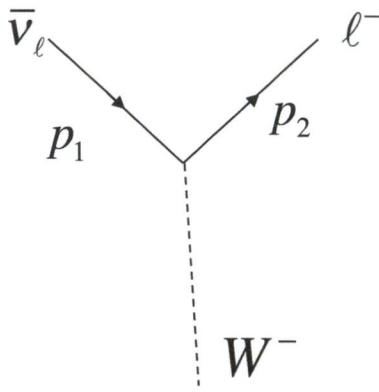

FIGURE 24.2
Decay diagram for $W^- \rightarrow \ell^- + \overline{\nu}_\ell$

calculate since it is a two body decay. We have

$$\mathcal{M} = (-i)\left(\frac{g_w}{2\sqrt{2}}\right)\left[\overline{u}(p_2)\gamma^\nu\left(1-\gamma^5\right)v(p_1)\right] \tag{24.6}$$

and neglecting the masses of the leptons in the final state decay rate in the rest frame of the W is

$$\Gamma\left(W^- \rightarrow \ell^- + \overline{\nu}_\ell\right) = \frac{|\vec{p}|c}{8\pi\hbar\left(M_W c\right)^2}\overline{|\mathcal{M}|}^2 = \frac{1}{3}\left(\frac{g_w}{2\sqrt{2}}\right)^2\frac{M_W c^2}{2\pi\hbar} = \frac{G_F\left(M_W c^2\right)^3}{6\sqrt{2}\pi\hbar} \tag{24.7}$$

which is the same for each type of lepton! The results for the quarks are given by the same diagram, except for two things. First, the W is color-neutral and so when it decays into a quark-antiquark doublet the colors of the decay products must be equal and opposite; since our detectors are color-blind we must add over the three colors. Second, the couplings will be multiplied by the appropriate CKM matrix elements.

The results for the full set of decay widths is given in table 24.2. You can see that there are no entries for decays of the W into a top quark because of the

TABLE 24.2

Decay Widths for the W Boson

Channel	$\hbar\Gamma$	Value (MeV)
$e^- + \bar{\nu}_e$	$\dfrac{G_F\left(M_W c^2\right)^3}{6\sqrt{2}\pi\hbar}$	225
$\mu^- + \bar{\nu}_\mu$	$\dfrac{G_F\left(M_W c^2\right)^3}{6\sqrt{2}\pi\hbar}$	225
$\tau^- + \bar{\nu}_\tau$	$\dfrac{G_F\left(M_W c^2\right)^3}{6\sqrt{2}\pi\hbar}$	225
$\bar{u} + d$	$\dfrac{G_F\left(M_W c^2\right)^3}{2\sqrt{2}\pi\hbar}\left|V_{ud}\right|^2$	640
$\bar{u} + s$	$\dfrac{G_F\left(M_W c^2\right)^3}{2\sqrt{2}\pi\hbar}\left|V_{us}\right|^2$	35
$\bar{u} + b$	$\dfrac{G_F\left(M_W c^2\right)^3}{2\sqrt{2}\pi\hbar}\left|V_{ub}\right|^2$	.011
$\bar{c} + d$	$\dfrac{G_F\left(M_W c^2\right)^3}{2\sqrt{2}\pi\hbar}\left|V_{cd}\right|^2$	33
$\bar{c} + s$	$\dfrac{G_F\left(M_W c^2\right)^3}{2\sqrt{2}\pi\hbar}\left|V_{cs}\right|^2$	660
$\bar{c} + b$	$\dfrac{G_F\left(M_W c^2\right)^3}{2\sqrt{2}\pi\hbar}\left|V_{cb}\right|^2$	1.20

3 colors (handwritten, at left of $\bar{u}+d$ through $\bar{c}+b$ rows)

heavier mass of the latter, a fact not known at the time the W was discovered. Furthermore, the decays into b quarks are very rare due to suppression factors from the CKM matrix. The total width, obtained by summing over all of the partial widths, gives $\Gamma_W = \Gamma\left(W^- \to \text{anything}\right) = 2.04$ GeV

The decay of the Z is computed in a similar manner. The decay diagram, fig. 24.3, is just like the one for the W except now the couplings are given by table 22.2. This means the matrix element is

$$\mathcal{M} = (-i)\left(\frac{g_Z}{2}\right)\left[\bar{u}(p_2)\gamma^\nu\left(c_V^f - c_A^f\gamma^5\right)v(p_1)\right] \tag{24.8}$$

yielding

$$\Gamma\left(Z \to f + \bar{f}\right) = \frac{N_c\left|\vec{p}\right|c}{8\pi\hbar\left(M_Z c\right)^2}\overline{\left|\mathcal{M}\right|}^2 = \frac{N_c}{3}\left(\frac{g_Z}{2}\right)^2\frac{M_Z c^2}{2\pi\hbar}\left(\left(c_V^\ell\right)^2 + \left(c_A^\ell\right)^2\right)$$

$$= N_c\frac{G_F\left(M_Z c^2\right)^3}{6\sqrt{2}\pi\hbar}\left(\left(c_V^\ell\right)^2 + \left(c_A^\ell\right)^2\right) \tag{24.9}$$

where the relations $M_Z\cos\theta_W = M_W$ and $g_Z\cos\theta_W = g_w$ have been used and I have included the color factor N_c, which is 3 for quarks and 1 for leptons.

The results for the full set of decay widths of the Z is given in table 24.3. In this case all of the allowed decay channels (remember, top decay is forbidden because the top is too heavy) make an appreciable contribution to the width. The decay into neutrinos cannot be directly observed, but the total decay rate into three neutrino flavors is $\sum_\ell \Gamma\left(Z \to \nu_\ell + \bar{\nu}_\ell\right) = 495$ MeV. If there are any other unknown particles that couple to the Z but not to the photon or gluons – an example would be an additional flavor of neutrino – they will contribute

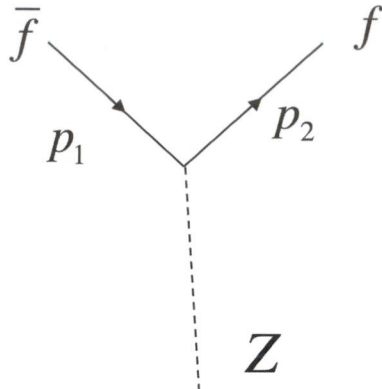

FIGURE 24.3

Decay of the Z boson into a fermion-antifermion pair

to the width of the Z if their masses are less than $\frac{M_Z}{2}$. Measurement of this "invisible" decay channel width provides an important check on the number of low-mass neutrinos. Likewise, it is not feasible experimentally to distinguish the different quark-antiquark channels (though sometimes clever methods can pick out the $c\bar{c}$ or $b\bar{b}$ ones), but we can obtain the total hadronic decay width, which is $\Gamma_Z = \Gamma\left(Z \rightarrow \text{hadrons}\right) = 2\Gamma\left(Z \rightarrow u + \bar{u}\right) + 3\Gamma\left(Z \rightarrow d + \bar{d}\right) = 1.67$ GeV. The total width is given by summing over the partial widths and is $\Gamma\left(Z \rightarrow \text{anything}\right) = 2.42$ GeV.

Of course these calculations neglect corrections due to loop diagrams, and can be done much more precisely. Today we know the empirical values of the decay widths to very good precision [1]

$$\Gamma_W = 2.141 \pm 0.041 \text{ GeV} \qquad \Gamma_Z = 2.4952 \pm 0.0023 \text{ GeV} \qquad (24.10)$$

as well as distinct channels

$$\Gamma(Z \rightarrow \text{invisible}) = 499.0 \pm\ 1.5 \text{ MeV}$$
$$\Gamma(Z \rightarrow l + l-) = 83.984 \pm 0.086 \text{ MeV} \qquad (24.11)$$
$$\Gamma(Z \rightarrow \text{hadrons}) = 1744.4 \pm\ 2.0 \text{ MeV}$$

and all are in excellent agreement with the Standard Model! An important constraint is that there cannot be any more than three light (i.e. less than half the mass of the Z) neutrino species, placing an important constraint on fundamental particle physics models.

TABLE 24.3

Decay Widths for the Z Boson

Channel	$\hbar\Gamma$	Value (MeV)
$\nu_\ell + \overline{\nu}_\ell$	$\dfrac{G_F\left(M_Z c^2\right)^3}{12\sqrt{2}\pi\hbar}$	165
$\ell^- + \ell^-$	$\dfrac{G_F\left(M_Z c^2\right)^3}{6\sqrt{2}\pi}\left(\frac{1}{2} - 2\sin^2\theta_W + 4\sin^4\theta_W\right)$	83
$\begin{matrix} \overline{u} + u \\ \overline{c} + c \end{matrix}$	$\dfrac{G_F\left(M_Z c^2\right)^3}{2\sqrt{2}\pi\hbar}\left(\frac{1}{2} - \frac{4}{3}\sin^2\theta_W + \frac{16}{9}\sin^4\theta_W\right)$	280
$\begin{matrix} \overline{d} + d \\ \overline{s} + s \\ \overline{b} + b \end{matrix}$	$\dfrac{G_F\left(M_Z c^2\right)^3}{2\sqrt{2}\pi\hbar}\left(\frac{1}{2} - \frac{2}{3}\sin^2\theta_W + \frac{4}{9}\sin^4\theta_W\right)$	370

24.2 Lepton Universality and Running Coupling

The kinematic criteria that allowed for unambiguous identification of W production and subsequent decay into $e\nu_e$ also applies for subsequent decay into $\mu\nu_\mu$ and $\tau\nu_\tau$. Cross-sections for each process can be separately calculated and measured, making a test of lepton universality possible. The results indicated [1]

$$\frac{g_\mu}{g_e} = 1.00 \pm 0.07(\text{stat}) \pm 0.04(\text{syst}) \qquad \frac{g_\tau}{g_e} = 1.00 \pm 0.10(\text{stat}) \pm 0.06(\text{syst})$$

$$(24.12)$$

showing that the $\mathfrak{E}_L$, $\mathfrak{M}_L$, and $\mathfrak{T}_L$ doublets all couple in the same way and with the same strength to the W boson.

This is in accord with the Standard Model prediction, which also states that there is absolute conservation of three separate lepton numbers: electron number, muon number, and tau number. These conservation laws continue to be tested in a variety of ways. For example, the best limits on the conversion of one charged-lepton type to another come from $\mu \to e\gamma$ and $\mu \to 3e$, for which [1]

$$\frac{\Gamma(\mu \to e\gamma)}{\Gamma(\mu \to \text{all possible})} < 1.2 \times 10^{-11} \qquad \frac{\Gamma(\mu \to 3e)}{\Gamma(\mu \to \text{all possible})} < 1.0 \times 10^{-12}$$

$$(24.13)$$

For the τ lepton the corresponding limits are not as stringent

$$\frac{\Gamma(\tau \to \mu\gamma)}{\Gamma(\tau \to \text{all possible})} < 6.8 \times 10^{-8} \qquad \frac{\Gamma(\tau \to e\gamma)}{\Gamma(\tau \to \text{all possible})} < 1.1 \times 10^{-7}$$

$$(24.14)$$

though there is still no evidence for any violation of charged lepton number. Semileptonic decays and decays of charged leptons into other charged antileptons yield similar limits, again in accord with the Standard Model.

What about the neutrinos? We do have firm evidence that the different flavors of neutrinos can change into each other. I'll postpone discussion of this subject until the next chapter.

Another early test of electroweak theory was in the value and evolution of the weak mixing angle θ_W. You might recall from Chapter 23 that this parameter has a clear relationship to the electroweak coupling constants and the electromagnetic coupling

$$g_{\mathrm{w}} \sin \theta_W = g_e \qquad g_Z = \frac{g_e}{\sin \theta_W \cos \theta_W} \qquad (24.15)$$

and so we have another prediction from the Standard Model that can be experimentally checked. It is difficult to directly measure the weak couplings, but fortunately we don't have to. We can instead make use of the relation $G_F = \frac{\sqrt{2}}{8} \left(\frac{g_{\mathrm{w}}}{M_W c^2} \right)^2$, which gives

$$M_W c^2 = \sqrt{\frac{\sqrt{2} g_{\mathrm{w}}^2}{8 G_F}} = \sqrt{\frac{\sqrt{2} g_e^2}{8 \sin^2 \theta_W G_F}} = \frac{1}{\sin \theta_W} \sqrt{\frac{\pi \alpha}{\sqrt{2} G_F}} = \frac{37.3}{\sin \theta_W} \text{ GeV} \qquad (24.16)$$

yielding a relationship between the measureable W boson mass, the fine structure constant, Fermi's constant, and $\sin \theta_W$. Similarly the relationship between the W and Z boson masses

$$\sin^2 \theta_W = 1 - \frac{M_W^2}{M_Z^2} \qquad (24.17)$$

provides an important cross-check on the validity of electroweak theory, since each quantity can be measured independently, though in the case of $\sin^2 \theta_W$ not directly. A variety of low-energy processes provide such checks, illustrated in figure 24.4.

1. **Parity violation in atoms**. Since electrons can exchange both photons and Z bosons with the quarks in the nucleus, there will be a parity-violating shift in the energy levels of atoms due to the different helicity couplings the electron has to the Z. The effect is only a few parts per million, but can be measured and provides a test of electroweak theory at the KeV scale [228].

2. **Forward-backward asymmetry in $e^+ e^-$ annihilation**. There is an interference between the virtual photon and Z diagrams, a process we looked at in Chapter 22. This asymmetry has been measured in a wide energy range [231] between 10 and 200 GeV, and is large for energies near the mass of the Z as we have seen.

3. **Scattering of ν_μ and $\bar{\nu}_\mu$ off of electrons.** This process is very clean due to the absence of hadronic interactions and provides the most direct means of determining the weak mixing angle. However, the cross-section is very low – four orders of magnitude smaller than neutrino-nucleus scattering. Nevertheless the CHARM2 experiment first provided the most precise measurements of this process, yielding the experimental result [229]

$$\sin^2 \theta_W = 0.2324 \pm .0083 \qquad (24.18)$$

results superseded by yet more accurate measurements [1].

Corrections from loop diagrams to the masses of the W and Z bosons modify the relationship they have with $\sin^2 \theta_W$. The most important diagrams are when the particle in the loop is either a Higgs boson or a top/bottom quark pair. The correction from the Higgs loop depends logarithmically on the mass of the Higgs boson, but the correction from the quark loop is proportional to

$$G_F \left(m_t^2 - m_b^2 \right) \simeq G_F m_t^2 \qquad (24.19)$$

which is quadratic in the mass of the top quark. A precise measurement of the mass of the W thus allows for a prediction of the mass of the top as long as the mass of the Higgs is not too large.

Experiments in the 1990s at LEP measured with increasing accuracy the masses of the W and Z and $\sin^2 \theta_W$ by a variety of methods. In 1993 a bound of the top mass was obtained

$$m_t = 166 \pm 27 \text{ GeV} \qquad (24.20)$$

provided 60 GeV$< m_\text{H} < 700$ GeV. The mass of the top was measured less than two years later (as we saw in Chapter 18) and found to be within this bound.

The coupling constants g_w and g_Y in the $\mathbf{SU}(2) \times \mathbf{U}(1)$ Model both undergo renormalization, and hence "run" – that is change with energy (or distance) – the way that the electromagnetic and strong couplings do. Since

$$\frac{g_Y}{g_\text{w}} = \tan \theta_W \qquad (24.21)$$

the quantity $\sin^2 \theta_W$ will also be a function of the momentum transfer between two particles that scatter due to weak interactions [230]. The dependence is more complicated than that for the electromagnetic and strong couplings because both numerator and denominator vary. The change is very small, only a few percent over more than six orders of magnitude. The predicted behavior and experimental values are given in figure 24.3. Fermion loops and boson loops contribute to the renormalization but with opposite sign. At low energies the fermion loops are important due to the large mass of the W, but at high energies loops due to W and Z self-couplings set in. However, these contribute to g_w (since it is a non-Abelian theory) but not to g_Y, changing the slope of the curve near the mass of the W.

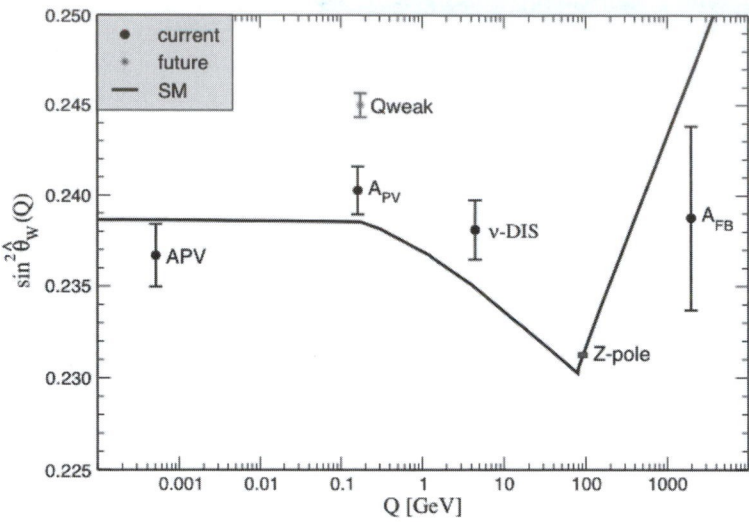

FIGURE 24.4

Running of the weak mixing angle as a function of energy, showing comparison with different experimental tests. The acronyms are APV: Atomic Parity Violation; A_{PV}: Asymmetry in Polarized Moller scattering; Z: Measurements taken when CM energy is the mass of the Z boson (the "Z-pole" measurements [232]); ν-DIS: Deep inelastic scattering of neutrinos from isoscalar targets [233]; AFB: Forward-backward asymmetry at LEP. The width of the curve reflects theoretical uncertainty from strong interaction effects [230]. Image courtesy of the Particle Data Group [1].

24.3 The Search for the Higgs

The only particle in the Standard Model that has not been observed is the Higgs particle. It obeys the equation

$$D_\mu D^\mu \Phi - \mu^2 \Phi + \lambda^2 \left(\Phi^\dagger \Phi\right) \Phi = \sum_{I=1}^{3} G_I \overline{\chi}_{IR}^{\mathfrak{L}} \chi_{IL}^{\mathfrak{L}} + \sum_{I,J=1}^{3} \overline{\chi}_{IR}^{\mathfrak{d}} G_{IJ} \chi_{JL}^{\tilde{\mathfrak{s}}}$$

$$+ \sum_{I,J=1}^{3} \overline{\chi}_{IR}^{\mathfrak{u}} H_{IJ} \tilde{\chi}_{JL}^{\tilde{\mathfrak{s}}} \qquad (24.22)$$

where the covariant derivative is

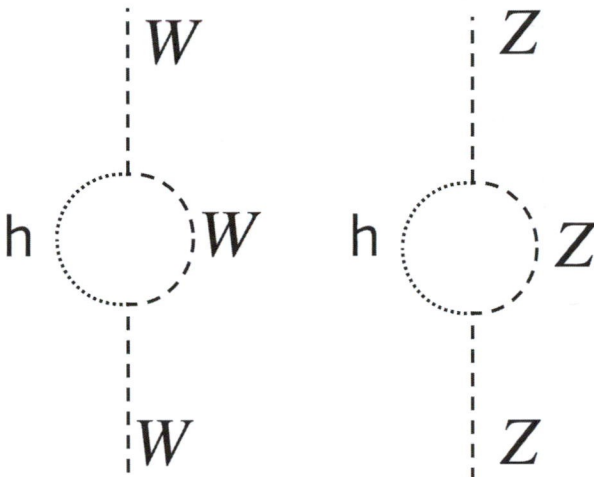

FIGURE 24.5
One-loop correction to the W and Z propagators (and hence their masses)
due to the Higgs boson.

$$D_\mu \Phi = \partial_\mu \Phi - i\frac{g_w}{2}\sigma^a \cdot W_\mu^a \Phi + i\frac{g_Y}{2}B_\mu \Phi \qquad (24.23)$$

and we remember that $\Phi = \begin{pmatrix} 0 \\ v + \mathsf{h}(x) \end{pmatrix}$ – the actual Higgs particle is described by the wavefunction $\mathsf{h}(x)$. If we insert this into the equation for Φ
we find after some algebra that its lower component gives

$$\partial^2 \mathsf{h} + 2\mu^2 \mathsf{h} + 3\lambda\mu \mathsf{h}^2 + \lambda^2 \mathsf{h}^3$$
$$-\frac{(g_w)^2}{4}(v + \mathsf{h})W_\mu^- W^{+\mu} - \left(\frac{g_w}{2\cos\theta_w}\right)^2 (v + \mathsf{h})Z_\mu Z^\mu$$
$$= \sum_{I=1}^{3} G_I \bar{\chi}_{IR}^{\mathfrak{L}} \chi_{IL}^{\mathfrak{L}} + \sum_{I,J=1}^{3} \bar{\chi}_{IR}^{\mathfrak{d}} G_{IJ} \chi_{JL}^{\mathfrak{d}} + \sum_{I,J=1}^{3} \bar{\chi}_{IR}^{\mathfrak{u}} H_{IJ} \tilde{\chi}_{JL}^{\mathfrak{d}} \qquad (24.24)$$

which shows that the mass m_h of the Higgs particle is

$$m_\mathsf{h} = \sqrt{2}\mu \qquad (24.25)$$

The Standard Model makes no prediction as to what its m_h is – the parameter μ must be input into the model. This is arbitrariness is not particularly
attractive, and theorists have wanted to do better. One idea for predicting
the Higgs mass from the Standard Model was to set $\mu = 0$ at the outset

and then compute the effective potential $V_{\text{eff}}(\mathsf{h})$ for the Higgs boson due to its interactions with all of the other particles in the Standard Model [234]. You might think that the only minimum of the effective potential would be at $\mathsf{h} = 0$ since we set $\mu = 0$, but quantum loop diagrams generate an effective potential $V_{\text{eff}}(\mathsf{h})$ that has a new minimum, from which a nonzero Higgs mass can be computed, determined from the other parameters in the Standard Model. This approach is clearly more attractive, since the arbitrary constant μ no longer appears. For a number of years it was thought to be viable until the discovery of the top quark with its enormous mass (and therefore large coupling to the Higgs) appeared to rule out this mechanism [235]. However, recent considerations that systematically sum over higher order corrections [236] have shown that perhaps this mechanism is viable after all, and that it predicts a Higgs mass of $m_{\mathsf{h}} \simeq 220 - 225$ GeV.

It is possible to obtain bounds on m_{h} indirectly from the Standard Model using precision measurements of electroweak observables. As I noted above, the masses of both the W and Z bosons get 1-loop correction terms (see figure 24.5) that depend on the logarithm of the Higgs boson mass. Since their masses are known so precisely, we can place a bound on the Higgs mass, although not a very stringent one due to the logarithmic dependence. A global fit to precision electroweak data, accumulated at LEP, SLC, the Tevatron and elsewhere implies that $m_{\mathsf{h}} < 182$ GeV at 95% confidence level [1]. This is somewhat less than the 220 GeV prediction above, but not so much less as to rule out the effective potential mechanism.

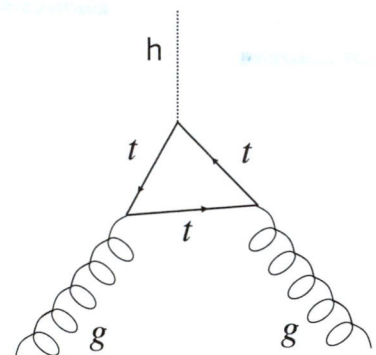

FIGURE 24.6
Gluon fusion produces a Higgs boson.

It's clear we will have to wait for LHC to do its job. At present, the best lower bound on m_h is from LEP [237]:

$$m_h > 114.4 \text{ GeV} \quad \text{at} \quad 95\% \text{c.l.} \tag{24.26}$$

which involved colliding electron-positron pairs to produce Z bosons at energies larger than M_Z. The virtual Z can emit a Higgs boson, which will preferentially decay to $b\bar{b}$ quark pairs due to the heavy mass of the bottom quark. The limit that can be reached on the Higgs mass is the difference between the center-of-mass collision energy E_{cm} and M_Z. LEP was able to attain E_{cm} of 209 GeV by installing additional superconductive radiofrequency cavities in the ring, providing additional power compensate for the large loss of energy in synchrotron radiation. This resulted in the upper limit above.

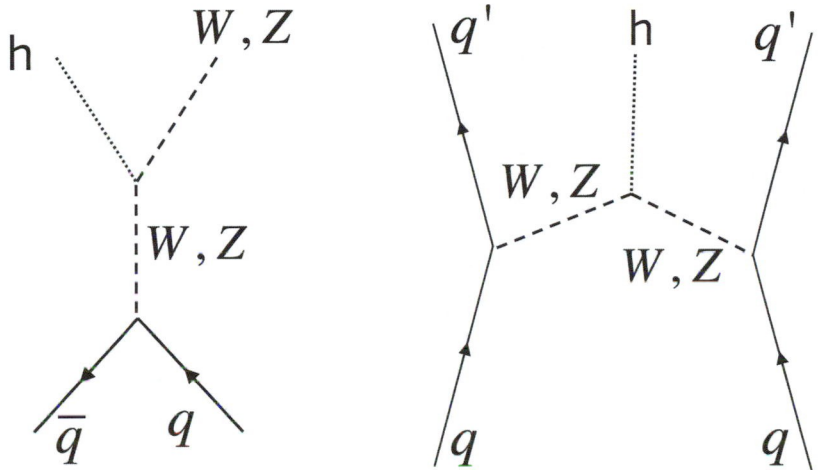

FIGURE 24.7
Other modes of Higgs production: Associate production via W/Z brehmsstrahlung (left) and WW or ZZ fusion (right).

It's possible to do a bit better. Precision electroweak measurements at LEP indirectly constrain m_h to be lower than 182 GeV at the 95% C.L. This is done by computing how the Higgs particle modifies the ratio $\frac{M_W}{M_Z}$ due to loop diagrams that arise from quantum field theory. The ratio $\frac{M_W}{M_Z}$ has been measured to high precision, and yields the above upper limit once the lower bound of 114.4 GeV is taken into account. The most recent experimental result for the Higgs mass is from Fermilab [238]. By combining all known data from proton-antiproton collisions at $E_{cm} = 1.96$ TeV, researchers have excluded a Higgs boson with mass $m_h = 170$ GeV at 95% c.l., a direct restriction, albeit within a very narrow energy range.

At the LHC the dominant mechanism for producing the Higgs boson is via gluon fusion, $g + g \rightarrow$ h, as in figure 24.6. Since the top is the heaviest quark, the Higgs couples most strongly to it, and so this quark loop dominates all other possible loops. It is also possible to produce the Higgs via fusion of W or Z bosons or via a brehmsstrahlung (perhaps it should be called "Higgsstrahlung") process as shown in figure 24.7, but these contribute less strongly.

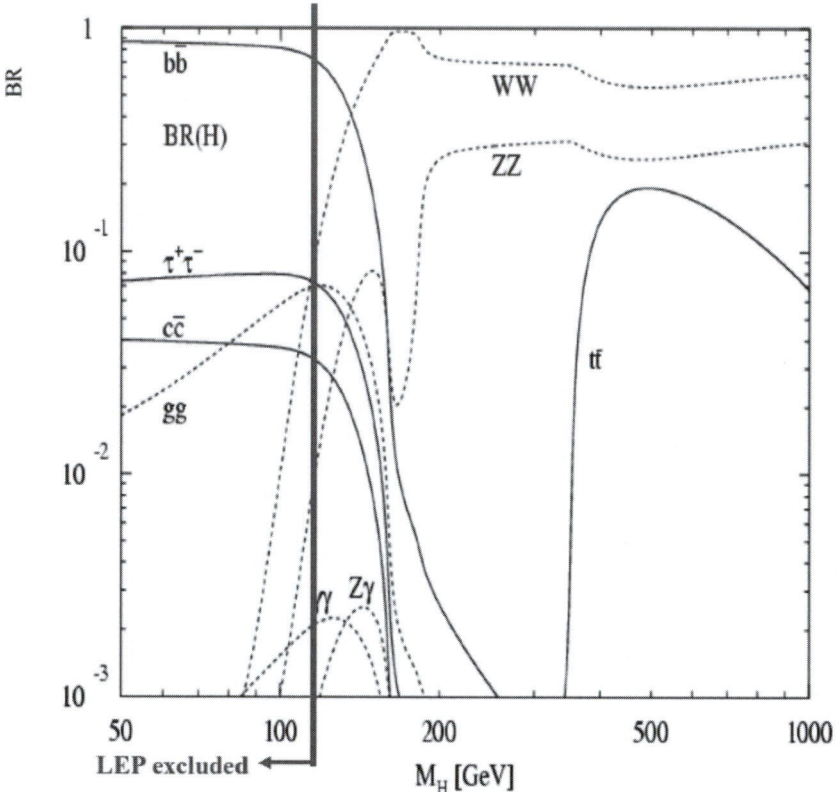

FIGURE 24.8

Branching Ratios for Higgs decay as a function of Higgs Mass. The solid vertical line is the LEP bound of 114 GeV. Reprinted figure from A. Djouadi, J. Kalinowski, and M. Spira, *HDECAY: a program for Higgs boson decays in the Standard Model and its supersymmetric extension* Comput. Phys. Commun. **108**, 56 (1998) [239]; Copyright (1998), with permission from Elsevier.

How the Higgs particle decays will depend on its mass. If it is less than half the W mass ($m_h \lesssim 140$ GeV) then it will decay into $b\bar{b}$ pairs. Above this mass up to 160 GeV it will predominantly decay into W pairs, and then above that into Z pairs, which fully dominate if $m_h > 180$ GeV. Only if the Higgs is very heavy – $m_h \lesssim 360$ GeV – will it decay into top pairs, and even then this is not the dominant process. There are other possibilities once loop diagrams are taken into account: the time-reversal of the diagram in figure 24.6 can yield decay into gluon pairs ($h \rightarrow g + g$) and also the less likely decay into photon pairs, which is easily seen by replacing the gluon lines with photons in figure 24.6.

A diagram of the all possible branching ratios for the decay of the Higgs boson as a function of its mass appears in figure 24.8.

The main purpose of the Large Hadron Collider is to carry out a direct search for the Higgs boson. If it is found it will put in place the last piece of the puzzle in the Standard Model. If it is not found, then we will have firm evidence of physics beyond the Standard Model – and undoubtedly another revolution in particle physics.

24.4   Questions

1. Consider the Higgs mechanism in the Abelian case. Write down the equations for the Higgs wavefunction and the gauge wavefunction if the Higgs field is written as $\varphi = v + h(x)$.

2. Suppose we used the $\mathbf{SU}(2)$ symmetry to write the Higgs field as

$$\Phi = \begin{pmatrix} v + h(x) \\ 0 \end{pmatrix}$$

What would the physical theory be in this case?

3. Show that the equation for the Higgs wavefunction

$$D_\mu D^\mu \Phi - \mu^2 \Phi + \lambda^2 \left(\Phi^\dagger \Phi\right) \Phi = \sum_{I=1}^{3} G_I \overline{\chi}_{IR}^\ell \, \chi_{IL}^\ell + \sum_{I,J=1}^{3} \overline{\chi}_{IR}^\mathfrak{d} G_{IJ} \chi_{JL}^\mathfrak{f}$$

$$+ \sum_{I,J=1}^{3} \overline{\chi}_{IR}^\mathfrak{u} H_{IJ} \tilde{\chi}_{JL}^\mathfrak{f}$$

   becomes

   $$\partial^2 h + 2\mu^2 h + 3\lambda\mu h^2 + \lambda^2 h^3$$

$$-\frac{(g_{\mathrm{w}})^2}{4}(v+\mathrm{h})\,W_\mu^-W^{+\mu}-\left(\frac{g_{\mathrm{w}}}{2\cos\theta_{\mathrm{w}}}\right)^2(v+\mathrm{h})\,Z_\mu Z^\mu$$

$$=\sum_{I=1}^{3}G_I\overline{\chi}_{IR}^{\,\ell}\,\chi_{IL}^{\ell}+\sum_{I,J=1}^{3}\overline{\chi}_{IR}^{\,\partial}G_{IJ}\chi_{JL}^{\tilde{\Im}}+\sum_{I,J=1}^{3}\overline{\chi}_{IR}^{\,u}H_{IJ}\tilde{\chi}_{JL}^{\tilde{\Im}}$$

when $\Phi=\begin{pmatrix}0\\v+\mathrm{h}(x)\end{pmatrix}$. Hint: remember that the gauge currents are conserved.

4. The Higgs potential is

$$\mathcal{V}(\Phi)=V_0-\frac{1}{2}\mu^2\Phi^\dagger\Phi+\frac{1}{4}\lambda^2\left(\Phi^\dagger\Phi\right)^2$$

(a) Suppose $V_0=0$. How much energy per unit volume does the Higgs potential contribute to the vacuum energy density of the universe if we assume $\lambda=\frac{1}{10}$?

(b) The actual energy density of empty space is smaller than 10^{-4} GeV/cm^3. What value would V_0 have to be set to in order to ensure that your answer in (a) does not exceed this value?

5. Compute the decay rate
$$\mathrm{h}\to f+\bar{f}$$
to leading order in the Standard Model, where f is a lepton or a quark.

6. Compute the decay rates

$$(\mathrm{a})\ \mathrm{h}\to W^++W^-$$
$$(\mathrm{b})\ \mathrm{h}\to Z+Z$$

to leading order in the Standard Model.

(c) Plot the ratio
$$\frac{\Gamma(\mathrm{h}\to W^++W^-)}{\Gamma(\mathrm{h}\to Z+Z)}$$
as a function of h.

7. Someone comes to you and proposes to find the Higgs using an e^+e^- machine by examining the process

$$e^++e^-\to\mathrm{h}\to W^++W^-$$

so that the existence of the Higgs particle can be inferred by finding deviations relative to other intermediate states (like the photon or Z) in producing the W^+W^- pair. Is this a good idea?

25

Beyond the Standard Model

The Standard Model – the comprehensive theory that describes our current understanding of the electromagnetic, strong, and weak interactions is not expected to be the final theory of physics. There are three main reasons underlying this expectation.

1. **It has too many adjustable parameters.** There are six different quark masses, three different lepton masses (since neutrino masses are assumed to be zero), one Higgs mass, four mixing angles from the CKM matrix, one mixing angle associated with ground state of QCD (as we shall see), and three coupling constants (one for each force). This is a total of 18 arbitrary parameters that are not predicted by the Standard Model itself, which is mathematically valid and theoretically self-consistent for any numerical values of these parameters*. If one allows for the possibility of neutrino masses then an additional seven parameters must be added to this list – three for the masses of the neutrinos and four associated with the mixing angles for a CKM-type matrix in the lepton sector – bringing the total to 25. If one includes gravitation, then two additional parameters – the gravitational coupling constant G and the cosmological vaccum energy (parametrized by a cosmological constant Λ) – must also be included bringing the total number of unexplained adjustable parameters to 27. It is generally believed (or at least preferred) that a more complete theory [240] will have fewer adjustable parameters – ideally only one!

2. **Gravity is not included.** The problem with including gravity is not simply a matter of adding Einstein's equations of general relativity with its two additional constants (G and Λ) to the model. If we neglect quantum mechanics, writing down the standard model in a self-consistent way that includes Einstein's theory of general relativity is well-known and understood. However, no one knows how to include gravity in a

---

*I should qualify this statement, since there are mathematical issues concerning the existence of quantum field theories that are not fully resolved. What I mean by this statement is that all of the mathematical manipulations performed to extract physical predictions from the Standard Model would be just as valid if other values for these 18 parameters had been used besides the ones measured in experiments.

quantum framework [241]. There are three major difficulties that must be overcome.

- *General Relativity is not a renormalizable theory.* Not only are quantum corrections to gravitational processes infinite (as they are for gauge theories), but the number of these infinite corrections is also infinite (unlike the situation in gauge theories). The renormalization procedure works in the Standard Model because we only ever need to redefine a finite number of parameters (such as mass and charge) to eliminate these infinities. However, in gravity, *every* possible physical quantity would need to be redefined to eliminate the infinite corrections. Doing this renders the quantum theory useless for predicting anything because there is always another adjustable infinite parameter that could be renormalized to agree with experiment.

- *Time is treated differently in gravity.* In quantum theories time plays the role of an ordering parameter that tells us how a quantum system evolves from the past into the future. However, in general relativity you can change your coordinate definitions of time and space pretty much any way you want, with only a few restrictions due to mathematical self-consistency. This means that two different observers in a gravitational field can have very different definitions of how time advances, and it is not clear which definition(s) is(are) the right ones to use in incorporating the quantum theory.

- *Causality becomes uncertain.* The light-cone structure of special relativity ensures a rigid separation of past and future for any observer at any place and time. The Standard Model – indeed all quantum gauge theories – rely upon this structure for making predictions. For example the scattering theory we developed in chapter 9 depends upon a clear definition of past quantum states that are emitted by sources and future quantum states that are absorbed by detectors. However, in general relativity gravity differentially slows down time, making the light-cone structure depend on the location of the observer. Quantum gravitational corrections are therefore expected to introduce some quantum-mechanical uncertainty into the structure of the light-cone (and our notions of causality) at each point in space and time. This leads to a strange conundrum – we won't know how to compute this kind of light cone without first having a quantum theory of gravity, but we won't be able to construct a quantum theory of gravity without first understanding this kind of light cone!

Theorists worldwide have been trying to make progress on these problems for more than 40 years, and many ideas have been proposed. Yet there has been no clear path forward, and a quantum theory of gravity appears as elusive as ever. Uniting gravity and quantum theory is a very deep problem, regarded by many as the most challenging and significant problem in physics today.

3. **The universe is sharply biophilic.** This means that our universe is not a typical specimen out of all possible universes one might consider, but instead has rather particular features that permit life. Of course it is clear that we could not live in a universe that did not permit life to exist. However, what has been slowly discovered over the past 50 years is that even rather minor adjustments of the parameters Standard Model and the structure of our cosmos would result in a universe that could not support carbon-based life as we know it. For example, if the mass of the up-quark differed by less than a percent from its experimental value, then neutrons would not decay into protons as observed. Either they would decay too quickly (if the up quark were a bit lighter), leaving pretty much only protons in the early universe, or they would decay too slowly (if the up quark were a bit heavier), making too much helium in the early universe, with little or no hydrogen to fuel the fusion processes in stars. In either case one of the conditions necessary for life as we know it would no longer hold true. Such arguments can be constructed for many different Standard Model parameters, including the masses of the light particles, the coupling constants of all of the forces, and the strength of the cosmological vacuum. It is rather peculiar that the existence of life is so sensitively correlated with the actual values of these parameters of the Standard Model. The reason for this situation goes beyond the explanatory power of the Standard Model [242].

There has been much effort expended in the past 15 years to find what physics lies beyond the Standard Model. In this chapter I will discuss some of the progress that has been made so far, and what the prospects are for further discovery.

25.1 Neutrino Oscillation

The best experimental evidence we have for new physics beyond the Standard Model is in the neutrino sector. A solid body of evidence has accumulated through a number of experiments indicating that the different flavors of neutrino can change into each other, a phenomenon called *neutrino oscillation* or *neutrino mixing.*

The idea of neutrino mixing goes back to 1957, when Bruno Pontecorvo suggested [243] that the neutrino-antineutrino system could oscillate in a manner analogous to the $K^0\bar{K}^0$ system. At that time only one neutrino flavor was known. When the ν_μ was discovered ten years later he pointed out that flavor oscillations between ν_μ and ν_e were not ruled out by any known experiments [244]. In the meantime two different Japanese groups had also proposed that neutrino flavors could mix.

The favored explanation for this phenomenon is that neutrinos are massive. If this is so then there will also be a 3×3 CKM-type mixing matrix in the lepton sector of the Standard Model, in which case we have physics beyond what the Standard Model predicts. As with the quarks, this will imply that the neutrino weak-eigenstates (which we call ν_e, ν_μ and ν_τ) are not the same as the neutrino mass-eigenstates. This means that an electron neutrino produced at a given point will, after propagating outward, have some probability for changing into one of the other two neutrino types.

To illustrate how neutrino oscillation works, I'll begin with a 2-flavor model. As we will see later, this will turn out to be a good approximation for atmospheric neutrinos. Since the physical eigenstates must have definite masses, we have the relationship

$$\Psi^m \equiv \begin{pmatrix} |\nu_2\rangle \\ |\nu_3\rangle \end{pmatrix} = V \begin{pmatrix} |\nu_\mu\rangle \\ |\nu_\tau\rangle \end{pmatrix} \equiv V\Psi^F \tag{25.1}$$

between the mass eigenstates and the flavor eigenstates, where V is a 2×2 matrix. Note that I have labeled the mass eigenstate wavefunctions with the numbers 2 and 3, using the convention of labeling in order of (anticipated) decreasing electron-neutrino content. Since probabilities are conserved, we have

$$(\Psi^m)^\dagger \Psi^m = (\Psi^F)^\dagger \Psi^F \quad \Rightarrow V^\dagger V = I \tag{25.2}$$

and so V is unitary. We can actually make it orthogonal by redefining the phases of the neutrino wavefunctions, as we did for the down-type quarks. Hence the most general form of V is

$$V = \begin{pmatrix} \cos\vartheta_{23} & \sin\vartheta_{23} \\ -\sin\vartheta_{23} & \cos\vartheta_{23} \end{pmatrix} = \begin{pmatrix} c_{23} & s_{23} \\ -s_{23} & c_{23} \end{pmatrix} \tag{25.3}$$

where I have used a notation that anticipates a full 3×3 CKM-type matrix in the neutrino sector, and for convenience I have written $\cos\vartheta_{23} = c_{23}$, etc.

Working non-relativistically, since the Ψ^m wavefunction contains states of definite mass-energy, we have

$$i\hbar\frac{\partial\Psi^m}{\partial t} = H\Psi^m = \begin{pmatrix} E_2 & 0 \\ 0 & E_3 \end{pmatrix}\Psi^m \tag{25.4}$$

which has the solution

$$|\nu_i(t)\rangle = e^{-iE_i t/\hbar}|\nu_i(0)\rangle \Rightarrow \Psi^m(t) = \exp\left[-iHt/\hbar\right]\Psi^m(0) \tag{25.5}$$

and so the flavor eigenstates have the time-dependence

$$\Psi^F(t) = V^{-1}\Psi^m(t) = \left(V^{-1}\exp\left[-iHt/\hbar\right]V\right)\Psi^F(0) \tag{25.6}$$

Multiplying the matrices out (and dropping the $\hbar$s) gives

$$\begin{pmatrix} |\nu_\mu(t)\rangle \\ |\nu_\tau(t)\rangle \end{pmatrix} = \begin{pmatrix} e^{-iE_2t}c_{23}^2 + e^{-iE_3t}s_{23}^2 & c_{23}s_{23}\left(e^{-iE_2t}-e^{-iE_3t}\right) \\ c_{23}s_{23}\left(e^{-iE_2t}-e^{-iE_3t}\right) & e^{-iE_2t}s_{23}^2 + e^{-iE_3t}c_{23}^2 \end{pmatrix} \begin{pmatrix} |\nu_\mu(0)\rangle \\ |\nu_\tau(0)\rangle \end{pmatrix} \tag{25.7}$$

Atmospheric neutrinos (that come from pion decays in cosmic rays) and laboratory neutrinos (from accelerator experiments) are essentially pure ν_μ sources. This means that we should set $\nu_\tau(0) = 0$ as one of our initial conditions. After a time t some ν_τ's will appear, with amplitude

$$|\nu_\tau(t)\rangle = c_{23}s_{23}\left(e^{-iE_2t}-e^{-iE_3t}\right)|\nu_\mu(0)\rangle \tag{25.8}$$

The probability of flavor conversion is therefore proportional to

$$\begin{aligned} |\langle\nu_\tau(t)\,|\nu_\mu(0)\rangle|^2 &= c_{23}^2 s_{23}^2 \left|e^{-iE_2t}-e^{-iE_3t}\right|^2 |\nu_\mu(0)|^2 \\ &= \frac{1}{4}\sin^2(2\vartheta_{23})\left(2-2\cos\left((E_2-E_3)\,t\right)\right)|\nu_\mu(0)|^2 \end{aligned} \tag{25.9}$$

and so we obtain (reinstating the appropriate factor of $\hbar$)

$$P(\nu_\mu \longrightarrow \nu_\tau, t) = \sin^2(2\vartheta_{23})\sin^2\left(\frac{(E_2-E_3)\,t}{\hbar}\right) \tag{25.10}$$

where the latter result is actually the relative probability, since plane-wave states can't be normalized. Note that the only wave to conserve lepton flavor is if the mixing angle $\vartheta_{23} = 0$ – any deviation away from this value indicates physics beyond the Standard Model.

Now I will make an approximation that goes beyond what I have assumed in using the non-relativistic Schroedinger equation (25.4), but which is valid for high-energy neutrinos, namely that rest-mass energy of the neutrinos is small compared to their total energy. For high-energy neutrinos all of the same momenta we therefore have

$$\begin{aligned} (E_2 - E_3) &= \sqrt{p^2 + m_2^2} - \sqrt{p^2 + m_3^2} \simeq p\left(1 + \frac{m_2^2}{2p^2} - \left(1 + \frac{m_3^2}{2p^2}\right)\right) \\ &= \frac{\Delta m^2}{2p} \simeq \frac{\Delta m^2}{2E} \end{aligned} \tag{25.11}$$

where $\Delta m^2 = m_2^2 - m_3^2$ and E is the average energy of either neutrino. The path-length traveled in a time t is approximately L/c (since they are nearly moving at the speed of light), so

$$P(\nu_\mu \longrightarrow \nu_\tau, t) = \sin^2(2\vartheta_{23})\sin^2\left(\frac{(\Delta mc^2)^2 L}{4E\hbar c}\right) \tag{25.12}$$

for neutrinos of energy E moving through a distance L. The probability that the ν_μ survives is then

$$P\left(\nu_\mu \longrightarrow \nu_\mu, t\right) = 1 - \sin^2\left(2\vartheta_{23}\right) \sin^2\left(\frac{\left(\Delta mc^2\right)^2 L}{4E\hbar c}\right) \qquad (25.13)$$

Laboratory experiments are best suited to measuring $P\left(\nu_\mu \longrightarrow \nu_\mu, t\right)$. A device capable of detecting ν_μ is placed a distance L away from a source of ν_μ of known flux and energy E, and the data are used to constrain the value of ϑ_{23} and $\left(\Delta mc^2\right)^2$. The expected flux and energy at the detector are calculated, and survival probablity is the ratio between the measured flux at the detector and the calculated survival flux.

In practice there will be a spread of energies in the flux of neutrinos emerging from the source. This will modify the simple trigonometric dependence above and therefore change the survival flux. It's not too hard to anticipate how. Suppose we have muon neutrinos at two similar (but not the same) energies emerging from the source. For a short period of time there will not be much difference in the survival probability for each energy. However, as time increases the low-energy ν_μ will more rapidly change flavor. Since the survival probability is minimized when $E = \frac{\left(\Delta mc^2\right)^2 L}{4\pi \hbar c}$, the lower-energy neutrino flux reaches its smallest value before the higher-energy one does. After this the low-energy ν_μ flux will increase while the high-energy ν_μ flux continues to decrease. The two fluxes soon become out of phase, with high-energy flux maximized when the low-energy flux is minimized and vice versa, with the average flux oscillating about a mean value dependent on $\sin^2\left(2\vartheta_{23}\right)$. Averaging over a spread of energies, the ν_μ flux will decrease rapidly from unity and after several oscillations settle down to a mean value about which it oscillates with small amplitude.

25.2 Neutrino Experiments

Neutrino oscillation is of much current interest in particle physics, more so than ever since experimental evidence at the Sudbury Neutrino Observatory (SNO) has definitively confirmed the phenomenon. There are three main sources of empirical evidence for this effect.

25.2.1 Solar Neutrinos

Standard nuclear astrophysics indicates that our sun should be emitting a flux of neutrinos that is three times larger than has been observed. Much effort has gone into trying to find out whether the discrepancy is due to some flaw in our understanding of the solar interior, or whether we need to modify our theory of neutrino physics [245]. The results from the SNO experiment have definitively adjudicated on this issue [42], indicating that the different neutrino flavors do indeed change into one another.

Our sun burns hydrogen into helium via a thermonuclear process, in which the average thermal energy is a few tens of KeV. This is small compared to the Coulomb barriers of the interacting nuclei. However, the core density of the sun (10^5 kg/m^3) is large enough that the reaction cross-sections converting hydrogen into helium can proceed.

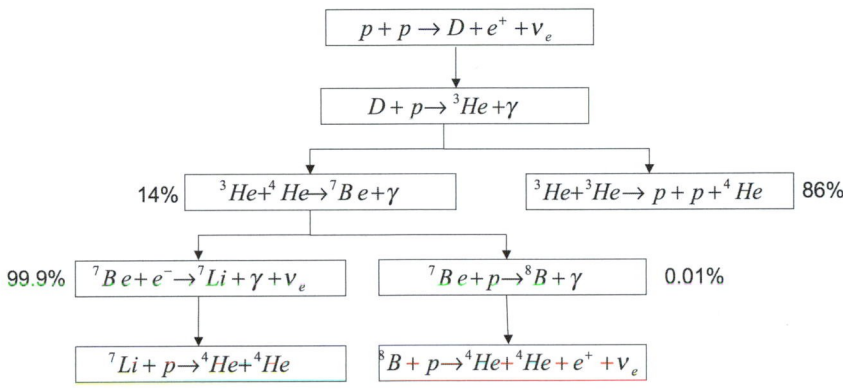

FIGURE 25.1

The pp-cycle of the Standard Solar Model.

About 95% of the energy produced by the sun comes via the conversion of four protons into a Helium nucleus, through a chain of reactions shown in fig. 25.1. Neutrinos of varying energy are produced as a by-product. These stream through the sun with essentially no absorption[†]. A "Standard Solar Model" (SSM) has been developed over the past 60 years that reliably describes the details of this chain of reactions and the neutrino flux produced.

Most of the ν_e flux produced by the sun is due to the conversion of two protons into deuterium. Detecting this is an experimental challenge because all the neutrinos produced are of low energy (less than 420 KeV). The highest energy neutrinos – up to 14 MeV – are produced in the Boron conversion to

---

[†]The photons produced in these same reactions follow a different path. They interact with the rest of the material in the sun, producing new photons of lower energy that in turn further react with this material. This sequence continues until eventually photons are produced at the surface of the sun that stream off into space, carrying energy that was produced from the original reaction several thousand years earlier.

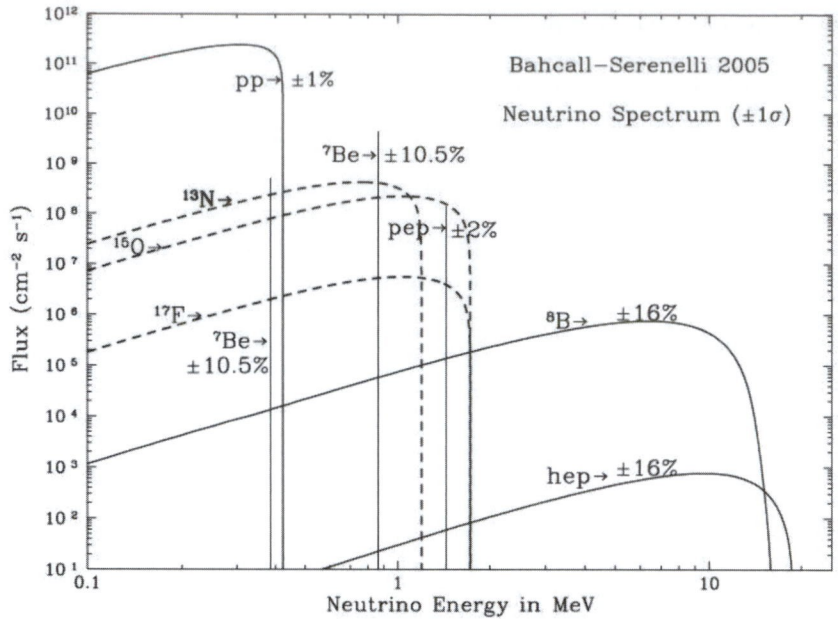

FIGURE 25.2

The energy spectrum of solar neutrinos. Image reprinted from J. Bahcall, A.M. Serenelli, and S. Basu Ap. J. **621**, L85 (2005) [?]. Reproduced by permission of the American Astronomical Society.

two Helium nuclei. This is a very rare process since it depends on a rare production channel from the $Be + p$ interaction. The more common Be interaction produces a Lithium nucleus along with a photon and a neutrino, which emerges at either 400 or 860 KeV. These predictions of the SSM are illustrated in fig. 25.2.

Pontecorvo proposed that solar electron neutrinos could be detected via the reaction

$$\nu_e + {}^{37}\text{Cl} \rightarrow e^- + {}^{37}\text{Ar} \tag{25.14}$$

in 1946, long before neutrinos were actually detected [246]. John Bahcall and collaborators computed the reaction rate for this process in 1962, using the solar model that he and his collaborators were beginning to construct [247]. While the inital reaction rate appeared to be too low for detection, a superallowed transition at 5 MeV was discovered afterward, raising the reaction rate by a factor of 20 to experimentally feasible levels.

Even so, the rate was still very tiny, and to observe it would mean using a huge volume of Chlorine in a region that was shielded as much as possible from all radioactive sources and from cosmic rays. An experiment was set up in the Homestake mine in South Dakota, USA, beginning in 1964 led by Ray

Davis, using 615 tonnes of C_2Cl_4 as the target "detector" [248]. The expected rate of Argon production is about 1 atom per day. This would be extracted every few weeks using a stream of helium gas and then put into a counter to detect the argon decays. The energy threshold for the experiment was 814 KeV, which meant that only the higher-energy Be and B neutrinos could be detected.

The experiment ran for 24 years (1970-94) and yielded the following result after 108 runs [249]

$$\Gamma\left(\nu_e +^{37}Cl \to e^- +^{37}Ar\right) = \begin{cases} 2.56 \pm 0.16 \pm 0.16 & \text{SNU (exp't)} \\ 8.1 \pm 1.3 & \text{SNU (SSM)} \end{cases} \quad (25.15)$$

expressed in Solar Neutrino Units (SNUs), which correspond to 1 capture per 10^{36} atoms per second. There is a clear discrepancy by a factor of 3 as compared to the SSM.

Of course discordant results were coming in from this experiment much earlier than 1994, and physicists began to explore other ways to measure the solar neutrino flux Φ. One attempt was the Kamiokande experiment, which used a water Cerenkov detector in the Kamioka underground observatory in Japan to measure the neutrino flux due to the elastic-scattering reaction

$$\nu_f + e^- \to \nu_f + e^- \quad (25.16)$$

which is sensitive to all neutrino flavors. However, this reaction can only proceed via Z-exchange for $f = \mu, \tau$, whereas it proceeds via both W and Z-exchange for electron neutrinos. Since the latter reaction has a cross-section six times that of the former, it is possible to check if the flux of solar neutrinos is purely ν_e – a diminished cross-section is indicative of a ν_μ and/or ν_τ flux component. The measured flux [250] – confirmed by super-Kamiokande (super-K), its successor – was found to be half the expected value. In particular

$$\Phi = \begin{cases} (2.35 \pm 0.02 \pm 0.08) \times 10^6 \text{ /cm}^2\text{/s (exp't)} \\ (5.69 \pm 0.91\ ) \times 10^6 \text{ /cm}^2\text{/s} \qquad \text{(SSM)} \end{cases} \quad (25.17)$$

Super-K was able to measure the direction of motion of the final state electron, and one could infer from this that the flux of neutrinos was definitely from the sun (and not some other source) [251].

Two other experiments were developed to measure the low-energy neutrino flux due to the pp reactions: SAGE in Russia [252] and GALLEX (later upgraded to GNO) in Italy [253]. These experiments respectively used 60 and 30 tonnes of gallium, which was used to measure the rate of the reaction

$$\nu_e +^{71}Ga \to e^- +^{71}Ge \quad (25.18)$$

at a threshold energy of 233 KeV. SAGE is still running, but in 2003 the GALLEX/GNO experiment ended. The results were

$$\Gamma\left(\nu_e +^{71} \text{Ga} \to e^- +^{71} \text{Ge}\right) = \begin{cases} 69.3 \pm 4.1 \pm 3.6 & \text{SNU (GALLEX/GNO)} \\ 70.8 \pm 5.3 \pm 3.7 & \text{SNU (SAGE)} \\ 126 \pm 10 & \text{SNU (SSM)} \end{cases}$$

(25.19)

again indicating a clear discrepancy between theory and experiment.

Could the SSM be wrong? This question troubled theorists and experimentalists alike for many years until 1995, when the GALLEX experiments reached sufficient precision to almost certainly rule this option out. By then the super-K experiments had measured the flux due to the Boron neutrinos and the flux inferred from solar luminosity due to the dominant pp contribution was well known. The sum of these was greater than the flux GALLEX had measured. But since the GALLEX experiment is sensitive to not only these two sources of neutrino flux but also to the Beryllium neutrinos, there is no reasonable way to modify the SSM because the Boron nucleus is a product of the Beryllium reaction.

Something was causing a large fraction of electron neutrinos to vanish on their way from the sun to the earth. The Sudbury Neutrino Observatory (SNO) was designed to see if neutrino physics was the cause. This experiment uses 1000 tons of heavy water (D_2O) located 2000 meters deep in the Creighton mine near Sudbury, Ontario. It began in 1999 and continued until 2002.

SNO measured the flux of solar neutrinos from the decay of ^8B via the charged current (CC) reaction on deuterium

$$\nu_e + D \to p + p + e^-$$

(25.20)

mediated by W bosons and sensitive exclusively to ν_e's, as well as the neutral current (NC) reaction

$$\nu_f + D \to p + n + \nu_f$$

(25.21)

mediated by the Z boson and sensitive to all flavors. SNO also measured the elastic scattering of all neutrino flavors ν_f with electrons [254].

Since the CC rate directly gives the ν_e flux and the NC rate gives the sum of all three fluxes (with all being equal according to the Standard Model), it is possible to infer the flux $\Phi_{\mu\tau}$ that is due to the other two flavors. In 2002 SNO confirmed that this flux was significant, as illustrated in fig. 25.3. The latest results from SNO in 2008 yielded [255]

$$\Phi^{\text{NC}} = 5.54 \pm 0.33 \pm 0.34 \times 10^6 / \text{cm}^2 / \text{s}$$

(25.22)

$$\Phi^{\text{CC}} = 1.67 \pm 0.05 \pm 0.08 \times 10^6 / \text{cm}^2 / \text{s}$$

(25.23)

$$\Phi^{\text{ES}} = 1.77 \pm 0.24 \pm 0.10 \times 10^6 / \text{cm}^2 / \text{s}$$

(25.24)

for the flux rates for all three processes. SNO has also precisely measured the total flux of Boron neutrinos from the sun independently from previous methods and finds agreement with SSM calculations, further indicating that new physics in the neutrino sector is responsible for the solar neutrino deficit.

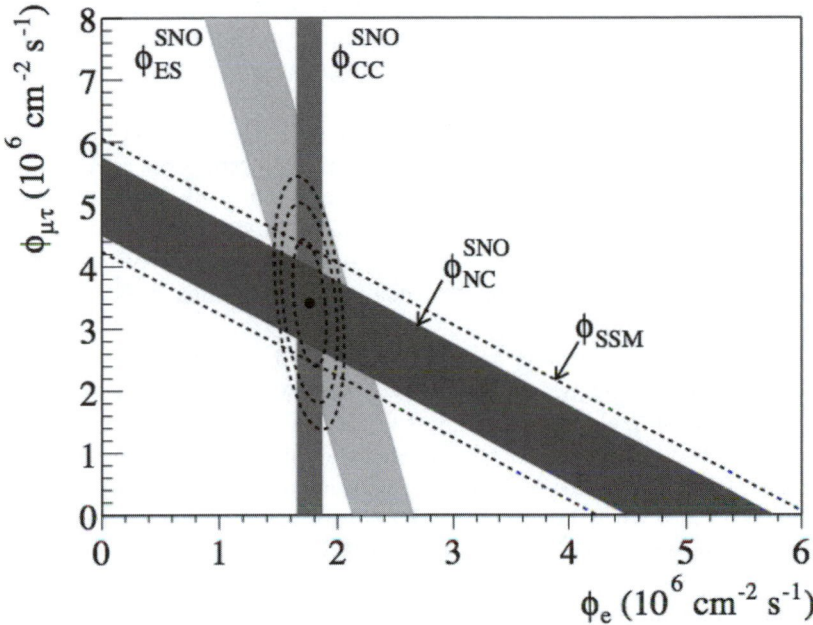

FIGURE 25.3

Fluxes of Boron neutrinos deduced from SNO data [42]. The bands represent uncertainties of 1 standard deviation. Image reprinted from Q.R. Ahmad et al. *Phys. Rev. Lett.* **89**, 011301 (2002); used with permission from the Sudbury Neutrino Observatory; copyright (2002) American Physical Society.

25.2.2 Atmospheric Neutrinos

Cosmic rays contain a large number of pions which (as we saw in Chapter 21) preferentially decay into muons and their associated (anti)neutrinos. The muons in turn decay into electrons (or positrons for the antimuon) and their associated antineutrinos (neutrinos). Hence one expects to see a flux of ν_μ that is twice that of ν_e. However, the observed ratio of fluxes is only 60% of the expected amount.

Atmospheric neutrino oscillations were first discovered at Super-K in 1998 [256]. This experiment consists of 22,500 tons (fiducial mass) of water in the Kamioka underground observatory. The water is used as a Cerenkov detector, which is sensitive to the processes

$$\nu_\mu + N \to \mu^- + N' \qquad \nu_e + N \to e^- + N' \qquad (25.25)$$

where N is the nucleus of either hydrogen or oxygen in the water. The electrons and muons each leave a single Cherenkov ring on the detector. These rings provide important information: the ring produced by a muon is expected to have a sharp shape, whereas the electron shape is fuzzy due to an electromagnetic shower resulting from interactions the electron has with the water in the detector. Fully contained muons and electrons are identified by a pattern recognition algorithm exploiting the maximum likelihood method to compare the measured distribution of hits with the expected shapes. Sometimes a particle is not fully contained but exits the central volume into the outer detector – in this case it is identified as a muon since only muons have such great penetrating power.

The neutrino energies range from a few hundred MeV to several GeV. The cross-sections are strongly forward-peaked at these energies, and so the direction of the outgoing electron or muon (which can determined from the direction of the ring) is strongly correlated with the direction of the neutrino. The amount of Cherenkov light is a measure of the energy of the lepton, which in turn is correlated with the energy of the neutrino.

This information is enough to determine the distance the neutrino has traveled from its production point in the atmosphere. The flight length can vary from about 10 km (if they are produced in the atmosphere above Japan, where the zenith angle $\vartheta = 0$) to 12,000 km (if they are produced on the other side of the earth, where $\vartheta = \pi$).

Events are divided into low-energy (less than 1 GeV) and high-energy (up to several GeV) types, and are also divided into e-type and μ-type events. Both "fully contained" (FC) events – when all of the visible energy is contained within the inner detector and "partially contained" (PC) events – when particles deposit visible energy in the outer detector were considered. The ν_e data show no indication of oscillation whereas the low-energy ν_μ data show a flux deficit that increases with increasing zenith angle [257], which is the same as increasing distance from the detector. The high-energy ν_μ have no deficit for small zenith angles but beyond a certain zenith angle the flux becomes about half of its expected value. These results are illustrated in fig. 25.4. Experiments at the Gran Sasso lab in Italy and SOUDAN2 in the USA confirmed the ν_μ flux deficit seen at super-K [258]. Since only the ν_μ flux is affected, this is almost certainly indicative of ν_μ/ν_τ oscillations. In the context of the 2-flavor model described above, the value of the distance can be used to determine Δm^2 and the value of about $\frac{1}{2}$ in the flux reduction suggests that $\vartheta_{23} \simeq \frac{\pi}{4}$.

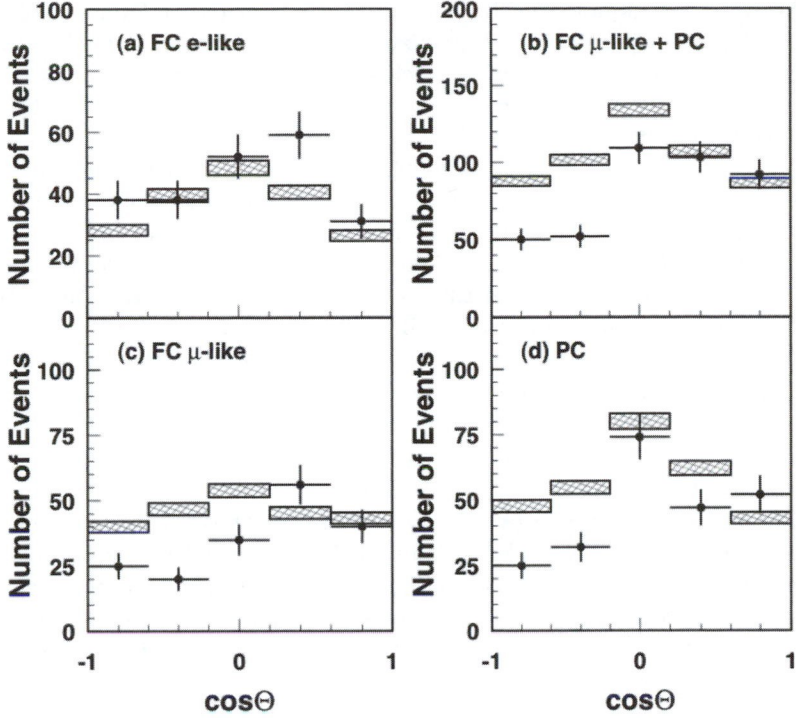

FIGURE 25.4

Zenith angle distributions for: (a) FC electron-like events, (b) FC muon-like and PC events, (c) FC muon-like events and (d) PC events. Down-going is when the cosine equals 1. The histograms with the shaded error bars show the Monte Carlo predictions with their statistical uncertainties [257]. Image reprinted from Y. Fukuda et al. *Study of the atmospheric neutrino flux in the multi-GeV energy range* Phys. Lett. **B436**, 33 (1998) fig. 5; Copyright (1998), with permission from Elsevier.

25.2.3 Laboratory Neutrinos

Another way to search for neutrino oscillations is by producing a source of neutrinos and then see if they disappear somewhere further along their trajectory away from the source. Disappearance experiments for ν_μ were carried out in both Japan at the KEK laboratory in Tsukuba and at Fermilab in the USA. In each case two detectors were used, one near the source to measure the initial flux and energy, and the other far away to measure the surviving flux. In Japan the far detector was super-K at 250 km away. In the USA MINOS was the far detector at 735 km from the source. These experiments also found a ν_μ flux deficit consistent with super-K.

The Kamioka Liquid scintillator Anti-Neutrino Detector (KamLAND) experiment provided important confirmation for the SNO experiments. This experiment used several Japanese nuclear power plants as a source for electron antineutrinos. Each source had a different power and was at a different distance (typically about 180 km) from the KamLAND detector. This detector, an ulta-pure scintillation detector, could accurately detect and measure the energy of recoil electrons from elastic scattering with the antineutrinos. This information can be used to infer the electron antineutrino flux and energy spectrum, which was found to be 0.658 ± 044(stat) ± 047 (syst) of the value expected if no electron antineutrinos had disappeared. Plotting the ratio of the observed spectrum to the no-oscillation spectrum as a function of $1/E$, KamLAND found that there was flux oscillation consistent with the sinusoidal L/E dependence that is characteristic of neutrino oscillation [259].

In 1995 the Liquid Scintillator Neutrino Detector (LSND) group at Los Alamos presented evidence that muon antineutrinos could change into electron antineutrinos [260]. The LSND detector consists of an approximately cylindrical tank 8.3 m long by 5.7 m in diameter, whose center is 30 m from the neutrino source. The tank is filled with 167 metric tons of liquid scintillator consisting of mineral oil and 0.031 g/l of b-PBD, which affords detection of both Cerenkov light and scintillation light. The Cerenkov cone and the time distribution of the light give excellent particle identification. Cosmic ray muons going through the detector are tagged by a veto shield surrounding it.

The experiment searches for the reaction

$$p + \bar{\nu}_e - > e^+ + n \tag{25.26}$$

followed by $n + p - > D + \gamma(2.2 \text{ MeV})$. The 2.2 MeV photon γ is determined to be correlated with a positron (or not) using a likelihood ratio that depends on the number of hit phototubes for the photon, the reconstructed distance between the positron and the photon, and the relative time between their production. The experiment found a total excess of $51.0 \pm 20.2 \pm 8.0$ events, which (if due to neutrino oscillations) corresponds to an oscillation probability of $(0.31 \pm 0.12 \pm 0.05)\%$.

This experiment, while intriguing, has yet to be confirmed. The related KArlsruhe Rutherford Medium Energy Neutrino (KARMEN) experiment searched for such an effect but saw no oscillation [261]. However,, this experiment differed from LSND in that the neutrinos travel a distance of 18m at KARMEN but 30m at LSND. The KARMEN results do not cover the full region of parameters that LSND does, and so it is possible that both experiments are correct.

While atmospheric experiments all suggest that ν_μ's predominantly decay into ν_τ's there is as yet no direct test of this expectation. The CNGS experiment (for CERN Neutrinos to Gran Sasso) is a proposal to make such a test [262]. In this experiment a ν_μ beam from the CERN SPS machine is aimed through the earth's crust to the OPERA detector at the Gran Sasso laboratory 737 km away. This detector is sensitive to the reaction

$$\nu_\tau + N \to \tau^- + N' \tag{25.27}$$

which is achieved as follows. The neutrino energy must be above 10 GeV to ensure that the above reaction is above threshold. However, this means the flight length is small compared to the oscillation length, and so only a small number of ν_τ's are expected, which in turn means a small number of τ^- events. This means the OPERA detector must have a large mass (to maximize cross section) and a high resolution (to distinguish the decay products of the τ^-). Using a combination of emulsion and electronic techniques, OPERA will achieve micrometer scale resolution with a mass of 2000 tons.

25.3 Neutrino Masses and Mixing Angles

From a particle physics viewpoint, there are two possible explanations for neutrino oscillations. The most favored is that neutrinos have mass. A flux of electron neutrinos produced inside the sun could, if sufficiently massive, oscillate into muon neutrinos which would go undetected in the experiments which have been carried out to date. In 1985 Mikayhev and Smirnov [263], building off work done by Wolfenstein [264], showed that this neutrino oscillation process could be greatly enhanced inside the core of the sun because of the high density of nucleons there. This effect remains the favored mechanism for explaining solar neutrino oscillations because it enhances only electron neutrino oscillations into the other two flavors.

Another possibility is that gravity is responsible [265]. Neutrinos, instead of being massive, might couple differently to gravity relative to the manner in which baryons couple to gravity. According to the equivalence principle, all forms of mass-energy couple in the same way to gravity, so one is postulating here that the equivalence principle is violated in (at least) the neutrino sector of the Standard Model. If so, then there is a discrepancy between the weak eigenstates and the gravitational eigenstates of neutrinos, permitting neutrino oscillations.

These two mechanisms can be distinguished from one another: the mass-mechanism has an oscillation probability which varies inversely with energy, whereas for the gravitational one the probability is proportional to the energy. So neutrino oscillation experiments act as a test of the equivalence principle [266]. The data from K2K have indicated that in the ν_μ-ν_τ sector mass oscillations appear to be the dominant mechanism [267], though both could be operative – perhaps neutrinos are massive particles that couple unequally to gravity!

Today all world experiments indicate the following for the neutrino mixing matrix, which has come to be called the Maki-Nakagawa-Sakata or MNS matrix [268]. It relates the mass eigenstates (labeled 1,2,3) to the weak eigenstates

$$
\begin{pmatrix} |\nu_e\rangle \\ |\nu_\mu\rangle \\ |\nu_\tau\rangle \end{pmatrix} \equiv \begin{pmatrix} V_{e1} & V_{e2} & V_{e3} \\ V_{\mu1} & V_{\mu2} & V_{\mu3} \\ V_{\tau1} & V_{\tau2} & V_{\tau3} \end{pmatrix} \begin{pmatrix} |\nu_1\rangle \\ |\nu_2\rangle \\ |\nu_3\rangle \end{pmatrix} \tag{25.28}
$$

The MNS matrix has the same form as the CKM matrix (21.55), but with an additional feature. Writing $\cos\vartheta_{ij} = c_{ij}$ and $\sin\vartheta_{ij} = s_{ij}$ the MNS matrix is

$$
\begin{pmatrix} V_{e1} & V_{e2} & V_{e3} \\ V_{\mu1} & V_{\mu2} & V_{\mu3} \\ V_{\tau1} & V_{\tau2} & V_{\tau3} \end{pmatrix} \tag{25.29}
$$

$$
= \begin{pmatrix} c_{12}c_{13} & -s_{12}c_{13} & s_{13}e^{-i\delta} \\ -s_{12}c_{23} - c_{12}s_{23}s_{13}e^{i\delta} & c_{12}c_{23} - s_{12}s_{23}s_{13}e^{i\delta} & s_{23}c_{13} \\ s_{12}s_{23} - c_{12}c_{23}s_{13}e^{i\delta} & -c_{12}s_{23} - s_{12}c_{23}s_{13}e^{i\delta} & c_{23}c_{13} \end{pmatrix} \begin{pmatrix} e^{i\alpha_1} & 0 & 0 \\ 0 & e^{i\alpha_2} & 0 \\ 0 & 0 & 1 \end{pmatrix}
$$

and we see that there is an additional diagonal phase matrix on the right. For quarks this phase matrix can be absorbed into redefinitions of the quark wavefunctions. However, it is possible that neutrinos are identical to their own antiparticles, in which case we call them Majorana particles. We have already observed antiquarks to be distinct particles from quarks, but we have no corresponding evidence that the same situation is true for neutrinos and antineutrinos.

If neutrinos are their own antiparticles then the $e^{i\alpha_k}$ phases cannot be absorbed into a redefinition of neutrino wavefunctions. These phases influence neutrinoless double beta decay and other processes, but do not affect flavor oscillation of neutrinos. Each of α_1, α_2, and δ contribute to $\mathbb{CP}$ violation – at present we have no empirical information as to the values of these parameters. Note that the size of $\mathbb{CP}$ violation in neutrino oscillation will depend on s_{13}, since δ always occurs in combination with this parameter.

We've seen above that the characteristic time (or length) for conversion from one flavor into another in a 2-flavor model is inversely proportional to $\left(\Delta mc^2\right)^2$. In a full 3-flavor model the same kind of relationship would hold, but with Δmc^2 replaced with much more complicated expression depending on the neutrino masses. However, the atmospheric neutrino data are are very well described by the hypothesis that the oscillation is purely a 2-flavor one between ν_μ and ν_τ. Furthermore, the empirical value of ϑ_{23} suggests that these are maximally mixed, i.e. the conversion between the two flavors is as large as possible. This greatly simplifies the interpretation of the data. The SNO and KamLAND data, in conjunction with atmospheric and reactor data, suggest that two neutrino mass eigenstates are significantly involved in solar neutrino evolution. SNO and KamLAND also establish that the mixing angle for this process is large. Finally, bounds on the short-distance oscillation of reactor electron-antineutrinos imply that $|V_{e3}|^2 < 0.032$, which in turn means that $s_{13}^2 < 0.032$.

If we put together all current experimental information on neutrino oscillations we find the following experimental values [1]

$$\vartheta_{12} \approx \vartheta_\odot = 33.9^o \; {}^{+2.4^o}_{-2.2^o} \qquad \vartheta_{23} \approx \vartheta_{\text{atm}} = 45^o \pm 8^o \qquad \vartheta_{13} < 3.2^o \quad (25.30)$$

$$\Delta m_{21}^2 \approx \Delta m_\odot^2 = 8.01^{+0.6}_{-0.4} \times 10^{-5} \text{ eV}^2 \quad (25.31)$$

$$\Delta m_{32}^2 \approx \Delta m_{31}^2 \approx \Delta m_{\text{atm}}^2 = 2.4^{+0.6}_{-0.5} \times 10^{-3} \text{ eV}^2 \quad (25.32)$$

I've used the convention here that the numbers 1,2,3 represent the order of decreasing ν_e content, and used notation to take into account that observed results indicate that essentially ϑ_{12} is the "solar mixing angle" and ϑ_{23} is the "atmospheric mixing angle."

The current situation is depicted in 25.5. This picture shows that two of

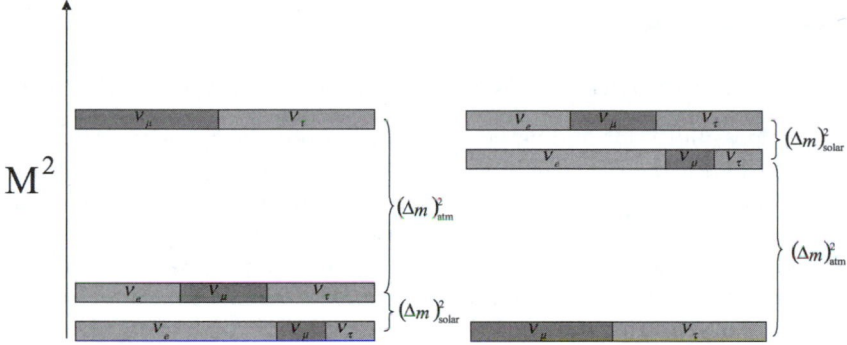

FIGURE 25.5

The currently understood possibilities for the mass spectrum of the weak neutrino eigenstates. Notice that the standard weak-flavor combinations significantly mix together for each mass eigenstate.

the three mass eigenstates are close together in mass, and that there is a 3rd mass eigenstate that is somewhat heavier – or lighter! At this point in time we don't know which is the case. Unlike the quarks, each mass eigenstate has lots of mixing between the different flavors of neutrinos. For example, the middle-mass eigenstate on the left (or the top one on the right) is an almost equal superposition of all three neutrino flavors.

There is another intriguing possibility suggested by the KARMEN/LSND results. If both of these experiments are correct then the best fit to the data implies that

$$0.2 \text{ eV}^2 < \Delta m_{\text{LSND}}^2 < 1 \text{ eV}^2 \quad (25.33)$$

which is a huge mass splitting compared to the solar and atmospheric cases. However, if there are only three mass eigenstates we must have

$$\Delta m_{21}^2 + \Delta m_{32}^2 + \Delta m_{13}^2 = 0 \tag{25.34}$$

and it is not possible for $\Delta m_{\mathrm{LSND}}^2, \Delta m_\odot^2$, and $\Delta m_{\mathrm{atm}}^2$ to satisfy this equation. So if all experiments so far are yielding reliable results, there must be at least one more flavor of neutrino! Confirming this would definitely indicate that there is a lot of new physics beyond the Standard Model.

Neutrino physics should yield some of the most interesting information about what lies beyond the Standard Model over the next few years. As experiments (such as LSND/KARMEN) are refined and as new observatories (such as SNOlab, the successor to SNO) collect data, we should learn much more about the behavior of the neutrino sector of the Standard Model.

25.4 Axions and the Neutron Electric Dipole Moment

Another possibility for finding new physics beyond the Standard Model consists of investigating the structure of the neutron. Recall from Chapter 6 that under time-reversal $\mathbb{T}$, the magnetic dipole moment $\vec{\mu}$ of a particle changes sign, but the electric dipole moment $\vec{\mathbf{d}}$ does not. Since the only direction that a subatomic particle at rest "picks out" is the one given by its spin, dipole moments must either be aligned or antialigned with the spin direction. But under $\mathbb{T}$ the spin flips sign but $\vec{\mathbf{d}}$ doesn't, violating time-reversal. Assuming that $\mathbb{CPT}$ is a symmetry, then this means that $\mathbb{CP}$ is violated for any elementary particle having a nonzero electric dipole moment, or EDM.

If an elementary fermion had an electric dipole moment then there would be an additional vertex in the Standard Model as illustrated in fig. 25.6 where $\Sigma^{\mu\nu} = \frac{i}{4}\left([\gamma^\mu, \gamma^\nu]\right)$ and the constant g_d is a coupling constant (having units of charge times length) parametrizing the strength of the EDM. Except for the factor of γ^5, this vertex looks just like that for the anomalous magnetic moment interaction induced by loop corrections in QED.

This vertex violates both $\mathbb{T}$ and $\mathbb{P}$ and so it does not exist in QED. Purcell and Ramsey in 1950 [269] considered a rather unconventional model in which the EDM of the neutron respected $\mathbb{P}$, which was thought to be a symmetry of the world at that time. They obtained the limit

$$g_d^{\mathrm{neutron}} < 10^{-13} \text{ e-cm} \tag{25.35}$$

for the neutron. Once parity was found to be violated by the weak interactions it was still thought that the EDM of the neutron (if it had one) was $\mathbb{T}$-invariant and thus also $\mathbb{CPT}$-invariant by virtue of the CPT-theorem. However, when

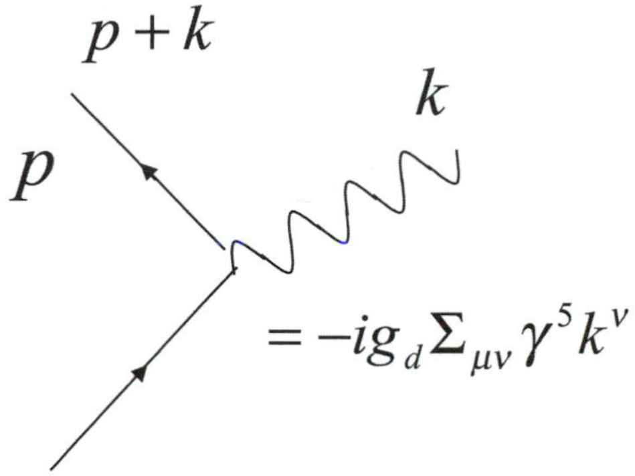

$$p+k$$

$$k$$

$$p$$

$$= -ig_d \Sigma_{\mu\nu} \gamma^5 k^\nu$$

FIGURE 25.6

Vertex for a fermion with an electric dipole moment.

$\mathbb{CP}$-violation was observed in 1964, the subject of the EDM of the neutron became of particular importance in physics, since it afforded a new test of $\mathbb{CP}$ physics.

QED is invariant under each of $\mathbb{C}$, $\mathbb{P}$, and $\mathbb{T}$ and as a consequence treats both right-handed and left-handed quarks on the same footing. It was originally thought that QCD respected the same symmetries since gluons coupled with the same strength to both right-handed and left-handed quarks. However, in the 1970s a subtlety in QCD was discovered that indicated it did not satisfy $\mathbb{CP}$ symmetry. The reason for this has to do with the vacuum (or ground state) structure of a non-abelian gauge theory.

A gauge particle (like a photon or gluon) that is in a configuration that has vanishing field strength at large distance must be a pure gauge artifact. For the photon this just means that its wavefunction is $A_\mu \to \partial_\mu \alpha$, where α is some function that vanishes sufficiently rapidly at large distance. Another gauge transformation can be made to set $A_\mu = 0$ everywhere. This is clearly the lowest energy state of the photon, and is regarded as the vacuum state of QED.

However, for gluons the situation is different. There are distinct non-vanishing configurations of the gluon wavefunction that are pure gauge at large distance. These configurations all have vanishing field strength at large distance but cannot be set to zero (or equal to each other) by another gauge transformation. Since each of these distinct configurations has a field strength that vanishes at large distance, each can be regarded as a possible vacuum state of QCD.

What distinguishes these configurations? The answer has to do with their topology. Two field configurations of a gauge particle are said to be topologically equivalent if there is a mathematical transformation that continuously changes one into the other without making the energy infinite in the process.

For example, suppose there were a theory in which there were two potential energy minima for two different types of field strengths. If the barrier between these minima were infinite then there is no continuous transformation we could make to change a field configuration whose energy was in one minimum to that in another minimum. In this case we would say the two configurations were topologically inequivalent, even though their minimum energy was the same.

A helpful analogy consists of considering a rubber band. Suppose we have a pole of finite length. We could either wrap the rubber band once around the pole or not. Suppose the band is wrapped around the pole once. Since the pole is of finite length we can easily remove the band from the pole by lifting it off of one end. This is a continuous set of operations (a continuous mathematical transformation) and so the configuration with the band wrapped around the finite pole is topologically equivalent to the one where it is not around the pole. We can do this for any configuration of the band: whether it is wrapped once, twice, or many times (as long as the rubber can be stretched).

Now suppose the pole has infinite length. A rubber band that is not wrapped around this pole is topologically distinct from one that is wrapped once around this pole, since the latter cannot be taken off of the pole by any continuous transformation. Since the pole is infinite we can't lift the band off. We could of course break the band and then glue it back together away from the pole, but this is not a continuous transformation since the band is no longer a band for part of this procedure. So we say in this case that the two configurations are topologically distinct. In fact, any configuration in which the band is wrapped around the pole a given number of times will be distinct from one where it is wrapped a different number of times. If we imagine having a band that won't break no matter how much it is twisted, then there will be a countably infinite number of different configurations that are topologically distinct. We can label these by the number of times the band is wrapped around the infinite pole – a quantity called the winding number.

The situation with gluons in QCD is similar to that of the rubber band and the infinite pole. In QCD the gauge group is $\mathbf{SU}(3)$ and the gluon can be in a configuration that at large distance has vanishing field strength and so approaches a pure $\mathbf{SU}(3)$ gauge transformation. However, this transformation can't be set to zero the way that we can for QED because the gauge configuration is "wrapped around" all of space analogous to the way that the rubber band is wrapped around the infinite pole. The configuration can be unwrapped (zero gluon field strength everywhere), wrapped once, twice, or many times around all of space[‡]. The different configurations are distin-

---

[‡]These wrappings are described mathematically by something called a homotopy group. The homotopy group in this case is $\pi_3\left(\mathbf{SU}(3)\right) = \mathbb{Z}$, which means that there are countably infinitely many topologically distinct maps from the group $\mathbf{SU}(3)$ to the three-dimensional sphere, each map labeled by an integer. These distinct maps are what I am calling "wrappings".

guished by a winding number. Since each of these "wrapped" configurations has vanishing field strength at infinity, they qualify as possible vacuum states (or ground states) of QCD. These configurations all violate $\mathbb{CP}$, except for the configuration where the winding number is zero.

Which one of these is the "correct" ground state? You might think that we should just choose the one with zero winding number, but this is wrong because of a subtle non-perturbative quantum effect. Here is where the quarks enter the picture. Recall that in order to get sensible quark masses, we had to separately transform the set of right-handed and left-handed quark wavefunctions to ensure that the Yukawa coupling matrix to the Higgs wavefunction was diagonal with all elements real. This can be done, of course, but one transformation required to ensure this involves simultaneously transforming all left-handed quark wavefunctions with one phase and all right-handed quark wavefunctions with the opposite phase:

$$\psi_L' = e^{i\chi}\psi_L \qquad \psi_R' = e^{-i\chi}\psi_R \qquad (25.36)$$

There is a non-perturbative quantum effect that induces from this transformation one of the wrapped configurations of the gluon vacuum that has nonzero winding number. So even if we started with a vacuum of zero winding number, this quantum effect would give us a vacuum that had a non-zero winding number once we made the above transformation. There is a non-zero probability to tunnel from a vacuum with zero winding number to one with nonzero winding number.

So why not include all the winding numbers, and say that the actual vacuum of QCD is some linear combination of them all? In fact, this is what is (and should be) done. Since each wrapping is labeled by a winding number n, its quantum state can be denoted by $|n\rangle$. Each vacuum state is separated from those with a different winding numbers by an energy barrier. Using methods beyond the scope of this text [270], it is possible to show that quantum mechanical transitions can take place between these different vacua, with amplitudes

$$\langle m| e^{-iHt} |n\rangle \propto \exp\left(-\frac{8\pi \left(m - n\right)^2}{g_s}\right) \qquad (25.37)$$

which is a non-perturbative effect (there is no good series expansion of the right-hand side of eq. (25.37) for small g_s). The amplitude is small for small strong coupling g_s, and large for large g_s. Since we have tunneling between the different vacua, the true vacuum state must be a linear superposition of all of them. As with periodic potentials in quantum mechanics (whose true ground state is the Bloch wave), we define the true vacuum to be

$$|\theta\rangle = \sum_n e^{-in\theta} |n\rangle \qquad (25.38)$$

and it is possible to show that this vacuum is invariant under all possible $\mathbf{SU}(3)$ gauge transformations. However, it is not $\mathbb{CP}$-invariant.

The parameter θ labels the different possible physically inequivalent sectors of QCD. We could choose any value of θ that we want and compute physical processes in a gauge–invariant manner. Unlike the vacuum states $|n\rangle$, which can interact with one another via quantum tunneling, each $|\theta\rangle$-world cannot communicate with any other $|\theta\rangle$-world. Consequently θ becomes another constant of nature (telling us what vacuum we live in), one of the undetermined parameters of the Standard Model.

Can we determine θ by experiment? Yes – this is again where the quarks come in. We can make another transformation of the form in eq. (25.36) on the quark wavefunctions to eliminate θ entirely, setting it equal to zero. This sounds good, except that now our quark mass terms have a complex phase that induces the $\mathbb{CP}$-violating vertex shown in fig. 25.6. We can further redefine the phases of all of the quark wavefunctions to get rid of this phase everywhere in the equations of motion except for the mass term of one quark, which we might as well take to be the up quark (though any quark would do). This phase is of the form $\exp(i\theta/N_f)$ where N_f is the number of quark flavors. Hence if one quark mass were zero, there would be no $\mathbb{CP}$-violation of this type in QCD.

However, experiment tells us that all quark masses appear to be nonzero. The net effect of this phase is to induce the vertex shown in fig. 25.6 for the up quark only. We could go to some trouble to compute the amplitude for the neutron, by constructing an effective neutron vertex for the electric dipole moment, but we don't need to in order to estimate how large this effect is. The coupling g_d must be proportional to $|\theta|$ and the mass m_u and charge of the up-quark. This has units of (charge)$\times$(mass), but on dimensional grounds the dipole moment must have units of charge times length, which is the same as charge/mass, since the Compton wavelength of a particle is inversely proportional to its mass. Hence we need to divide by the square of a mass. The only mass scale for the neutron is the mass of the neutron m_n, so on dimensional grounds

$$g_d^{\text{neutron}} \sim |\theta| \, e \frac{m_u}{m_n^2} = 10^{-16} \, |\theta| \text{ e-cm} \tag{25.39}$$

a result that is supported by more detailed calculations. The best experimental limits today imply [1]

$$|g_d^{\text{neutron}}| < 10^{-25} \text{ e-cm} \tag{25.40}$$

which is very tiny, and sets a bound of $|\theta| < 10^{-9}$.

This suggests that perhaps $\theta = 0$, and that the actual vacuum of our world really does conserve $\mathbb{CP}$ after all. The problem with this is that we have no mechanism within the Standard Model for setting $\theta = 0$ any more than we have a mechanism for setting one of the parameters of the CKM matrix to a particular value. However, the near-vanishing value of θ is suggestive that perhaps some dynamical mechanism is at work to naturally make $\theta = 0$.

Peccei and Quinn in 1977 proposed such a mechanism [271] in which θ became a dynamical variable that relaxed to the minimum of an effective potential where $\mathbb{C}$ and $\mathbb{P}$ were both conserved. This mechanism requires the existence of a new pseudoscalar particle called the axion.

The original axion from the Peccei-Quinn model was ruled out by experiment, but new versions of the model appeared in which the axion couples weakly enough to ordinary matter to have thus far escaped detection [272]. The common feature of all such models is that there is some $\mathbf{U}(1)$ symmetry that is spontaneously broken at some high energy and is slightly broken by the non-trivial QCD vacuum, providing a small mass for the axion.

If there is an axion, it will most likely be observed as result of its interactions with the light quarks, or in other words with the low-mass hadrons. However, because axions couple to gluons, they indirectly couple to pions. The most general model for coupling the axion to up and down quarks leads to a theory (that I won't go into detail about here) that describes interaction of the axion with pions. To leading order in the pion decay constant f_π, this theory leads to the result [273]

$$m_{\text{axion}} = \frac{f_\pi}{f_a} \frac{\sqrt{m_u m_d}}{(m_u + m_d)} m_\pi = \frac{6 \times 10^{-3} \ (\text{MeV})^2}{f_a} \tag{25.41}$$

where f_a is the "axion decay constant" with units of energy that parametrizes the energy scale of the spontaneous breaking of the $\mathbf{U}(1)$ symmetry.

The interactions that axions can have with fermions would modify predicted cross-sections for a whole range of accelerator and reactor experiments. By setting f_a to be large enough we can easily make these modifications tiny enough to have escaped detection so far. Since axions couple to pions they also couple to photons. Stars can therefore "cool off" by emitting axions, and limits on the rate of stellar cooling of red giant stars force $f_a > 10^7 \text{GeV}$. A stronger limit [274] comes from the supernova SN 1987A: the observed neutrino pulse from confirmed the theoretically expected cooling speed young neutron stars to be a few seconds, which means that excessive cooling by axions cannot take place, forcing $f_a > 10^{10} \text{GeV}$.

This last limit means that axions are very light – about 6×10^{-3} eV. You might imagine that searching for such a low mass particle in the lab would be very challenging. The most promising approaches exploit the fact that axions and photons will convert into each other in the presence of an external electric or magnetic field. An external magnetic field will cause axion–photon oscillation in a manner similar to neutrino-flavor oscillations, whereas the electric field due to a charged particle (like a nucleus) will make the conversion an effective scattering process of the form $\gamma + Ze \to a + Ze$, a phenomenon called the Primakoff effect [275]. If a photon is moving in a direction perpendicular to an external magnetic field it will convert into axions moving in the same direction as the beam [276]. This phenomenon can be searched for by sending a laser beam down the bore of a two long superconducting dipole magnets (like the bending magnets in high-energy accelerators). Axions will be generated in the first magnet provided the oscillation length is comparable to the length of the magnet. Putting a barrier to the laser beam in between the two magnets will cause only axions (if there are any) to move through the second magnet, which will convert them to photons that can then be detected. So far such searches have yielded null results [277], and have not yet placed limits on axion couplings that are better than what we have from stellar observations.

It is also possible to get an upper bound on f_a from cosmological considerations [278]. Axions will be produced in the early universe, contributing to the overall energy density of the universe. However, the axion energy density doesn't rapidly dissipate into other particles because axions couple very weakly to known matter. This "invisible" energy density will exceed the amount that would cause the universe to be closed (contrary to observation) unless $f_a < 10^{12}$GeV. So perhaps there really are axions out there!

$$
\begin{array}{cc}
\text{QED} & \text{QCD} \\[4pt]
\psi' = \underbrace{e^{i\alpha(x)}}_{U}\,\psi & \psi' = U(3\times3)\,\psi \\[8pt]
& U_1 U_2 \neq U_2 U_1 \\[4pt]
U_1 U_2 = U_2 U_1 & \text{Non-Abelian} \\[4pt]
\text{Abelian} &
\end{array}
$$

25.5   Frontiers

Neutrino physics and axions represent only a small portion of the effort being expended today to understand what lies beyond the Standard Model. In the remaining part of this chapter I will sketch out a few of the many other approaches being taken to extend the frontiers of our knowledge.

25.5.1   Dark Matter

Axions and neutrinos both form examples of what might be called "dark matter" – matter that interacts so feebly with known matter that its presence in the universe on cosmological scales can almost certainly be detected only via its gravitational interactions with known matter. The latter half of the 20th century led to the construction of a standard cosomological model of the universe that describes all known cosmological data in a single framework. The total mass-energy of the universe is now known to be within about 1% of the critical value needed for the universe to be closed and finite (the so-called critical density) [1]. Within the context of this model it is known that only 4% of this consists of known matter, which is inferred from the relative abundance of hydrogen, helium, and other light nuclei. Most of this – 3.6% of the total – is in the form of interstellar gas, with the remaining 0.4% in the form of stars and planets.

Another 23% of the energy density is dark matter – a form of mass-energy whose presence is known only via its gravitational attraction to other luminous (and known) matter. This kind of matter neither emits nor absorbs photons (or if it does, it does so at a rate so feeble as to have so far escaped detection). The composition of this matter is an outstanding puzzle for both particle physics and cosmology. It was once thought that neutrinos could be the dark matter, but it is now known that they contribute at best 1% to this amount. It is generally thought that dark matter is some new kind of particle not described by the Standard Model. In view of the limits above, axions are a viable candidate today, but there are several other possibilities. A new kind of "sterile neutrino" – one that can convert into known neutrinos but does not participate in the weak interactions – is another possibility. Yet another is a WIMP, short for Weakly Interacting Massive Particle. WIMPs experience weak interactions, but have escaped detection in accelerator experiments so far due to their presumed large mass, which is perhaps 10s or 100s of GeV. Supersymmetry (see below) provides a theoretical framework for describing such particles [280].

Trillions of WIMPs must be passing through the Earth each second if they indeed exist. Experimental searches for WIMPs are ongoing. Some attempt to directly detect them with a laboratory detector, whereas other experiments search for indirectly for WIMPs via the products of their decays and annihilations. Some experiments have claimed detection (for example EGRET [281], CDMS [282] and DAMA/NaI [283]) but without confirmation, and others have found only null results. A number of experiments are currently in progress, and still others are in development. The LHC counts as one of these – it could detect WIMPs as a form of missing energy and momentum provided their coupling to ordinary matter were sufficiently strong and their masses not too large [284].

25.5.2 Dark Energy

Normal matter – including dark matter – is gravitationally attractive. It therefore imposes a deceleration on the expansion of the universe, slowing down the expansion in a manner analogous to the way the earth's gravity slows down a projectile launched from its surface. An outstanding goal in cosmology for a number of years was to measure the amount of this deceleration. In 1999 this goal was achieved by looking at supernova data. Supernovae of type 1a can be used as reliable distance indicators (standard candles) in astronomy because their absolute luminosity is known. They are visible at very large distances, and so can be used to infer the deceleration rate. Distant supernovae should appear brighter and closer than their redshifts would otherwise suggest.

When the measurements were completed, most supernovae were found to be dimmer and further away than expected. In other words instead of *de*celerating, the universe was *ac*celerating in its expansion! This means that there must be some other form of matter with a kind of gravitationally repulsive character, and it must be in enough abundance to overcome the attractive decelerating effects of the known and dark matter. Further measurements indicate that the remaining 73% of the energy density of the universe is of this type: it is a very peculiar form of matter that has negative pressure.

The simplest explanation of this form of matter is that it is a relic vacuum energy, parameterized by what is called a cosmological constant [285]. This means that the dark energy is the same at all times and places. This introduces yet another parameter into the Standard Model as noted earlier. However, this parameter is very problematic for physics. All subatomic particles contribute to the vacuum energy density, and an estimate based on known methods from quantum field theory estimate a value for this vacuum energy density that is 10^{120} times the amount that is observed. This is perhaps the largest discrepancy between theory and experiment known in science [15].

As you might have guessed, many physicists have attempted to provide some other explanation for dark energy by making it depend on space and time in some way. These approaches typically suggest the existence of new particles. Others have attempted to modify General Relativity, our established theory of gravity, at large distances and/or over long cosmological times. On general grounds there is no reason at this point in time to prefer a cosmological constant over any of these other models, and the question of the composition of dark energy is wide open [8]. At present, cosmologists and astronomers are in the process of mapping out the historical development of our universe to see if the dark energy is indeed constant by measuring thousands of more supernovae and the distribution of mass on very large scales.

25.5.3 Grand Unification

The fact that weak and electromagnetic interactions can be unified into a single theory suggests that the strong interactions should somehow be included as well. Such theories were called "Grand Unified Theories", or GUTs, since they would unify all non-gravitational forces into one grand theoretical framework, with one coupling constant governing them all.

How might this work? We've seen in chapters 14, 18 and 23 that each of the coupling constants is a function of energy. The electromagnetic coupling g_e grows as a function of energy whereas the other two decrease. So it is not unreasonable to imagine that they might all have the same value at some energy. A rough estimate puts this energy scale at 10^{16} GeV – the grand unification scale.

The simplest GUT was a model with $\mathbf{SU}(5)$ as the gauge symmetry group [286]. It's simplest in the sense that this is the group with the smallest number of generators that can include the groups of the strong and electroweak theories. All of the quarks and leptons in a given generation were assigned to be in irreps of this group. There are fifteen such particles: the right- and left-handed electron, the electron-neutrino, and up and down quarks of 3 colors each, each right and left handed. The smallest two irreps – the quintet and the decuplet of $\mathbf{SU}(5)$ – were used: three left-handed charge-conjugate down quarks, the left handed electron and the electron-neutrino were placed in the quintet – so they would be regarded as one kind of "particle," and the rest (charge-conjugated as needed) were placed in the decuplet, the other kind of "particle." That all the particles did not fit into a single irrep was regarded as somewhat unattractive, and a different GUT, based on the group $\mathbf{SO}(10)$ allows all known particles (plus the right-handed neutrino) to fit into a single irrep, the 16, of this group.

Some kind of Higgs mechanism is presumed to break the $\mathbf{SU}(5)$ symmetry down into the $\mathbf{SU}(3) \times \mathbf{SU}(2) \times \mathbf{U}(1)$ symmetry of the Standard Model (with the $\mathbf{SU}(2) \times \mathbf{U}(1)$ in turn broken by the Higgs mechanism we looked at in chapters 22 and 23). But in the absence of symmetry breaking all the leptons and quarks in a given generation have the same mass, and interact identically, attracting/repelling one another under a single force, which in $\mathbf{SU}(5)$ would be governed by the exchange of what are called *leptoquark bosons*, a generalization of gluons. There are 24 of them. Twelve are really just the photon, $W^{\pm}$, Z and gluons in disguise; the rest are new bosons that change leptons into (anti)quarks and vice versa – hence the name, analogous to the way that a W boson changes an electron into its neutrino partner, or a gluon changes a blue quark into a green one. Leptoquark bosons come in two types (X with charge $+\frac{4}{3}$ and Y with charge $+\frac{1}{3}$, along with their antiparticles) and couple antiquarks to leptons and quarks to antiquarks; some examples of vertices in the $\mathbf{SU}(5)$ model appear in figure 25.7. Under symmetry breaking these leptoquark bosons become very massive, about 10^{15} GeV.

This is very heavy – but not infinitely heavy. Every now and then a down quark could emit one of these objects and change into a positron. The leptoquark boson can then collide with one of the remaining up quarks in the proton, changing it into an anti-up quark. This in turn binds with the other up quark to produce a π^0. These kinds of theories therefore predict that baryons are unstable, and that the proton can decay via the process

$$p \rightarrow e^+ + \pi^0 \tag{25.42}$$

as shown in figure 25.8, albeit very slowly since the leptoquark boson is so heavy.

The original estimates for this process put a bound on the range of lifetime of the proton [287] as $10^{31} > \tau_p > 10^{30}$ years, twenty orders of magnitude longer than the age of the universe! This sounds nigh impossible to detect, but if a sufficiently large number of protons are collected together – say about 10^{30}, roughly the amount in a swimming pool of water – then about 1 per year should decay. Since the decay channel (25.42) is so unique, there is a chance of actually observing this process.

Well, people looked hard and carefully over a series of increasingly precise searches, but didn't see any evidence for proton instability [288]. The current lower bound is [1]

$$\tau_p > 10^{33} \text{years} \tag{25.43}$$

ruling out the original $\mathbf{SU}(5)$ model and putting constraints on all GUTs.

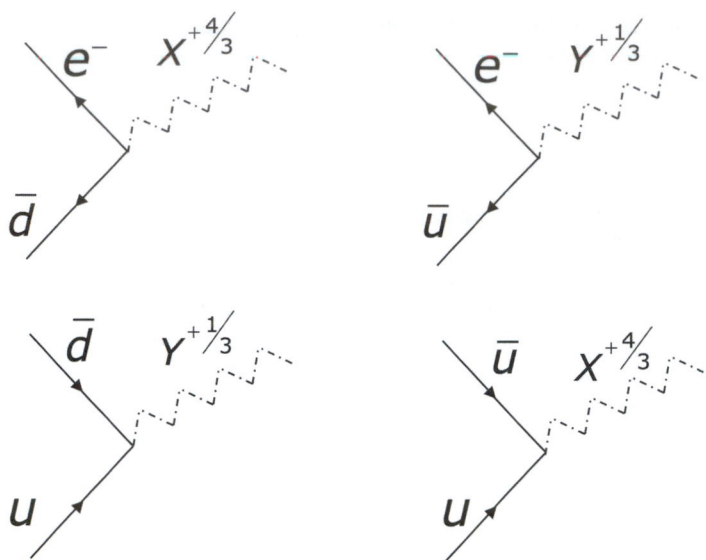

FIGURE 25.7

Examples of leptoquark boson vertices in **SU**(5). The vertex rules carry with them appropriate combinations of color to ensure color conservation.

The reluctance of the proton to decay as hoped for put a damper on GUT models. To make matters worse, the three coupling constant were also found to not actually coalesce at a single value of the energy. This was due to precision measurements at LEP. Prior to these measurements there was enough uncertainity in the values of the constants that within limits of error they all had the same value at an energy of 6×10^{14} GeV (which led to the prediction in eq. (25.42) for the proton lifetime). However, it is now known that the three constants do *not* meet at a single point in the original **SU**(5) model [289] – see figure 25.10.

So if there is to be a Grand Unified Theory, it must include some kind of new physics beyond the leptoquark bosons that must appear in such a scheme. Either there must be new (unstable) particles with masses in between the GUT-scale of 10^{16} GeV and the electroweak scale of 100 GeV, or new gravity effects, or new energy scales less than the GUT scale at which the GUT symmetry (whatever it is) is broken.

25.5.4 Supersymmetry and Superstrings

A vexing problem of the Standard Model (known as the hierarchy problem) is connected with the Higgs particle. The mass of any subatomic particle will

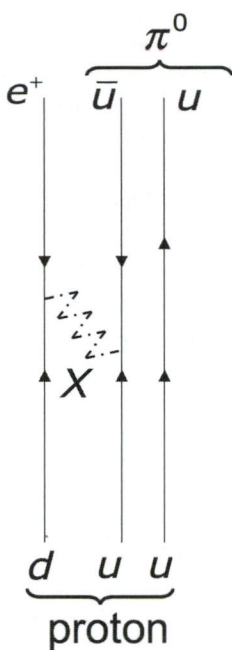

FIGURE 25.8

A possible means for the proton decay process in eq. (25.42) to take place in **SU**(5).

be corrected by loops that consist of the fermions and gauge bosons from the Standard Model that couple to that particle. These corrections are generally under control for pretty much all of the particles in the Standard Model except for the Higgs particle, where the loop corrections to the mass tend to induce corrections that are about 10^{16} GeV in the context of any Grand Unified Theory. It's possible to adjust parameters to get this value to be between 100 and 250 GeV, but only by a very delicate cancellation to 13 decimal place precision, and this is just at one loop order. A two-loop correction to the Higgs mass would entail cancellation to 26 decimal place precision. What would cause such cancellations?

One possible explanation – favored by a very large number of theorists – is that supersymmetry is responsible [290]. Recall that fermion loops contribute a minus sign whenever they appear. A loop with bosons contributes positively. This suggests that we could impose a new symmetry on a given theory, one in which every fermion loop is cancelled by a corresponding boson loop. This will only work if the boson in the loop has exactly the same properties as the fermion except for spin. A supersymmetric theory is one in which each fermion has a corresponding boson partner with identical properties (mass, charge, etc.) except for spin. It is possible to formally transform the fermion wavefunction into its boson partner and vice versa – such a transformation is called a supersymmetry transformation. From the perspective of supersymmetry, these fermion and boson particles are really just two different states of one "superparticle," analogous to the way that the electron and its neutrino are different states of one weak doublet.

Superstring theories go a step further [11]. In such theories particles are regarded not as pointlike, but as line-like, or as a string. The string can be open – like a shoelace – or closed, like a rubber band, as illustrated in figure 25.9. The idea is that the most fundamental elementary particle in nature is a string, and the known particles we see – the quarks, the leptons, the photon, the weak bosons, the gluon, even the graviton – are different kinds of vibrations of this fundamental object. It can be shown that these kinds of theories predict states of every possible allowed spin. One of these is a spin 3/2 particle, referred to as the gravitino since it transforms with the graviton under supersymmetry. Many theorists view the string approach as attractive because it incorporates gravity into the theoretical scheme that describes the interactions of all of the other particles. String theory is presumed to be

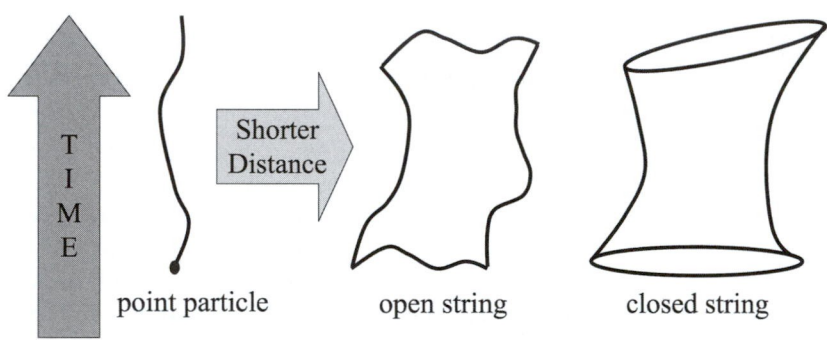

FIGURE 25.9

In string theory, point particles are large-distance approximations to strings.

valid at very high energies, energies at which quantum gravity is expected to become important. This is higher than the GUT scale, so high that the the Compton wavelength of a particle of mass M becomes comparable to its Schwarzschild radius $2GM/c^2$, the radius of a black hole of the same mass. This energy scale is called the Planck scale, the scale at which all particle interactions are effectively indistinguishable from one another, and at which perhaps even our standard concepts of time and space break down. Some kind of symmetry breaking mechanism is presumed to take place that reduces the theory to general relativity and a Grand Unified Theory of some kind at lower energies.

It is clear that none of the known particles in the Standard Model can be the superpartners of any other particle since their properties are all so very different. Hence if supersymmetry is valid there must be at least twice as many particles as observed, since each known particle will have a superpartner. For example the electron will have a partner called the "selectron" that will have the same charge and mass as the electron but zero spin. This symmetry will ensure that the corrections to the Higgs mass are zero, since every loop contributing positively from a boson will be cancelled by one from its fermion superpartner.

So where are these superpartners? None have yet been observed, and it is easy to see why. If a selectron really existed with the above properties it would bind to nuclei to form atoms. Since the selectron is a boson, the Pauli principle is not operative – many selectrons would Bose-condense in atoms, causing normal matter to implode. Supersymmetry therefore must be a broken symmetry if it is to play a role in the description of particle physics. A symmetry-breaking mechanism must be in place that ensures all unobserved superpartners gain masses that are too large to have been detected so far. This effect will spoil the perfect loop cancellations that correct the Higgs mass, but if the symmetry breaking scale is small enough (say about 1 TeV) then the imbalance between the boson and fermion loops can be kept small enough to ensure the corrections to the mass of the Higgs are not too large.

This leads to a very attractive feature of supersymmetry – it restores the unification of the forces at high energy! Provided supersymmetry is broken at about 1 TeV, all three coupling constants become the same at GUT scale energies. The actual calculations from the running coupling constants in the minimal supersymmetric extension of the Standard Model give [289]

$$M_{\text{SUSY}} = 10^{3.4 \pm 1.3} \text{GeV}$$
$$M_{\text{GUT}} = 10^{15.8 \pm 0.4} \text{GeV} \tag{25.44}$$
$$\alpha_{\text{GUT}}^{-1} = 26.3 \pm 2.9$$

in order to ensure unification, as illustrated in figure 25.10. The coupling constants are

$$\alpha_1 = \frac{5}{3} \frac{e^2}{4\pi\hbar c} \qquad \alpha_2 = \alpha_W = \frac{(g_W)^2}{4\pi\hbar c} \qquad \alpha_3 = \alpha_s = \frac{g_s^2}{4\pi\hbar c}$$

where the factor of $5/3$ in the electromagnetic case is a convention chosen to normalize its generator relative to those of the strong and electroweak gauge groups.

The numbers in eq. (25.45) are tantalizing, given that the LHC will explore the TeV energy scale. Yet the mechanism for supersymmetry breaking is an outstanding problem. The simplest supersymmetric generalization to

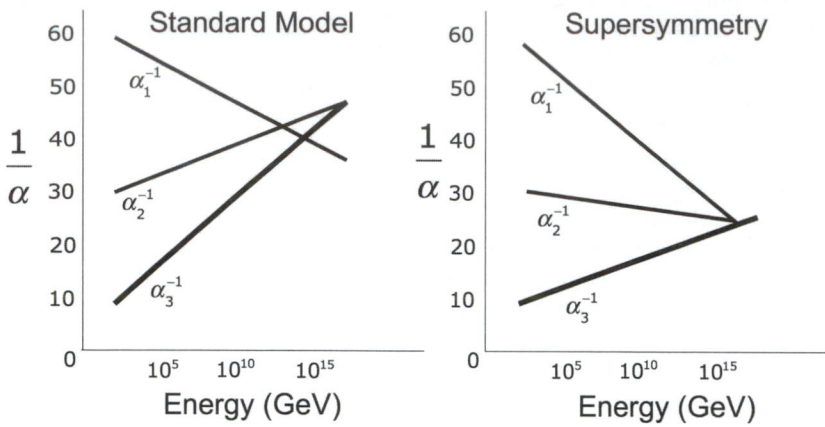

FIGURE 25.10

Running of the electromagnetic, weak, and strong coupling constants in the Standard Model (left), and in its minimal Supersymmetric (SUSY) extension (right). The thickness of the lines is indicative of the errors.

the Standard Model has over 125 adjustable parameters, and there is neither a compelling approach to breaking supersymmetry nor a clear way to predict these parameters. However, all supersymmetric theories have the feature that the superpartners cannot all decay into known particles. There must be a lightest supersymmetric particle (LSP) whose presence has so far gone undetected. This particle could perhaps be a dark matter WIMP referred to above.

One of the major hopes of the LHC is to find some of these superpartners. If none are seen, new bounds on supersymmetry breaking and its role in physics will be established. Conversely, if definitive evidence for superpartners is obtained then our picture of particle physics will be revolutionized.

25.6 Summing Up

There are many new ideas for extensions to the Standard Model that are being explored in various laboratories around the world, and I have mentioned only a few of the most popular ideas here. Keep in mind that there are lots of other ideas out there for both theoretical generalizations and experimental study, and there is a need to generate more of each. Perhaps you will someday contribute in the quest to find the fundamental theory of physics.

25.7 Questions

1. (a) Consider neutrino oscillations in the full 3-flavor case. Show that the probability for a neutrino flavor ν_α to change into a flavor ν_β in time t is given by

$$P\left(\nu_\alpha \to \nu_\beta\right) = \left| \sum_I V_{\alpha I} V_{\beta I}^* \exp\left(-i\frac{m_I^2 t}{2E}\right) \right|$$

using the same approximations as for the 2-flavor case, where E is energy of the neutrino.

(b) Suppose now that we assume $\Delta m_{21}^2 \ll \Delta m_{23}^2 \simeq \Delta m_{31}^2$. Show that

$$P\left(\nu_e \to \nu_\mu\right) = \sin^2 \vartheta_{23} \sin^2 2\vartheta_{13} \sin^2 \left(\frac{\Delta m_{23}^2 t}{2E}\right)$$

$$P\left(\nu_e \to \nu_\tau\right) = \cos^2 \vartheta_{23} \sin^2 2\vartheta_{13} \sin^2 \left(\frac{\Delta m_{23}^2 t}{2E}\right)$$

$$P\left(\nu_\mu \to \nu_\tau\right) = \cos^2 \vartheta_{13} \sin^2 2\vartheta_{13} \sin^2 \left(\frac{\Delta m_{23}^2 t}{2E}\right)$$

where the ϑ_{ij} are the parameters of the MNS matrix.

2. Assuming ϑ_{13} is small, find $P\left(\nu_e \to \nu_e\right)$.

3. Consider writing the CKM matrix in the same form as the MNS matrix:

$$\begin{pmatrix} V_{ud} & V_{us} & V_{ub} \\ V_{cd} & V_{cs} & V_{cb} \\ V_{td} & V_{ts} & V_{tb} \end{pmatrix}$$

$$= \begin{pmatrix} c_{12}c_{13} & -s_{12}c_{13} & s_{13}e^{-i\delta} \\ -s_{12}c_{23} - c_{12}s_{23}s_{13}e^{i\delta} & c_{12}c_{23} - s_{12}s_{23}s_{13}e^{i\delta} & s_{23}c_{13} \\ s_{12}s_{23} - c_{12}c_{23}s_{13}e^{i\delta} & -c_{12}s_{23} - s_{12}c_{23}s_{13}e^{i\delta} & c_{23}c_{13} \end{pmatrix}$$

as in eq. (21.55).

Off-diagonal elements of the CKM matrix become increasingly smaller as the generation number increases. Suppose we define

$$s_{12} = \lambda \qquad s_{23} = A\lambda^2 \qquad s_{13}e^{-i\delta} = A\lambda^3 \left(\rho - i\eta\right)$$

and consider λ to be a small parameter.

(a) Find the approximate form of the CKM matrix, retaining in each entry the relevant power of λ.

(b) Why isn't this approximation valid for the MNS matrix?

4. A "large" gauge transformation U acts on the vacuum $|n\rangle$ as follows

$$U\,|n\rangle = |n+1\rangle$$

Show that the $|\theta\rangle$-vacuum is eigenstate of U.

5. Show that the $|\theta\rangle$-vacuum is unique. In other words, show that the amplitude for a transition from $|\theta\rangle$ to $|\theta'\rangle$ is zero unless $\theta = \theta'$, where H is the Hamiltonian of the system.

6. Find two other diagrams that can yield the decay process

$$p \rightarrow e^+ + \pi^0$$

in $\mathbf{SU}(5)$.

7. Find the energy at which the Compton wavelength of a particle of mass M is equal to its Schwarzschild radius. How much larger is it than the GUT scale? How large is it in kilograms? Convert it to units of time (called the Planck time) and distance (called the Planck length).

A

Notation and Conventions

A.1 Natural Units

The fundamental constants of nature that appear in all studies of particle physics are Planck's constant $\hbar$ and the speed of light c. These have values

$$\hbar = 1.05457266 \times 10^{-34} \text{ Js} \tag{A.1}$$

$$c = 2.99792458 \times 10^8 \text{ ms}^{-1} \tag{A.2}$$

which are rather cumbersome to use. It is much more convenient to work in *natural units*[*] where $\hbar = \frac{h}{2\pi} = 1$ and $c = 1$. This also allows one to set the permittivity of free space, $\varepsilon_0 = 1$, provided all charges are rescaled in units of $(\hbar c)^{-1/2}$. I will typically adopt these conventions, except on occasions where it is useful to illustrate the explicit units. This will typically be when I display a result that can be directly compared to experiment (such as a decay rate or a cross-section), in which case the factors of $\hbar$ and c are useful.

The easiest way to work with this in particle physics is to (a) express all velocities as a fraction of the speed of light and all times in terms of the light-travel distance (b) convert distances into units of inverse energy (or vice versa) as appropriate, using the conversion factor $\hbar c = 197$ MeV-fm, and (c) express charges, masses and momenta in units of energy.

So we use the following prescriptions:

Physical Quantity	Notation	Units	Natural$\rightarrow$ Physical
velocity	$\vec{\beta}$	unitless	$\vec{\beta} \rightarrow \vec{v}/c$
time	t	$\hbar/\text{MeV}$	$t \rightarrow t/\hbar$
length	d	$\hbar c/\text{MeV}$	$d \rightarrow d/\hbar c$
mass	m	MeV/c^2	$m \rightarrow mc^2$
momentum	$\vec{p}$	MeV/c	$\vec{p} \rightarrow \vec{p}c$
charge	q	unitless	$q \rightarrow q/\sqrt{\hbar c}$
energy	E	MeV	

---

[*]The reason why they are called natural units is because their definition is a consequence of the properties of nature rather than any human construct. It is common to include Newton's constant of gravity G, setting $G = 1$ in which case natural units are often called Planck units.

So in other words, if we see a given expression that depends on mass, time, momentum and energy, to convert it to standard units we apply the conversion factors in the right-hand column. I have given a set of examples in table A.1.

Units of charge follow differing conventions. Particle physicists like to use Heaviside-Lorentz units, and I shall adopt these here. This means that the fine-structure constant is

$$\alpha = \frac{e^2}{4\pi\hbar c}$$

though I will typically use natural units and set $\hbar = c = 1$.

Coulomb's Law	System	Units
$F = \frac{q_1 q_2}{4\pi\epsilon_0 r^2}$	*SI*	Coulombs
$F = \frac{q_1 q_2}{r^2}$	*Gaussian*	Electrostatic units (esu)
$F = \frac{q_1 q_2}{4\pi r^2}$	*Heaviside-Lorentz*	$\frac{1}{\sqrt{4\pi}}$ Electrostatic units (esu)

A.2 Relativistic Notation

A 4-vector a_μ will generally be noted in component form as $a_\mu = (a_0, \vec{a})$, where $\vec{a}$ is a 3-vector. Repeated indices are almost always summed over all of their values – when this is not the case it will be explicitly indicated. So for example

$$A_\mu \cdot B^\mu \equiv \sum_{\mu=0}^{3} \left(A_\mu \cdot B^\mu \right) = A_0 B^0 + A_1 B^1 + A_2 B^2 + A_3 B^3$$

The reason for this convention is that the cumbersome summation symbols can be suppressed in writing expressions and doing calculations. I will use Greek letters $(\alpha, \beta, \gamma, \ldots)$ to denote the components of 4-vectors (and 4-tensors), and Latin letters $(a, b, c, \ldots)$ to denote components of 3-vectors.

A.2.1 Metric

A very important quantity is the **metric** $g_{\mu\nu}$, which has the associated line element (or interval)

$$ds^2 = g_{\mu\nu} dx^\mu dx^\nu = \left(dx^0 \right)^2 - d\vec{x} \cdot d\vec{x} = \left(dx^0 \right)^2 - \left(dx^1 \right)^2 - \left(dx^2 \right)^2 - \left(dx^3 \right)^2$$

TABLE A.1
Units Conversion

Quantity	Particle Physics Expression	Standard Expression
Rest mass	m	mc^2
Compton Wavelength	$\lambda = \frac{1}{m}$	$\frac{\lambda}{\hbar c} = \frac{1}{mc^2} \Rightarrow \lambda = \frac{\hbar}{mc}$
Time Dilation Factor	$\gamma = \frac{1}{\sqrt{1-\beta^2}}$	$\gamma = \frac{1}{\sqrt{1-\left(\frac{v}{c}\right)^2}}$
Relativistic Momentum	$\vec{p} = m\gamma\vec{\beta}$	$\vec{p}c = m\gamma c^2 \frac{\vec{v}}{c} \Rightarrow \vec{p} = \frac{m\vec{v}}{\sqrt{1-\left(\frac{v}{c}\right)^2}}$
Momentum Operator	$\vec{P} = -i\vec{\nabla}$	$\vec{P}c = \left(-i\hbar c\vec{\nabla}\right) \Rightarrow \vec{P} = -i\hbar\vec{\nabla}$
Angular Momentum Operator	$\vec{L} = -i\vec{r}\times\vec{\nabla}$	$\vec{L}c = -i\vec{r}\times\left(\hbar c\vec{\nabla}\right) \Rightarrow \vec{L} = -i\hbar\left(\vec{r}\times\vec{\nabla}\right)$
Coulomb Energy	$E_c = \frac{q_1 q_2}{r}$	$E_c = \frac{(q_1/\sqrt{\hbar c})(q_2/\sqrt{\hbar c})}{r/\hbar c} \Rightarrow E_c = \frac{q_1 q_2}{r}$

where in matrix form the metric is

$$g_{\mu\nu} = \begin{pmatrix} +1 & & & \\ & -1 & & \\ & & -1 & \\ & & & -1 \end{pmatrix}$$

and its inverse is

$$g^{\mu\nu} = \begin{pmatrix} +1 & & & \\ & -1 & & \\ & & -1 & \\ & & & -1 \end{pmatrix}$$

which means that

$$g_{\mu\nu}g^{\mu\sigma} = g_{\nu\mu}g^{\sigma\mu} = \delta_\nu^\sigma = \begin{pmatrix} 1 & & & \\ & 1 & & \\ & & 1 & \\ & & & 1 \end{pmatrix}$$

and δ_ν^σ is the relativistic Kronecker-delta function. From this perspective, the metric generalizes the notion of the 3-dimensional Kronecker-delta function δ_{ij} to special relativity. It is the object that tells us how to measure distances and time intervals, and takes the dot-product between vectors. For any 4-vector we define

$$a_\mu \ : \text{covariant} \qquad\qquad a^\mu : \text{contravariant} \qquad (A.3)$$
$$a^\mu = g^{\mu\nu}a_\nu = (a_0, -\overrightarrow{a}) \iff a_\mu = (a_0, \overrightarrow{a}) = g_{\mu\nu}a^\nu \qquad (A.4)$$

and the relativistic version of the dot-product is

$$a \cdot b = g_{\mu\nu}a^\mu b^\nu = a^\mu b_\mu = a_\mu b^\mu = g^{\mu\nu}a_\mu b_\nu = a_0 b_0 - \overrightarrow{a} \cdot \overrightarrow{b}$$

In general the magnitude of a 4-vector is the dot-product of the vector with itself

$$(\text{4-vector})^2 = (\text{timepart})^2 - (\text{3-vector})^2 \qquad (A.5)$$
$$a^2 = a^\mu a_\mu = a_0^2 - |\overrightarrow{a}|^2 \qquad (A.6)$$

The negative sign means that a given 4-vector can be one of three possible types:

Spacelike	$a^2 < 0$
Timelike	$a^2 > 0$
Null	$a^2 = 0$

A.2.2 Momentum and Energy

The most important 4-vector we will be using is the 4-Momentum $p_\mu = (E, \overrightarrow{p})$, which has the following property

$$p \cdot p \equiv g^{\mu\nu} p_\mu\, p_\nu = E^2 - \overrightarrow{p} \cdot \overrightarrow{p} = m^2 \quad \text{(where m is a constant)} \quad \text{(A.7)}$$

$$\text{or} \qquad E^2 = |\overrightarrow{p}|^2 + m^2 \quad \text{(or, converting to standard} \qquad \text{(A.8)}$$

$$\text{notation, } E^2 = |\overrightarrow{p}|^2\, c^2 + (mc^2)^2 \text{)}$$

where we interpret $E = p_0$ as the energy of a particle with 4-momentum p_μ and mass m. Depending on the physical situation, a 4-momentum in a given process can have any one of the possible norms:

Spacelike	$p^2 < 0$	(e.g. for a mediator in a scattering process)
Timelike	$p^2 > 0$	(e.g. (rest mass)^2 of a physical particle)
Null	$p^2 = 0$	(e.g. a physical photon)

A.2.3 Lorentz Transformations

A *Lorentz-transformation* is

$$a'_\mu = \Lambda_\mu{}^\nu a_\nu \qquad a'^\mu = \Lambda^\mu{}_\nu a^\nu \qquad \text{(A.9)}$$

$$a'_\mu \cdot b'{}^\mu = a_\mu \cdot b^\mu \iff g^{\alpha\beta}\Lambda_\alpha{}^\mu\Lambda_\beta{}^\nu = g^{\mu\nu} \iff \Lambda^\sigma{}_\mu\Lambda_\sigma{}^\nu = \delta_\mu^\nu = \Lambda_\sigma{}^\nu\Lambda^\sigma{}_\mu$$

The matrix Λ performs rotations and boosts:

(1) Boosts ($\gamma = 1/\sqrt{1-\beta^2}$)

along x-axis: $\quad \Lambda^\mu{}_\nu = \begin{pmatrix} \gamma & -\beta\gamma & 0 & 0 \\ -\beta\gamma & \gamma & 0 & 0 \\ 0 & 0 & 1 & 0 \\ 0 & 0 & 0 & 1 \end{pmatrix} \implies \begin{cases} E' = \gamma(E - \beta p_x) \\ p'_x = \gamma(p_x - \beta E) \\ p'_y = p_y \\ p'_z = p_z \end{cases}$

along y-axis: $\quad \Lambda^\mu{}_\nu = \begin{pmatrix} \gamma & 0 & 0 & 0 \\ 0 & 1 & -\beta\gamma & 0 \\ 0 & -\beta\gamma & \gamma & 0 \\ 0 & 0 & 0 & 1 \end{pmatrix} \implies \begin{cases} E' = \gamma(E - \beta p_z) \\ p'_x = p_x \\ p'_y = \gamma(p_y - \beta E) \\ p'_y = p_y \end{cases}$

along z-axis: $\quad \Lambda^\mu{}_\nu = \begin{pmatrix} \gamma & 0 & 0 & -\beta\gamma \\ 0 & 1 & 0 & 0 \\ 0 & 0 & 1 & 0 \\ -\beta\gamma & 0 & 0 & \gamma \end{pmatrix} \implies \begin{cases} E' = \gamma(E - \beta p_z) \\ p'_x = p_x \\ p'_y = p_y \\ p'_z = \gamma(p_z - \beta E) \end{cases}$

(2) Rotations

about x-axis: $\quad \Lambda^\mu{}_\nu = \begin{pmatrix} 1 & 0 & 0 & 0 \\ 0 & 1 & 0 & 0 \\ 0 & 0 & \cos\theta & \sin\theta \\ 0 & 0 & -\sin\theta & \cos\theta \end{pmatrix} \implies \begin{cases} E' = E \\ p'_x = p_x \\ p'_y = \cos\theta\, p_y + \sin\theta\, p_z \\ p'_z = -\sin\theta\, p_y + \cos\theta\, p_z \end{cases}$

about y-axis: $\Lambda^{\mu}{}_{\nu} = \begin{pmatrix} 1 & 0 & 0 & 0 \\ 0 & \cos\theta & 0 & -\sin\theta \\ 0 & 0 & 1 & 0 \\ 0 & \sin\theta & 0 & \cos\theta \end{pmatrix} \implies \begin{cases} E' = E \\ p'_x = \cos\theta\, p_x - \sin\theta\, p_z \\ p'_y = p_y \\ p'_z = \sin\theta\, p_x + \cos\theta\, p_z \end{cases}$

about z-axis: $\Lambda^{\mu}{}_{\nu} = \begin{pmatrix} 1 & 0 & 0 & 0 \\ 0 & \cos\theta & \sin\theta & 0 \\ 0 & -\sin\theta & \cos\theta & 0 \\ 0 & 0 & 0 & 1 \end{pmatrix} \implies \begin{cases} E' = E \\ p'_x = \cos\theta\, p_x + \sin\theta\, p_y \\ p'_y = -\sin\theta\, p_x + \cos\theta\, p_y \\ p'_z = p_z \end{cases}$

From the above, we see that Λ is the generalization of a rotation matrix to special relativity. It obeys the following properties

(a) $g_{\alpha\beta}\Lambda^{\alpha}{}_{\mu}\Lambda^{\beta}{}_{\nu} = g_{\mu\nu}$
(b) $|\det\Lambda| = 1$
(c) $(\Lambda_1)^{\alpha}{}_{\mu}\,(\Lambda_2)^{\mu}{}_{\beta} = (\Lambda_{12})^{\alpha}{}_{\beta}$ (2 Lorentz-tmfs $\rightarrow$ Lorentz tmf)
(d) $\left|(\Lambda)^{0}{}_{0}\right| \geq 1$

A.3   Greek Alphabet

TABLE A.2
Greek Alphabet

alpha	α	A	nu	ν	N
beta	β	B	omicron	o	O
gamma	γ	Γ	xi	ξ	X
delta	δ	Δ	pi	π	Π
epsilon	ε, ϵ	E	rho	ρ, ϱ	P
zeta	ζ	Z	sigma	σ, ς	Σ
eta	η	H	tau	τ	T
theta	θ, ϑ	Θ	upsilon	υ	Υ
iota	ι	I	phi	ϕ, φ	Φ
kappa	$\kappa, \varkappa$	K	chi	χ	Ξ
lambda	λ	Λ	psi	ψ	Ψ
mu	μ	M	omega	ω, ϖ	Ω

B

Kronecker Delta and Levi-Civita Symbols

B.1 Kronecker Delta

The identity matrix in any N-dimensional space is denoted in index form as δ_I^J, where $I, J, = 1, \ldots, N$. It is defined as

$$\delta_I^J = \begin{cases} 0 & I \neq J \\ 1 & I = J \end{cases} = \mathrm{diag}(1, 1, 1, \ldots, 1)$$

or in other words as a matrix with 1's along the diagonal and zeroes everywhere else. The trace of an $N \times N$ matrix A_{IJ} is

$$\mathrm{Tr}[A] = A_{IJ}\delta^{IJ} = A^{IJ}\delta_{IJ} = A_I^J\delta_J^I$$

The indices on δ_I^J can be raised or lowered without penalty except in the case of relativistic spacetime, where the metric raises and lowers indices. In general, the indices in the Kronecker delta function take on the values and notation of the indices in the space under consideration.

For example, in spacetime the indices of the Kronecker-delta function would be Greek indices running over the values $0, 1, 2, 3$, and we have

$$\delta_\nu^\sigma = g_{\mu\nu}g^{\mu\sigma} = g_{\nu\mu}g^{\sigma\mu} = \begin{pmatrix} 1 & 0 & 0 & 0 \\ 0 & 1 & 0 & 0 \\ 0 & 0 & 1 & 0 \\ 0 & 0 & 0 & 1 \end{pmatrix}$$

Here only δ_ν^σ has meaning; the quantities $\delta^{\nu\sigma}$ and $\delta_{\nu\sigma}$ are meaningless.

In three-dimensional space the Kronecker-delta function is denoted δ_{ij} where

$$\delta_{ij} = \delta^{ij} = \delta_j^i = \begin{pmatrix} 1 & 0 & 0 \\ 0 & 1 & 0 \\ 0 & 0 & 1 \end{pmatrix}$$

and here the indices can be raised lowered as is convenient.

In the two-dimensional space of Pauli matrices

$$\delta_{ab} = \delta^{ab} = \delta_b^a = \begin{pmatrix} 1 & 0 \\ 0 & 1 \end{pmatrix}$$

where a, b have the values $1, 2$.

In the four-dimensional space of Dirac matrices

$$\delta_{ab} = \delta^{ab} = \delta^a_b = \begin{pmatrix} 1 & 0 & 0 & 0 \\ 0 & 1 & 0 & 0 \\ 0 & 0 & 1 & 0 \\ 0 & 0 & 0 & 1 \end{pmatrix}$$

where a, b have the values $1, 2, 3, 4$. The range and labeling of the indices is context-dependent.

B.2 Levi-Civita Symbol

The N-dimensional Levi-Civita symbol (or epsilon-tensor) is denoted $\varepsilon_{J_1 \cdots J_N}$, where $J_1, J_2, \ldots J_N = 1, \ldots, N$. It has as many indices as the dimension of the space in which it is defined, and is defined by

$$\varepsilon_{J_1 \cdots J_N} = \begin{cases} +1 & \text{if } J_1, \ldots, J_N \text{ is an even permutation of } 1, 2, 3, \ldots, N \\ -1 & \text{if } J_1, \ldots, J_N \text{ is an odd permutation of } 1, 2, 3, \ldots, N \\ 0 & \text{otherwise (when any two indices are equal)} \end{cases}$$

or in other words the $\varepsilon_{J_1 \cdots J_N}$ symbol vanishes unless all indices are different and is otherwise ± 1 depending on the permutation of the indices. The determinant of an $N \times N$ matrix A_{IJ} is

$$\det[A] = A_{I_1 J_1} A_{I_2 J_2} \cdots A_{I_N J_N} \varepsilon^{I_1 \cdots I_N} \varepsilon^{J_1 \cdots J_N} \tag{B.1}$$

$$= A^{I_1 J_1} A^{I_2 J_2} \cdots A^{I_N J_N} \varepsilon_{I_1 \cdots I_N} \varepsilon_{J_1 \cdots J_N} \tag{B.2}$$

$$= A^{J_1}_{I_1} A^{J_2}_{I_2} \cdots A^{J_N}_{I_N} \varepsilon^{I_1 \cdots I_N} \varepsilon_{J_1 \cdots J_N} \tag{B.3}$$

$$= A_{11} A_{22} A_{33} \cdots A_{NN} - A_{12} A_{21} A_{33} \cdots A_{NN} + \cdots \tag{B.4}$$

The ordering of the indices of $\varepsilon_{J_1 \cdots J_N}$ matters but (as with the Kronecker delta) they can be raised or lowered without penalty except in the case where the space under consideration is spacetime, in which case the metric raises/lowers indices. In general, the indices in the Levi-Civita symbol take on the values and notation of the indices in the space under consideration.

For example, in spacetime the indices of the Levi-Civita symbol would be Greek indices running over the values $0, 1, 2, 3$, and we have

$$\varepsilon^{\mu\nu\alpha\beta} = g^{\mu\rho} g^{\nu\lambda} g^{\alpha\sigma} g^{\beta\tau} \varepsilon_{\rho\lambda\sigma\tau} \tag{B.5}$$

$$\varepsilon^{\mu\nu\alpha\beta} = \begin{cases} -1 & \text{(if } \mu\nu\alpha\beta \text{ is an even permutation of } 0123) \\ +1 & \text{(if } \mu\nu\alpha\beta \text{ is an odd permutation of } 0123) \end{cases} \tag{B.6}$$

$$\varepsilon^{0123} = -1 \qquad \varepsilon_{0123} = +1$$

$$\varepsilon_{\mu\nu\alpha\beta} = \begin{cases} +1 & \text{(if } \mu\nu\alpha\beta \text{ is an even permutation of } 0123) \\ -1 & \text{(if } \mu\nu\alpha\beta \text{ is an odd permutation of } 0123) \end{cases} \tag{B.7}$$

The following identities hold in spacetime:

$$\varepsilon^{\mu\nu\alpha\beta}\varepsilon_{\mu\rho\tau\varkappa} = -\delta^\nu_\rho\delta^\alpha_\tau\delta^\beta_\varkappa - \delta^\alpha_\rho\delta^\beta_\tau\delta^\nu_\varkappa - \delta^\beta_\rho\delta^\nu_\tau\delta^\alpha_\varkappa$$
$$+\delta^\nu_\rho\delta^\beta_\tau\delta^\alpha_\varkappa + \delta^\beta_\rho\delta^\alpha_\tau\delta^\nu_\varkappa + \delta^\alpha_\rho\delta^\nu_\tau\delta^\beta_\varkappa \qquad (B.8)$$

$$\varepsilon^{\mu\nu\alpha\beta}\varepsilon_{\mu\nu\tau\varkappa} = -2\left(\delta^\alpha_\tau\delta^\beta_\varkappa - \delta^\alpha_\varkappa\delta^\beta_\tau\right) \qquad (B.9)$$

$$\varepsilon^{\mu\nu\alpha\beta}\varepsilon_{\mu\nu\alpha\varkappa} = -6\delta^\beta_\varkappa \qquad (B.10)$$

$$\varepsilon^{\mu\nu\alpha\beta}\varepsilon_{\mu\nu\alpha\beta} = -24 \qquad (B.11)$$

In three dimensions

$$\varepsilon^{ijk} = \varepsilon_{ijk} = \begin{cases} +1 & (\text{if } ijk \text{ is an even permutation of } 123) \\ -1 & (\text{if } ijk \text{ is an odd permutation of } 123) \end{cases} \qquad (B.12)$$

$$\varepsilon_{123} = \varepsilon^{123} = +1 \qquad (B.13)$$

In two dimensions

$$\varepsilon^{ij} = \varepsilon_{ij} = \begin{cases} +1 & (\text{if } ij \text{ is an even permutation of } 12) \\ -1 & (\text{if } ijk \text{ is an odd permutation of } 12) \end{cases} \qquad (B.14)$$

$$= \begin{pmatrix} 0 & 1 \\ -1 & 0 \end{pmatrix} \qquad (B.15)$$

$$\varepsilon_{12} = \varepsilon^{12} = +1 \qquad (B.16)$$

C

Dirac Delta-Functions

The Dirac δ-function is defined as

$$\delta(x) = \begin{cases} 0 & x \neq 0 \\ \int_{-\infty}^{\infty} dx \delta(x) = 1 \end{cases}$$

and is mathematically a distribution, since it is undefined at $x = 0$. It is kind of a functional generalization of the Kronecker-delta.

Any continuous function $f(x)$ obeys the rule

$$f(x)\delta(x) = f(0)\delta(x)$$

since $\delta(x)$ vanishes when $x \neq 0$. This means that in general

$$\int_{-\infty}^{\infty} dx f(x)\delta(x - a) = f(a)$$

which can easily be shown by writing $x' = x - a$ in the above integral.

More generally

$$\int_{-\infty}^{\infty} dx f(x)\delta(g(x)) = \sum_{i=1}^{n} \frac{f(x_i)}{|g'(x_i)|}$$

with $g'(x_i) = \frac{dg}{dx}\big|_{x=x_i}$, where

$$g(x_i) = 0 \qquad i = 1, \ldots n$$

are the places where the function $g(x)$ vanishes. This is straightfowardly shown by letting $y = g(x)$, so that $dx = \frac{dy}{g'(x)}$ and $y = 0$ when $x = x_i$. The range of integration flips sign if $g'(x_i) < 0$, which gives a negative value for the answer in all such cases, hence leading to the absolute-value $|g'(x_i)|$ in the denominator.

The notation $\delta^{(n)}$ refers to a product over n variables:

$$\delta^{(n)}(x) = \delta(x_1)\delta(x_2)\cdots\delta(x_n)$$

so that

$$\delta^{(4)}(p - p') = \delta((p - p')_1)\delta((p - p')_2)\delta((p - p')_3)\delta((p - p')_4)$$

The δ-function can also be understood as the derivative of the Heaviside step function $\Theta(x)$:

$$\Theta(x) = \begin{cases} 0 & x < 0 \\ 1 & x > 0 \end{cases}$$

We can see this as follows. Instead of integrating the δ-function over all values of x, integrate it up to an arbitrary value of x. This gives

$$\int_{-\infty}^{x} dy\,\delta(y) = \begin{cases} 0 & x < 0 \\ 1 & x > 0 \end{cases}$$

because clearly if $x < 0$ then the δ-function vanishes everywhere in the range of integration, and if $x > 0$ we have included the singular point $x = 0$, and so we can continue to integrate out to $x = \infty$ (since $\delta(x) = 0$ for $x > 0$), yielding a value of 1 for the integral. Consequently

$$\int_{-\infty}^{x} dy\,\delta(y) = \Theta(x) \quad \Rightarrow \quad \frac{d\Theta(x)}{dx} = \delta(x)$$

D

Pauli and Dirac Matrices

D.1 Pauli Matrices

The 3-Pauli matrices are Hermitian, unitary, traceless 2×2 matrices:

$$\{\sigma^x, \sigma^y, \sigma^z\} = \{\sigma^1, \sigma^2, \sigma^3\} = \left\{ \begin{pmatrix} 0 & 1 \\ 1 & 0 \end{pmatrix}, \begin{pmatrix} 0 & -i \\ i & 0 \end{pmatrix}, \begin{pmatrix} 1 & 0 \\ 0 & -1 \end{pmatrix} \right\}$$

We don't distinguish between upper and lower indices, so that

$$\sigma^1 = \sigma_1, \sigma^2 = \sigma_2, \sigma^3 = \sigma_3$$

We have the product rule

$$\sigma_i \sigma_j = \delta_{ij} I + i\epsilon_{ijk}\sigma_k = \delta_{ij} + i\epsilon_{ijk}\sigma_k$$

where the 2×2 unit matrix I is often suppressed, as in the second part of the expression above. This rule implies

$$
\begin{aligned}
(\sigma_1)^2 = (\sigma_2)^2 = (\sigma_3)^2 &= 1 \\
\sigma_1 \sigma_2 = i\sigma_3 \quad &\text{and cyclic} \\
[\sigma_i, \sigma_j] = 2i\epsilon_{ijk}\sigma_k \quad &\{\sigma_i, \sigma_j\} = 2\delta_{ij}
\end{aligned}
\tag{D.1}
$$

and for any two vectors $\vec{a}$ and $\vec{b}$:

$$(\vec{a} \cdot \vec{\sigma})\left(\vec{b} \cdot \vec{\sigma}\right) = \vec{a} \cdot \vec{b} + i\left(\vec{a} \times \vec{b}\right) \cdot \vec{\sigma}$$

We also have the exponential relation

$$\exp\left[\vec{\theta} \cdot \vec{\sigma}\right] = \sum_{n=0}^{\infty} \frac{\left(\vec{\theta} \cdot \vec{\sigma}\right)^n}{n!} = \cos\theta + i\widehat{\theta} \cdot \vec{\sigma} \sin\theta$$

where $\vec{\theta} = \theta\widehat{\theta}$.

D.2   Dirac Matrices

The Dirac γ-matrices are four unitary traceless 4×4 matrices defined by the relation

$$\{\gamma^\mu, \gamma^\nu\} = 2g^{\mu\nu}$$

and are most commonly represented in the form

$$\gamma^0 = \begin{pmatrix} I & 0 \\ 0 & -I \end{pmatrix} \qquad \gamma^i = \begin{pmatrix} 0 & \sigma^i \\ -\sigma^i & 0 \end{pmatrix}$$

where I is the 2×2 identity matrix and the σ^i are the Pauli matrices.

The indices of the Dirac matrices are raised/lowered using the metric

$$\gamma_\mu = g_{\mu\nu}\gamma^\nu \qquad \Rightarrow \gamma_0 = \gamma^0 \quad \gamma_i = -\gamma^i$$

For any 4-vector a^μ we have

$$\slashed{a} = a^\mu \gamma_\mu = a_\mu \gamma^\mu$$

and the "bar" of any 4×4 matrix M is

$$\bar{M} = \gamma^0 M^\dagger \gamma^0$$

From the Dirac matrices the additional matrices

$$\gamma^5 = i\gamma^0\gamma^1\gamma^2\gamma^3 = \begin{pmatrix} 0 & I \\ I & 0 \end{pmatrix} \tag{D.2}$$

$$\Sigma^{\mu\nu} = \frac{i}{4}\left([\gamma^\mu, \gamma^\nu]\right) \tag{D.3}$$

$$\Sigma^{0i} = -\frac{i}{2}\begin{pmatrix} 0 & \sigma^i \\ \sigma^i & 0 \end{pmatrix} \qquad \Sigma^{ij} = \frac{1}{2}\epsilon_{ijk}\begin{pmatrix} \sigma^k & 0 \\ 0 & \sigma^k \end{pmatrix} \tag{D.4}$$

are in common usage, γ^5 for parity transformations and helicity projections and $\Sigma^{\mu\nu}$ for Lorentz transformations. We also have

$$\{\gamma^\mu, \gamma^5\} = 0$$

D.3   Identities and Trace Theorems

The Dirac matrices obey the following useful identities

$$\left(\gamma^0\right)^\dagger = \gamma^0 \quad \left(\gamma^i\right)^\dagger = -\gamma^i \qquad \overline{\gamma}^\mu = \gamma^\mu$$

$$\left(\gamma^0\right)^2 = I \quad \left(\gamma^i\right)^2 = -I \qquad \gamma_\alpha \gamma^\alpha = 4$$

$$\gamma_\alpha \gamma^\nu \gamma^\alpha = -2\gamma^\nu \qquad \gamma_\alpha \not{a} \gamma^\alpha = -2\not{a}$$

$$\gamma_\alpha \gamma^\mu \gamma^\nu \gamma^\alpha = 4g^{\mu\nu} \qquad \gamma_\alpha \not{a} \not{b} \gamma^\alpha = 4a \cdot b$$

$$\gamma_\alpha \gamma^\mu \gamma^\nu \gamma^\lambda \gamma^\alpha = -2\gamma^\lambda \gamma^\mu \gamma^\nu \quad \gamma_\alpha \not{a} \not{b} \not{c} \gamma^\alpha = -2\not{c} \not{b} \not{a}$$

where the bar of a matrix Γ is $\overline{\Gamma} = \gamma^0 \Gamma^\dagger \gamma^0$. These can be used to prove the following trace theorems

$$\mathrm{Tr}\,[I] = 4 \qquad\qquad \mathrm{Tr}\,[\text{odd } \# \text{ of } \gamma\text{-matrices}] = 0$$

$$\mathrm{Tr}\,[\gamma^\mu \gamma^\nu] = 4g^{\mu\nu} \qquad\qquad \mathrm{Tr}\,[\not{a} \not{b}] = 4a \cdot b$$

$$\mathrm{Tr}\,[\gamma^\mu \gamma^\nu \gamma^\alpha \gamma^\beta] = 4\left(g^{\mu\nu} g^{\alpha\beta} - g^{\mu\alpha} g^{\nu\beta} + g^{\mu\beta} g^{\nu\alpha}\right)$$

$$\mathrm{Tr}\,[\not{a} \not{b} \not{c} \not{d}] = 4\left(a \cdot b \, c \cdot d - a \cdot c \, b \cdot d + a \cdot d \, b \cdot c\right)$$

$$\mathrm{Tr}\,[\gamma^5] = 0 \qquad\qquad \mathrm{Tr}\,[\gamma^5 \gamma^\mu \gamma^\nu] = 0$$

$$\mathrm{Tr}\,[\gamma^5 \gamma^\mu \gamma^\nu \gamma^\alpha \gamma^\beta] = 4i\varepsilon^{\mu\nu\alpha\beta} \quad \mathrm{Tr}\,[\gamma^5 \not{a} \not{b} \not{c} \not{d}] = 4i\varepsilon^{\mu\nu\alpha\beta} a_\mu b_\nu c_\alpha d_\beta$$

E

Cross-Sections and Decay Rates

E.1 Decays

For a single particle of 4-momentum p^μ decaying into m particles of 4-momenta $p_1'^\mu \cdots p_m'^\mu$ via the process

$$1 \rightarrow 1' + 2' + \cdots m'$$

the decay rate is

$$d\Gamma = \frac{\mathcal{S}c^2}{2\hbar E} \left[\prod_{i=1}^{m} \frac{c\left(\Delta p_i'\right)^3}{2E_i'\left(2\pi\right)^3} \right] |\langle p_1' \cdots p_m' |\mathsf{M}| p\rangle|^2 \left(2\pi\right)^4 \delta^{(4)} \left(\sum_{i=1}^{m} p_i' - p \right)$$

where E is the energy of the decaying particle. If this particle is at rest and has mass M then $E = Mc^2$. The quantity $\mathcal{S}$ is a statistical factor — it equals $1/n!$ for each group of n identical particles in the final state.

If $m = 2$ then there are only 2 particles in the final state, in which case the decay rate simplifies to

$$\Gamma_{2\text{-Body}} = \frac{\mathcal{S}|\vec{p}\,'|c}{8\pi\hbar\left(Mc\right)^2} |\mathcal{M}|^2$$

when the initially decaying particle is at rest, and where $\vec{p}\,'$ is the momentum of either of the outgoing momenta in the final state, with $\langle p_1' \cdots p_m' |\mathsf{M}| p\rangle = \mathcal{M}$

$$|\vec{p}\,'| = \frac{c\sqrt{M^4 + m_{1'}^4 + m_{2'}^4 - 2m_{1'}^2 m_{2'}^2 - 2M^2 m_{2'}^2 - 2M^2 m_{1'}^2}}{2M}$$

and $\mathcal{S} = 1/2!$ if the outgoing particles are identical.

E.2 Cross-Sections

For two particles of 4-momenta p_1^μ and p_2^μ that collide and produce m particles of 4-momenta $p_1'^\mu \cdots p_m'^\mu$ via the process

$$1 + 2 \rightarrow 1' + 2' + \cdots m'$$

the cross-section is

$$d\sigma = \frac{\hbar^2 \mathcal{S}}{4\sqrt{(p_1 \cdot p_2)^2 - p_1^2 p_2^2}} \left[\prod_{i=1}^{m} \frac{c\,(\Delta p_i')^3}{2E_i'\,(2\pi)^3} \right] |\langle p_1' \cdots p_m' |\mathsf{M}| p_1 p_2 \rangle|^2 \quad \text{(E.1)}$$

$$\times (2\pi)^4\, \delta^{(4)} \left(\sum_{i=1}^{m} p_i' - p_2 - p_1 \right) \quad \text{(E.2)}$$

where now $\langle p_1' \cdots p_m' |\mathsf{M}| p_1 p_2 \rangle = \mathcal{M}$ and as before $\mathcal{S}$ is a statistical factor that equals $1/n!$ for each group of n identical particles in the final state.

E.2.1   2-Body CMS

If $m = 2$ then there are only 2 particles in the final state, and in a reference frame where the spatial momenta of the colliding particles are equal and opposite (the CMS, or center-of-momentum system) so that $\vec{p}_1 = -\vec{p}_2$, the cross-section simplifies to

$$\left(\frac{d\sigma}{d\Omega} \right)_{\text{2-Body CMS}} = \left(\frac{\hbar c}{8\pi} \right)^2 \frac{\mathcal{S}\,|\mathcal{M}|^2}{(E_1 + E_2)^2} \frac{|\vec{p}\,'|}{|\vec{p}|}$$

where the outgoing momenta obey the relation

$$\vec{p}_1' = -\vec{p}_2' = \vec{p}\,'$$

and where

$$\sqrt{(p_1 \cdot p_2)^2 - p_1^2 p_2^2} = (E_1 + E_2)\,|\vec{p}_1|\,/c$$

E.2.2   2-Body Lab Frame

By definition the lab frame is the frame where one particle is at rest, which can be taken to be particle # 2 so that $\vec{p}_2 = 0$ and so $E_2 = M_2 c^2$.

In general, the formula is messy, even if there are only two particles in the final state. However, if the two outgoing particles are massless ($M_{1'} = M_{2'} = 0$) then the cross-section simplifies to

$$\left(\frac{d\sigma}{d\Omega} \right)_{\text{2-Body inelastic-LAB}} = \left(\frac{\hbar}{8\pi} \right)^2 \frac{\mathcal{S}\,|\mathcal{M}|^2\,|\vec{p}_1'|}{M_2\,|\vec{p}_1|\,[(E_1 + M_2 c^2) - |\vec{p}_1|\,c\cos\theta]}$$

If the particles scatter elastically, so that the outgoing particles are the same as the two particles in the initial state ($1 + 2 \longrightarrow 1 + 2$) then the differential cross section simplifies to

$$\left(\frac{d\sigma}{d\Omega} \right)_{\text{2-Body elastic-LAB}} = \left(\frac{\hbar}{8\pi} \right)^2 \frac{\mathcal{S}\,|\mathcal{M}|^2\,|\vec{p}_1'|^2}{M_2\,|\vec{p}_1|\,[|\vec{p}_1'|\,(E_1 + M_2 c^2) - |\vec{p}_1|\,E_1'\cos\theta]}$$

If $M_1 = 0$ (so that the incoming particle is massless) then this formula further simplifies to

$$\left(\frac{d\sigma}{d\Omega}\right)_{\text{2-Body elastic-LAB}} = \left(\frac{\hbar E_1'}{8\pi M_2 c E_1}\right)^2 |\mathcal{M}|^2$$

Alternatively, if $M_2 c^2 \gg E_1$ (when the target recoil can be neglected), it simplifies to

$$\left(\frac{d\sigma}{d\Omega}\right)_{\text{2-Body elastic-LAB}} = \left(\frac{\hbar}{8\pi M_2 c}\right)^2 |\mathcal{M}|^2$$

and in both cases $\mathcal{S} = 1$ because the two particles in the final state cannot be identical.

In all of these cases

$$\sqrt{(p_1 \cdot p_2)^2 - p_1^2 p_2^2} = M_2 c |\vec{p}_1|$$

F

Clebsch-Gordon Coefficients

Clebsch-Gordon tables, reproduced here from the Particle Data Book [1], contain explicit formulae for all the $\mathcal{C}_m^j$ (j_1, j_2; $m_1 m_2$) coefficients given in eq. (5.29). The total spins (j_1, j_2) being combined are given in the upper left of one of the sub-tables. The respective (m_1, m_2)-values (or z-components) of these spins are given in the lower-left boxes in a subtable, and the possible output $|j\, m\rangle$ wavefunctions are in the upper right boxes in the same sub-table. The $\mathcal{C}_m^j$'s are the square roots of the numbers in the relevant middle boxes, where the minus sign (if there is one) goes outside of the square root. Additional formulae are for lowest order spherical harmonics.

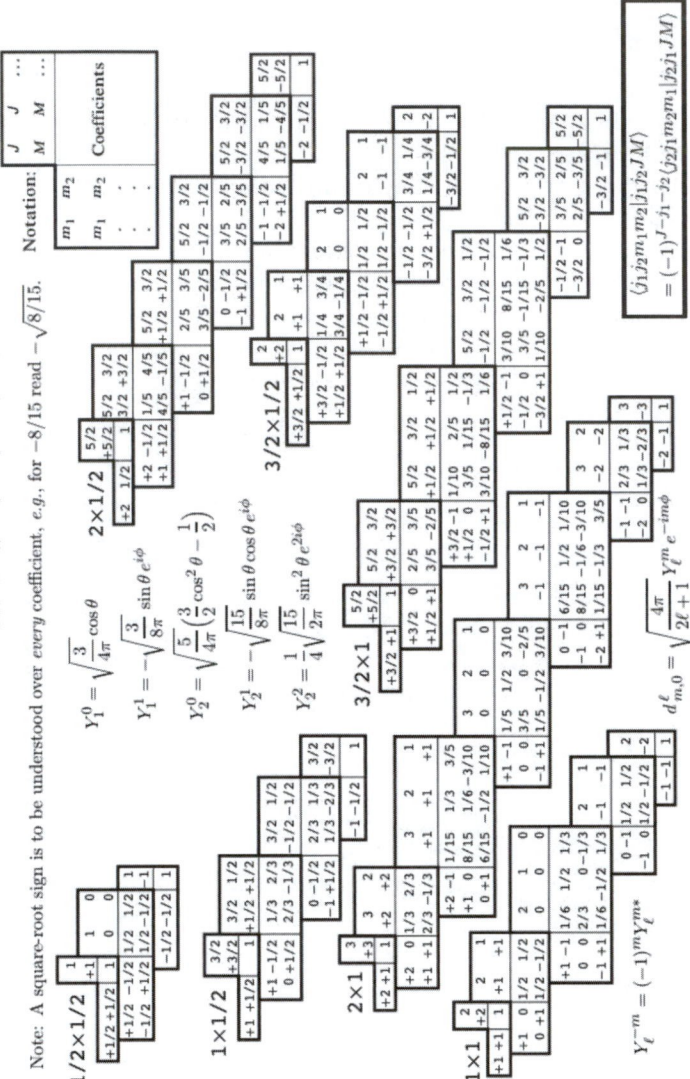

FIGURE F.1

Table of Clebsch-Gordon Coefficients

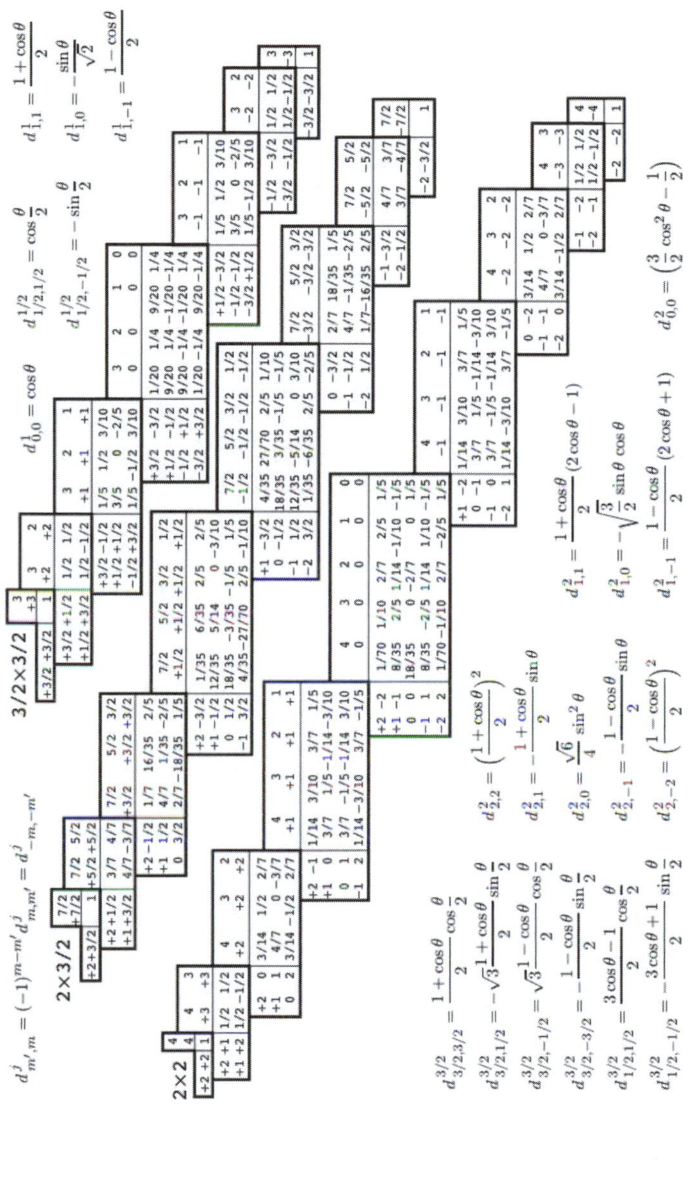

FIGURE F.2

Table of Clebsch-Gordan Coefficients (cont'd)

G

Fundamental Constants

The constants of nature are taken from CODATA (Committee on Data for Science and Technology), and current values can be found at

http://physics.nist.gov/cuu/Constants/index.html

and from the Particle Data Group, where current values are at

http://pdglive.lbl.gov/listings1.brl?exp=Y

Quantity	Symbol	Value	Units	Uncertainty
Speed of Light in Vacuum	c	$2.99792458 \times 10^8 \mathrm{ms}^{-1}$	LT^{-1}	
Planck constant	h	$6.62606896(33) \times 10^{-34} \mathrm{Js}$	$\mathrm{ML}^2\mathrm{T}^{-1}$	50 ppb
		$= 4.13566733(10) \times 10^{-15} \mathrm{eVs}$	$\mathrm{ML}^2\mathrm{T}^{-1}$	25 ppb
Reduced Planck constant	$\hbar = \frac{h}{2\pi}$	$1.054571628(53) \times 10^{-34} \mathrm{Js}$	$\mathrm{ML}^2\mathrm{T}^{-1}$	50 ppb
		$= 6.58211899(16) \times 10^{-16} \mathrm{eVs}$	$\mathrm{ML}^2\mathrm{T}^{-1}$	25 ppb
conversion constant	$\hbar c$	$197.3269631(49) \mathrm{\ MeV\ fm}$	$\mathrm{ML}^3\mathrm{T}^{-2}$	25 ppb
electron charge	e	$-1.602176487(40) \times 10^{-19} \mathrm{C}$	Q	25 ppb
electron mass	m_e	$9.10938215(45) \times 10^{-31} \mathrm{kg}$	M	50 ppb
		$0.510998910(13) \mathrm{\ MeV}$	$\mathrm{ML}^2\mathrm{T}^{-2}$	25 ppb
electron magnetic moment	g	$-9.28476377(23) \times 10^{-24} \mathrm{JT}^{-1}$	$\mathrm{ML}^2\mathrm{T}^{-3}$	25 pbb
proton mass	m_p	$1.672621637(83) \times 10^{-27} \mathrm{kg}$	M	50 ppb
		$938.272013(23)\mathrm{MeV}$	$\mathrm{ML}^2\mathrm{T}^{-2}$	25ppb
Bohr magneton	$\mu_B = \frac{eh}{2m_e}$	$5.7883817555(79) \times 10^{-5} \mathrm{eVT}^{-1}$	$\mathrm{QL}^2\mathrm{T}^{-1}$	1.4 ppb
Nuclear Magneton	$\mu_N = \frac{eh}{2m_p}$	$3.1524512326(45) \times 10^{-8} \mathrm{eVT}^{-1}$	$\mathrm{QL}^2\mathrm{T}^{-1}$	1.4 ppb
Bohr Radius	$a = \frac{4\pi\hbar^2}{m_e e^2}$	$0.52917720859(36) \times 10^{-10} \mathrm{m}$	L	0.68 ppb
Rydberg Constant	$E_R = \frac{m_e c e^4}{2\hbar}$	$13.60569193(34)\mathrm{eV}$	$\mathrm{ML}^2\mathrm{T}^{-2}$	25 ppb
Inverse Fine Structure constant	α^{-1}	$137.035999679(94)$		0.68 ppb
Fermi Constant	G_F	$1.16637(1) \times 10^{-5} \mathrm{GeV}^{-2}$	$\mathrm{M}^{-2}\mathrm{L}^{-4}\mathrm{T}^4$	8600 ppb
Weak Mixing Angle	$\sin^2\theta_W$	$0.22255(56)$		2500000 ppb
W boson mass	M_W	$80.398(25) \mathrm{\ GeV}$		3.6×10^5 ppb
Z boson mass	M_Z	$91.1876(21) \mathrm{\ GeV}$		2.3×10^4 ppb
Strong Coupling	$\alpha_s(M_Z)$	$0.1176(20)$		1.7×10^7 ppb
Avagadro Number	N_A	$6.02214179(30) \times 10^{23} \mathrm{mol}^{-1}$		50 ppb
Boltzman's constant	k_B	$1.3806504(24) \times 10^{-23} \mathrm{JK}^{-1}$	$\mathrm{ML}^2\mathrm{T}^{-2}\mathrm{K}^{-1}$	1700 ppb
		$= 8.617343(15) \times 10^{-5} \mathrm{eVK}^{-1}$	$\mathrm{ML}^2\mathrm{T}^{-2}\mathrm{K}^{-1}$	1700 ppb
Newton's constant	G	$= 6.67428(67) \times 10^{-11} \mathrm{m}^3 \mathrm{kg}^{-1}\mathrm{s}^{-2}$	$\mathrm{M}^{-1}\mathrm{L}^3\mathrm{T}^{-2}$	1.0×10^5 ppb

H

Properties of Elementary Particles

The following tables list some basic properties of the elementary particles –
the gauge bosons, leptons, and quarks – as well as of some of the properties
of the lowest-energy quark bound states (mesons and baryons).

TABLE H.1
Gauge Bosons (Spin-1)

Force	Type	Mass (MeV)	Charge	Color	Lifetime(sec)	Main Decays
Electromagnetism	photon: γ	$< 6 \times 10^{-17}$	0	0	∞	none
Strong	gluon: g	0 (assumed)	0	octet	∞	none
Weak	charged: $W^\pm$	80,398	± 1	0	3.11×10^{-25}	$\ell\nu_\ell,\, q\bar{q}$
	neutral: Z	91,187.6	0	0	2.64×10^{-25}	$\ell^+\ell^-,\, q\bar{q}$

TABLE H.2
Leptons* (Spin-1/2)

Generation	Flavor	Mass (MeV)	Charge	Lifetime (sec)	Main Decays
1st	electron: e	0.51099	-1	∞	none
	e-neutrino: ν_e	$< 2 \times 10^{-6}$	0	∞	none
2nd	muon: μ	105.659	-1	2.19703×10^{-6}	$e\,\nu_\mu\bar{\nu}_e$
	μ-neutrino: ν_μ	< 0.17	0	∞	none
3rd	tau: τ	1776.99	-1	2.906×10^{-13}	$e\,\nu_\tau\bar{\nu}_e,\ \mu\,\nu_\mu\bar{\nu}_\mu,\ \pi^-\nu_\tau$
	τ-neutrino: ν_τ	< 1.8	0	∞	none

* Neutrino masses are not associated with the particular neutrino flavors, but rather with a linear combination of them as discussed in Chapter 25.

TABLE H.3
Quarks** (Spin-1/2)

Generation	Flavor	Mass (MeV)	Charge
1st	up: u	$3-7$	$+2/3$
	down: d	$1.5-3.0$	$-1/3$
2nd	charm: c	1270	$+2/3$
	strange: s	95	$-1/3$
3rd	top: t	171,200	$+2/3$
	bottom: b	4200	$-1/3$

** Light quark masses are not precisely determined as discussed in chapter 16.

TABLE H.4
Baryons (Spin-1/2)

Symbol	Quark Content	Mass (MeV)	Charge	Lifetime (sec)	Main Decays
p	uud	938.272	1	∞	none
n	udd	939.565	0	885.7	$p\,e^-\bar{\nu}_e$
Λ	uds	1115.68	0	2.63×10^{-10}	$p\,\pi^-$, $n\,\pi^0$
Σ^+	uus	1189.37	1	8.02×10^{-11}	$p\,\pi^0$, $n\,\pi^+$
Σ^0	uds	1192.64	0	7.4×10^{-20}	$\Lambda\gamma$
Σ^-	dds	1197.45	-1	1.48×10^{-10}	$n\,\pi^-$
Ξ^0	uds	1314.8	0	2.90×10^{-10}	$\Lambda\pi^0$
Ξ^-	dds	1321.3	-1	1.64×10^{-10}	$\Lambda\pi^-$
Λ_c	udc	2286.5	1	2.00×10^{-13}	$\Lambda\pi\pi$, $\Sigma\pi\pi$, $p\,K\pi$
Σ_c	$uuc,\ udc,\ ddc$	2452	2, 1, 0	3.3×10^{-22}	$\Lambda_c\pi$
Ξ_c	$usc,\ dsc$	2469	1, 0	1.1×10^{-13}	$S = -2$ Hadrons
Ω_c	ssc	2697	0	6.9×10^{-14}	$S = -3$ Hadrons
Λ_b	udb	5624	0	1.23×10^{-12}	$\Lambda_c + \cdots$

TABLE H.5

Baryons (Spin-3/2)

Symbol	Quark Content	Mass (MeV)	Charge	Lifetime (sec)	Main Decays
Δ	$uuu,\ uud,\ udd,\ ddd$	1232	2,1,0,−1	5.6×10^{-24}	$p\ \pi,\ n\ \pi$
Σ^*	$uus,\ uds,\ dds$	1385	1, 0, −1	1.8×10^{-23}	$\Lambda\pi^0,\ \Sigma\pi$
Ξ^*	$uss,\ dss$	1533	0, −1	6.9×10^{-23}	$\Xi\pi$
Ω^-	sss	1672	1	8.2×10^{-11}	$\Lambda K^-,\ \Sigma\pi$

TABLE H.6
PseudoScalar Mesons* (Spin-0)

Symbol	Quark Content	Mass (MeV)	Charge	Lifetime (sec)	Main Decays
$\pi^\pm$	$u\bar{d}, d\bar{u}$	139.57	1, −1	2.6×10^{-8}	$\mu\nu_\mu$
π^0	$\frac{u\bar{u}-d\bar{d}}{\sqrt{2}}$	134.977	0	8.4×10^{-17}	$\gamma\gamma$
$K^\pm$	$u\bar{s}, s\bar{u}$	493.68	1, −1	1.24×10^{-8}	$\mu\nu_\mu, \pi\pi, \pi\pi\pi$
$K^0, \bar{K}^0$	$d\bar{s}, s\bar{d}$	497.65	0	$\begin{cases} K_S^0 : 8.95 \times 10^{-11} \\ K_L^0 : 5.11 \times 10^{-8} \end{cases}$	$\pi\pi$ $\pi e\nu_e, \pi\mu\nu_\mu, \pi\pi\pi$
η	$\frac{u\bar{u}+d\bar{d}-2s\bar{s}}{\sqrt{6}}$	547.51	0	5.1×10^{-19}	$\gamma\gamma, \pi\pi\pi$
η'	$\frac{u\bar{u}+d\bar{d}+s\bar{s}}{\sqrt{3}}$	957.78	0	3.2×10^{-21}	$\gamma\gamma, \pi\pi\pi$
$D^\pm$	$c\bar{d}, d\bar{c}$	1869.3	1, −1	1.04×10^{-12}	$K\pi\pi, K\mu\nu_\mu, Ke\nu_e$
$D^0, \bar{D}^0$	$c\bar{u}, u\bar{c}$	1864.5	0	4.1×10^{-13}	$K\pi\pi, K\mu\nu_\mu, Ke\nu_e$
$D_s^\pm$	$c\bar{s}, s\bar{c}$	1968.2	1, −1	5.0×10^{-13}	$\eta\rho, \phi\pi\pi, \phi\rho$
$B^\pm$	$u\bar{b}, b\bar{u}$	5279.0	1, −1	1.6×10^{-12}	$D^*\ell\nu_\ell, D\ell\nu_\ell, D^*\pi\pi\pi$
$B^0, \bar{B}^0$	$d\bar{b}, b\bar{d}$	5279.4	0	1.5×10^{-12}	$D^*\ell\nu_\ell, D\ell\nu_\ell, D^*\pi\pi$

TABLE H.7

Vector Mesons* (Spin-1)

Symbol	Quark Content	Mass (MeV)	Charge	Lifetime (sec)	Main Decays
ρ	$u\bar{d}$, $\frac{u\bar{u}-d\bar{d}}{\sqrt{2}}$, $d\bar{u}$	775.5	1, 0, −1	4×10^{-24}	$\pi\pi$
K^*	$u\bar{s}$, $d\bar{s}$, $s\bar{d}$, $s\bar{u}$	894	1, 0, 0, −1	1×10^{-23}	$K\pi$
ω	$\frac{u\bar{u}+d\bar{d}}{\sqrt{2}}$	782.6	0	8×10^{-23}	$\pi\pi\pi$, $\pi\gamma$
ϕ	$s\bar{s}$	1019.5	0	2.0×10^{-22}	$^+K^-$, $K^0\bar{K}^0$
$J\psi$	$c\bar{c}$	3097	0	7×10^{-21}	e^+e^-, $\mu^+\mu^-$, 5π, 7π
D^*	$c\bar{d}$, $c\bar{u}$, $u\bar{c}$, $d\bar{c}$,	2008	1, 0, 0, −1	3×10^{-21}	$K\pi$
Υ	$b\bar{b}$	9460	0	1×10^{-20}	e^+e^-, $\mu^+\mu^-$, $\tau^+\tau^-$

I

Feynman Rules for the Standard Model

External Lines

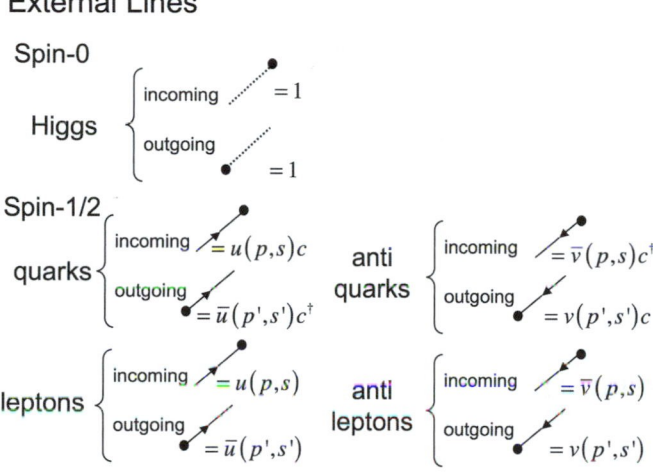

Spin-0

Higgs
- incoming $\cdots = 1$
- outgoing $= 1$

Spin-1/2

quarks
- incoming $= u(p,s)c$
- outgoing $= \bar{u}(p',s')c^\dagger$

anti quarks
- incoming $= \bar{v}(p,s)c^\dagger$
- outgoing $= v(p',s')c$

leptons
- incoming $= u(p,s)$
- outgoing $= \bar{u}(p',s')$

anti leptons
- incoming $= \bar{v}(p,s)$
- outgoing $= v(p',s')$

Spin-1

photons
- incoming μ $\quad = \varepsilon^{\mu}(p)$
- outgoing μ' $\quad = \varepsilon^{\mu'}(p')$

gluons
- incoming μ,a $\quad = \varepsilon^{\mu}(p)a^{a}$
- outgoing μ',a' $\quad = \varepsilon^{\mu'}(p')a^{a'}$

W or Z
- incoming μ,a $\quad = \varepsilon^{\mu}(p)$
- outgoing μ',a' $\quad = \varepsilon^{\mu'}(p')$

Propagators

Higgs
$$\frac{i}{q^2 - m^2} = \quad\cdots\cdots\cdots \quad q$$

fermion
$$\frac{i(\gamma \cdot q + m)}{q^2 - m^2}\delta^{ab} = \quad a \xrightarrow{\quad} b \quad q$$

photon
$$-\frac{i}{q^2}g_{\mu\nu} = \quad \mu \sim\!\!\sim\!\!\sim\!\!\sim \nu \quad q$$

gluon
$$-\frac{i}{q^2}g_{\mu\nu}\delta^{ab} = \quad \mu,a \,\,\text{\Large\textcircled{}} \,\, \nu,b \quad q$$

W or Z
$$-\frac{i\left(g_{\mu\nu} - \dfrac{q_{\mu}q_{\nu}}{M_a^2}\right)\delta^{ab}}{q^2 - M_a^2 + iM_a\Gamma_a} = \quad \mu,a \,\text{-----}\, \nu,b \quad q$$

Vertices

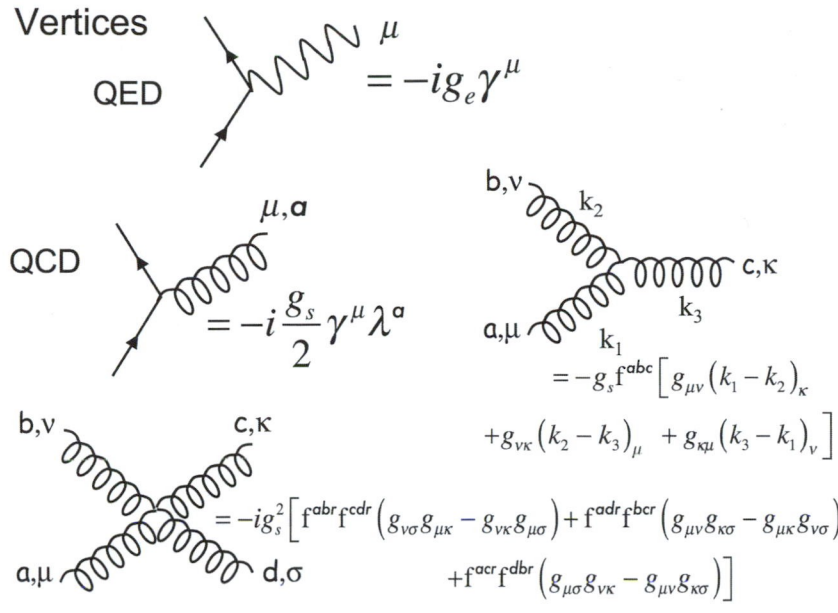

QED

$$= -ig_e\gamma^\mu$$

QCD

μ,a

$$= -i\frac{g_s}{2}\gamma^\mu\lambda^a$$

b,ν k_2 c,κ k_3 a,μ k_1

$$= -g_s f^{abc}\Big[g_{\mu\nu}\big(k_1 - k_2\big)_\kappa$$
$$+ g_{\nu\kappa}\big(k_2 - k_3\big)_\mu + g_{\kappa\mu}\big(k_3 - k_1\big)_\nu\Big]$$

b,ν c,κ a,μ d,σ

$$= -ig_s^2\Big[f^{abr}f^{cdr}\big(g_{\nu\sigma}g_{\mu\kappa} - g_{\nu\kappa}g_{\mu\sigma}\big) + f^{adr}f^{bcr}\big(g_{\mu\nu}g_{\kappa\sigma} - g_{\mu\kappa}g_{\nu\sigma}\big)$$
$$+ f^{acr}f^{dbr}\big(g_{\mu\sigma}g_{\nu\kappa} - g_{\mu\nu}g_{\kappa\sigma}\big)\Big]$$

Electroweak

ν_ℓ W^- ℓ^-

$$= -\frac{ig_w}{2\sqrt{2}}\gamma^\mu\big(1 - \gamma^5\big)$$

u_I W^- ∂_J

$$= -\frac{ig_w}{2\sqrt{2}}\gamma^\mu\big(1 - \gamma^5\big)V_{IJ}$$

$\partial_J = d, s$ or b

$u_I = u, c$ or t

V_{IJ} is the CKM matrix

f Z f

$$= -\frac{ig_z}{2\sqrt{2}}\gamma^\mu\big(c_V^f - c_A^f\gamma^5\big)$$

f	c_V	c_A
ν_ℓ	$\dfrac{1}{2}$	$\dfrac{1}{2}$
ℓ	$-\dfrac{1}{2} + 2\sin^2\theta_w$	$-\dfrac{1}{2}$
u,c,t	$\dfrac{1}{2} - \dfrac{4}{3}\sin^2\theta_w$	$\dfrac{1}{2}$
d,s,b	$-\dfrac{1}{2} + \dfrac{2}{3}\sin^2\theta_w$	$-\dfrac{1}{2}$

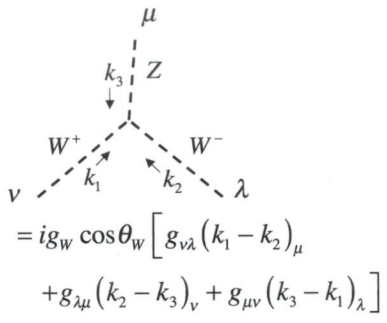

$$= ig_w \cos\theta_w \Big[g_{v\lambda} (k_1 - k_2)_\mu$$
$$+ g_{\lambda\mu} (k_2 - k_3)_v + g_{\mu v} (k_3 - k_1)_\lambda \Big]$$

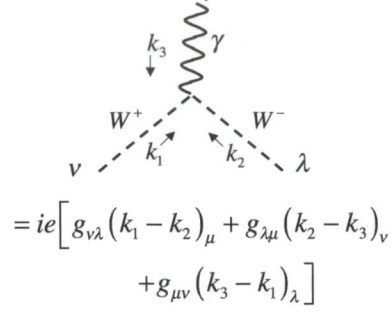

$$= ie\Big[g_{v\lambda} (k_1 - k_2)_\mu + g_{\lambda\mu} (k_2 - k_3)_v$$
$$+ g_{\mu v} (k_3 - k_1)_\lambda \Big]$$

$$= -i\frac{3g_w m_h^2}{2M_w}$$

$$= -\frac{ig_w m_f}{2M_w}$$

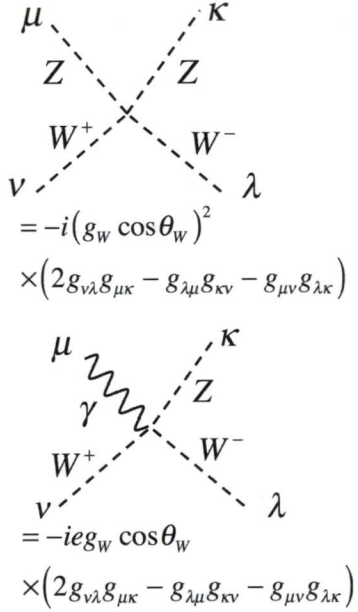

$$= -i\big(g_w \cos\theta_w \big)^2$$
$$\times \big(2g_{v\lambda}g_{\mu\kappa} - g_{\lambda\mu}g_{\kappa v} - g_{\mu v}g_{\lambda\kappa} \big)$$

$$= -ig_w^2 \big(2g_{\mu v}g_{\lambda\kappa} - g_{v\lambda}g_{\mu\kappa} - g_{\lambda\mu}g_{\kappa v} \big)$$

$$= -ieg_w \cos\theta_w$$
$$\times \big(2g_{v\lambda}g_{\mu\kappa} - g_{\lambda\mu}g_{\kappa v} - g_{\mu v}g_{\lambda\kappa} \big)$$

$$= -ie^2 \big(2g_{v\lambda}g_{\mu\kappa} - g_{\lambda\mu}g_{\kappa v} - g_{\mu v}g_{\lambda\kappa} \big)$$

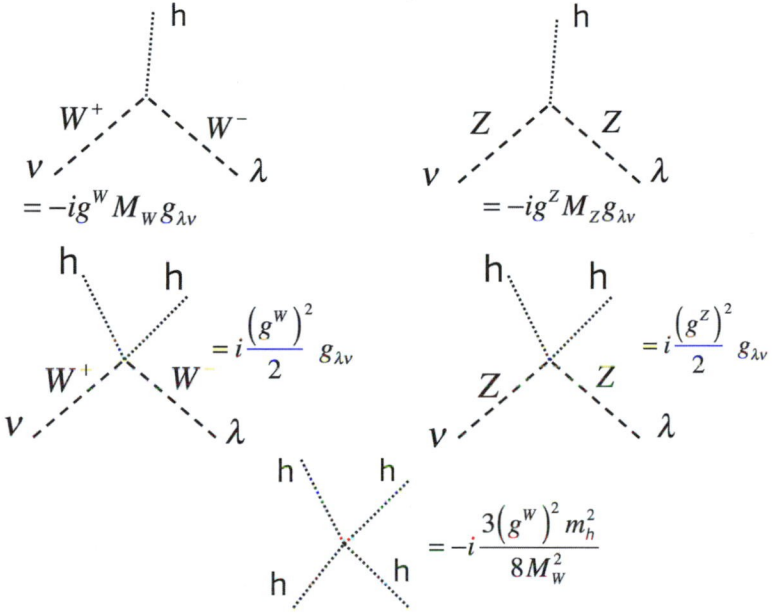

J

The Large Hadron Rap

Lyrics by Kate McAlpine; used with permission

Twenty-seven kilometers of tunnel under ground
Designed with mind to send protons around
A circle that crosses through Switzerland and France
Sixty nations contribute to scientific advance
Two beams of protons swing round, through the ring they ride
'Til in the hearts of the detectors, they're made to collide
And all that energy packed in such a tiny bit of room
Becomes mass, particles created from the vacuum
And then

LHCb sees where the antimatter's gone
ALICE looks at collisions of lead ions
CMS and ATLAS are two of a kind
They're looking for whatever new particles they can find.
The LHC accelerates the protons and the lead
And the things that it discovers will rock you in the head.

We see asteroids and planets, stars galore
We know a black hole resides at each galaxy's core
But even all that matter cannot explain
What holds all these stars together something else remains
This dark matter interacts only through gravity
And how do you catch a particle there's no way to see?
Take it back to the conservation of energy
And the particles appear, clear as can be

You see particles flying, in jets they spray
But you notice there ain't nothin', goin' the other way
You say, "My law has just been violated – it don't make sense!
There's gotta be another particle to make this balance."
And it might be dark matter, and for first
Time we catch a glimpse of what must fill most of the known 'Verse.
Because

LHCb sees where the antimatter's gone
ALICE looks at collisions of lead ions
CMS and ATLAS are two of a kind
They're looking for whatever new particles they can find.

Antimatter is sort of like matter's evil twin
Because except for charge and handedness of spin
They're the same for a particle and its anti-self
But you can't store an antiparticle on any shelf
Cuz when it meets its normal twin, they both annihilate
Matter turns to energy and then it dissipates.

When matter is created from energy
Which is exactly what they'll do in the LHC
You get matter and antimatter in equal parts
And they try to take that back to when the universe starts
The Big Bang back when the matter all exploded
But the amount of antimatter was somehow eroded
Because when we look around we see that matter abounds
But antimatter's nowhere to be found.
That's why

LHCb sees where the antimatter's gone
ALICE looks at collisions of lead ions
CMS and ATLAS are two of a kind
They're looking for whatever new particles they can find.
The LHC accelerates the protons and the lead
And the things that it discovers will rock you in the head.

The Higgs Boson – that's the one that everybody talks about.
And the one sure thing that this machine will sort out
If the Higgs exists, they ought to see it right away
And if it doesn't, then the scientists will finally say
"There is no Higgs! We need new physics to account for why
Things have mass. Something in our Standard Model went awry."

But the Higgs I still haven't said just what it does
They suppose that particles have mass because
There is this Higgs field that extends through all space
And some particles slow down while other particles race
Straight through like the photon it has no mass
But something heavy like the top quark, it's draggin' its ***
And the Higgs is a boson that carries a force
And makes particles take orders from the field that is its source.

They'll detect it.

LHCb sees where the antimatter's gone
ALICE looks at collisions of lead ions
CMS and ATLAS are two of a kind
They're looking for whatever new particles they can find.

Now some of you may think that gravity is strong
Cuz when you fall off your bicycle it don't take long
Until you hit the earth, and you say, "Dang, that hurt!"
But if you think that force is powerful, you're wrong.
You see, gravity – it's weaker than Weak
And the reason why is something many scientists seek
They think about dimensions we just live in three
But maybe there are some others that are too small to see
It's into these dimensions that gravity extends
Which makes it seem weaker, here on our end.
And these dimensions are "rolled up" curled so tight
That they don't affect you in your day to day life
But if you were as tiny as a graviton
You could enter these dimensions and go wandering on
And they'd find you...

When LHCb sees where the antimatter's gone
ALICE looks at collisions of lead ions
CMS and ATLAS are two of a kind
They're looking for whatever new particles they can find.
The LHC accelerates the protons and the lead
And the things that it discovers will rock you in the head.

References

[1] C. Amsler, et al., (Particle Data Group), *Phys. Lett.* **B667**, 1 (2008).

[2] For a readable account of the history of particle physics see S. Weinberg, *Subatomic Particles*, Scientific American Library, New York, 1983.

[3] Accelerators have many applications beyond particle physics. For a comprehensive overview see E.J.N. Wilson *An Introduction to Particle Accelerators*, Oxford University Press, 2001. For a technical introduction at the undergraduate level see H. Weidemann *Particle Accelerator Physics*, Springer, 2007.

[4] The most reliable source of information about the LHC comes directly from CERN, which has a document *CERN - FAQ: LHC – the guide*, CERN-Brochure-2008-001-Eng that you can obtain from the LHC website (http://www.cern.ch/lhc).

[5] There are no books at present about the Sudbury Neutrino Observatory. Your best bet for learning the history of this facility and its experiments is to go to the SNO website:
http://www.sno.phy.queensu.ca/sno/sno2.html .

[6] For the latest information on the comparsion between observation and the concordance model of cosmology see E. Komatsu, et. al., *Ap. J. Supp.* (to appear) [e-print: arXiv: 0803.0547v2] (2008); for an introduction to our modern understanding of cosmology see J. Levin, *How the Universe Got Its Spots*, Anchor Books, 2003; G.F.R. Ellis *Before the Beginning: Cosmology Explained*, Marion Boyars Publishers Ltd., 2000.

[7] See the first chapter of K. Zuber, *Neutrino Physics*, CRC Press, 2004.

[8] A recent exposition of this subject appears in U. Sarkar, *Particle and Astroparticle Physics*, Taylor & Francis, 2007.

[9] I. A. D'Souza and C. S. Kalman, *Preons: Models of Leptons, Quarks and Gauge Bosons as Composite Objects*, World Scientific, 1992.

[10] To learn about the origins of this subject see K.G. Wilson, *Nucl. Phys. Proc. Suppl.* **140** 3 (2005).

[11] For a readable introduction to this vast technical subject, see P. C. W. Davies and J.R. Brown, *Superstrings: A Theory of Everything?*, Cambridge University Press, 1992.

[12] For a comprehensive account of the history of particle physics, see A. Pais, *Inward Bound*, Clarendon Press, Oxford, 1986.

[13] M. Fierz *Physica Acta* **12**, 3 (1939); W. Pauli, *Phys. Rev.* **58**, 716 (1940); *Prog. Theo. Phys.* **5** no. 4 (1950); For a readable introduction see I. Duck and E.C.G. Sudarshan, *Am. J. Phys.* **66**, 284 (1998).

[14] J.M. Leinaas, and J. Myrheim, *Nuovo Cim.* **B37**, 1 (1977); F. Wilczek, *Phys. Rev. Lett.* **49**, 957 (1982).

[15] For a readable explanation, see R.J. Adler, B. Casey, and O.C. Jacob, *Am. J. Phys.* **63**, 620 (1995).

[16] An engaging discussion of this and other interesting phenomena in modern physics appears in A. Hey and P. Walters, *The New Quantum Universe*, Cambridge University Press, 2003.

[17] J.J. Thomson, *Phil. Mag.* **44**, 293(1897).

[18] E. Rutherford, *Phil. Mag., Series 6*, **21**, 669 (1911).

[19] S.A. Goudschmidt and G.H. Uhlenbeck, *Die Naturwissenschaften* **13**, 953 (1925).

[20] P. Zeeman, *Phil. Mag.* **43**, 226 (1897); *Phil. Mag.* **44**, 55 (1897); *Nature* **55**, 347 (1897).

[21] W. Gerlach and O. Stern, *Z. Physik* **9**, 349 (1922).

[22] For a good introduction at a technical level, see the second chapter of S. Weinberg, *Gravitation and Cosmology, Principles and Applications of the General Theory of Relativity*, John Wiley & Sons, New York, 1972.

[23] There are many good books on special relativity. A nice introduction appears in the second chapter of R.K. Pathria, *The Theory of Relativity*, Dover Publications, 2003. The book by A.P. French, *Special Relativity (M.I.T. Introductory Physics Series)*, W.H. Norton, 1968, provides a comprehensive introduction at a basic level.

[24] There are about as many books on group theory as there are on special relativity, each tailored to particular interests and perspectives. A good introduction is by Harry J. Lipkin, *Lie Groups for Pedestrians*, Dover, New York, 2002. A treatment that is a bit more advanced is B.G. Wybourne, *Classical Groups for Physicists*, John Wiley & Sons, New York, 1974.

[25] I won't prove this here. If you want to find out more, see ref. [24], as well as M. Tinkham, *Group Theory and Quantum Mechanics*, Dover, New York, 2003.

[26] A very complete treatment that proves this result and many others of group theory at an advanced level is by J.F. Cornwell, *Group Theory in Physics*, vols I - III, Academic Press, New York, 1989.

[27] If you want to see a full treatment of how Group Theory is used in particle physics, see R. Slansky, *Group Theory for Unified Model Building*, *Phys. Rep.* **79**, 1 (1981).

[28] For an introduction discussing the application of Lie Algebras in particle physics, see H. Georgi, *Lie Algebras in Particle Physics*, Benjamin/Cummings Publishing Co., Reading, MA, 1982.

[29] For a thorough treatment of action principles in classical mechanics, see H. Goldstein, C.P. Poole and J.L. Safko, *Classical Mechanics (3rd Edition)*, Addison Wesley, 2001.

[30] For a general treatment see R.M. Wald and A. Zoupas, *Phys. Rev.* **D61**, 084027 (2000); for a discussion of action principles in general relativity at an introductory level, see R.M. Wald, *General Relativity*, University of Chicago Press, 1984.

[31] See F. Halzen and A.D. Martin, *Quarks and Leptons*, John Wiley & Sons, New York, 1984.

[32] The idea of using symmetries to classify particles dates back to Wigner; see E.P. Wigner, *Annals of Mathematics* (1) **40**, 149 (1939).

[33] H. Dehmelt, *Physica Scripta* **T22**, 102 (1988).

[34] E.J. Eichten, K.D. Lane, and M. Peskin, *Phys. Rev. Lett.* **50**, 1 (1983).

[35] There are many books that discuss how to obtain the irreps of the rotation group. For a formal treatment from a quantum-mechanical viewpoint, see Chapter 9 of K. Hannabuss, *An Introduction to Quantum Theory*, Oxford University Press, 1999.

[36] W. Pauli, *Zeit. fur Phys.* **43**, 601 (1927).

[37] G. Herzberg, *Atomic Spectra and Atomic Structure*, Dover Publications (1944).

[38] For a thorough discussion, see Chapter 3 of E.U. Condon and G.H. Shortley, *The Theory of Atomic Spectra*, Cambridge University Press, 1970.

[39] C.S. Wu, E. Ambler, R.W. Hayward, D.D. Hoppes and R.P. Hudson, *Phys. Rev.* 105, 1413 (1957).

[40] H. Enge, *Introduction to Nuclear Physics*, Addison Wesley, 1966.

[41] *Kamiokande Collaboration:* Y. Fukuda et al., *Phys. Rev. Lett.* **77**, 1683 (1996); *Super-Kamiokande Collaboration:* Y. Fukuda et al., *Phys. Rev. Lett.* **81**, 1158 (1998); ibid. **82**, 1810 (1999); ibid. **86**, 5656 (2001); *Phys. Rev.* **D 69**, 011104 (2004).

[42] *SNO Collaboration:* Q.R. Ahmad et al., *Phys. Rev. Lett.* **87**, 071301 (2001); ibid. **89**, 011301 (2002); ibid. **89**, 011302 (2002); ibid. **92**, 181301 (2004).

[43] The question of the arrow of time and its origin has implications across a broad range of fields in particle physics, cosmology, thermodynamics, and more. For further reading, see O.E. Overseth, *Scientific American* **88**, 89 (October 1969). R.P. Feynman, *The Character of Physical Law*, Chapter 5, BBC Publications, 1965. H. Price, *Time's Arrow & Archimedes' Point: New Directions for the Physics of Time*, Oxford University Press, 1992; R. Le Poidevin *Travels in Four Dimensions: The Enigmas of Space and Time*, Oxford University Press, 2003.

[44] See R.C. Tolman, *Principles of Statistical Mechanics*, Oxford University Press, Oxford, 1938; J.S. Thomsen, *Phys. Rev.* **91**, 1263 (1953).

[45] B. Regan, et. al., *Phys. Rev. Lett.* **88**, 071805 (2002).

[46] P.G. Harris, et. al., *Phys. Rev. Lett.* **82**, 904 (1999).

[47] J. McDonough, et. al., *Phys. Rev.* **D38**, 2121 (1988).

[48] Y. Tomozawa, *Ann. Phys.* **128**, 463 (1980).

[49] G.S. Adkins, R.N. Fell, and J. Sapirstein, *Phys. Rev. Lett.* **84** (2000) 5086.

[50] R.S. Vallery, P.W. Zitzewitz, and D.W. Gidley, *Phys. Rev. Lett.* **90** 203402 (2003).

[51] D.W. Gidley, et. al., *Phys. Rev. Lett.* **49** , 525 (1982).

[52] C.I. Westbrook, et. al., *Phys. Rev. Lett.* **58**, 1328 (1987).

[53] C.I. Westbrook, et. al., *Phys. Rev.* **A58**, 5489 (1989).

[54] J.S. Nico, et. al., *Phys. Rev. Lett.* **65**, 3455 (1990).

[55] K. Neubeck, H. Schober, and H. Wäffler, *Phys. Rev.* **C10**, 320 (1974).

[56] O. Klein and Y. Nishina *Zeit. fur f. Phys.* **52**, 853 (1929).

[57] C. S. Wu and I. Shaknov, *Phys. Rev.* **77**, 136 (1950).

[58] J. Schwinger, *Phys. Rev.* **82** (1951) 914; G. Lüders, *Det. Kong. Danske Videnskabernes Selskab Mat.-fysiske Meddelelser* **28**, no. 5 (1954); W. Pauli, *Niels Bohr and the Development of Physics*, p. 30 (W. Pauli, ed.) McGraw-Hill, New York, 1955.

[59] S.W. Hawking and R. Penrose, *The Nature of Space and Time*, Princeton University Press, 2000.

[60] For a general overview of accelerator physics, see E.M. McMillan, in *Experimental Nuclear Physics*, p. 639, John Wiley & Sons, New York, 1959;

M.S. Livingstone and J.P. Blewett, *Particle Accelerators*, McGraw Hill, 1962; M.H. Blewett, *Ann. Rev. Nucl. Sci.* **17**, 427 (1967); C. Pellegrini, *Ann. Rev. Nucl. Sci.* **22**, 1 (1972); J.R. Rees *Sci. Am.* **261**, 58 (April 1990); S. Meyers and E. Picasso, *Sci. Am.* **263**, 54 (January 1990); L.M. Lederman, *Sci. Am.* **264**, 48 (March 1991); E.J.N. Wilson *An Introduction to Particle Accelerators*, Oxford University Press, 2001; S.Y. Lee *Accelerator Physics*, World Scientific, 2004.

[61] A nice overview of the history of DC acclerators is given by E. Cottereau, CERN Accelerator School course Particle Accelerators for Medicine & Industry, Pruhonice, Czech Republic, May 9-17 (2001), CERN report.

[62] A. Tollenstrup, *Nature* **404**, 350 (2000).

[63] J.D. Cockcroft et al, *Nature* **129**, 242 (1932).

[64] R.J. Van de Graaff, et. al., *Phys. Rev.* **43**, 149 (1933).

[65] L. Alvarez, *Rev. Sci. Inst.* **22**, 705 (1951).

[66] Nicholas Christofilos never published his work, but it was acknowledged by E.D. Courant, M.S. Livingston and H.S. Snyder, *Phys. Rev.* **91**, 202 (1953).

[67] E.D. Courant, M.S. Livingston, and H.S. Snyder, *Phys. Rev.* **88**, 1190 (1952).

[68] E. O. Lawrence and N. E. Edlefsen, *Science* **72**, 276 (1930).

[69] V. Veksler, *Comptes Rendus (Doklady), Acad. Sci. U.S.S.R.* **43**, No. 8, 444 IX (1944); V. Veksler, *Comptes Rendus (Doklady), Acad. Sci. U.S.S.R.* **44**, No. 9, 393 (1944); V. Veksler, *J. Phys. (U.S.S.R.)* **9**, No. 3, 153 (1945).

[70] E. McMillan, *Phys. Rev.* **68**, 143 (1945).

[71] J.D. Jackson, *Classical Electrodynamics*, John Wiley & Sons, 3rd ed. (1998).

[72] S. van der Meer, *Rev. Mod. Phys.* **57**, 689 (1985).

[73] C. Rubbia, *Rev. Mod. Phys.* **57**, 699 (1985).

[74] LEP Working Group, *Phys. Lett.* **B565**, 61 (2003).

[75] See CERN Courier, **48** no. 7 (September 2008) for the latest on the CLIC facility as of this writing.

[76] For a relatively recent comprehensive overview of detectors, see K. Kleinknecht, *Detectors for Particle Radiation*, Cambridge University Press, 1998.

[77] H. Bethe, *Ann. Phys.* (Leipz.) **5**, 325 (1930); F. Bloch, *Ann. Phys.* (Leipz.) **16**, 285 (1933).

[78] For a semi-classical derivation of eq. (8.2), see B. Rossi, *High Energy Particles*, p. 17, Prentice-Hall, Englewood Cliffs, N.J., 1961.

[79] For a general overview of how radiation interacts with matter see R.H. Sternheimer in *Methods of Experimental Physics*, p.1. C.L.Yuan and C.S. Wu eds., Academic Press, 1961.

[80] W. Crookes, *Chemical News* **87**, 241 (1903).

[81] H. Geiger and E. Marsden, *Proc. Roy. Soc.* (London) **A 82**, 495 (1909).

[82] I. Broser and H. Kallmann, *A. Naturforsch* **2a**, 439 (1947); G.T. Reynolds, et. al., *Phys. Rev.* **78**, 488 (1950); H. Kallmann, *Phys. Rev.* **78**, 621 (1950); M. Ageno, et. al., *Phys. Rev.* **79**, 720 (1950).

[83] C. T. R. Wilson, *Proc. Roy. Soc.* (London) **87**, 292 (1912).

[84] C.D. Anderson, *Phys. Rev.* **43**, 491 (1933).

[85] D. Glaser, *Phys. Rev.* **91**, 762 (1953).

[86] H. Greinacher, *Helv. Phys. Acta* **7**, 360 (1934).

[87] G. Charpak, R. Bouclier, T. Bressani, J. Favier, and C. Zupancic, *Nucl. Instrum. Methods* **62**, 202 (1968); G. Charpak, *Rev. Mod. Phys.* **65**, 591 (1993).

[88] D.R. Nygren, Proceedings of the 1975 PEP Summer Study, LBL, Berkley, (1975); J.N. Marx and D.R. Nygren, *Physics Today* **31**, 46 (October 1978).

[89] P.A. Cerenkov, *Doklady Akad. Nauk SSSR* **2**, 451 (1934).

[90] I.M. Frank and I.E. Tamm, *C.R. Acad. Sci. USSR* **14**, 107 (1937).

[91] SNO Collaboration, *Nucl. Inst. Meth.* **A449**,172 (2000).

[92] For a review, see A. Peisert in *Instrumentation in High Energy Physics*, F. Sauli, ed., World Scientific, 1992.

[93] SNO Collaboration (S.N. Ahmed et al.) *Phys. Rev. Lett.* **92**, 102004 (2004).

[94] H.O. Back, et. al., *Phys. Lett.* **B525**, 29 (2002).

[95] For the actual bounds, see information on the properties of the proton and the electron in C. Amsler et al. (Particle Data Group), *Phys. Lett.* **B667**, 1 (2008) (or alternatively go to the website http://pdg.lbl.gov/).

[96] G. Breit and E. Wigner *Phys. Rev.* **49** (1936) 519.

[97] The proper definition of the S-matrix – along with the decay rate and cross-section formulae – is within the context of quantum field theory. For a rigorous introduction, see S. Weinberg, *The Quantum Theory of*

Fields, Modern Applications vol. I, Cambridge University Press, Cambridge, 1996.

[98] See for example A. Beiser, *Perspectives of Modern Physics*, McGraw-Hill, 1969.

[99] E. Fermi, *Nuclear Physics*, University of Chicago Press, (1950).

[100] P.A.M. Dirac, *Proc. Roy. Soc.* (London) **A 114**, 243 (1927).

[101] See sec. 9.2.3 of D.J. Griffiths, *Introduction to Quantum Mechanics*, 2nd ed., Prentice Hall, Upper Saddle River, N.J., 2005.

[102] R. Durbin, H. Loar, and J. Steinberger, *Phys. Rev.* **83**, 646(1951); D.L. Clark, A. Roberts, and R. Wilson, *Phys. Rev.* **83**, 649 (1951) ; 85, 523 (1952); W.F. Cartwright, C. Richman, M.N. Whitehead, and H.A. Wilcox, *Phys. Rev.* **91** 677, (1953).

[103] R.P. Feynman, *Phys. Rev.* **76**, 749 (1949); **76**, 769 (1949)

[104] F. Dyson, *Phys. Rev.* **75**, 486 (1949); **75**, 1376 (1949); **82**, 428 (1951).

[105] To get a sense of the history behind the development of the diagrammatic formalism, see F. Dyson, *Disturbing the Universe*, New York, Harper & Row, 1979.

[106] If you want to develop a thorough understanding of the origin of the rules, you will have to work through them in a book such as S. Weinberg, *The Quantum Theory of Fields, Modern Applications* vols I and II, Cambridge Univrsity Press, Cambridge, 1996.

[107] A similar theory was presented by David Griffiths in Chapter 6 of D.J. Griffiths *Introduction to Elementary Particles*, John Wiley & Sons, 2008, in which three distinct spinless particles were introduced. I have chosen to stick to just two particles, since this toy theory will (as you will see in Chapter 13) model QED rather closely.

[108] S. Tomonaga, *Prog. Theor. Phys.* **1**, 27 (1946); *Phys. Rev.* **74**, 224 (1948).

[109] J. Schwinger, *Phys. Rev.* **75**, 651 (1949); *Phys. Rev.* **76**, 790 (1949).

[110] E. Schroedinger, *Ann. Phys.* **79**, 361 (1926); **81**, 109 (1926).

[111] To get a better understanding of the history of the first relativistic wave equation, see H. Kragh *Am. J. Phys.* **52**, 1024 (1984).

[112] P.A.M. Dirac, *Proc. Roy. Soc.* (London) **A 117**, 610 (1928).

[113] P.A.M. Dirac, *Proc. Roy. Soc.* (London) **A 126**, 360 (1930).

[114] J.C. Maxwell, *Phil. Trans. Roy. Soc.* **155**, 459 (1865); if you have trouble finding this paper, you might check J.C. Maxwell, *Scientific Papers* (Dover reprint) **1**, 526.

[115] O. Klein, *Zeit. fur Phys.* **37**, 895 (1926).

[116] V. Fock, *Zeit. fur Phys.* **38**, 242 (1926).

[117] To learn about the history of the gauge principle, see J.D. Jackson and L.B. Okun, *Rev. Mod. Phys.* **73**, 663 (2001).

[118] H. Weyl, *Proc. Nat. Acad. Sci.* **15**, 323 (1929); *Zeit. fur Phys.* **56**, 330 (1929).

[119] L. Lorenz, *Tidsckr. Fys. Chem.* **6**, 1 (1867); *Ann. der Physik und Chemie* **131**, 243 (1867); *Phil. Mag.* ser. 4, **34**, 287 (1867).

[120] H.A. Lorentz, *Encykl. Math. Wissen. Band V:2, Heft 1*, **13**, p. 63 (1904); *Encykl. Math. Wissen. Band V:2, Heft 1*, **14**, p. 145 (1904); G. B. Teubner, Leipzig and G. E. Stechert, *Theory of Electrons*, New York, 1909 (reprinted by Dover Publications, New York, 1952).

[121] For an historical account of the development of Quantum electrodynamics, see S.S. Schweber, *QED and the Men who Made it: Dyson, Feynman, Schwinger, and Tomonaga*, Princeton University Press, 1994.

[122] Evidently Hendrik Casimir was the first person to perform these manipulations. See A. Pais, *Inward Bound* p. 375, New York: Oxford, 1986.

[123] N.F. Mott, *Proc. Roy. Soc.* **A124**, 425 (1929).

[124] W. Pauli, *Zeit. fur Phys.* **31**, 373 (1925).

[125] J. Schwinger, *Phys. Rev.* **73**, 416 (1948).

[126] For an outline of the most recent calculation, see G. Gabrielse, D. Hanneke, T. Kinoshita, M. Nio, and B. Odom, *Phys. Rev. Lett.* **97**, 030802 (2006).

[127] If you want to get a picture of the details behind all of this, see P. J. Mohr and B. N. Taylor, *Rev. Mod. Phys.* **72**, 351 (2000); *Rev. Mod. Phys.* **77**, 1 (2005).

[128] G.W. Bennett, et. al., *Phys. Rev. Lett.* **89** 101804 (2002); Erratum *Phys. Rev. Lett.* **89** 129903 (2002); G.W. Bennett et al., *Phys. Rev. Lett.* **92** 161802 (2004); G.W. Bennett et al., *Phys. Rev.* **D73** 072003 (2006).

[129] A detailed treatment yielding this result is in Chapter 3 of J.J. Sakurai, Advanced Quantum Mechanics, Addison Wesley, 1967.

[130] H.A. Bethe, L.M. Brown, and J.R. Stehn *Phys. Rev.* **77**, 370 (1950).

[131] A. van Wijngaarden, F. Holuj, and G.W.F. Drake, *Can. J. Phys.* **76**, 95 (1998); Erratum **76**, 993 (1998).

[132] W.E. Lamb, Jr. and R.C. Retherford, *Phys. Rev.* **72**, 241 (1947).

[133] S. Treibwasser, E.S. Dayhoff, and W.E. Lamb, *Phys. Rev.* **89**, 98 (1953); **89**, 106 (1953); R.T. Robiscoe and T.W. Shyn, *Phys. Rev. Lett.* **24**, 559 (1970); T.W. Shyn, W.L. Williams, R.T. Robiscoe, and T. Rebane, *Phys. Rev.* **A3**, 116 (1971); K.A. Safinya, K.K. Chan, S.R. Lundeen, and F.M. Pipkin, *Phys. Rev. Lett.* **45**, 1934 (1980); B.L. Cosens and T.V. Vorberger, *Phys. Rev.* **A12**, 16 (1976); S.L. Kaufman, W.E. Lamb, Jr., K.R. Lea, and M. Leventhal, *Phys. Rev.* **A4**, 2128 (1971); G. Newton, D.A. Andrews, and P.J. Unworth, *Philos. Trans. R. Soc. London* **290**, 373 (1979); E.W. Hagley and F.M. Pipkin, *Phys. Rev. Lett.* **72**, 1172 (1994); S.R. Lundeen and F.M. Pipkin, *Metrologia* **22**, 9 (1986); V.G. Palchikov, L. Sokolov, and V.P. Yakovlev, *Sov. Phys. JETP Lett.* **38**, 347 (1983).

[134] If you want to get into all of the gory details for this and other QED calculations, see M.E. Peskin and H.D. Schroeder, *An Introduction to Quantum Field Theory*, Addison-Wesley, 1995.

[135] L.D. Landau, in *Niels Bohr and the Development of Physics*, ed. W. Pauli, Pergamon Press, London, 1955; L.D. Landau and I.J. Pomeranchuk, *Dokl. Akad. Nauk., SSSR* **102** (1955) 489.

[136] D.V. Shirkov, in *Quantum Field Theory: A 20th Century Profile*, p. 25, ed. by A.N. Mitra, Hindustan Book Agency and Indian National Science Academy, New Delhi, 2000.

[137] J. Chadwick, *Nature* **192**, 312 (1932).

[138] E.P. Wigner, *Phys. Rev.* **43** (1933) 252.

[139] W. Heisenberg, *Zeit. fur Phys.* **77** (1932) 1.

[140] C.M.G. Lattes, et. al., *Nature* **159**, 694 (1947).

[141] V.B. Flagin, et. al., *Sov. Phys. JETP* **35**, 592 (1959).

[142] For a more complete discussion see S. Gasiorowicz, *Elementary Particle Physics*, John Wiley & Sons, 1966.

[143] H.L. Anderson, E. Fermi, E.A. Long, and D.E. Nagle, *Phys. Rev.* **85**, 936 (1952).

[144] G.D. Rochester and C.C. Butler, *Nature* **160**, 855 (1947).

[145] A. Pais, *Phys. Rev.* **86**, 663 (1952).

[146] M. Gell-Mann, *Phys. Rev.* **92**, 883 (1953); *Nuovo Cim.* **4**, Suppl. 2 848 (1956); T. Nakano and K. Nishijima, *Prog. Theor. Phys.* **10**, 581 (1953).

[147] M. Gell-Mann, *Phys. Lett.* **8**, 214 (1964).

[148] G. Zweig, CERN preprint 8182/Th.401 (1964), reprinted in S.P. Rosen and D. Lichtenberg, (ed.) *Developments in Quark Theory of Hadrons*, vol. I, Hadronic Press, MA, 1980.

[149] V.E. Barnes, et. al., *Phys. Rev. Lett.* **12**, 204 (1964).

[150] O.W. Greenberg, *Phys. Rev. Lett.* **13**, 598 (1964).

[151] If you are interested, there are a number of reviews on this subject: M.L. Perl, E.R. Lee, and D. Lomba, *Mod. Phys. Lett.* **A19**, 2595 (2004); P.F. Smith, *Ann. Rev. Nucl. and Part. Sci.* **39**, 73 (1989); L. Lyons, *Phys. Reports* **129**, 225 (1985); M. Marinelli and G. Morpurgo, *Phys. Reports* **85**, 161 (1982).

[152] This was a long-term project and the full set of papers is too long to list here – the earliest work is in: N. Isgur and G. Karl, *Phys. Lett.* **B72**, 109 (1977); *Phys. Rev.* **D20**, 1191 (1979); N. Isgur, G. Karl and R. Koniuk, *Phys. Rev. Lett.* **41**, 1269 (1978); R. Koniuk and N. Isgur, *Phys. Rev. Lett.* **44**, 845 (1980).

[153] A lot of useful information about the quark model appears in S. Gasiorowicz and J.L. Rosner, *Am. J. Phys.* **49**, 954 (1981).

[154] S. Coleman and S.L. Glashow, *Phys. Rev. Lett.* **6**, 423 (1961).

[155] R. van Royen and V.F. Weisskopf, *Nuovo Cim.* **50**, 617 (1967); **51**, 583 (1967);

[156] K.G. Chetyrkin, et. al., *Nucl. Phys.* **B586**, 56 (2000); erratum **B634**, 413 (2002).

[157] M.L. Perl, et. al., *Phys. Rev. Lett.* **35**, 1489 (1975).

[158] M.N. Rosenbluth, *Phys. Rev.* **79**, 615 (1950).

[159] For a discussion of the physical significance of these quantities, see H. Frauenfelder and E.M. Henley, *Subatomic Physics*, Englewood Cliffs, N.J. Prentice-Hall, 1991.

[160] J.D. Bjorken, *Phys. Rev.* **163**, 767 (1967); *Phys. Rev.* **179**, 1547 (1969).

[161] C.G. Callen and D. Gross, *Phys. Rev. Lett.* **22**, 156 (1975).

[162] An in-depth study was carried out by A. Bodek, et. al., *Phys. Rev.* **20**, 1471 (1979).

[163] For a review of the process by which quarks were discovered and the Bjorken and Callen-Gross relations verified, see R.E. Taylor, *Rev. Mod. Phys.* **63**, (1991).

[164] K.G. Wilson, *Phys. Rev.* **D10**, 2445 (1974); K.G. Wilson and J. Kogut, *Rev. Mod. Phys.* **55**, 775 (1983).

[165] For a non-technical introduction, see D.H. Weingarten, *Sci. Am.* **274**, 116 (February 1996).

[166] E. Eitchen in *1983 Proceedings of the SLAC Summer Institute on Particle Physics: Dynamics and Spectroscopy at High Energy*, published as SLAC-R-267 (1983).

[167] K.Niu, et. al., *Prog. Theor. Phys.* **46**, 1644 (1971).

[168] For more details on this story see K. Niu, *Proc. 1st Inst. Workshop on Nuclear Emulsion Techniques*, Nagoya, preprint DPNU-98-39, 1998.

[169] J.E. Augustin, et. al., *Phys. Rev. Lett.* **33**, 1406 (1974).

[170] J.J. Aubert, et. al., *Phys. Rev. Lett.* **33**, 1404 (1974).

[171] B. Richter, *Physica Scripta* **15**, 23 (1977).

[172] S.L. Glashow, I. Iliopoulos, and L. Maiani, *Phys. Rev.* **D2**, 1285 (1970).

[173] T. Applequist and D. Politzer, *Phys. Rev. Lett.* **34**, 43 (1975); *Phys. Rev.* **D12**, 1404 (1975).

[174] G. Goldhaber, et. al., *Phys. Rev. Lett.* **37**, 255 (1976); I. Peruzzi, et. al., *Phys. Rev. Lett.* **37**, 569 (1976).

[175] G. Zweig, CERN preprint No. 8419 TH 412, (1964), reprinted in S.P. Rosen and D. Lichtenberg, (ed.) *Developments in Quark Theory of Hadrons*, vol. I, Hadronic Press, MA, 1980; S. Okubo, *Phys. Lett.* **5**, 165 (1963); *Phys. Rev.* **D16**, 2336 (1977); J. Iizuka, K. Okada, and O. Shito, *Prog. Th. Phys.* **35**, 1061 (1966); J. Iizuka, *Prog. Th. Phys. Suppl* **37-38**, 21 (1966).

[176] E. Bloom and G. Feldman, *Sci. Am.* **246**, 66 (May 1982).

[177] S.W. Herb, et. al., *Phys. Rev. Lett.* **39**, 252 (1977).

[178] L. Lederman, *Rev. Mod. Phys.* **61**, 547 (1989).

[179] F. Abe, et. al., *Phys. Rev. Lett.* **74**, 2626 (1995).

[180] For a nice readable account of the discovery of the top quark see T.M. Liss and P. Tipton, *Sci. Am.* **261**, 54 (September 1997).

[181] T. Aaltonen et al. (CDF Collaboration), Phys. Rev. **D79**, 092005 (2009).

[182] C.N. Yang and R. Mills, *Phys. Rev.* **96**, 191 (1954).

[183] For a review of experimental tests of the fifth force see E.G. Adelberger, *Class. Quant. Grav.* **18**, 2397 (2001).

[184] H.D. Politzer, *Phys. Rev. Lett.* **30**, 1346 (1973); *Phys. Rep.* **14C**, 120 (1974); D.J. Gross and F. Wilczek, *Phys. Rev. Lett.* **30**, 1343 (1973).

[185] H. Becquerel, *Comptes Rendus* **122**, 420 (1896).

[186] If you want to read about the history of radioactivity and its relationship to the rest of physics see M. F. L'Annunziata, *Radioactivity: Introduction and History*, Elsevier Science, 2007.

[187] A. P. Dickin, *Radiogenic Isotope Geology* Cambridge University Press (1997)

[188] E. Fermi, *Z. Phys.* **88**, 161 (1934); *Nuovo Cim.* **11**,1 (1934).

[189] For a discussion of the Coulomb factor see H. Enge, *Introduction to Nuclear Physics*, Addison Wesley, 1966.

[190] For a more complete discussion of the determination of Fermi's constant from nuclear decays see Chapter 6 of D.H. Perkins, *Introduction to High Energy Physics*, Cambridge University Press, 2000.

[191] W.J. Marciano and A. Sirlin, *Phys. Rev. Lett.* **61**, 1815 (1988); T. van Ritbergen and R.G. Stuart, *Phys. Rev. Lett.* **82**, 488 (1999).

[192] Ch. Kraus, et al., *Eur. Phys. J.* **C40**, 447 (2005); V.M. Lobashev, *Phys. Lett.* **B460**, 227 (1999).

[193] K. Assamagan, et al. *Phys. Rev.* **D53**, 6065 (1996)

[194] R. Barate, et al., *Eur. Phys. J.* **C2**,395 (1998).

[195] O. Elgaroy, et al., *Phys. Rev. Lett.* **89**, 061301 (2002).

[196] M. Colless, et al., *Mon. Not. R. Astron. Soc.* **328**, 1039 (2001); W. J. Percival, et al., *Mon. Not. R. Astron. Soc.* **327**, 1297 (2001).

[197] F. Reines and C.L. Cowan, Jr., *Phys. Rev.* **113**, 273 (1959); F. Reines, C.L. Cowan, Jr., F.B. Harrison, A.D. McGuire, and H.W. Kruse, *Phys. Rev.* **113**, 159 (1960).

[198] NOMAD Collaboration, J. Altegoer, et al., *Nucl. Instr. and Meth.* **A404**, 96 (1998); NOMAD Collaboration, P. Astier, et al., *Nucl. Phys.* **B611**, 3 (2001).

[199] T.D. Lee and C.N. Yang, *Phys. Rev.* **104**, 254 (1956).

[200] M. Goldhaber, L. Grodzins, and A. W. Sunyar, *Phys. Rev.* **109**, 1015 (1958).

[201] W. Pauli, *Writings on Physics and Philosophy*, eds. C.P. Enz and K.V. Meyenn, Springer, 1994.

[202] M. Gell-Mann and A. Pais, , et al., *Phys. Rev.* **97**, 1387 (1955).

[203] J.H. Christensen, et al., *Phys. Rev. Lett.* **13**, 138 (1964).

[204] L. Wolfenstein, *Phys. Rev. Lett.* **13**, 562 (1964).

[205] For a review of this situation see M. Trodden *Rev. Mod. Phys.* **71**, 1463 (1999).

[206] *BELLE Collaboration:* K. Abe et al., *Phys. Rev. Lett.* **87**, 091802 (2001).

[207] M.K. Gaillard, *Physics Today* **74** (April 1981).

[208] O. Klein "On the Theory of Charged Fields" at *New Theories in Physics* International Institute of Intellectual Cooperation, Paris (1939); for a review of the conference at which Klein presented this idea – decades ahead of its time – see D.J. Gross' article in *The Oskar Klein Centenary: proceedings* ed. U. Lindstroem, Singapore, World Scientific, 1995.

[209] S. Eidelman, et. al., *Phys. Lett.* **B592**, 1 (2004).

[210] D.B. Renner, et. al., *J. Phys. Conf. Ser.* **46**, 152 (2006).

[211] N. Cabbibo, *Phys. Rev. Lett.* **10**, 531 (1963).

[212] For an in-depth review of weak interactions in the quark model see J.F. Donoghue, et. al., *Phys. Rep.* **131**, 319 (1986).

[213] D. Ambrose, et. al., *Phys. Rev. Lett.* **84**, 1389 (2000).

[214] M. Kobayashi and T. Maskawa, *Prog. Theor. Phys.* **49**, 652 (1973).

[215] C. Jarlskog, *Phys. Rev. Lett.* **55**, 1039 (1985); *Z. Phys.* **C 29**, 491 (1985).

[216] S. Glashow, *Nucl. Phys.* **22**, 579 (1961).

[217] A. Salam, in *Elementary Particle Theory*, ed. N. Svartholm, Almquist and Forlag, Stockholm, 1968; S. Weinberg, *Phys. Rev. Lett.* **19**, 1264 (1967).

[218] G. 't Hooft, *Nucl. Phys.* **B33**, 173 (1971); *Nucl. Phys.* **B35**, 167 (1971); G. t Hooft and M. Veltman, *Nucl. Phys.* **B44**, 189 (1972); *Nucl. Phys.* **B50**, 318 (1972).

[219] B.W. Lee, *Phys. Rev.* **D5**, 823 (1972).

[220] S.A. Bludman, *Nuovo Cim.* **9**, 443 (1958).

[221] F.J. Hasert, et. al., *Phys. Lett.* **B46**, 121 (1973); **B46**, 138 (1973); *Nucl. Phys.* **B73**, 1 (1974).

[222] R.H. Heisterberg, et. al., *Phys. Rev. Lett.* **44**, 635 (1980); For data on $\nu_\mu + e^- \to \nu_e + \mu^-$ see F. Bergsma, et. al., *Phys. Lett.* **B122**, 465 (1983).

[223] P.W. Higgs, *Phys. Lett.* **12**, 132 (1964); *Phys. Rev. Lett.* **13**, 508 (1964); *Phys. Rev.* **145**, 1156 (1966).

[224] J. Goldstone, *Nuovo Cim.* **9**, 154 (1961); Y. Nambu, *Phys. Rev. Lett.* **4**, 380 (1960); J. Goldstone, A. Salam and S. Weinberg, *Phys. Rev.* **127**, 965 (1962).

[225] C. Rubbia, et. al., *Proc. Int. Neutrino Conf. Aachen*, eds. H. Faissner, H. Reithler and P. Zerwas, p. 683, Vieweg, Braunschweig, 1977.

[226] G. Arnison, et. al., *Phys. Lett.* **B122**, 103 (1983); **B126**, 398 (1983).

[227] M. Banner, et. al., *Phys. Lett.* **B122**, 476 (1983).

[228] The most precise measurements to date are by Carl Wieman and collaborators: C.S. Wood, et. al., *Science* **275**, 1759 (1997); S.C. Bennett and C.E. Wieman, *Phys. Rev. Lett.* **82**, 2484 (1999).

[229] P. Villain, et. al., (CHARM II collaboration) *Phys. Lett.* **B335**, 246 (1994).

[230] J. Erler and M.J. Ramsey-Musolf *Phys. Rev.* **D72**, 073003 (2005); A. Czarnecki and W.J. Marciano, *Int. J. Mod. Phys.* **A15**, 2365 (2000).

[231] D. Acosta, et. al., *Phys. Rev.* **D71**, 052002 (2005).

[232] J. Alcarez, et. al., *ALEPH, DELPHI, L3, OPAL, and LEP Electroweak Working Group*, CERN-PH-EP-2006-042 (hep-ex/0612034), 2006.

[233] See the article by F. Perrier in *Precision Tests of the Standard Electroweak Model*, ed. P. Langacker, World Scientific, Singapore, 1995.

[234] S. Coleman and E. Weinberg, *Phys. Rev.* **D7**, 1888 (1973).

[235] For a review see M. Sher, *Phys. Rep.* **179**, 273 (1989).

[236] V. Elias, R.B. Mann, D.G.C. McKeon, T.G. Steele, *Phys. Rev. Lett.* **91**, 251601 (2003); V. Elias, R.B. Mann, D.G.C. McKeon, T.G. Steele, *Nucl. Phys.* **B678**, 147 (2004); Err. **B703**, 413 (2004).

[237] R. Barate, et. al., *Phys.Lett.* **B565**, 61 (2003).

[238] Announced at 34th International Conference on High Energy Physics (ICHEP 2008).

[239] A. Djouadi, J. Kalinowski, and M. Spira, Comput. Phys. Commun. **108**, 56 (1998).

[240] If you want to read more about the kind of theory that might supercede the Standard Model see S. Weinberg, *Dreams of a Final Theory: The Scientist's Search for the Ultimate Laws of Nature*, Vintage, 1994.

[241] For an overview of the different approaches to Quantum Gravity see L. Smolin *Three Roads to Quantum Gravity*, Basic Books, 2002.

[242] There is a wide divergence of opinion as to how to respond to the biophilic properties of our universe. For an overview of approaches, see *Universe or Multiverse?* ed. B. Carr, Cambridge University Press, 2007. An earlier book that lays out the situation is by J. Barrow and F. Tipler, *The Anthropic Cosmological Principle*, Oxford University Press, 1988, and a recent book that presents the problem at current readable level is by P. Davies, *Cosmic Jackpot: Why Our Universe Is Just Right for Life*, Houghton Mifflin Harcourt, 2007. Another interesting perspective is given by J. Polkinghorne in *John Marks Templeton, Evidence of Purpose*, p. 111, New York: The Continuum Publishing Company, 1996.

[243] B. Pontecorvo, *Zh. Eksp. Teor. Fiz.* **33**, 549 (1957); [*Sov. Phys. JETP* **6**, 429 (1958)]; *Zh. Eksp. Teor. Fiz.* **34**, 247 (1957); [*Sov. Phys. JETP* **7**, 172 (1958)].

[244] B. Pontecorvo, *Zh. Eksp. Teor. Fiz.* **53**, 1717 (1967); [*Sov. Phys. JETP* **26**, 984 (1968)];

[245] If you want to see the major papers on the solar neutrino problem, look at *Solar Neutrinos: The First 30 Years*, ed. J.N. Bahcall, Westview, Boulder, CO, 2002. For a readable summary of the problem see A.N. MacDonald, J.R. Klein, and D.L. Wark, *Sci. Am.* **288**, 40 (April 2003).

[246] B. Pontecorvo, Chalk River Lab PD-205 report.

[247] J.N. Bahcall, *Astr. J.* **137**, 334 (1963).

[248] R. Davis Jr., D.S. Harmer and K.C. Hoffman, *Phys. Rev. Lett.* **20**, 1209 (1986).

[249] B.T. Cleveland, *Astr. J.* **496**, 505 (1998).

[250] K.S. Hirata, et. al., *Phys. Rev. Lett.* **63**, 16 (1989).

[251] J. Hosaka, et. al., *Phys. Rev.* **D73**, 112001 (2006).

[252] J.N. Abdurashitov, et. al., *J.E.T.P.* **95**, 181 (2002).

[253] P. Anselmann, et. al., *Phys. Lett.* **B285**, 375 (1992); **B285**, 390 (1992); **B357**, 237 (1995).

[254] B. Aharmim, et. al., *Phys. Rev.* **C72**, 055502 (2005).

[255] B. Aharmim, et. al., *Phys. Rev. Lett.* **101**, 111301 (2008).

[256] Y. Fukuda, et. al., *Phys. Lett.* **B433**, 9 (1998); *Phys. Rev. Lett.* **81**, 1562 (1998).

[257] Y. Fukuda, et. al., *Phys. Lett.* **B436**, 33 (1998).

[258] Y. Ashie, et. al., *Phys. Rev.* **D71**, 112005 (2005).

[259] T. Araki, et. al., *Phys. Rev. Lett.* **94**, 081801 (2005).

[260] A. Aguilar, et. al., *Phys. Rev.* **D64**, 112007 (2001).

[261] B. Armbruster, et. al., *Phys. Rev.* **D65**, 112001 (2002).

[262] Andrea Longhin (OPERA Collaboration) *Proceedings of 28th Physics in Collision (PIC 2008)*, Perugia, Italy, 2008.

[263] S.P. Mikheyev and A. Smirnov, *Yad. Fiz.* **42**, 1441 (1985) [*Sov. J. Nucl. Phys.* **42**, 913 (1986)].

[264] L. Wolfenstein, *Phys. Rev.* **D17**, 2369 (1978).

[265] M. Gasperini, *Phys. Rev.* **D38**, 2635 (1988).

[266] R.B. Mann and U. Sarkar, *Phys. Rev. Lett.* **76**, 865 (1996).

[267] G. L. Fogli, E. Lisi, A. Marrone and D. Montanino, *Phys. Rev.* **D6**, 093006 (2003).

[268] This nomenclature is in honor of the work of Z. Maki, M. Nakagawa and S. Sakata, *Prog. Theor. Phys.* **28**, 870 (1962), which predated the discovery of neutrino oscillations.

[269] E.M. Purcell and N.F. Ramsey, *Phys. Rev.* **78**, 807 (1950).

[270] C.G. Callan, R. Dashen and D. Gross, *Phys. Lett.* **B63**, 334 (1976); R. Jackiw and C. Rebbi, *Phys. Rev. Lett.* **37**, 172 (1976).

[271] R.D. Peccei and H.R. Quinn, *Phys. Rev. Lett.* **38**, 1440 (1977); *Phys. Rev.* **D16**, 1791 (1977).

[272] For a review of such models see Chapter 7 of U. Sarkar, *Particle and Astroparticle Physics*, Taylor & Francis, 2007.

[273] D. Demir and E. Ma, *Phys. Rev.* **D62**, 111901 (R) (2000); *J. Phys. G: Nucl. Part. Phys.* **27**, L87 (2001); D. Demir, E. Ma and U. Sarkar, *J.Phys. G: Nucl. Part. Phys.* **26**, L117 (2000).

[274] R. Brinkmann and M. Turner, *Phys. Rev.* **D38**, 2338 (1988); G. Raffelt and D. Seckel, *Phys. Rev. Lett.* **60**, 1793 (1988); M.S. Turner, *Phys. Rev. Lett.* **60**, 1797 (1988); K. Choi, K. Kang and J.E. Kim, *Phys. Rev. Lett.* **62**, 849 (1989).

[275] P. Sikivie, D.B. Tanner, K. van Bibber, *Phys. Rev. Lett.* **98**, 172002 (2007).

[276] P. Sikivie, *Phys. Rev. Lett.* **51**, 1415 (1983); *Phys. Rev.* **D32**, 2988 (1985).

[277] G.G. Raffelt and L. Stodolsky, *Phys. Rev.* **D37**, 1237 (1988).

[278] J. Preskill, M. Wise and F. Wilczek, *Phys. Lett.* **B120**, 127 (1983); L.F. Abbott and P. Sikivie, *Phys. Lett.* **B120**, 133 (1983); M. Dine and W. Fishler, *Phys. Lett.* **B120**, 137 (1983); J.E. Kim, *Phys. Rep.* **150**, 1 (1987); M.S. Turner, *Phys. Rev.* **D33**, 819 (1986).

[279] K. Abazajian, G.M. Fuller and M. Patel, *Phys. Rev.* **D64**, 023501 (2001).

[280] J. Ellis, J. Hagelin, D. Nanopoulos, K. Olive and M. Srednicki, *Nucl. Phys.* **B238**, 453 (1984); K. Griest, *Phys. Rev.* **D38**, 2357 (1988); J. Ellis and R. Flores, *Phys. Lett.* **B300**, 175 (1993); A. Bottino et al., *Astroparticle Phys.* **2**, 67; (1994) T. Falk, K. Olive and M. Srednicki, *Phys. Lett.* **B339**, 248 (1994); V. Bednyakov, H.V. Klapdor-Kleingrothaus and S.G. Kovalenko, *Phys. Lett.* **B329**, 5 (1994); *Phys. Rev.* **D55**, 503 (1997).

[281] P. Sreekumar, et. al. *Astr. J.* **494** (1998) 523.

[282] *CDMS Collaboration:* R. Abusaidi et al., *Phys. Rev. Lett.* **84**, 5699 (2000); D.S. Akerib et al., *Nucl. Instrum. Meth.* **A20**, 105 (2004). *Edelweiss Collaboration:* L. Chabert et al., *Eur. Phys. J.* **C33**, S965 (2004).

[283] R. Bernabei, et. al. *Phys. Lett.* **B389**, 757 (1996); **408**, 439 (1996); **424**, 195 (1996); *AIP Conf. Proc.* **698**, 328 (2003); *Tokyo 2004, Neutrino oscillations and their origin*, p. 378; *Neutrinoless double beta decay*, ed. V.K.B. Kota and U. Sarkar, Narosa, Delhi, 121 (2007); A. Incicchitti et al., *Proc. TAUP 2005*, Zaragoza, Spain (2005).

[284] M.E. Peskin, *J. Phys. Soc. Japan* **76**, 111017 (2007).

[285] For a review of this and other possibilities, see M.S. Turner and D. Huterer, *J. Phys. Soc. Japan* **76**, 111015 (2007).

[286] H. Georgi and S. Glashow, *Phys. Rev. Lett.* **32**, 438 (1974).

[287] H. Georgi, H. Quinn and S. Weinberg, *Phys. Rev. Lett.* **33**, 451 (1974).

[288] Super-Kamiokande Collaboration: Y. Hayato et al., *Phys. Rev. Lett.* **83**, 1529 (1999); K.S. Hirata et al., *Phys. Lett.* **B220**, 308 (1989); *IMB Collaboration:* W. Gajewski et al., *Nuc. Phys.* **A28**, 1610164 (1992); *FREJUS Collaboration:* C. Berger et al., *Phys. Lett.* **B245**, 305 (1991).

[289] U. Amaldi, W. de Boer and H. Furstenau, *Phys. Lett.* **B 260**, 447 (1991).

[290] For a readable account see G.L. Kane *Supersymmetry: Unveiling the Ultimate Laws of Nature*, Perseus, Cambridge, MA, 2000.

Index